D0643957

The Fortran 2003 Handbook

Jeanne C. Adams · Walter S. Brainerd ·
Richard A. Hendrickson · Richard E. Maine ·
Jeanne T. Martin · Brian T. Smith

The Fortran 2003 Handbook

The Complete Syntax, Features and Procedures

Jeanne C. Adams
National Center for Atmospheric Research
Boulder, Colorado, USA

Walter S. Brainerd
The Fortran Company
Tucson, Arizona, USA

Richard A. Hendrickson
Spackman & Hendrickson, Inc.
Minneapolis, Minnesota, USA

Richard E. Maine
NASA Dryden, Edwards AFB
California, USA

Jeanne T. Martin,
Retired: Lawrence Livermore National
Security, California, USA

Brian T. Smith
University of New Mexico
Albuquerque, New Mexico, USA

ISBN: 978-1-84628-378-9 e-ISBN: 978-1-84628-746-6
DOI: 10.1007/978-1-84628-746-6

British Library Cataloguing in Publication Data
A catalogue record for this book is available from the British Library

Library of Congress Control Number: 2008934286

Printed on acid-free paper

Springer Science+Business Media
springer.com

Contents

Preface

Fortran has been the premier language for scientific computing since its introduction in 1957. Fortran originally was designed to allow programmers to evaluate formulas—FORmula TRANslation—easily on large computers. Fortran compilers are now available on all sizes of machines, from small desktop computers to huge multi-processors.

The *Fortran 2003 Handbook* is a definitive and comprehensive guide to Fortran 2003. Fortran 2003, the latest standard version of Fortran, has many modern features that will assist the programmer in writing efficient, portable, and maintainable programs that are useful for everything from "hard science" to text processing.

The *Fortran 2003 Handbook* is an informal description of Fortran 2003, developed to provide not only a readable explanation of features, but also some rationale for the inclusion of features and their use. In addition, "models" give the reader better insight as to why the language is the way it is.

Target Audience

This handbook is intended for anyone who wants a comprehensive survey of Fortran 2003, including those familiar with programming language concepts but unfamiliar with Fortran. Experienced Fortran 95 programmers will be able to use this volume to assimilate quickly those features in Fortran 2003 that are not in Fortran 95 (Fortran 2003 contains all of the features of Fortran 95).

Although the handbook is written for use in conjunction with the standard, it is also designed as a practical stand-alone description of Fortran 2003. The syntax rules have been recast into more readable form. On the other hand, in places where the standard is not completely clear, a reasonable interpretation is often given, together with ways to implement and program that will avoid potential problems. Of course, if information is being sought to understand a fine point of compiler implementation, settle a bet, resolve a court case, or determine the answer to a Fortran trivia question, the standard itself should be considered the final authority.

Organization

Chapters 1–16 correspond to Sections 1–16 in the standard. (The standard is the complete official description of the language, but it is written in a legally airtight, formal style without tutorial material and can be difficult to understand in places.) The handbook and the standard can be read in parallel for insights into the Fortran language. This makes it feasible to use this handbook to "decipher" the standard, and this is an ideal use of this book.

Specific information can be found in the following places:

- A brief list of references can be found at the end of Chapter 1.
- Each chapter begins with a summary of the main terms and concepts described in the chapter.
- Each of the standard intrinsic procedures is described in detail in Appendix A; a general discussion of the intrinsic functions is in Chapter 13.
- The IEEE module procedures are described in detail in Appendix B and Chapter 14.
- Appendix C contains a listing of the new, obsolescent, and deleted features.
- The index is unusually comprehensive.

Style of the Programming Examples

In order to illustrate many features of the language and as many uses of these features as possible, no single particular style has been used when writing the examples. In many cases, the style illustrated is not necessarily one that the authors recommend.

Jeanne Adams

It is with deep regret that we acknowledge the passing in 2007 April of Jeanne Adams—our coauthor and longtime colleague and friend. Among her many contributions to computing and Fortran standardization, she is best known for her chairmanship of the committee that developed Fortran 90.

Walter S. Brainerd

Richard A. Hendrickson

Richard E. Maine

Jeanne T. Martin

Brian T. Smith

USA, 2008 May

1 Introduction

For a programming language, Fortran has been around a long time. It was one of the first widely used "high-level" languages, as well as the first programming language to be standardized. Although Fortran has been enhanced many times, the enhancements almost always have been upward compatible; old programs continue to work with new compilers. It is still the premier language for scientific and engineering computing applications.

The purpose of this handbook is to describe the latest version of this language, Fortran 2003. This chapter sets the stage by providing relevant background and describing the notation used to specify the syntax of Fortran 2003.

1.1 History

1.1.1 Initial Development of Fortran

In 1954 a project was begun under the leadership of John Backus at IBM to develop an "automatic programming" system that would convert programs written in a mathematical notation to machine instructions for the IBM 704 computer. Many were skeptical that the project would be successful because, at the time, computer memories were so small and expensive and execution time so valuable that it was believed necessary for the compiled program to be almost as good as that produced by an assembly language programmer.

This project produced the first Fortran compiler, which was delivered to a customer in 1957. It was a great success by any reasonable criterion. The efficiency of the code generated by the compiler surprised even some of its authors. A more important achievement, but one that took longer to realize, was that programmers could express their computations in a much more natural way. This increased productivity and permitted the programmer to write a program that could be maintained and enhanced much more easily than an assembly language program.

About one year after the introduction of the first Fortran compiler, IBM introduced Fortran II. One of the most important changes in Fortran II was the addition of subroutines that could be compiled independently. Thus, Fortran changed substantially even during its first year; it has been changing continually ever since.

1.1.2 Standardization

By the early 1960s, many computer vendors had implemented a Fortran compiler. They all included special features not found in the original IBM compiler. These features usually were included to meet needs and requests of the users and thus provide an inducement for the customer to buy computer systems from the vendor providing the best features. Because the language was very young, a special added feature could be

J.C. Adams et al., *The Fortran 2003 Handbook*,
DOI: 10.1007/978-1-84628-746-6_1,

tested to see if it was a good long-term addition to the language. Unfortunately, the profusion of dialects of Fortran prevented programs written for one computer from being transported to a different computer system.

1.1.2.1 Fortran 66

At about this time, the American Standards Association (ASA), which became the American National Standards Institute (ANSI) and is now the National Committee for Information Technology Standards (NCITS), began a project of standardizing many aspects of data processing. Someone had the daring idea of standardizing programming languages. A committee, which became X3J3, then J3, and was renamed INCITS/PL22.3 in 2007, was formed to develop a standard for Fortran. This standard was adopted in 1966 [3]; after the adoption of Fortran 77, it became known as Fortran 66 to distinguish the two versions.

1.1.2.2 Fortran 77

The language continued to develop after 1966, along with general knowledge in the areas of programming, language design, and computer design. Work on a revision of Fortran 66 was completed in 1977 (hence the name Fortran 77) and officially published in 1978 [4]. The most significant features introduced in this version were the character data type, the IF-THEN-ELSE construct, and many new input/output facilities, such as direct access files and the OPEN statement. Except for the character data type, most of these features had been implemented in many compilers or preprocessors. During this revision, Hollerith data was removed because the character data type is a far superior facility. Although this idea of removing features did not seem very controversial when Fortran 77 was introduced, it proved to be controversial later—so much so that no Fortran 77 features were removed in Fortran 90.

Fortran 77, developed by X3J3, was an ANSI standard—an American National Standard. At about this time the International Standards Organization (ISO) began to mature in the computing language area and adopted Fortran 77 as an international standard; the ISO standard was identical to the ANSI standard, and in fact consisted of one page that referenced the ANSI standard.

1.1.2.3 Fortran 90

As soon as the technical development of Fortran 77 was completed, X3J3 and its ISO counterpart Working Group 5 (SC22/WG5) teamed up for the next revision, which was called Fortran 90. Fortran 90 was an ISO standard first [11], which the US adopted, word for word, as an ANSI standard. Although X3J3 did the technical work on Fortran 90, and produced the standard document, the torch had been passed as to the "owner" of the Fortran standard; that "owner", for Fortran 90 and for the foreseeable future, is ISO.

Fortran 90 was a major advance over Fortran 77. It included: a greatly liberalized source form, a complete set of iteration and selection control structures, enhanced numeric facilities (e.g., the environmental intrinsic functions), a comprehensive data-parallel array language, data structures (including dynamic structures), user-defined types and operators, procedure extensions (e.g., recursion, internal procedures, explicit procedure interfaces, user-defined generic procedures), module encapsulation (with powerful data hiding features), kind type parameters (e.g., to regularize the different "kinds" of reals, provide the corresponding kinds of complex, accommodate different kinds of character, and to resolve overloads in a simple way), dynamic objects (e.g., allocatable arrays and pointers), and some I/O extensions (e.g., NAMELIST and non-advancing I/O). The concept of "obsolescent" features was introduced, and a handful of Fortran 77 features were so identified. But removal of significant numbers of archaic features was controversial and so no features were actually removed. A standard-conforming Fortran 77 program is a standard-conforming Fortran 90 program with the same interpretation.

1.1.2.4 Fortran 95

Fortran 95 [10], specified by WG5 and produced by X3J3, represented a minor revision to Fortran 90. Most of the changes corrected and clarified what was in Fortran 90. However, a few significant features, such as pure functions and the FORALL construct and statement, were added because they were considered important contributions from High Performance Fortran [17]. A few (quite insignificant) features designated as obsolescent in Fortran 90 were removed from Fortran 95. These features are:

1. Real and double precision DO variables
2. Branching to an END IF from outside the block
3. PAUSE statement
4. ASSIGN statement, assigned GO TO statement, and related features
5. *n*H edit descriptor

1.1.2.5 Fortran 2003

Fortran 2003 [7], while not the major advance that Fortran 90 represented, still added considerably more features than did Fortran 95, which was a minor revision of Fortran 90. The most important features introduced in Fortran 2003 are:

- interoperability with the C programming language [15], permitting easy portable access to the low-level facilities of C from Fortran programs and the portable use of Fortran libraries by programs written in C
- support for exceptions and IEEE arithmetic [13] in so far as it does not conflict with existing Fortran arithmetic rules
- support for object-oriented programming, including inheritance (type extension), polymorphism (dynamic typing), and type-bound procedures
- data-type enhancements, such as parameterized derived types, allocatable components, and finalizers

- input/output enhancements, such as user-defined derived-type input/output, asynchronous input/output, stream input/output, and the FLUSH statement to empty buffers
- support for international usage, including the ISO 10646 character set [16] and choice of a comma or period for the decimal symbol in numeric formatted input/output
- other features, such as procedure pointers, the PROTECTED and VOLATILE attributes, the IMPORT statement, access to environment variables and command-line arguments, better error handling, and better rounding control

1.2 The Fortran 2003 Language Standard

The Fortran 2003 standard [7] describes the syntax and semantics of the Fortran programming language but only certain, not all, aspects of the Fortran processing system. When specifications are not covered by the standard, the interpretation is processor dependent; that is, the processor defines the interpretation, but the interpretation for any two processors need not be the same. Programs that rely on processor-dependent interpretations typically are not portable.

The specifications that are included in the standard are:

1. the syntax of Fortran statements and forms for Fortran programs
2. the semantics (meaning) of Fortran statements and the semantics of Fortran programs
3. interoperability requirements between Fortran and C programs
4. requirements for IEEE floating-point support
5. specifications for input data
6. appearance of output data

The specifications that are not defined in the standard are:

1. the way in which a Fortran compiler is written
2. operating system facilities defining the computing system
3. methods used to transfer data to and from peripheral storage devices and the nature of the peripheral devices
4. behavior of extensions implemented by vendors
5. the size and complexity of a Fortran program and its data
6. the hardware or firmware used to run the program

7. the way values are represented and the way numeric values are computed
8. the physical representation of data
9. the characteristics of disks and other storage media

1.2.1 Program Conformance to the Standard

A program conforms to the standard if the statements are all syntactically correct, execution of the program causes no violations of the standard (e.g., referencing an element outside the bounds of an array), and the input data is all in the correct form. A program that uses a vendor extension is not standard conforming and may not be portable. In particular, a program that uses intrinsic procedures or modules provided by the vendor is not standard conforming.

1.2.2 Processor Conformance to the Standard

In the Fortran 2003 standard, the term "processor" means the combination of a Fortran compiler and the computing system that executes the code. A processor conforms to the standard if it correctly processes any standard-conforming program, provided the Fortran program is not too large or complex for the computer system in question. Except for certain restrictions in format specifications, the processor must be able to flag any nonstandard syntax (described by the syntax rules and constraints) used in the program. This includes the capability to flag any extensions available in the vendor software (including deleted features) and used in the program. Note that the compiler is not required to scan a character string used as a format. The standard also requires that the processor detect, with appropriate explanation, the following:

1. obsolescent features (see C)
2. intrinsic type kind values not supported
3. characters not permitted by the processor
4. illegal source form
5. violations of the scope rules for names, labels, operators, and assignment symbols

The standard does not require the processor to detect nonstandard intrinsic modules, but most processors probably will detect their use.

Rules for the form of the output are less stringent than for other features of the language in the sense that the processor may have some options about the format of the output and the programmer may not have complete control over which of these options is used.

A processor may include extensions not in the standard; if it processes standard-conforming programs according to the standard, it is considered to be a standard-conforming processor.

1.2.3 Portability

One of the main purposes of a standard is to describe how to write portable programs. However, there are some things that are standard conforming, but not portable. An example is a program that computes a very large number. Certain computing systems will not accommodate a number this large. Thus, such a number could be a part of a standard-conforming program, but may not run on all systems and thus may not be portable. Another example is a program that uses a deeper nesting of control constructs than is allowed by a particular compiler.

1.2.4 A Permissive Standard

The primary purpose of the Fortran standard is to describe a language with the property that, if a programmer uses the language, the difficulties of porting programs from one computer system to another will be minimized. But to handle the somewhat contradictory goal of permitting experimentation and development of the language, the standard is *permissive*; that is, a processor can conform to the standard even if it allows features that are not described in the standard. This has its good and bad aspects.

On the positive side, it allows implementors to experiment with features not in the standard; if they are successful and prove useful, they can become candidates for standardization during the next revision. Thus, a vendor of a compiler may choose to add some features not found in the standard and still conform to the standard by correctly processing all of the features that are described in the standard.

On the negative side, the burden is on the programmer to know about and avoid these extra features when the program is to be ported to a different computer system. The programmer is given some help with this problem in that a Fortran processor is required to recognize and warn the programmer about syntactic constructs in a program that do not conform to the standard. A good Fortran programmer's manual also will point out nonstandard features with some technique, such as shading on the page. But there is no real substitute for knowledge of the standard language itself. This handbook provides this knowledge.

1.3 Notation Used in this Book

When a word or words are in **bold font**, this indicates that the term is being defined.

Fortran keywords, such as CALL and IF, are capitalized when discussed in text.

Examples in a `monospaced font` should be compilable when incorporated into a complete program unit.

Braces { } are used in Appendices A and B to indicate optional intrinsic procedure arguments.

In this book, a simplified form (compared to that used in the standard) is used to describe the syntax of Fortran 2003 programs. The forms consist of program text in the same font used to display program examples (such as `END DO`) and syntactic terms that must be replaced with correct Fortran source for those terms, which are printed using a sans serif font (such as input-item-list). Optional items are enclosed in brackets; items enclosed in brackets followed by ellipses (...) may occur any number (including zero) of times. The ampersand (&) is used to continue a line, just as it is used to continue a line in a Fortran 2003 program. Use of one of the syntactic forms always produces a syntactically correct part of a Fortran 2003 program. These syntactic forms indicate how to construct most of the correct Fortran 2003 statements, but in some cases are incomplete in that they do not describe all of the possible forms. For example, specifiers in some input/output statements are listed in order, but may be written in any order.

The following syntax form occurs in 9. It describes one form that can be used to construct a direct access formatted WRITE statement. The general syntax for the WRITE statement is quite complex and gives no hint as to which options are allowed for direct access formatting. On the other hand, this rule is overly restrictive in that it indicates a particular order for the options, which is not required by the standard. Nevertheless, using this form always will produce a correct WRITE statement.

```
WRITE([UNIT=]scalar-integer-expression&
  ,[FMT=]format&
  ,REC=scalar-integer-expression&
  [,IOSTAT=scalar-default-integer-variable]&
  [,ERR=label]&
)[output-item-list]
```

Another property of the syntactic forms is that the terms used are informal. They are not necessarily defined precisely anywhere in the book and are not always the same as those in the standard; they are often longer and more descriptive. If you need to know the precise syntax allowed, refer to Fortran 2003 standard [7].

A general restriction on all syntax rules is that, for forms with lists of keywords, any particular keyword may appear at most once. For example, there may be at most one IOSTAT in a WRITE statement.

In the text near many syntax rules is a reference, such as (R*nnn*). This indicates that the syntax rule is related to syntax rule *nnn* of the Fortran 2003 standard.

The syntax rules use abbreviations for some common forms. These are listed in Table 1-1.

Table 1-1 Syntax form abbreviations

char	character
dtio	derived-type input/output
expr	expression
id	identifier
io	input/output
spec	specification

An occurrence of abc-list is shorthand for a list of one or more things of form abc, separated by comma; that is abc [, abc] ....

The standard categorizes some restrictions as constraints. The difference between constraints and the other restrictions is that the compiler must be able to detect violations of the constraints during compilation. In this book, there is no distinction between restrictions that the compiler must detect and those it need not detect and also between compile-time and run-time restrictions.

1.4 Approximations to Real and Complex Values

Most real (and hence complex) values cannot be represented exactly in a computer. For example, when

```
X = 1.23
```

is executed, the value stored for X might not be exactly 1.23, but the nearest approximation that can be represented in the computer, which usually will be a binary representation.

Whenever real and complex values and operations are discussed in this book, it is to be understood that the values will be approximate when represented in the computer. For examples, when this book indicates that the symbol "+" represents addition, it really means that it represents an approximation to the sum of two values.

1.5 References

1. Adams, Jeanne C., Walter S. Brainerd, Jeanne T. Martin, Brian T. Smith, and Jerrold L. Wagener, *The Fortran 90 Handbook*, McGraw-Hill, 1992.

2. Adams, Jeanne C., Walter S. Brainerd, Jeanne T. Martin, Brian T. Smith, and Jerrold L. Wagener, *The Fortran 95 Handbook*, MIT Press, 1997.

3. *American National Standard FORTRAN, X3.9-1966,* United States of America Standards Institute, 1966 (Fortran 66).

4. American National Standards Institute, *American National Standard Programming Language FORTRAN, ANSI X3.9-1978,* New York, 1978 (Fortran 77).

5. Brainerd, Walter S., Ed., Fortran 77, *Communications of the ACM,* Vol. 21, No. 10, October 1978, pp. 806–820.

6. Greenfield, Martin H., History of FORTRAN standardization, *Proceedings of the 1982 National Computer Conference,* AFIPS Press, Arlington, VA, 1982.

7. International Standards Organization, *ISO/IEC 1539-1 : 2004, Information technology—Programming languages—Fortran—Part 1: Base language,* Geneva, 2004 (Fortran 2003).

8. International Standards Organization, *ISO/IEC 1539-2 : 1994, Information technology—Programming languages—Fortran—Part 2: Varying length character strings,* Geneva, 1994.

9. International Standards Organization, *ISO/IEC 1539-3 : 1998, Information technology—Programming languages—Fortran—Part 3: Conditional compilation,* Geneva, 1998.

10. International Standards Organization, *ISO/IEC 1539-1 : 1997, Information technology—Programming languages—Fortran—Part 1: Base language:* Geneva, 1997 (Fortran 95).

11. International Standards Organization, *ISO/IEC 1539 : 1991, Information technology—Programming languages—Fortran,* Geneva, 1991 (Fortran 90).

12. International Standards Organization, *ISO/IEC 646:1991. Information technology—ISO 7-bit coded character set for information interchange.* ASCII.

13. International Standards Organization, *IEC 60559 (1989-01), Binary floating-point arithmetic for microprocessor systems.* The original was IEEE 754-1985 and is often called the floating-point standard.

14. International Standards Organization, *ISO 8601:1988, Data elements and interchange formats—Information interchange—Representation of dates and times.*

15. International Standards Organization, *ISO/IEC 9989:1999, Information technology—Programming languages—C.* The C standard.

16. International Standards Organization, *ISO/IEC 10646-1:2000, Information technology—Universal multiple-octet coded character set (UCS)—Part 1: Architecture and basic multilingual plane.*

17. Koelbel, Charles H., David B. Loveman, Robert S. Schreiber, Guy L. Steel Jr., and Mary E. Sisal, *High Performance Fortran Handbook,* MIT Press, Cambridge, MA, 1993.

2 Fortran Concepts and Terms

- A **Program** is an organized collection of program units. There must be exactly one main program, and in addition there may be modules, external subprograms, and block data units. Elements described by means other than Fortran may be included.
- A **Module** provides a means of packaging related data and procedures, and hiding information not needed outside the module. There are several intrinsic modules.
- The **Data Environment** consists of the data objects upon which operations will be performed to create desired results or values. These objects may have declared and dynamic types; they may have type parameters, and they may possess attributes such as dimensionality. They need not exist for the whole execution of the program. Allocatable objects and pointer targets may be created when needed and released when no longer needed.
- **Program Execution** begins with the first executable construct in the main program and continues with successive constructs unless there is a change in the flow of control. When a procedure is invoked, its execution begins with its first executable construct. On normal return, execution continues where it left off. Execution may occur simultaneously with input/output processes.
- The **Definition Status** of a variable indicates whether or not the variable has a value; the value may change during execution. Most variables are initially undefined and become defined when they acquire a value. The status also may become undefined during execution. Pointers have both an association status and a definition status. Allocatable objects have both an allocation status and a definition status.
- **Scope** and **Association** determine where and by what names various entities are known and accessible in a program. These concepts form the information backbone of the language.

This chapter introduces the basic concepts and fundamental terms needed to understand Fortran. Some terms are defined implicitly by the syntax rules. Others, such as "associated" or "present" are ordinary English words, but they have a specific Fortran meaning.

One of the major concepts involves the organization of a program. A program consists of program units; program units consist of Fortran statements. Some statements are executable; some are not. In general, the nonexecutable statements define the data environment, and the executable statements specify the actions taken. This chapter presents the high-level syntax rules for a Fortran program. It also describes the order in which constructs and statements may appear in a program and concludes with an example of a short, but complete, Fortran program.

J.C. Adams et al., *The Fortran 2003 Handbook*,
DOI: 10.1007/978-1-84628-746-6_2, © Springer-Verlag London Limited 2009

While there is some discussion of language features here to help explain various terms and concepts, Chapters 3–16 contain the complete description of all language features.

2.1 Program Organization

A collection of program units constitutes an executable program. A Fortran program must have one main program and may have any number of the other program units. Program units may serve as hosts for smaller scoping units. Information may be hidden within part of a program or communicated to other parts of a program by various means. The programmer may control the parts of a program in which information is accessible.

With the introduction of C interoperability in Fortran 2003, it is possible to include, with much greater ease and portability, external procedures and other entities defined by a means other than Fortran. A processor has one or more companion processors. A **companion processor** is a processor-dependent mechanism by which global data and procedures may be referenced or defined. It may be the Fortran processor itself, or it may be another Fortran processor. If a procedure is defined by means of a companion processor that is not the Fortran processor itself, the standard refers to the C function that defines the procedure. Although the procedure need not be defined by means of the C programming language, the interoperability mechanisms are designed to mesh well with C.

2.1.1 Program Units

A Fortran program unit is one of the following:

main program
module
external subprogram
block data

A Fortran program may consist of only a main program, although usually there are also modules and/or external subprograms which may be subroutine or function subprograms. These program units contain constructs and statements that define the data environment and the steps necessary to perform calculations. Each program unit has an END statement to terminate the program unit. Each has a special initial statement as well, but the initial statement for a main program is optional. For example, a program might contain a main program, a module, and a subroutine:

```
program task
   . . .
   call calc (z)
   . . .
end program task
```

```
module info
   . . .
end module info

subroutine calc(x)
   use info
      . . .
end subroutine calc
```

An ideal Fortran program would consist of a main program and several modules; that is, there would be no external subprograms. This is the best model for packaging and encapsulation (2.2.5). Subroutine and function subprograms are a fundamental part of the language. They may be module, internal, or external subprograms.

The interface of a procedure supplies information about the name and type (if a function) of the procedure, as well as information about its arguments. A program is more robust if the interfaces of procedures are known when the procedures are invoked. This is inherently the case for internal procedures, module procedures, and all of the intrinsic procedures. In addition, the interfaces of procedures defined in other languages must be described to the Fortran system as C function interfaces (15.6).

The main program could be defined in a language other than Fortran, but it is usually the language of the main program that determines the program's primary nature. For example, a Fortran main program with some elements specified in another language is still a Fortran program; whereas, if the main program is specified in C but there is access to Fortran elements, the program is generally considered to be a C program. Interlanguage communication is described in 15.

Because all except the most trivial of programs will make use of subroutines and functions in some form, it might be expected that subroutines and functions would be described earlier, but that is not the case. Chapter 12 describes them in detail. Chapter 11 describes all program units—the main program, modules, external subprograms, and block data program units.

Internal procedures and module procedures gain access to information in their hosts by host association. A USE statement specifying a module can appear in a main program, a subprogram, a module, an interface body, or a block data subprogram to gain access to the module's public information. This method of access is called use association. Association is described in 16.

Figure 2-1 illustrates the organization of a sample Fortran program. The lines with thin arrows represent internal and external subprogram references with the arrow pointing to the subprogram. The thick solid arrows represent access by use association with the arrow pointing to the position of a USE statement.

2.1.1.1 Main Program

The main program is required; if there are other program units, the main program acts as a controller; that is, it takes charge of the program and controls the order in which procedures are executed.

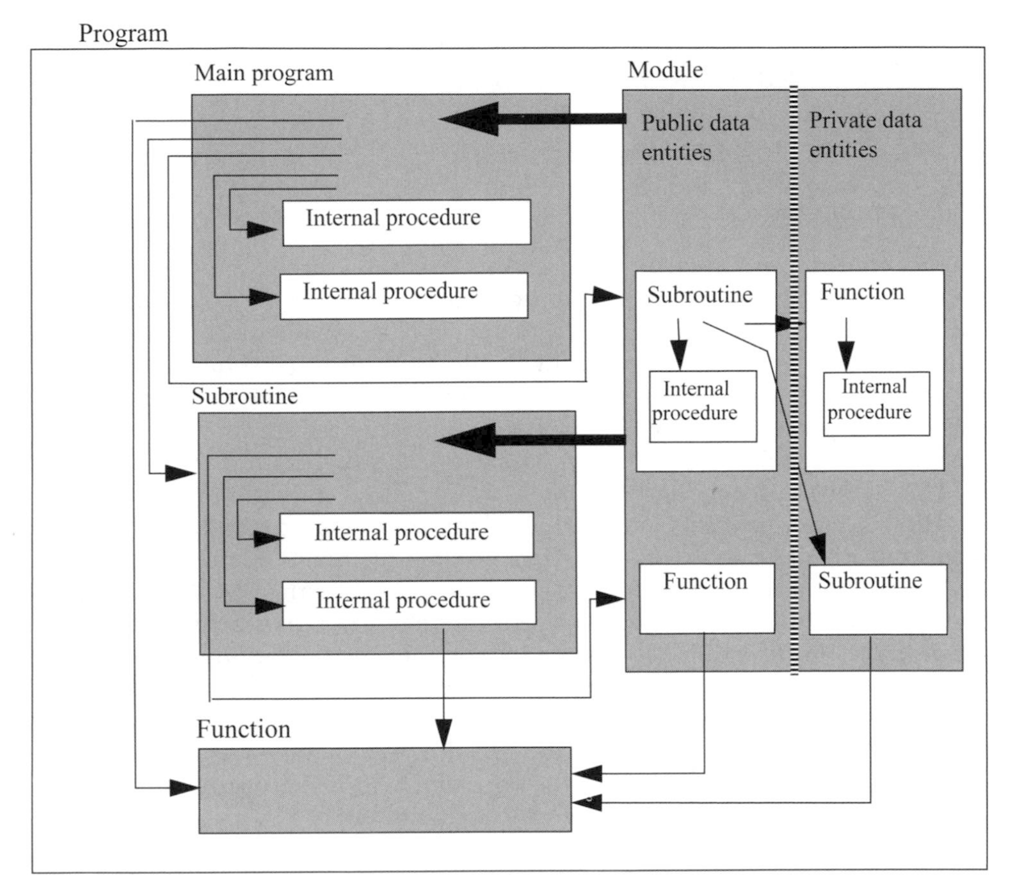

Figure 2-1 Example of program packaging. The thick arrows represent use association; the thin arrows represent procedure references.

2.1.1.2 Module

A module contains definitions that can be made accessible to other program units by use association. These definitions include data definitions, type definitions, definitions of procedures known as module procedures, and specifications of procedure interfaces. A module procedure may be invoked by another module procedure in the module or by other program units that access the module. Fortran 2003 introduced intrinsic modules; there were no intrinsic modules in earlier standards. These are the ISO_FORTRAN_ENV module (13.6.1) that provides public entities relating to the environment such as input/output units and storage sizes, the ISO_C_BINDING module (15.3) that provides access to named constants representing kind values that are compatible with C types, and three IEEE modules (14.3) that provide support for exceptions and IEEE arithmetic.

2.1.1.3 External Subprogram

An external subprogram (a function or a subroutine) may be used to perform a task or calculation on entities available to the external subprogram. These entities may be the arguments to the subprogram that are provided in the reference, entities defined in the subprogram, or entities accessible from modules or common blocks. A CALL statement is used to invoke a subroutine. A function is invoked when its value is needed in an expression. The computational process that is specified by a function or subroutine subprogram is called a **procedure**. An external subprogram provides one way to define a procedure. It may be invoked from other program units of the Fortran program. Unless it is a pure procedure, a subroutine or function may change the program state by changing the values of data objects accessible to the procedure.

2.1.1.4 Block Data Program Unit

A block data program unit (11.5) contains data definitions only and is used to specify initial values for a restricted set of data objects.

2.1.1.5 Compilation

Prior to the introduction of modules into Fortran, program units could be compiled independently with no need for information from any other program unit. Any information needed in more than one program unit had to be replicated wherever it was needed. The compiled program unit could be used in a number of applications without the necessity of recompiling; this is called **independent compilation**.

If a program unit contains a USE statement, the referenced module must be available in some form when that program unit is compiled.

There are many ways to implement modules; however, most implementations require compilation of modules prior to compilation of any program units that use the modules. The compilation often produces a file containing encoded or summarized information about the module, which is accessed when a program using the module is compiled.

The situation regarding the availability of include files is somewhat similar, but because include files are simply inserted as text in a program, they are not usually preprocessed in any way.

2.1.2 Procedures

A procedure specifies a task or a calculation, usually one that can be separated out from the main flow or one that is needed in different parts of the program. A procedure may take the form of a subroutine or a function. Every procedure has an interface that must be unique in some way, A set of generic procedures may be identified by the same name or symbol, but made unique by their arguments. A procedure may be defined by means other than the Fortran language.

2.1.2.1 Internal Procedures

Main programs, module subprograms, and external subprograms may contain internal subprograms, which may be either subroutines or functions. The procedures they define are called **internal procedures**. Internal subprograms must not themselves contain internal subprograms, however. The main program, external subprogram, or module subprogram that contains an internal subprogram is referred to as the internal subprogram's **host**. Entities known in a host are available to an internal procedure by host association. Internal procedures may be invoked within their host or within other internal procedures in the same host. Internal procedures are described in 12.

There is also an obsolescent feature, the statement function (12.4.4), which specifies a function by a single statement.

2.1.2.2 Procedure Interfaces

An interface provides the procedure name, the number of arguments, their types, attributes, names, and the type and attributes of a function result. This information is required in some cases, such as for a dummy argument, which assumes the shape of its actual argument (12.5.1.2). The information also allows the processor to check the validity of an invocation.

If a procedure interface is not inherently available, it may be specified in an interface block. All program units, except block data, may contain procedure interface blocks. A procedure interface block contains one or more interface bodies that are used to describe the interfaces of procedures that would otherwise be unknown. Interface blocks are used for external procedures, dummy procedures, procedure pointers, abstract procedures, or type-bound procedures. An interface block with a generic specification may be used to describe generic procedures or user-defined operators, assignment, or input/output. Procedure interfaces are described in 12.

2.1.2.3 Generic Procedures

Fortran has the concept of a generic procedure, that is, one that can accept arguments that have different types in different invocations. If the procedure is a function, in most cases the type of the result is the same as that of the arguments. An example is the intrinsic SIN (the sine function), which can accept a real, double precision, or complex argument. A user-defined procedure also can be generic. A user defines several specific procedures, and either collects their interfaces in an interface block with a generic specification or lists them in a GENERIC statement in the type definition. The identifier that appears in the generic specification or the GENERIC statement may be used to reference the specific procedure whose arguments match those of the reference.

2.1.2.4 Procedures Defined by Other Languages

Chapter 15 describes how procedures defined by means of the C programming language can be accessed from Fortran and how procedures defined in Fortran can be accessed from C programs. Other languages may be accommodated by these same mechanisms. The mechanisms are not limited to C, but are described in terms of C protocols. Some of the additions to Fortran 2003 to facilitate this process are useful in themselves to strictly Fortran programs, such as the VALUE attribute for dummy arguments (5.9.2), enumerations (4.6), and stream input/output (9.1.5.3).

2.2 Data Environment

Before a calculation can be performed, its data environment must be established. The data environment consists of data objects that possess properties, attributes, and values. The steps in a computational process generally specify operations that are performed on operands (or objects) to create desired results or values. Operands may be constants, variables, constructors, function references, or more complicated expressions made up of these items; each operand has a data type (which may be dynamic); it may have type parameters; and, if it is defined, it has a value. A data object has attributes in addition to type. Chapter 4 discusses data type in detail; Chapter 5 discusses how program entities and their attributes are declared; and Chapters 6 and 7 describe how data objects may be used.

2.2.1 Data Type

The Fortran language provides five intrinsic data types—real, integer, complex, logical, and character—and allows users to define additional types. Sometimes it is natural to organize data in combinations consisting of several components of different types. Because the data describe one object, it is convenient to have a means to refer to this aggregation of data by a single name. In Fortran, an aggregation of such data values is called a **structure**. To use a structure, a programmer must first define the type of the structure. Once the new type is defined, any number of structures (or objects) of that type may be declared.

Some applications require related objects, such as a basic line plus a line of a certain style (dotted or dashed), or of a certain color, or both style and color. A base type may be defined and then extended by adding different components. When a type is defined, it is not necessary to specify that it may be extended. Generic procedures may be defined (such as DRAW or ADD_TO_FIGURE) that accept as an actual argument an object of the base type or any extension of it. Such an argument that may be of any of these types is polymorphic.

2.2.2 Type Parameters

Both intrinsic and user-defined types may have parameters. For the intrinsic types, a kind type parameter specifies a particular representation. In addition, the character type has a length parameter.

Each of the intrinsic types may have more than one representation (specified by a KIND parameter). The Fortran standard requires at least two different representations for each of the real and complex types that correspond to "single precision" and "double precision", and permits more.

A type parameter for a user-defined type is also either a kind type parameter or a length type parameter. Type parameters for user-defined types are specified in the type definition.

Portable mechanisms for specifying precision are provided so that numerical algorithms that depend on a minimum numeric precision can be programmed to produce reliable results regardless of the processor's characteristics. Fortran permits more than one representation for the integer, logical, and character types as well. Alternative representations for the integer type permit different ranges of integers. Alternative representations for the logical type might include a "packed logical" type to conserve memory space and an "unpacked logical" type to increase speed of access. The large number of characters required for ideographic languages, such as those used in Asia with thousands of different graphical symbols, cannot be represented as concisely as alphabetic characters and require "more precision". For international usage Fortran 2003 encourages support of the ISO 10646 character set (1.5).

A kind type parameter value must be known at compile time and may be used to resolve generic procedure references. A length type parameter value need not be known at compile time; it may be used for character lengths, array dimensions, or other sizes. If it is a deferred type parameter, indicated by a colon (:), it may change during execution. If it is an assumed type parameter, indicated by an asterisk (*), it assumes its value from another entity, such as an actual argument.

Examples of type declarations with parameters are:

```
complex (kind = HIGH) x
integer (kind = SHORT) days_of_week
character (kind = ISO_10646, len = 500) HAIKU
type MY_ARRAY (pick_kind, rows, cols)        ! Type definition
   integer, kind :: pick_kind
   integer, len :: rows, cols
   real (pick_kind) :: VALUES (rows, cols)
end type MY_ARRAY
type(MY_ARRAY) AA(HIGH, i, j)
```

where HIGH, SHORT, and ISO_10646 are named integer constants given appropriate values by the programmer. The length parameter for the character string HAIKU has the value 500. AA is of type MY_ARRAY; its single component, VALUES, is a real array of kind HIGH and dimension (i, j), where i and j are specification expressions.

2.2.3 Dimensionality

Single objects, whether intrinsic or user-defined, are scalar. Even though a structure has components, it is technically a scalar. A set of scalar objects, all of the same type, may be arranged in patterns involving columns, rows, planes, and higher-dimensioned configurations to form arrays. It is possible to have arrays of structures. An array may have up to seven dimensions. The number of dimensions is called the **rank** of the array. It is declared when the array is declared and cannot change. The **size** of the array is the total number of elements and is equal to the product of the extents in each dimension. The **shape** of an array is the list of its extents. Two arrays that have the same shape are said to be **conformable**. A scalar is conformable with any array. Examples of array declarations are:

```
real :: coordinates (100, 100)
integer :: distances (50)
type(line) :: mondrian(10)
```

An array is an object and may appear in an expression or be returned as a function result. Intrinsic operations involving arrays of the same shape are performed element-by-element to produce an array result of the same shape. There is no implied order in which the element-by-element operations are performed.

A portion of an array, such as an element or section, may be referenced as a data object. An array element is a single element of the array and is scalar. An array section is a subset of the elements of the array and is itself an array.

2.2.4 Dynamic Data

Data objects may be dynamic in size, shape, type, or length type parameters, but not rank or kind type parameters. The dynamic data objects are:

polymorphic objects
pointers
allocatable objects
automatic objects

The type of a polymorphic object (5.2) may change during program execution. Objects that may have both a dynamic type as well as a dynamic size and shape are data pointers, allocatable variables, and dummy arguments. Automatic objects appear in subprograms and come into existence when the subprogram is invoked.

Dynamic type was introduced in Fortran 2003. An entity that is not polymorphic has both a declared and a dynamic type, but they are the same. The dynamic type of a polymorphic object that is not allocated (6.7.1) or associated (7.5.5.1) is its declared type. The CLASS keyword is used to declare polymorphic entities. An object declared with CLASS (*) is an unlimited polymorphic object with no declared type.

Procedures and data objects in Fortran may be declared to have the POINTER attribute. A procedure pointer must be a procedure entity. A data pointer must be associated with a target before it can be used in any calculation. This is accomplished by allocation (6.7.1.2) of the space for the target or by assignment of the pointer to an existing target (7.5.5.1). A pointer assignment statement is provided to associate a pointer with a target (declared or allocated). It makes use of the symbol pair => rather than the single character =; otherwise, it is executed in the same way that an ordinary assignment statement is executed, except that instead of assigning a value it associates a pointer with a target. For example,

```
real, target :: VECTOR(100)
real, pointer :: ODDS(:)
   . . .
ODDS => VECTOR(1:100:2)
```

The pointer assignment statement associates ODDS with the odd elements of VECTOR. The assignment statement

```
ODDS=1.5
```

defines each odd element of VECTOR with the value 1.5. Later in the execution sequence, the pointer ODDS could become associated with a different target by pointer assignment or allocation, as long as the target is a one-dimensional, default real array. Chapter 7 describes the pointer assignment statement.

If a pointer object is declared to be an array, its size and shape may change dynamically, but its rank is fixed by the declaration. If a pointer target is polymorphic, the pointer must be of a type that is compatible with the target, or both the pointer and target must be declared unlimited polymorphic. An example of pointer array declaration and allocation is:

```
real, pointer :: lengths (:)
allocate (lengths (200))
```

A variable may be declared to have the ALLOCATABLE attribute. Space must be allocated for the variable before it can be used in any calculation. The variable may be deallocated and reallocated with a different type, length type parameters, and shape as the program executes. As with a pointer, the rank is fixed by the declaration. An allocatable variable cannot be made to point to an existing named object; the object always must be created by an ALLOCATE statement. An example of allocatable array declaration and allocation is:

```
real, allocatable :: lengths (:)
allocate (lengths (200))
```

The similarities of these examples reflect the similarity of some of the uses of allocatable arrays and pointers, but there are differences. Pointers may be used to create dynamic data structures, such as linked lists and trees. The target of a pointer can be changed by reallocation or pointer assignment; the new target must be of the same rank but may have different extents in each dimension. The attributes of an allocatable variable can be changed only by deallocating and reallocating the variable. There is a MOVE_ALLOC intrinsic function that can be used if the values of the elements of an allocatable array are to be preserved when its size is changed. Use of allocatable variables generally leads to more efficient execution than use of the more flexible pointers.

Only pointers and allocatable objects may be allocated or deallocated. It is possible to inquire whether an object is currently allocated. Chapter 5 describes the declaration of pointers and allocatable objects; Chapter 6 covers the ALLOCATE and DEALLOCATE statements; Chapter 13 and Appendix A describe the ASSOCIATED intrinsic inquiry function for pointers and the ALLOCATED intrinsic inquiry function for allocatable variables. Chapter 15 describes dynamic interoperable objects.

Automatic data objects, either arrays or character strings (or both), may be declared in a subprogram. These local data objects are created on entry to the subprogram and disappear when the execution of the subprogram completes. These are useful in subprograms for temporary arrays and characters strings whose sizes are different for each reference to the subprogram. An example of a subprogram unit with an automatic array TEMP is:

```
subroutine SWAP_ARRAYS (A, B)
   real, dimension (:) :: A, B
   real, dimension (size (A)) :: TEMP

   TEMP = A
   A = B
   B = TEMP
end subroutine SWAP_ARRAYS
```

A and B are assumed-shape array arguments; that is, they take on the shape of the actual arguments. TEMP is an automatic array that is created the same size as A on entry to subroutine SWAP_ARRAYS. SIZE is an intrinsic function.

2.2.5 Packaging and Encapsulation

The packaging of a fair-sized program is an important design consideration when a new Fortran application is planned. The most important benefit of packaging is information hiding. Entities can be kept inaccessible except where they are actually needed. This provides some protection against inadvertent misuse or corruption, thereby improving program reliability. Packaging can make the logical structure of a program more apparent by hiding complex details at lower levels. Programs are therefore easier to comprehend and less costly to maintain. The Fortran features that provide these benefits are

- user-defined types
- internal procedures
- modules

The accessibility of a user-defined type in a module may be public, private, or protected. In addition, even if the type is public, it may have private components. A type definition has a type-bound procedure part in which the procedures bound to that type are specified.

Internal procedures may appear in main programs, module subprograms, and external subprograms; they are known only within their host. The name of an internal procedure must not be passed as an argument. The Fortran standard further restricts internal procedures in that an internal procedure must not itself be the host of another internal procedure. However, statement functions may appear within an internal procedure.

Modules provide the most comprehensive opportunities to apply packaging concepts including several levels of organization and hiding (5.8). The entities specified in a module (types, data objects, procedures, interfaces, etc.) may be made available to other scoping units; may be made available, but protected from corruption outside the module; or may be kept private to the module. Thus modules provide flexible encapsulation facilities for entities in an object-oriented application. The procedures, mentioned in a type definition (4.4.2) and referred to as type-bound procedures (4.4.11), generally appear as module procedures in the module that contains the type definition.

In addition to the usual capabilities of procedures, these type-bound procedures may specify

- defined operators
- defined assignment
- defined input/output
- finalization

Finalization is accomplished by a final procedure that is invoked automatically just before an object of the type is destroyed by deallocation, the execution of a RETURN or END statement, or some other means.

Of course, more than one type definition may appear in a module, so if there is a need for communication among separate but related objects, the module provides the appropriate means for permitting and controlling access to information.

2.3 Program Execution

During program execution, constructs and statements are executed in a prescribed order. Variables become defined with values and may be redefined later in the execution sequence. Procedures are invoked, perhaps recursively. Space may be allocated and later deallocated. The targets of pointers may change. The types of polymorphic variables may change.

2.3.1 Execution Sequence

Program execution begins with the first executable construct in the main program. An executable construct is an instruction to perform one or more of the computational actions that determine the behavior of the program or control the flow of the execution of the program. These actions include performing arithmetic, comparing values, branching to another construct or statement in the program, invoking a procedure, or reading from or writing to a file or device. Examples of executable statements are:

```
          read (5, *) z, y
          x = (4.0 * z) + base
          if (x > y) go to 100
          call calculate (x)
     100  y = y + 1
```

When a procedure is invoked, its execution begins with the first executable construct after the entry point in the procedure. On normal return from a procedure invocation, execution continues where it left off in the invoking procedure.

Unless a control statement or construct that alters the flow of execution is encountered, program statements are executed in the order in which they appear in a program unit until a STOP, RETURN, or END statement is executed. Branch statements specify a change in the execution sequence and consist of the various forms of GO TO statements, a procedure reference with alternative return specifiers, EXIT and CYCLE statements in DO constructs, and input/output statements with branch label specifiers, such as ERR, END, and EOR specifiers. The control constructs (IF, CASE, DO, and SELECT TYPE) can cause internal branching implicitly within the structure of the construct. The SELECT TYPE construct chooses a block of code based on the dynamic type of its polymorphic selector. Chapter 8 discusses in detail control flow within a program.

Another feature of Fortran 2003 is asynchronous input/output. It allows computation to occur in parallel with an input/output process if the processor supports parallel processing. A WAIT statement may be used to synchronize the processes. This and other new input/output features are described in 9.

Normal termination of execution occurs if the END statement of a main program or a STOP statement is executed. Normal termination of execution also may occur in a procedure defined by means other than Fortran. If a Fortran program includes procedures executed by a companion processor, the normal termination process will include the effect of executing the C exit function.

2.3.2 Definition and Undefinition

Unless initialized, variables have no value initially; uninitialized variables are considered to be **undefined**. Variables may be initialized in type declaration statements, type declarations, DATA statements, or by means other than Fortran; initialized variables are considered to be **defined**. Some variables initialized by default initialization, such as that specified in a type definition, are initialized when the variables come into existence, whereas other variables such as those initialized in a DATA statement are initialized when execution begins.

A variable may acquire a value or change its current value, typically by the execution of an assignment statement or an input statement. Thus it may assume different values at different times, and under some circumstances it may become undefined. This is part of the dynamic behavior of program execution. Defined and undefined are the Fortran terms that are used to specify the definition status of a variable. The events that cause variables to become defined and undefined are described in 16.

A variable is considered to be defined only if all parts of it are defined. For example, all the elements of an array, all the components of a structure, or all characters of a character string must be defined; otherwise, the array, structure, or string is undefined. Fortran permits zero-sized arrays and zero-length strings; these are always considered to be defined.

Pointers have both a definition status and an association status. When execution begins, the association status of all pointers is undefined, except for data or default initialized pointers given the disassociated status. During execution a pointer may become disassociated, or it may become associated with a target. At some point the association status may revert to undefined. Even when the association status of a pointer is defined, the pointer is not considered to be defined unless the target with which it is associated is defined. Pointer targets become defined in the same way that any other variable becomes defined, typically by the execution of an assignment or input statement.

Allocatable variables have a definition status and an allocation status. The allocation status is never undefined.

2.3.3 Scope

The scope of a program entity is the part of the program in which that entity is known, is available, and can be used. A scoping unit is

1. a program unit or subprogram, excluding any scoping units in it
2. a derived-type definition
3. an interface body, excluding any scoping units in it

Some entities have scopes that are something other than a scoping unit. For example, the scope of a name, such as a variable name, can be any of the following:

1. an executable program
2. a scoping unit
3. a construct
4. a statement or part of a statement

The scope of a label is a scoping unit. The scope of an input/output unit is a program.

2.3.4 Association

Association is the concept that is used to describe how different entities in the same scoping unit or different scoping units can share values and other properties. Argument association allows values to be shared between a procedure and the program that calls it. Use association and host association allow entities described in one part of a program to be used in another part of the program. Use association makes entities defined in modules accessible, and host association makes entities in the containing environment available to a contained procedure. The IMPORT statement (12.5.2), introduced in Fortran 2003, makes entities in a host scoping unit available in an interface body by host association.

Additional forms of association are inheritance association (between the entities in an extended type and its parent), linkage association (between corresponding Fortran and C entities), and construct association (relevant to the ASSOCIATE and SELECT TYPE constructs). The complete description of association may be found in 16.

An old form of association, storage association, which allows two or more variables to share storage, can be set up by the use of EQUIVALENCE, COMMON, or ENTRY statements. It is best avoided.

2.4 Terms

Frequently used Fortran terms are defined in this section. Definitions of less frequently used terms may be found by referencing the index of this handbook or Annex A of the Fortran 2003 standard.

Entity	This is the general term used to refer to any Fortran "thing", for example, a program unit, a procedure, a common block, a variable, an expression value, a constant, a statement label, a construct, an operator, an interface, a type, an input/output unit, a namelist group, etc.
Name	A name is used to identify many different entities of a program such as a program unit, a named variable, a named constant, a common block, a construct, a formal argument of a subprogram (dummy argument), or a user-defined type (derived type). The rules for constructing names are given in 3.
Named entity	A named entity is referenced by a name without any qualification such as an appended subscript list or substring range.
Data object	A data object is a constant, a variable, or a subobject of a constant. It may be a scalar or an array. It may be of intrinsic or derived type.
Constant	A constant is a data object whose value cannot be changed. A named entity with the PARAMETER attribute is called a **named constant**. A constant without a name is called a **literal constant**. A constant may be a scalar or an array.
Variable	A variable is a data object whose value can be defined and redefined. A variable may be a scalar or an array.
Local variable	A variable that is in a main program, module, or subprogram and is not associated by being: a dummy argument, in COMMON, a BIND(C) variable, or accessed via host or USE association. A subobject of a local variable is also a local variable.
Subobject of a constant	A subobject of a constant is a portion of a constant. The portion referenced may depend on the value of a variable, in which case it is neither a constant nor a variable.
Data entity	A data entity is a data object or the result of the evaluation of an expression. A data entity has a type, possibly type parameters, and a rank (a scalar has rank zero). It may have a value.
Expression	An expression may be a simple data reference or it may specify a computation and thus be made up of operands, operators, and parentheses. The type, type parameters, value, and rank of an expression result are determined by the rules in 7.
Function reference	A function reference invokes a function. It is made up of the name of a function followed by a parenthesized list of arguments, which may be empty. The type, type parameters,

and rank of the result are determined by the interface of the function and the reference.

Data type

A data type provides a means for categorizing data. Each intrinsic and user-defined data type has—a name, a set of values, a set of operators, and a means to represent values of the type in a program. For each data type there is a type specifier that is used to declare objects of the type.

Type parameter

There are two categories of type parameters for types: kind and length. For intrinsic types a kind type parameter indicates the range for the integer type, the decimal precision and exponent range for the real type and parts of the complex type, and the machine representation method for the character and logical types. The length type parameter indicates a length for the intrinsic character type. For a derived type, the type parameters are defined in its type definition.

Derived type

A derived type (or user-defined type) is a type that is not intrinsic; it requires a type definition to name the type and specify its parameters and components. The components may be of intrinsic or user-defined types. An object of derived type is called a structure. For each derived type, a structure constructor is available to specify values. Operations on objects of derived type must be defined by a function. Assignment for derived-type objects is defined intrinsically, but may be redefined by a subroutine. Finalizers may be specified for derived-type objects. Data entities of derived type may be used as procedure arguments and function results, and may appear in input/output lists and other places. Derived types may be extended by inheritance.

Ultimate component

The ultimate components of a derived type entity are the lowest-level components that have storage in the entity. They are a) any components that are of an intrinsic type, b) any components that have the ALLOCATABLE or POINTER attribute (the entity has storage for the pointer or allocation descriptor, but the object or target does not, itself, have storage in the entity), and c) the ultimate components of any derived type components that have neither the ALLOCATABLE nor POINTER attribute. The ultimate components are subject to, for example, storage association rules.

There is a distinction between a component of derived type and an allocatable or pointer component of the same type. In the first case, the elements of the derived type component are ultimate components; in the other cases only the descriptor or pointer is an ultimate component

Inheritance
: Inheritance is the process of automatically acquiring entities (parameters, components, or procedure bindings) from a parent.

Polymorphism
: Polymorphism is the ability to change type during program execution. Dummy arguments, pointers, and allocatable objects may be polymorphic.

Scalar
: A scalar is a single object of any intrinsic or derived type. A structure is scalar even if it has a component that is an array. The rank of a scalar is zero.

Array
: An array is an object with the dimension attribute. It is a collected set of scalar data, all of the same type and type parameters. The rank of an array is at least one and at most seven. An array of any rank may be of zero size. An array of size zero or one is not a scalar. Data entities that are arrays may be used as expression operands, procedure arguments, and function results, and may appear in input/output lists, as well as other places.

Subobject
: A subobject is a portion of a data object. Portions of a data object may be referenced and defined (if the object is a variable) separately from other portions of the object. Array elements and array section are portions of arrays. Substrings are portions of character strings. Structure components are portions of structures. Portions of complex objects are the real and imaginary parts. Subobjects are referenced by designators or intrinsic functions and are considered to be data objects themselves.

Designator
: Sometimes it is convenient to reference only part of an object, such as an element or section of an array, a substring of a character string, or a component of a structure. This requires the use of a designator which is the name of the object followed by zero or more selectors that select a part of the object.

Selector
: This term is used in several different ways. A selector may designate part of an object (array element, array section, substring, or structure component) or the set of values for which a CASE block is executed, or the dynamic type for which a SELECT TYPE block is executed, or the object associated with the name in an ASSOCIATE construct.

Declaration
: A declaration is a nonexecutable statement that specifies the attributes of a program element. For example, it may be used to specify the type of a variable or function or the shape of an ar-

ray. It may indicate that an entity is a data pointer or a procedure pointer. Attributes that were introduced in Fortran 2003 are: ASYNCHRONOUS, which indicates that the value of the variable may change outside the execution flow due to a possibly simultaneous input/output process; BIND (C), which is used to indicate data and functions that interoperate with C; PROTECTED, which prohibits any change to the value of the variable or the association status of the pointer outside the module in which it is declared; VALUE, which, when applied to a dummy argument, specifies an argument passing mechanism useful in C interoperability; and VOLATILE, which indicates that the value of the variable may change by means other than the normal execution sequence

Definition

This term is used in several ways. A data object is said to be defined when it has a valid or predictable value; otherwise it is undefined. It may be given a valid value by execution of statements such as assignment or input. Under certain circumstances described in 16, it may subsequently become undefined.

Procedures and derived types are said to be defined when their descriptions have been supplied by the programmer and are available in a program unit.

The association status of a pointer is defined when the pointer is associated or disassociated; otherwise, it is undefined.

Statement keyword

A statement keyword is part of the syntax of a statement. Each statement, other than an assignment statement, pointer assignment statement, or statement function definition, begins with a statement keyword. Some statement keywords appear in internal positions within statements. Examples of these keywords are THEN, KIND, and INTEGER. Statement keywords are not reserved; they may be used as names.

List keyword

A list keyword is a name that is used to identify an item in a list (rather than its position) such as an argument list, type parameter list, or structure constructor list. Keywords for the argument lists of all of the intrinsic procedures are specified by the standard (A). Keywords for user-supplied external procedures may be specified in a procedure interface block. Keywords for structure constructors and user-defined type parameters are specified in the type definition.

Sequence

A sequence is a set ordered by a one-to-one correspondence with the numbers 1, 2, through *n*. The number of elements in the sequence is *n*. A sequence may be empty, in which case it contains no elements.

Operator — An operator indicates a computation involving one or two operands. Fortran defines a number of intrinsic operators; for example, +, −, *, /, ** with numeric operands, and .NOT., .AND., .OR. with logical operands. In addition, users may define operators for use with operands of intrinsic or derived types.

Construct — A construct is a sequence of statements starting with an ASSOCIATE, DO, FORALL, IF, SELECT CASE, SELECT TYPE, or WHERE statement and ending with the corresponding terminal statement.

Executable construct — An executable construct is an action statement (such as a READ statement) or a construct (such as a DO or CASE construct).

Procedure — A procedure is defined by a sequence of statements that expresses a computation that may be invoked as a subroutine or function during program execution. It may be an intrinsic procedure, an external procedure, an internal procedure, a module procedure, a dummy procedure, or a statement function. If a subprogram contains an ENTRY statement, it defines more than one procedure.

Procedure interface — A procedure interface is a sequence of statements that specifies the name and characteristics of a procedure, the name and attributes of each dummy argument, and the generic specifier by which it may be referenced, if any.

Reference — A data object reference is the appearance of the object designator in a statement requiring the value of the object.

A procedure reference is the appearance of the procedure designator, operator symbol, or assignment symbol in an executable program requiring execution of the procedure.

A module reference is the appearance of the module name in a USE statement.

Intrinsic — Anything that is defined by the Fortran processor is intrinsic. There are intrinsic data types, procedures, modules, operators, and assignment. These may be used freely in any scoping unit. The Fortran programmer may define types, procedures, modules, operators, and assignment; these entities are not intrinsic.

Companion processor — A companion processor is a processor that provides mechanisms by which global data and procedures may be referenced or defined—perhaps by means other than Fortran, such as the C programming language.

Scoping unit — A scoping unit is a portion of a program in which a name has a fixed meaning. A program unit or subprogram generally defines a scoping unit. Type definitions and procedure interface bodies also constitute scoping units. Scoping units are non-overlapping, although one scoping unit may contain another in the sense that it surrounds it. If a scoping unit contains another scoping unit, the outer scoping unit is referred to as the **host scoping unit** of the inner scoping unit.

Association — In general, association permits an entity to be referenced by different names in a scoping unit or by the same or different names in different scoping units. There are several kinds of association: the major ones are name association, pointer association, inheritance association, and storage association. Name association is argument association, use association, host association, linkage association, and construct association.

Inheritance association — Inheritance association occurs between the inherited entities of an extended type and the corresponding entities of its parent.

Linkage association — Linkage association occurs between a module variable with the BIND(C) attribute and the relevant C variable or between a Fortran common block and the relevant C variable. It has the scope of the program.

Construct association — Construct association occurs between the selector in an ASSOCIATE or SELECT TYPE construct and the associate name of the construct. It has the scope of the construct.

2.5 High-Level Syntax Forms

The form of a program (R201) is:

```
program-unit
  [ program-unit ] . . .
```

The forms for Fortran program units are shown in the first section below. The constructs that may appear in a program unit are shown in the subsequent sections. All program units may have a specification part. The main program and the three forms of subprogram (module, external, and internal) may have an execution part.

The notation used in this chapter is the same as that used to show the syntax in all the remaining chapters; it is described in 1.3 along with an assumed syntax rule and some frequently used abbreviations for syntax terms. This is not the complete set of rules; many lower-level rules are missing. Many of these rules may be found in the following chapters. The Fortran 2003 standard [7] contains the complete syntax rules.

2.5.1 Fortran Program Units

The forms of a program unit (R202) are:

```
main-program
module
external-subprogram
block-data
```

The form of a main program (R1101) is:

```
[ PROGRAM program-name ]
    [ specification-part ]
    [ execution-part ]
[ CONTAINS
    internal-subprogram
    [ internal-subprogram ] ... ]
END [ PROGRAM [ program-name ] ]
```

The form of a module (R1104) is:

```
MODULE module-name
    [ specification-part ]
[ CONTAINS
    module-subprogram
    [ module-subprogram ] ...]
END [ MODULE [ module-name ] ]
```

The form of a module subprogram (R1108) and an external subprogram (R203) is:

```
subprogram-heading
    [ specification-part ]
    [ execution-part ]
[ CONTAINS
    internal-subprogram
    [ internal-subprogram ] ...]
subprogram-ending
```

The form of an internal subprogram (R211) is:

```
subprogram-heading
    [ specification-part ]
    [ execution-part ]
subprogram-ending
```

The forms of a subprogram heading (R1224, R1232) are:

```
[ prefix ] [ declaration-type-spec ] FUNCTION function-name &
      ( [ dummy-argument-list ] ) [ suffix ]
[ prefix ] SUBROUTINE subroutine-name [ ( [dummy-argument-list ] ) ] [ binding-spec ]
```

A prefix (R1228) is any combination of the keywords:

```
RECURSIVE
PURE
ELEMENTAL
```

A suffix (R1229) is one of the forms:

```
RESULT ( result-name ) [ binding-spec ]
binding-spec [ RESULT ( result-name ) ]
```

A binding specification (R509) is:

```
BIND ( C [ , NAME = scalar-char-initialization-expr ] )
```

The forms of a subprogram ending (R1230, R1234) are:

```
END [ FUNCTION [ function-name ] ]
END [ SUBROUTINE [ subroutine-name ] ]
```

The form of a block data program unit (R1116) is:

```
BLOCK DATA [ block-data-name ]
   [ specification-part ]
END [ BLOCK DATA [ block-data-name ] ]
```

2.5.2 The Specification Part

The form of the specification part (R204) is:

```
[ use-statement ] ...
IMPORT [ [ : : ] import-name-list ] ] ...
[ implicit-part ]
[ declaration-construct ] ...
```

The forms of a USE statement (R1109) are:

```
USE [ [ , module-nature ] : : ] module-name [ , rename-list ]
USE [ [ , module-nature ] : : ] module-name , ONLY : [ only-list ]
```

The form of the implicit part (R206) is:

```
[ implicit-part-statement ] ...
IMPLICIT implicit-spec-list
```

The forms of an implicit part statement (R205) are:

```
IMPLICIT implicit-spec-list
PARAMETER ( named-constant = initialization-expr &
      [ , named-constant = initialization-expr ] ... )
entry-statement
format-statement
```

The forms of an implicit specification (R550) are:

```
NONE
declaration-type-spec ( letter-spec-list )
```

The forms of a declaration construct (R207) are:

```
declaration-type-spec [ [ , attribute-spec ] ... : : ] entity-declaration-list
specification-statement
derived-type-definition
interface-block
enumeration-definition
entry-statement
format-statement
statement-function-statement
```

The forms of a declaration type specification (R502) are:

```
INTEGER [ kind-selector ]
REAL [ kind-selector ]
DOUBLE PRECISION
COMPLEX [ kind-selector ]
CHARACTER [ character-selector ]
LOGICAL [ kind-selector ]
TYPE ( derived-type-spec )
CLASS ( derived-type-spec )
CLASS ( * )
```

The form of a kind selector (R404) is:

```
( [ KIND = ] kind-value )
```

The forms of a character selector (R424) are:

```
( length-value [ , [ KIND = ] kind-value ] )
( LEN = length-value [ , KIND = kind-value ] )
( KIND = kind-value [ , LEN = length-value ] )
* character-length [ , ]
```

A length value (R402) has one of the forms:

```
scalar-integer-expression
*
:
```

A kind value (R404) has the from:

```
scalar-integer-initialization-expr
```

A character length (R426) has one of the forms:

```
( length-value )
scalar-integer-literal-constant
```

A derived-type specification (R455) has the from:

```
type-name [ ( type-parameter-spec-list ) ]
```

A type parameter specification (R456) has the from:

```
[ keyword = ] length-value
```

The forms of an attribute specification (R503) are:

```
ALLOCATABLE
ASYNCHRONOUS
BIND ( C [ , NAME = scalar-char-initialization-expr ] )
DIMENSION ( array-spec )
EXTERNAL
INTENT ( intent-spec )
INTRINSIC
OPTIONAL
PARAMETER
POINTER
PRIVATE
PROTECTED
PUBLIC
SAVE
TARGET
VALUE
VOLATILE
```

The form of an entity declaration (R504) is:

```
object-name [ ( array-spec ) ] [ * character-length ] [ initialization ]
```

The forms of initialization (R506) are:

```
= initialization-expr
=> function-reference
```

The forms of specification statements (R212) are:

```
ALLOCATABLE [ : : ] object-name [ ( deferred-shape-spec-list ) ] &
      [ , object-name [ ( deferred-shape-spec-list ) ] ] ...
ASYNCHRONOUS [ [ : : ] variable-name-list ]
BIND ( C [ , NAME = scalar-char-initialization-expr ] ) [ : : ] bind-entity-list
```

```
COMMON [ / [ common-block-name ] / ] common-block-object-list
DATA data-statement-object-list / data-value-list / &
     [ [ , ] data-statement-object-list / data-value-list / ]
DIMENSION [ :: ] array-name ( array-spec ) [ , array-name ( array-spec ) ] ...
EQUIVALENCE equivalence-set-list
EXTERNAL [ :: ] external-name-list
INTENT ( intent-spec ) [ :: ] dummy-argument-name-list
INTRINSIC [ :: ] intrinsic-procedure-name-list
NAMELIST / namelist-group-name / namelist-group-object-list
OPTIONAL [ :: ] dummy-argument-name-list
POINTER [ :: ] pointer-declaration-list
PARAMETER ( named-constant = initialization-expr &
     [ , named-constant = initialization-expr ] ... )
PROCEDURE ( [ procedure-interface ] ) [ [ , procedure-attribute-spec ] ... :: ] &
     procedure-declaration-list
PROTECTED [ :: ] entity-name-list
PUBLIC [ [ :: ] access-id-list ]
PRIVATE [ [ :: ] access-id-list ]
SAVE [ [ :: ] saved-entity-list ]
TARGET [ :: ] object-name [ ( array-spec ) ] [ , object-name [ ( array-spec ) ] ] ...
VALUE [ :: ] dummy-argument-name-list
VOLATILE [ :: ] variable-name-list
```

The forms of a procedure interface (R1212) are:

```
interface-name
declaration-type-spec
```

The forms of a procedure attribute specification (R1213) are:

```
BIND ( C [ , NAME = scalar-char-initialization-expr ] )
INTENT ( intent-spec )
OPTIONAL
POINTER
PRIVATE
PUBLIC
SAVE
```

The form of a procedure declaration (R1214) is:

```
procedure-entity-name [ => function-reference ]
```

The form of a derived-type definition (R429) is:

```
TYPE [ [ , type-attribute-list ] :: ] type-name [ ( type-parameter-name-list ) ]
   [ type-parameter-definition-statement ] ...
   [ private-or-sequence-statement ]
   [ component-definition-statement ] ...
[ CONTAINS
```

```
    [ PRIVATE ]
    procedure-binding-statement
    [ procedure-binding-statement ] ... ]
END TYPE [ type-name ]
```

The forms of a type attribute (R431) are:

```
ABSTRACT
BIND ( C )
EXTENDS ( parent-type-name )
PRIVATE
PUBLIC
```

The form of a type parameter definition statement (R435) is:

```
INTEGER [ kind-selector ] , type-parameter-attribute-spec :: &
      type-parameter-declaration-list
```

The forms of a type parameter attribute specification (R437) are:

```
KIND
LEN
```

The form of a type parameter declaration (R436) is:

```
type-param-name [ = scalar-integer-initialization-expr ]
```

The forms of a component definition statement (R439) are:

```
declaration-type-spec [ [ , component-attribute-spec-list ] :: ] &
      component-declaration-list
PROCEDURE ( [ procedure-interface ] ) , procedure-component-attribute-spec-list :: &
      procedure-declaration-list
```

The forms of a component attribute specification (R441) are:

```
ALLOCATABLE
DIMENSION ( component-array-spec )
POINTER
PRIVATE
PUBLIC
```

The form of a component declaration (R442) is:

```
component-name [ ( component-array-spec ) ] [ * character-length ] [ initialization ]
```

The forms of a procedure component attribute specification (R446) are:

```
NOPASS
PASS [ ( argument-name ) ]
POINTER
PRIVATE
PUBLIC
```

The form of an interface block (R1201) is:

```
[ ABSTRACT ] INTERFACE [ generic-spec ]
   [ subprogram-heading
      [ specification-part ]
   subprogram-ending ] ...
   [ [ MODULE ] PROCEDURE procedure-name-list ] ...
END INTERFACE [ generic-spec ]
```

The forms of a generic specification (R1207) are:

```
generic-name
OPERATOR ( defined-operator )
ASSIGNMENT ( = )
derived-type-io-generic-spec
```

The form of an enumeration definition (R460) is:

```
ENUM , BIND ( C )
   ENUMERATOR [ :: ] enumerator-list
   [ ENUMERATOR [ :: ] enumerator-list ] . . .
END ENUM
```

2.5.3 The Execution Part

The form of the execution part (R208) is:

```
execution-part-construct
   [ execution-part-construct ] ...
```

The forms of an execution part construct (R209) are:

```
executable-construct
entry-statement
format-statement
```

The forms of an executable construct (R213) are:

```
action-statement
associate-construct
case-construct
do-construct
forall-construct
if-construct
select-type-construct
where-construct
```

The forms of an action statement (R214) are:

```
variable = expression
data-pointer-object [ ( bounds-list ) ] => data-target
data-pointer-object ( bounds-remap-list ) => data-target
```

```
procedure-pointer-object => procedure-target
ALLOCATE [ declaration-type-spec :: ] ( allocation-list [ , allocate-option-list ] )
BACKSPACE scalar-integer-expression
BACKSPACE ( position-spec-list )
CALL subroutine-name [ ( [ actual-argument-spec-list ] ) ]
CLOSE ( close-spec-list )
CONTINUE
CYCLE [ do-construct-name ]
DEALLOCATE ( allocate-object-list [ , deallocate-option-list ] )
ENDFILE scalar-integer-expression
ENDFILE ( position-spec-list )
EXIT [ do-construct-name ]
FLUSH scalar-integer-expression
FLUSH ( flush-spec-list )
FORALL ( forall-triplet-specification-list [ , scalar-logical-expression ] ) &
      forall-assignment-statement
GO TO label
GO TO ( label-list ) [ , ] scalar-integer-expression
IF ( scalar-logical-expression ) action-statement
IF ( scalar-numeric-expression ) label , label , label
INQUIRE ( inquire-spec-list )
INQUIRE ( IOLENGTH = scalar-integer-variable ) output-item-list
NULLIFY ( pointer-object-list )
OPEN ( connection-spec-list )
PRINT format [ , output-item-list ]
READ ( io-control-spec-list ) [ input-item-list ]
READ format [ , input-item-list ]
RETURN [ scalar-integer-expression ]
REWIND scalar-integer-expression
REWIND ( position-spec-list )
STOP [ scalar-character-constant ]
STOP digit [ digit [ digit [ digit [ digit ] ] ] ]
WAIT ( wait-spec-list )
WHERE ( logical-expression ) where-assignment-statement
WRITE ( io-control-spec-list ) [ output-item-list ]
```

The form of the ASSOCIATE construct (R816) is:

```
[ associate-construct-name : ] ASSOCIATE ( association-list )
   block
END ASSOCIATE [ associate-construct-name ]
```

The form of the CASE construct (R808) is:

```
[ case-construct-name : ] SELECT CASE ( case-expression )
[ CASE ( case-value-range-list ) [ case-construct-name ]
   block ] ...
```

```
[ CASE DEFAULT [ case-construct-name ]
   block ]
END SELECT [ case-construct-name ]
```

The form of the DO construct (R825) is:

```
[ do-construct-name : ] DO [ label ] [ loop-control ]
   block
[ label ] END DO [ do-construct-name ]
```

The form of the FORALL construct (R752) is:

```
[ forall-construct-name : ] &
            FORALL ( forall-triplet-spec-list [ , scalar-logical-expression ] )
   [ forall-body-construct ] ...
END FORALL [ forall-construct-name ]
```

The form of the IF construct (R802) is:

```
[ if-construct-name : ] IF ( scalar-logical-expression ) THEN
   block
[ ELSE IF ( scalar-logical-expression ) THEN [ if-construct-name ]
   block ] ...
[ ELSE [ if-construct-name ]
   block ]
END IF [ if-construct-name ]
```

The form of the SELECT TYPE construct (R821) is:

```
[ select-construct-name : ] SELECT TYPE ( [ associate-name => ] selector )
   [ type-guard [ select-construct-name ]
      block ] . . .
END SELECT [ select-construct-name ]
```

The form of the WHERE construct (R744) is:

```
[ where-construct-name : ] WHERE ( logical-expression )
   [ where-body-construct ] ...
[ ELSEWHERE ( logical-expression ) [ where-construct-name ]
   [ where-body-construct ] ... ] ...
[ ELSEWHERE [ where-construct-name ]
   [ where-body-construct ] ... ]
END WHERE [ where-construct-name ]
```

2.6 Ordering Requirements

Within program units, subprograms, and interface bodies there are ordering requirements for statements and constructs. The syntax rules above do not fully describe the ordering requirements. Therefore, they are illustrated in Table 2-1. In general, data declarations and specifications must precede executable constructs and statements, although FORMAT, DATA, and ENTRY statements may appear among the executable statements. Placing DATA statements among executable constructs is now an obsolescent feature. USE statements, if any, must appear first. Internal or module subprograms, if any, must appear last following a CONTAINS statement.

In Table 2-1 a vertical line separates statements and constructs that can be interspersed; a horizontal line separates statements that must not be interspersed.

Table 2-1 Requirements on statement ordering

<table>
<tr><td colspan="3">PROGRAM, FUNCTION, SUBROUTINE, MODULE, or BLOCK DATA statement</td></tr>
<tr><td colspan="3">USE statements</td></tr>
<tr><td colspan="3">IMPORT statements[3]</td></tr>
<tr><td rowspan="4">FORMAT[5]
and
ENTRY[4]
statements</td><td colspan="2">IMPLICIT NONE</td></tr>
<tr><td>PARAMETER statements</td><td>IMPLICIT statements</td></tr>
<tr><td>PARAMETER and
DATA statements[6]</td><td>Derived-type definitions,
interface blocks,[7]
type declaration statements,
enumeration statements,
procedure statements,
statement function statements,[2,5]
and specification statements</td></tr>
<tr><td>DATA statements[1]</td><td>Executable constructs[5]</td></tr>
<tr><td colspan="3">CONTAINS statement[8]</td></tr>
<tr><td colspan="3">Internal subprograms or module subprograms</td></tr>
<tr><td colspan="3">END statement</td></tr>
<tr><td colspan="3">1. Placing DATA statements among executable constructs is obsolescent.
2. Statement function statements are obsolescent.
3. Can appear only in interface bodies.
4. Can appear only in modules and external procedures.
5. Cannot appear in module specification parts, interface bodies, and block data subprograms.
6. Cannot appear in interface bodies.
7. Cannot appear in block data subprograms.
8. Cannot appear in internal subprograms, interface bodies, and block data subprograms.</td></tr>
</table>

2.7 Example Fortran Program

Illustrated below is a very simple Fortran program consisting of one program unit, the main program. Three data objects are declared: H, T, and U. These become the loop indices in a triply-nested loop construct (8.7) containing a logical IF statement (8.4.2) that conditionally executes an input/output statement (9.4).

```
program sum_of_cubes
! This program prints all 3-digit numbers that
! equal the sum of the cubes of their digits.
implicit none
integer :: H, T, U
do H = 1, 9
      do T = 0, 9
            do U = 0, 9
                  if (100*H + 10*T + U == H**3 + T**3 + U**3) &
                        print "(3I1)", H, T, U
            end do
      end do
end do
end program sum_of_cubes
```

This Fortran program is standard conforming and should be compilable and executable on any standard Fortran computing system, producing the following output:

```
153
370
371
```

3 Language Elements and Source Form

- **The Fortran Character Set** consists of the uppercase letters and lowercase letters of the English alphabet, the decimal digits, underscore, and special characters. Lower case letters are considered the same as the corresponding upper case letters except in character contexts and input/output records.
- **Lexical Tokens** are constructed from characters in the Fortran character set. They include statement keywords, names, constants, and operators.
- The **Processor Character Set** consists of the Fortran character set plus additional characters with or without graphics. (Control characters generally do not have graphics.) A processor may support other character sets as well, such as Greek or Japanese. Any of these characters may appear in character strings, comments, or input/output.
- **Free Source Form** has no position restrictions. Lines may contain up to 132 characters. Blanks are significant and cannot be used within tokens, particularly names, keywords, literal constants, and multicharacter operators. A semicolon may be used to separate statements on the same line. Comments beginning with an exclamation (!) may appear on a separate line or at the end of a line. Lines can be continued by placing an ampersand at the end of the line to be continued.
- **Fixed Source Form** reserves positions 1 through 5 for labels, position 6 for continuation, positions 7 through 72 for Fortran statements, and positions 73-80 are unused. Comments are indicated by a C or asterisk in position 1. An exclamation may also be used in position 1 or to indicate an end of line comment. A semicolon can be used to separate statements on a line. Blanks are insignificant except in a character context.
- An **INCLUDE Line** specifies the location of text to be included in the source in place of the INCLUDE line.
- It is possible to prepare source so that it conforms to the rules for both free and fixed source form. Text in this restricted form can be included in either free form or fixed form source.

A program is made up of language elements consisting of lexical tokens that include names, keywords, operators, and statement labels. There are rules for forming lexical tokens from the characters in the Fortran character set. There are also rules (called source form) for placing these tokens on a line.

A processor must have a character set that includes the Fortran character set, but may permit other characters in certain contexts. The additional characters may include control characters (which may have no graphic representation, such as escape or new

J.C. Adams et al., *The Fortran 2003 Handbook*,
DOI: 10.1007/978-1-84628-746-6_3, © Springer-Verlag London Limited 2009

line) or may include characters with specified graphics such as those found in languages, like Greek, Arabic, Chinese, or Japanese. These characters are not required to be part of the character set for the default character type, but could be part of some optional, nondefault character type, permitted by the standard and supplied by a particular implementation.

There are two source forms in Fortran. One is oriented towards terminal input of source code. It is called **free source form**. The other is oriented towards the Hollerith punched card common in the 1960s and is restricted to 80 positions. It is called **fixed source form**. Fixed source form is an obsolescent feature. There is a convenient way to place the same text in several places in a program; it makes use of an **INCLUDE line**.

3.1 The Processor Character Set

The **processor character set** contains:

- the Fortran character set of Table 3-1.
- as an extension, a processor-dependent set of control characters that have no graphic representation, such as "new line" or "escape"
- as an extension, a set of characters with graphics (such as lowercase letters, Greek letters, Japanese ideographs, or characters in the shape of a heart or a diamond)

It is recommended that the programmer consult the implementor's documentation describing the processor-dependent features of each particular Fortran implementation.

3.1.1 The Fortran Character Set

Characters in the **Fortran character set** are shown in Table 3-1.

Rules and restrictions:

1. Lowercase letters are considered the same as uppercase letters except within a character constant, a quote or apostrophe edit descriptor, or input/output records, where uppercase and lowercase letters are different data values in character data. The following two statements are equivalent:

   ```
   PRINT *, N
   Print *, n
   ```

   Whether uppercase and lowercase letters are distinguished in the FILE= or NAME= specifier in an OPEN or an INQUIRE statement is processor dependent.

2. The digits are assumed to be decimal numbers when used to describe a numeric value, except in binary, octal, and hexadecimal (BOZ) literal constants or input/output records corresponding to B, O, or Z edit descriptors. For example, consider the following DATA statement:

Table 3-1 The Fortran character set

Alphanumeric characters	
Letters	A B C D E F G H I J K L M N O P Q R S T U V W X Y Z a b c d e f g h i j k l m n o p q r s t u v w x y z
Digits	0 1 2 3 4 5 6 7 8 9
Underscore	_

Special characters			
Graphic	Name of character	Graphic	Name of character
	Blank	;	Semicolon
=	Equals	!	Exclamation point
+	Plus	"	Quotation mark or quote
-	Minus	%	Percent
*	Asterisk	&	Ampersand
/	Slash	~	Tilde
\	Backslash	<	Less than
(	Left parenthesis	>	Greater than
)	Right parenthesis	?	Question mark
[	Left square bracket	’	Apostrophe
]	Right square bracket	’	Grave accent
{	Left curly bracket (brace)	^	Circumflex accent
}	Right curly bracket (brace)	|	Vertical bar
,	Comma	$	Currency symbol
.	Decimal point or period	#	Number sign
:	Colon	@	Commercial at

```
DATA  X, I, J / 4.89, B'1011', Z'BAC91' /
```

The digits of the first constant are decimal digits, those of the second constant are binary digits, and those of the third are hexadecimal digits.

3. The underscore is used to make names more readable. For example, in the identifier NUMBER_OF_CARS, each underscore is used to separate the obvious English words. It is a significant character in any name. It cannot be used as the first character of a name; however, it may be the last character. An underscore is also used

to separate the kind value from the actual value of a literal constant (for example, 123_SHORT is a literal constant with value 123 and of integer type with kind SHORT).

4. Except for the currency symbol ($), the graphic for each character must be the same as in Table 3-1; however, any style, font, or printing convention may be used.

5. The special characters, \, {, }, ~, ?, ', ^, |, $, #, and @, are used only in a character context or a comment.

The special characters are used for operators like multiply and add, and as separators or delimiters in Fortran statements. Separators and delimiters make the form of a statement unambiguous.

3.1.2 Other Characters

In addition to the Fortran character set, other characters may be included in the processor character set. These are either control characters with no graphics or additional characters with graphics. The selection of the other characters and where they may be used is processor dependent. However, wherever they are permitted, the other characters are restricted in use to character constants, quote and apostrophe edit descriptors, comment lines, and input/output records. All characters of the Fortran character set may be used in character constants, quote and apostrophe edit descriptors, comment lines, and input/output records.

A processor is required to support the Fortran character set as part of a character set referred to as the **default character set**. A processor is allowed to support more than one character set, each set using a different kind value of the intrinsic character type (4.3.5); each such character set is a **nondefault character set**. The choice of characters in such sets is processor dependent except that each such set must contain a character that can be used as a blank. This specially designated character is used where blank padding is required.

The choice of the representable characters beyond the Fortran character set is expected to be dependent on the particular implementation. It is recommended that the implementor's documentation be consulted for specific details.

3.1.3 The Tab Character

The tab character is not in the Fortran character set and is an example of an optional control character that may be permitted by the processor in the source forms and typically is used as a blank separator. When it appears as the first character in a fixed source form line, it often represents at least six blank characters, so that the next character may begin the body of a statement that must appear in columns 7-72. However, this is not standard and its use may make it difficult to port the program; therefore, its use is not recommended.

This recommendation does not help the programmer who has code that uses tab characters. To use a Fortran file containing tabs with a compiler that accepts only standard-conforming programs, replace the tab with a blank for free source form and with six blanks for fixed source form. This conversion is not fool proof; replacing a tab in fixed source form may extend the line beyond position 72. In either form, a tab may be used in character context for output format control; in this case, the modification may lead to an undesirable layout of data in the output.

3.2 Lexical Tokens

A statement is constructed from low-level syntax. The low-level syntax describes the basic language elements, called **lexical tokens**, in a Fortran statement. A lexical token is the smallest meaningful unit of a Fortran statement and may consist of one or more characters. Tokens are names, keywords, literal constants (except for complex literal constants), labels, operator symbols, comma, =, =>, :, ::, ;, %, and delimiters. A complex literal (4.3.3) consists of several tokens. Examples of operator symbols are + and //.

Delimiters in Fortran are pairs of symbols that enclose parts of a Fortran statement. The delimiters are slashes (in pairs), left and right parentheses, left and right brackets, and the symbol pairs (/ and /).

```
/ ... /
( ... )
(/ ... /)
[ ... ]
```

In the statements:

```
DATA  X, Y/ 1.0, -10.2/
CALL PRINT_LIST (LIST, SIZE)
VECTOR = (/ 10, 20, 30, 40 /)
```

the slashes distinguish the value list from the object list in a DATA statement, the parentheses are delimiters marking the beginning and end of the argument list in the CALL statement, and the pairs (/ and /) and [and] mark the beginning and end of the elements of an array constructor.

3.2.1 Statement Keywords

Statement keywords appear in uppercase letters in the syntax rules. Some statement keywords also identify the statement, such as in the DO statement:

```
DO I = 1, 10
```

where DO is a statement keyword identifying the DO statement. Other keywords delimit parts of a statement such as ONLY in a USE statement, or WHILE in one of the forms of a DO construct, as, for example:

```
DO WHILE( .NOT. FOUND )
```

Others specify options in the statement such as IN, OUT, or INOUT in the INTENT statement.

There are two statements in Fortran that have no statement keyword. They are the assignment statement (7.5.1) and the statement function (12.4.4).

Some sequences of capital letters in the formal syntax rules are not statement keywords. For example, EQ in the lexical token **.EQ.** and EN as an edit descriptor are not statement keywords.

A dummy argument keyword, a different sort of keyword, is discussed in 12.1.2.

3.2.2 Names

Variables, named constants, program units, common blocks, procedures, arguments, constructs, derived types (types for structures), namelist groups, structure components, dummy arguments, and function results are among the elements in a program that have a name.

Rules and restrictions:

1. A name must begin with a letter and consist of letters, digits, and underscores. Note that an underscore must not be the first character of a name.

2. Fortran permits up to 63 characters in a name.

Examples of names:

```
A
CAR_STOCK_NUMBER
A__BUTTERFLY
Z_28
TEMP_
```

3.2.3 Constants

A **constant** is a syntactic notation for a value. The value may be of any intrinsic type, that is, a numeric (integer, real, or complex) value, a character value, or a logical value.

A value that does not have a name is a **literal constant**. Examples of literal constants are:

```
1.23
400
( 0.0, 1.0 )
"ABC"
B'0110110'
.TRUE.
```

No literal constant can be array-valued or of derived type. The forms of literal constants are given in more detail in 4.2.6.

A value that has a name is called a **named constant** and may be of any type, including a derived type. A named constant may also be array-valued. Examples of named constants are:

```
X_AXIS
MY_SPOUSE
```

where these names have been specified in a declaration statement as follows:

```
REAL, DIMENSION(2), PARAMETER :: X_AXIS = (/0.0, 1.0/)
TYPE(PERSON), PARAMETER :: MY_SPOUSE = PERSON( 39, 'PAT' )
```

Note, however, that the entity on the right of the equal sign is not itself a constant but an initialization expression (7.4.1). The forms for defining named constants are described in more detail in 5.7.1.

3.2.4 Operators

Operators are used with operands in expressions to produce other values. Examples of language-supplied operators are:

*	representing multiplication of numeric values
//	representing concatenation of character values
==	representing comparison for equality (same as .EQ.)
.OR.	representing logical disjunction
.NOT.	representing logical negation

The complete set of the intrinsic operators is given in 7.1.1.1.

Users may define operators (12.5.4.2) in addition to the intrinsic operators. User-defined operators begin with a period (.), followed by a sequence of up to 63 letters, and end with a period (.), except that the letter sequence must not be the same as any intrinsic operator or either logical constant. Note that, unlike names, underscores and digits are not allowed.

3.2.5 Statement Labels

A label may be used to identify a statement. A **label** consists of one to five decimal digits, one of which must be nonzero. If a Fortran statement has a label, it is uniquely identified and the label can be used in DO constructs, CALL statements, branching statements, and input/output statements. In most cases, two statements in the same program unit must not have the same label (there are exceptions because a program unit may contain more than one scoping unit, for example, several internal procedures). Leading zeros in a label are not significant so that the labels 020 and 20 are the same label. This means that there are 99999 different labels and the processor must accept any of them, but may limit the total number of labels allowed in a program unit.

Any statement may have a label, but a label is used only:

1. to designate to target of a branch
2. to specify a FORMAT statement
3. to indicate the termination of some DO loops

The cases in which duplicate labels can be used in the same program unit are explained in 16 as part of the general treatment of the scope of entities. Examples of statements with labels are:

```
100 CONTINUE
 21 X = X + 1.2
101 FORMAT (1X, 2F10.2)
```

The Fortran syntax does not permit a statement with no content, sometimes referred to as a blank statement in other programming languages. A label must have a statement so each of the following lines is nonstandard Fortran:

```
10
          X=0;101;
```

3.3 Source Form

A Fortran program is a sequence of one or more lines organized as Fortran statements, comments, and INCLUDE lines; this collection of statements, comments, and lines is called **source text**. A Fortran statement consists of one or more complete or partial lines of source text and is constructed from low-level syntax. A complete or partial line is a sequence of characters. The following examples illustrate how statements can be formed from partial or complete lines:

```
! This example is written for one of the source forms, called free
! source form (3.3.1).  It uses the & on the continued line to
! indicate continuation.  A ! after an & indicates the beginning
! of a comment.
10 FORMAT( 2X, I5 )            ! A statement on a complete line
13 FORMAT( 2X, &               ! A statement on two
           I5 )                !   complete lines

X = 5; 10 FORMAT( 2X, I5)      ! Two statements, each as part of a line

X = 5 +  &                     ! A statement consisting of a complete
    Y; 10 FORMAT( 2X, I5 )     !   line and a partial line

X = 5 +  &
    Y; 10 FORMAT( 2X,   &      ! A statement made up of
            I5);  READ &       !   two partial lines
          (5, 10)  A, B, C
```

The lines within a program unit (except comment lines) and the order of the lines are in general significant (Table 2-1), except that the order of the subprograms following a CONTAINS statement and before the END statement for the containing program unit is insignificant.

There are two source forms for writing source text: free source form and fixed source form, which is the traditional Fortran form. Programmers must use either fixed or free source form throughout a program unit, although different program units within the program may use different source forms. Each Fortran processing system must provide a way to indicate which source form is being used; for example, this might be indicated with a compiler option, file suffix (e.g., .f or .f03), or compiler directive, or the processor might assume one of the forms by default. Section 3.3.3 describes a way to write Fortran statements so that the source text is acceptable to both free and fixed source forms.

Characters that form the value of a character literal constant or a character string edit descriptor (quote or apostrophe edit descriptor) are said to be in a **character context**. Note that the characters in character context do not include the delimiters used to indicate the beginning and end of the character constant or string. Also, the ampersands in free source form, used to indicate that a character string is being continued and used to indicate the beginning of the character string on the continued line, are never part of the character string value and thus are not in character context (3.3.1).

The rules that apply to characters in a character context are different from the rules that apply to characters in other contexts. For example, blanks are always significant in a character context, but are never significant in other parts of a program written using fixed source form.

```
CHAR = CHAR1 // "Mary K. Williams"
! The blanks within the character string
! (within the double quotes) are significant.

! The next two statements are equivalent in fixed source form.
DO2I=1,N
DO 2 I = 1, N
```

Comments may contain any graphic character that is in the processor character set. For fixed source form, comments may contain, in addition, certain control characters as allowed by the processor—see the implementor's manual for the specific control characters allowed.

3.3.1 Free Source Form

In free source form, there are no restrictions limiting statements to specific positions on a Fortran line. The blank character is significant and may be required to separate lexical tokens.

Rules and restrictions:

1. Blank characters are significant everywhere except that a sequence of blank characters outside a character context is treated as a single blank character. They may be used freely between tokens and delimiters to improve the readability of the source text. For example, the two statements:

   ```
   SUM=SUM+A(I)
   SUM = SUM + A (I)
   ```

   are the same.

2. Each line may contain from 0 to 132 characters, provided that they are of default character kind. If any character is of a nondefault character kind, the processor may limit the number of characters to fewer than 132 characters. For example, a line such as

```
TEXT = GREEK_'This line has 132 characters and contains α'
```

may use exactly 132 graphic characters, but the implementation may require more space to represent this source line than 132 Fortran characters. The processor may thus limit how many graphic characters may be used on a line if any of them are of nondefault character kind.

3. The exclamation mark (!), not in character context, is used to indicate the beginning of a comment that ends with the end of the line. A line may contain nothing but a comment. Comments, including the !, are ignored and do not alter the interpretation of Fortran statements in any way. There is no language limit on the number of comments in a program unit, although the processor may impose such a limit. A line whose first nonblank character is an exclamation mark is called a **comment line**. An example of a Fortran statement with a trailing comment is:

```
ITER = ITER + 1    ! Begin the next iteration.
```

An example of a comment line is:

```
! Begin the next iteration.
```

A comment may appear before or after a program unit, but the standard does not indicate which program unit it belongs to if it is between program units.

A line with only blank characters or with no characters is treated as a comment line.

4. The ampersand (&) is used as the continuation symbol in free source form. If it is the last nonblank character after any comments are deleted and it is not in a character context, the statement is continued on the next line that does not begin with a comment. If the first nonblank character on the continuing line is an ampersand, the statement continues after the ampersand; otherwise, the statement continues with the first position of the line. The ampersand or ampersands used as the continuation symbols are not considered part of the statement. For example, the following statement takes two lines (one continuation line):

```
STOKES_LAW_VELOCITY = 2 * GRAVITY * RADIUS ** 2 *   &
      (DENSITY_1 - DENSITY_2) / (9*COEFF_OF_VISCOSITY)
```

The leading blanks on the continued line are included in the statement and are allowed in this case because they are between lexical tokens.

No more than 255 continuation lines are allowed in a Fortran statement. No line may contain an ampersand as the only nonblank character before an exclamation mark. Comment lines cannot be continued; that is, the ampersand as the last character in a comment is part of the comment and does not indicate continuation.

The double-ampersand convention must be used to continue a name, a character constant, or a lexical token consisting of more than one character split across lines. The following statement is the same statement as in the previous example:

```
STOKES_LAW_VELOCITY = 2 * GRAVITY * RADIUS * 2 * (DEN&
      &SITY_1 - DENSITY_2) / (9 * COEFF_OF_VISCOSITY)
```

However, splitting names across lines makes the code difficult to read and is not recommended.

Ampersands may be included in a character constant. Only the last ampersand on the line is the continuation symbol, as illustrated in the following example:

```
LAWYERS = "Jones & Clay & &
   &Davis"
```

The value of this constant is "Jones & Clay & Davis". The first two ampersands are in character context; they are part of the value of the character string.

5. More than one statement may appear on a line. The statement separator is the semicolon (;), provided it is not in a character context; multiple successive semicolons on a line with or without blanks intervening are considered as a single separator. The end of a line is also a statement separator, and any number of semicolons at the end of the line have no effect. For example:

```
! The semicolon is a statement separator.
   X = 1.0;  Y = 2.0

   ! However, the semicolon below at the end of a line
   ! is not treated as a separator and is ignored.
   Z = 3.0;

   ! Also, consecutive semicolons are treated as one
   ! semicolon, even if blanks intervene.
   Z = 3.0; ; W = 4.0

! Continuation lines and statement separators may be mixed.
   A = &
   B; C = D; E &
   = D
```

A semicolon must not be the first nonblank character on a line. Thus, the following is illegal:

```
A = B &
; C = D
```

but the following is legal:

```
A = B &
&; C = D
```

This rule does not seem to make much sense, but that is what the standard says.

6. A label may appear before a statement, provided it is not part of another statement, but it must be separated from the statement by at least one blank. For example:

```
10 FORMAT(10X,2I5)                     ! 10 is a label.
   IF (X == 0.0) 200 Y = SQRT(X)       ! Label 200 is not allowed.
```

7. Any graphic character in the processor character set may be used in character literal constants (4.3.5.5) and character string edit descriptors (10.2.3). Note that this excludes control characters; it is recommended that the implementor's manual be consulted for the specific details.

3.3.1.1 Blanks as Separators

Blanks in free source form may not appear within tokens, such as names or symbols consisting of more than one character, except that blanks may be freely used in format specifications. For instance, blanks may not appear between the characters of multi-character operators such as ** and **.NE.** Format specifications are an exception because blanks may appear within edit descriptors such as BN, SS, or TR in format specifications. On the other hand, a blank must be used to separate a statement keyword, name, constant, or label from an adjacent name, constant, or label. For example, the blanks in the following statements are required.

```
INTEGER SIZE
PRINT 10,N
DO I=1,N
```

Adjacent keywords require a blank separator in some cases (for example, CASE DEFAULT) whereas in other cases two adjacent keywords may be written either with or without intervening blanks (for example, BLOCK DATA); The following list gives the situations where blank separators are optional.

BLOCK DATA
DOUBLE PRECISION
ELSE IF
ELSE WHERE
END ASSOCIATE
END BLOCK DATA
END DO
END ENUM
END FILE
END FORALL
END FUNCTION
END IF

END INTERFACE
END MODULE
END PROGRAM
END SELECT
END SUBROUTINE
END TYPE
END WHERE
GO TO
IN OUT
SELECT CASE
SELECT TYPE

Thus both of the following statements are legal:

```
END IF
ENDIF
```

Despite these rules, blank separators between statement keywords make the source text more readable and clarify the statements. In general, if common rules of English text are followed, everything will be correct. For example, blank separators in the following statements make them quite readable, even though the blanks between the keywords DOUBLE and PRECISION and between END and FUNCTION are not required.

```
RECURSIVE PURE FUNCTION F(X)
DOUBLE PRECISION X
END FUNCTION F
```

3.3.1.2 Sample Program, Free Source Form

A sample program in free source form is:

```
 123456789.......
-----------------------------------------------------
|PROGRAM LEFT_RIGHT
| REAL  X(5), Y(5)
|        ! Print arrays X and Y
|        PRINT 100, X, Y
|        100 FORMAT (F10.1, F10.2, F10.3, F10.4,  &
|                    F10.5)
|    . . .
|END
```

3.3.2 Fixed Source Form

Fixed source form is position oriented on a line using the conventions for position that were used historically for Fortran written on punched cards. Currently, most programmers use Fortran systems that permit a less stilted style of source form; this is similar to or the same as the free source form described in the previous sections. Fixed source form is now obsolescent.

Rules and restrictions:

1. Fortran statements or parts of Fortran statements must be written between positions 7 and 72. Character positions 1 through 6 are reserved for special purposes.

2. Blanks are not significant in fixed source form except in a character context. For example, the two statements:

```
D O  10  I = 1, L O O P E N D
DO 10 I = 1, LOOPEND
```

   are the same.

   A C or * in position 1 identifies a comment. In this case, the entire line is a comment and is called a **comment line**. A ! in any position except position 6 and not in character context indicates that a comment follows to the end of the line. Comments are not significant, and there is no language limit on the number of comment lines. However, a processor may impose a limit. A comment line may appear before or after a program unit, but the standard does not indicate which program unit it belongs to if it is between program units.

3. A line with only blank characters or with no characters is treated as a comment line.

4. Multiple statements on a line are separated by one or more semicolons; semicolons may occur at the end of a line and have no effect. A semicolon must not be the first nonblank character in positions 7 through 72.

5. Any character (including ! and ;) other than blank or zero in position 6 indicates that the line is a continuation of the previous line. Such a line is called a **continuation line**. The text on the continuation line begins in position 7. There must be no more than 19 continuation lines for one statement in fixed source form. The first line of a continued statement is called the **initial line**.

6. Statement labels may appear only in positions 1 through 5. Labels may appear only on the first line of a continued statement. Thus, positions 1 through 5 of continuation lines must contain blanks.

7. An END statement must not be continued. END also must not be an initial line of a statement other than an END statement. For example, an assignment statement for the variable ENDLESS may not be written as

```
 END
+LESS = 3.0
```

   because the initial line of this statement is identical to an END statement.

8. Any character from the processor character set (including graphic and control characters) may be used in a character literal constant and character edit descriptors, except that the processor is permitted to limit the use of some of the control

characters in such character contexts. Consult the implementor's documentation for such limitations.

3.3.2.1 Sample Program, Fixed Source Form

A sample program in fixed source form is:

```
12345678901234567890123.....
--------------------------------------------
|      PROGRAM LEFT_RIGHT
|      REAL  X(5), Y(5)
|C     Print arrays X and Y
|      PRINT 100, X, Y
|  100 FORMAT (F10.1, F10.2, F10.3, F10.4,
|    1         F10.5)
|          . . .
|      END
```

3.3.3 Rules for Fixed/Free Source Form

For many purposes, such as an included file (3.4), it is desirable to use a form of the source code that is valid and equivalent for either free source form or fixed source form. Such a fixed/free source form can be written by obeying the following rules and restrictions:

1. Limit labels to positions 1 through 5, and statements to positions 7 through 72. These are the limits required in fixed source form.

2. Treat blanks as significant. Because blanks are ignored in fixed source form, using the rules of free source form will not impact the requirements of fixed source form.

3. Use the exclamation mark (!) for a comment, but don't place it in position 6, which indicates continuation in fixed source form. Do not use the C or * forms for a comment.

4. To continue statements, use the ampersand in both position 73 of the line to be continued, and in position 6 of the continuation line. Positions 74 to 80 must remain blank or have only a comment there. Positions 1 through 5 of the continuation line must be blank. The first ampersand continues the line after position 72 in free source form and is ignored in fixed source form. The second ampersand indicates a continuation line in fixed source form and in free source form indicates that the text for the continuation of the previous line begins after the ampersand.

3.3.3.1 Sample Program, Use with Either Source Form

A sample program that is acceptable for either source form is:

```
12345678901234567890123....                                   73
---------------------------------------.....----
|      PROGRAM LEFT_RIGHT
|      REAL  X(5), Y(5)
|!     Print arrays X and Y
|      PRINT 100, X, Y
|  100 FORMAT (F10.1, F10.2, F10.3, F10.4,              &
|    &        F10.5)
|         . . .
|      END
```

3.4 The INCLUDE Line

Source text may be imported from another file and included within a program file during processing. An **INCLUDE line** consists of the keyword INCLUDE followed by a character literal constant. For example,

```
INCLUDE  'MY_COMMON_BLOCKS'
```

The specified text is substituted for the INCLUDE line before compilation and is treated as if it were part of the original program source text. The location of the included text is specified by the value of the character constant in some processor-dependent manner. A frequent convention is that the character literal constant is the name of a file containing the text to be included. Use of the INCLUDE line provides a convenient way to include source text that is the same in several program units. For example, the specification of interface blocks or objects in common blocks may constitute a file that is referenced in the INCLUDE line.

The form for an INCLUDE line is:

INCLUDE   character-literal-constant

Rules and restrictions:

1. The character literal constant used must not have a kind parameter that is a named constant.
2. The INCLUDE line is a directive to the compiler; it is not a Fortran statement.
3. The INCLUDE line is placed where the included text is to appear in the program.
4. The INCLUDE line must appear on one line with no other text except possibly a trailing comment. There must be no statement label.
5. INCLUDE lines may be nested. That is, a second INCLUDE line may appear within the text to be included. The permitted level of nesting is not specified and is processor dependent. However, the text inclusion must not be recursive at any level; for example, included text A must not include text B, which includes text A.

6. A file intended to be referenced in an INCLUDE line must not begin or end with an incomplete Fortran statement. This means that the line before the INCLUDE line must not be continued and that the line after the INCLUDE line must not be a continuation line.

   An example of a program unit with an INCLUDE line follows:

```
PROGRAM MATH
REAL, DIMENSION (10,5,79) :: X, ZT
!  Some arithmetic
INCLUDE 'FOURIER'
!  More arithmetic
   . . .
END
```

The Fortran source text in the file FOURIER in effect replaces the INCLUDE line. The INCLUDE line behaves like a compiler directive.

4 Data Types

- A **Type** has a name, type parameters, a set of values, a set of operations and procedures, and a means to represent constants of the type.
- **Type Parameters** allow a type to have a family of representations.
- A **Type Specifier** is used to specify a particular type and type parameter values.
- The **Intrinsic Types** are integer, real, complex, logical, and character.
- **Derived Types** are defined by a user.
- A **Structure** is an object of derived type.
- A **Structure Constructor** creates values of derived type.
- An **Array Constructor** creates array values.
- **Operations** on objects of derived type are defined by functions supplied by the user.
- **Type Extension** is a means of defining a new type by building on a previously defined type. The new type **inherits** aspects of the previously defined one.
- A **Procedure Binding** is a relationship between a type and a procedure. It allows a procedure to be selected based on the type of an object.
- An **Enumeration** is a set of named integer constants with a declaration form intended to facilitate interoperation with C.

Data type is a fundamental concept in Fortran, as well as in many other programming languages. Every piece of data in a Fortran program has a data type, which determines what kinds of values it can take and what can be done with it. This chapter details what is meant by a type in Fortran. It then describes each of the types defined by the standard, plus the facilities for user-defined types.

The standard defines intrinsic types corresponding to the broad categories of computational tasks listed in Table 4-1. Additional types can be built of (or derived from) the intrinsic types and thus are called **derived types**. Most derived types are defined by the programmer using the facilities described in this chapter. A few derived types are defined in the standard intrinsic modules (14.3, 15.3). The Fortran types are categorized in Figure 4-1.

The type of a datum determines the operations that can be performed on it. Table 4-2 lists the intrinsically-defined operations. The user can define additional operations for any type.

J.C. Adams et al., *The Fortran 2003 Handbook*,
DOI: 10.1007/978-1-84628-746-6_4,

Table 4-1 Intrinsic types for computational tasks

Task	Type of data
Calculating typical numeric results	Real data
Calculating in the complex domain	Complex data
Counting	Integer data
Making decisions	Logical data
Explaining	Character data

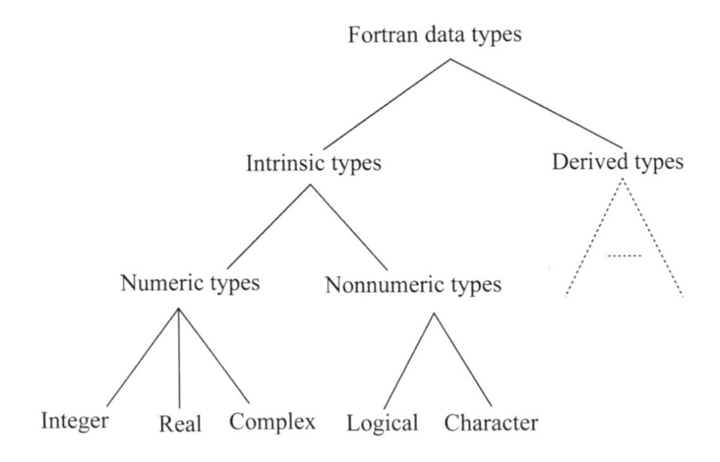

Figure 4-1 Fortran data types

Table 4-2 Intrinsically-defined operations

Type of data	Intrinsic operations
Real, complex, integer	Addition, subtraction, multiplication, division, exponentiation, negation, comparison, identity
Logical	Negation, conjunction, disjunction, equivalence and nonequivalence
Character	Concatenation, comparison
Derived	None

4.1 Data Type Selection

The appropriate intrinsic type for a data entity is often obvious. More careful thought is likely to be required for the selection of suitable kind parameter values, derived types and classes.

4.1.1 Kinds of Intrinsic Types

Once the type is decided, it may be necessary for the programmer to consider which "kind" of the type to use. "Kind" is a technical term in Fortran. Each of the intrinsic types has a **kind parameter** that selects a processor-dependent representation of objects of that type and kind. Each compiler supports a particular set of representations from which the programmer can select. If no kind parameter value is specified, the default kind is assumed.

Kinds are known to the processor as integer values, but if a program is to be portable, the actual numbers should not be used because the particular kind values for the intrinsic types are processor dependent. For portability, appropriate kind values should be determined using the procedures described in 4.3, 13.3.1.2, and 14.3.3.7 and assigned to named constants, which are then used in type specifiers, literal constants, and kind arguments of intrinsic functions. When named constants are used to designate kinds, only the value of the constant matters—not the names of the constant. If two different named constants have the same numerical value and are used as kind parameters, then they will represent the same kind.

The intrinsics for determining kind values all return negative values when appropriate kinds are not available for the particular compiler. Negative values are guaranteed to cause compilation error diagnostics when used as kind values, thus ensuring that programs will not accidentally be run with kinds that do not meet specified requirements.

The Fortran kind parameters for each of the intrinsic types serve the following purposes:

1. **Real.** The real kind parameter primarily selects the precision and range of the representation. There may also be multiple representations for the same precision and range, for example native and IEEE representations. The standard requires that each compiler support at least two real kinds which must have different precisions. The default real kind is the lower precision of these. Compilers may support additional real kinds to provide other precisions or other representations with the same precisions.

   Fortran 77 did not have kind parameters but did provide two kinds for the real type: REAL and DOUBLE PRECISION. REAL is often referred to as single precision. It treated double precision real as a separate type. Fortran 90 and later versions, while remaining compatible, treat double precision as a separate kind of real. That is, there are two ways to specify double precision real: one is with a REAL specifier with the kind corresponding to double precision, and the other is with a DOUBLE PRECISION specifier.

   Programs with default REAL and DOUBLE PRECISION declarations are not numerically portable across machine architectures with different word sizes. Each compiler vendor chooses a representation for the real type that is efficient on the host machine. For example, a representation that will fit into 32 bits might be chosen on a 32-bit-word machine while a representation that fits into 64 bits might be chosen for a 64-bit-word machine. If a 64-bit representation is required for the numerical stability of the algorithm, DOUBLE PRECISION declarations must be used on the 32-bit machine. When the program is moved to the 64-bit machine, the DOUBLE PRECISION declarations usually would be changed to REAL declarations because a 128-bit representation is not needed and probably would degrade the performance of the program. A programmer can use kind parameters in REAL declarations to specify a required minimum precision of perhaps 12 decimal digits. When the program is run on the 32-bit machine, it will use the kind corresponding to double precision. When the same program (without change) is run on the 64-bit machine, the kind corresponding to single precision will be used.

2. **Complex**. The kind parameter for complex selects the same representation as that for real. Every real kind has a corresponding complex kind and vice versa.

3. **Integer**. The integer kind parameter primarily selects the range of integer values that can be represented. Some representations might provide for a large range at the cost of large storage size, while others might provide for small storage size at the cost of correspondingly smaller range. In principle it is also possible for there to be multiple representations for the same range, perhaps differing in byte order or in the representation of negative values. Only one integer kind is required in a standard-conforming processor, but more are permitted. The default integer kind is represented in the same size as the default real kind.

4. **Character**. The character kind parameter selects a character set. The default character type usually has an underlying machine representation of a single byte (8 bits). This is adequate to represent 2^8 or 256 different characters, which is more than enough for alphabetic languages. However, ideographic languages, such as Japanese and Chinese, have several thousand graphic symbols that require at least a two-byte representation (16 bits). To accommodate this spectrum of users, Fortran makes provision for (although it does not require implementation of) different kinds of character data. Because these additional kinds of character data are not required for standard-conforming Fortran processors, many processors intended for English-speaking Fortran users might not support ideographic languages. Nevertheless, the character kind mechanism allows an implementation to support an alphabetic language or an ideographic language or both simultaneously. The standard makes particular provisions for the ASCII and ISO 10646 character sets, but support for them is optional.

5. **Logical**. Because the logical type has only two values (true and false), it could be represented in a single bit. Although efficient in terms of storage, single-bit representations can be inefficient in computation time. For storage association, the default logical type is represented in the same size as the default real type, but alternative representations of logical data are permitted; that is, a nondefault logical kind might be represented in a byte or in a bit for architectural efficiency or application requirements.

4.1.2 Derived Types

Sometimes it is easier to think about an essential element of a problem as several pieces of related data. Arrays can be used to collect homogeneous data (all of the same type) into a single variable. In contrast, a **structure** is a collection of possibly nonhomogeneous data in a single variable. To declare a structure, it is first necessary to define a type that has components of the desired types. The structure is then declared as an object of this user-defined (or derived) type. In the example below, first a type named PATIENT is defined, then two structures JOHN_JONES and SALLY_SMITH are declared.

```
TYPE PATIENT
   INTEGER                PULSE_RATE
   REAL                   TEMPERATURE
   CHARACTER (LEN=300)    PROGNOSIS
END TYPE PATIENT

TYPE (PATIENT)            JOHN_JONES, SALLY_SMITH
```

Type PATIENT has three components, each of a different intrinsic type (integer, real, and character). In practice, a type of this nature probably would have even more components, such as the patient's name and address, insurance company, room number in the hospital, etc. For purposes of illustration, three components are sufficient. JOHN_JONES and SALLY_SMITH are variables (or structures) of type PATIENT. A type definition indicates names, types, and attributes for its components; it does not declare any variables that have these components. Just as with the intrinsic types, a type declaration is needed to declare variables of this type. There can be any number of variables of type PATIENT; there can be subprogram arguments and function results of type PATIENT; there can be arrays of type PATIENT; and operations, such as .appendectomy., can be defined that manipulate objects of type PATIENT. Thus the derived-type definition can be used as a way to specify a pattern for a particular collection of related but nonhomogeneous data; but, because the user can define the pattern, a number of other capabilities are available.

4.1.3 Classes

Sometimes there are multiple derived types that share some components and operations, but have other components and operations that differ. It is desirable to take advantage of the commonalities in order to minimize code duplication and maintenance problems. In that case, a class of related types can be defined in a tree structure. The shared components and operations are defined for a base type. The other types are defined as extension types of the base type.

For example, consider a program that has multiple linked lists, each with a different type of object. Each of these linked lists has a corresponding node type. The basic linked list operations and the node components needed to support those operations are the same for all the linked lists. The object type and any operations related to it differ among the lists. In this case, a base type can be defined for a linked list node with no object. The types for the nodes are defined as extensions of the base type, each of which adds different components to the base type.

4.2 What Is Meant by "Type" in Fortran?

A data type provides a means to categorize data and thus determine which operations may be applied to the data. For each type there is:

1. a type name
2. a set of type parameters
3. a set of values
4. a set of operations and procedures
5. a form for constants of the type

4.2.1 Type Names

Each of the intrinsic types has a name supplied by the standard. The names of derived types must be supplied in type definitions. A derived-type name must not be DOUBLEPRECISION or the same as any of the intrinsic type names (INTEGER, REAL, COMPLEX, LOGICAL, or CHARACTER), even if the intrinsic type is never used; this is a rare exception to Fortran's general rule that the language has no reserved names.

4.2.2 Type Parameters

A type may have multiple representations, which are specified by type parameter values. Type parameters are classified as either kind or length type parameters. Each type parameter is itself of type integer and has a type parameter name. The type parameter names for the intrinsic types are specified by the standard. Any type parameters for derived types must be defined by the derived-type definition.

Kind type parameter values need to be known at compile time; the compiler is likely to need to generate different machine instructions for different kind type parameter values. Thus, anywhere a kind parameter value is required, it must be specified by a scalar integer initialization expression, which is described in 7.4.1. Each of the intrinsic types has a kind type parameter named KIND, which is a default integer. The set of valid values for the intrinsic kind type parameters and the representations specified by those kind values are defined by the compiler.

Invalid kind type parameter values for intrinsic types are guaranteed to give compiler error diagnostics. Examples in this book generally assume that the kind type parameter values used are valid.

Length type parameter values may, in some cases, be determined or change during program execution. As the name suggests, they are most commonly used for lengths or sizes, where a base representation form is repeated. The intrinsic character type has a length parameter named LEN, which is an integer of a processor-dependent kind.

4.2.3 Type Specifier

A type specifier is used in several contexts to specify a particular type and type parameter values. A type specifier (R401) is either an intrinsic type specifier or a derived-type specifier. The syntax of type specifiers is detailed in 4.3 and 4.4.4. The following examples illustrate the use of type specifiers in type declaration statements, an array constructor, and an allocate statement.

```
integer :: i
type(patient) :: jane_doe
names = [character(16):: "Lisa", "Pam", "Julie"]
allocate (real_node_type:: node)
```

A type parameter value (R402) in a type specifier is either a kind value or a length value. The form of a kind value is:

scalar-integer-initialization-expression

The form of a length value is one of:

scalar-integer-expression
*
:

An asterisk as a length value specifies that the type parameter is assumed, that is its value is copied from the corresponding type parameter value of something else. An assumed type parameter is allowed only in a type guard statement of a SELECT TYPE construct, in the allocation of a dummy argument, or in the declaration of a dummy argument, named character constant, or character function result.

A colon as a length value specifies that the type parameter is deferred; the value of the type parameter may be set and changed during execution. A deferred type parameter may be specified only in the declaration of a pointer or allocatable entity or component.

4.2.4 Type Values

Each type has a set of valid values, which usually depend on the type parameter values.

4.2.5 Type Operations and Procedures

An operator has either one or two operands. The definition of an operator depends on the types, type parameters, and ranks of the operands. For intrinsic types, a set of operations with corresponding operators is provided by the language as described in 7.

A user may specify new operators and define operations for the new operators. The form of a new operator is an alphabetic name of the user's choice delimited by periods. These new operators are analogous to intrinsic operators such as .GT., .AND., and .NEQV. For example, a user might specify and define the operations .PLUS., .REMAINDER., and .REVERSE. In defining the operation, the types of allowable operands must be specified. It is not possible to override the standard's definition of an intrinsic operation, but it is possible to define application of an intrinsic operator symbol to cases that are not defined by the standard. For example, consider the expression A + B. If both A and B are of numeric type, the operation is intrinsically defined. However, if either A or B is of derived type or nonnumeric type, then the plus operation between A and B is not intrinsically defined, and the user may provide a definition. The user defines an operation using a function and an interface block as described in 12.5.4.2.

The definition of assignment depends on the types, type parameters, and ranks of both the variable and the expression that provides the value. The language defines assignment for several combinations of intrinsic types, as described in 7.5.2. The language also defines assignment of a derived-type expression to a variable whose type and type parameter values are the same or can be allocated to be the same. A user may define assignment for cases not defined by the standard. A user also may override the standard's definition of assignment for a derived type, but not for intrinsic types. The user defines assignment using a subroutine and an interface block, as described in 12.5.4.3.

A user may define procedure bindings for derived types. These are discussed in 4.4.11.

4.2.6 Forms for Constants

A literal constant (R306) is one of:

```
integer-literal-constant
real-literal-constant
complex-literal-constant
logical-literal-constant
character-literal-constant
boz-literal-constant
```

Two aspects of this definition merit particular note. First is that it excludes the signed forms of the integer and real literal constants (4.3.1.4, 4.3.2.4); a sign can be used in most contexts, but it parses as an operator instead of as part of the constant. Second is that it includes BOZ literal constants (4.3.1.4) even though the allowed contexts for BOZ literal constants are so restricted that they might better be thought of as additional forms allowed in those contexts rather than as literal constants.

The language specifies the syntactic forms for literal constants of each of the intrinsic types. The form of a constant indicates the type, type parameters, and value (Table 4-3). There are no literal constant forms for derived types. However, a comparable role can be played by a structure constructor whose component values are all initialization expressions. For example, if the derived type PATIENT is defined as described in 4.1.2, patient(10.99.7,"Recovering") is such a constructor.

4.3 Intrinsic Types

Each of the intrinsic types is described below. The descriptions include simple examples to show how objects of these types may be declared. These simple examples do not give the complete story. The complete forms for declarations are in 5.1.

Table 4-3 Constant forms

Syntax	Type and parameters	Value
1	integer	1
103.1 or 1.031E2	real	103.1
(1.0, 1.0)	complex	1+ι
.TRUE.	logical	true
"Hello"	character(len=5)	Hello

4.3.1 Integer Type

4.3.1.1 Name, Type Parameters, and Type Specifier

The name of the integer type is INTEGER. It has a single kind type parameter named KIND. Only one kind of integer, referred to as default integer, is required by the standard, but a processor may provide more. The storage occupied by a default integer is called a numeric storage unit (13.6.1, 16.2.3.1); a default real or logical must occupy the same amount of storage as a default integer.

The form of the integer type specifier is:

```
INTEGER [ ( [ KIND = ] kind-value ) ]
```

If the kind value is omitted, default integer kind is implied.

Examples of type declaration statements using integer type specifiers are:

```
INTEGER X
INTEGER (KIND=LONG) COUNT, K, TEMPORARY_COUNT
INTEGER (SHORT) PARTS
INTEGER, DIMENSION (0:9) :: SELECTORS, IX
```

where LONG and SHORT are named integer constants.

4.3.1.2 Values

The integer type has values that represent a subset of the mathematical integers. The set of values varies from one processor to another. The intrinsic inquiry function RANGE provides the decimal exponent range for integers of the kind of its argument. The intrinsic function KIND can be used to determine the kind of its integer argument.

There is an intrinsic function SELECTED_INT_KIND that returns an integer kind based on a range requirement. For example:

```
INTEGER (KIND=SELECTED_INT_KIND(5)) I, J
```

declares I and J to be integer objects with a representation method that permits at least five decimal digits; that is, it includes at least all integers between -10^5 and 10^5.

Every integer kind has a single zero value, which is considered neither positive nor negative. If a processor has separate internal representations for positive and negative zero integers of a kind, they are considered to have the same value.

4.3.1.3 Operators

There are both binary and unary intrinsic operators for the integer type. Binary operators have two operands and unary operators have one. The binary arithmetic operations for the integer type are: +, −, *, /, and **. The unary arithmetic operations are + and −. The relational operations (all binary) are: <. <=, ==, /=, >=, and >. The result of an intrinsic arithmetic operation on integer operands is an integer value; the result of an intrinsic relational operation is a logical entity of default logical kind.

4.3.1.4 Form for Constant Values

An integer constant is a string of decimal digits, optionally followed by an underscore and a kind parameter.

The form of an integer literal constant (R406) is:

```
digit-string [ _ kind-parameter ]
```

where the kind parameter is one of:

```
digit-string
scalar-integer-constant-name
```

This syntax for a kind parameter in a literal constant is considerably more restrictive than that of a kind value used in a type specifier and other places; however, the kind parameter in a literal constant can be a named constant, which in turn can be defined by an initialization expression, so the same functionality exists. If a kind parameter is specified, the constant is of that kind; otherwise, it is of type default integer.

Examples of integer literal constants are:

```
42
999999999999999999999_LONG
```

where LONG is a named integer constant.

A signed integer constant (R405) is an integer constant preceded by an optional sign, which is either + or −. Contrary to what might be expected from the terminology, a signed integer constant is not in general an integer constant. Signed integer constants are used in only a few places in the language. In most contexts, a sign followed by an integer constant parses as an operator and an unsigned integer constant rather than as a signed integer constant. This distinction makes little difference in practice except in one special case: on machines where the most negative integer is larger in magnitude than the most positive one, overflow can result from trying to write a literal for the most negative integer in the obvious way.

Examples of signed integer constants are:

```
+64
10000000
-255_SHORT
```

where SHORT is a named integer constant.

Integer constants are interpreted as decimal values. However, in limited contexts, there are forms for unsigned binary, octal, or hexadecimal constants, collectively referred to as BOZ literal constants. These forms may be used only to initialize integer variables in DATA statements or as actual arguments of the intrinsic functions CMPLX, DBLE, INT, or REAL.

A binary constant (R412) has one of the forms:

```
B ' digit [ digit ] ... '
B " digit [ digit ] ... "
```

where a digit is restricted to 0 or 1.

An octal constant (R413) has one of the forms:

```
O ' digit [ digit ] ... '
O " digit [ digit ] ... "
```

where a digit is restricted to the values 0 through 7.

A hexadecimal constant (R414) has one of the forms:

```
Z ' digit [ digit ] ... '
Z " digit [ digit ] ... "
```

where a digit is 0 through 9 or one of the letters A through F (representing the decimal values 10 through 15). Although these forms use quotes or apostrophes, they are not character strings; lowercase letters are equivalent to uppercase in the hexadecimal forms.

Although the standard refers to these forms as integer constants, their interpretation when used as actual arguments of CMPLX, DBLE, and REAL is as a bit pattern rather than a numeric integer value. Where these constants are interpreted as numeric values, the binary, octal, and hexadecimal digits are interpreted according to the binary, octal, and hexadecimal number systems; the result is represented with the integer kind having the largest range supported by the compiler. Examples (all of which have a value equal to the decimal value 10) are:

```
B"1010"
O'12'
Z"a"
```

4.3.2 Real Type

4.3.2.1 Name, Type Parameters, and Type Specifier

The name of the real type is REAL. It has a single kind type parameter named KIND. A processor must provide at least two kinds for the real type. One of the kinds is for the default real type and the other is for the double precision real type, which must have more precision than the default real type. The default real type must occupy the same amount of storage as default integer; double precision real type must occupy twice as much.

The forms of the real type specifier are:

```
REAL [ ( [ KIND = ] kind-value ) ]
DOUBLE PRECISION
```

If the kind value is omitted from the form with the REAL keyword, default real kind is implied. The form with the DOUBLE PRECISION keyword is an alternate form for specifying a real with the kind for double precision.

Examples of type declaration statements using real type specifiers are:

```
REAL X, Y
REAL (KIND =  HIGH), SAVE :: XY(10, 10)
REAL, POINTER :: A, B, C
DOUBLE PRECISION DD, DXY, N
```

where HIGH is a named integer constant.

4.3.2.2 Values

The values of the real data type approximate the mathematical real numbers. The set of values varies from processor to processor.

Intrinsic functions are available to inquire about the representation methods provided on a processor. The intrinsic function KIND can be used to determine the kind of its real argument. The intrinsic functions PRECISION and RANGE return the decimal precision and exponent range of the approximation method used for the kind of the argument. The intrinsic function SELECTED_REAL_KIND returns a real kind value based on precision and range requirements. For example:

```
REAL (SELECTED_REAL_KIND (5)) X
```

declares X to have at least five decimal digits of precision and no specified minimum range.

```
REAL (SELECTED_REAL_KIND (8, 70)) Y
```

declares Y to have at least eight decimal digits of precision and a range that includes values between 10^{-70} and 10^{70} in magnitude.

Every real kind has a zero value. If a processor has separate internal representations for positive and negative zeros of a kind, those representations are both treated as numerically equivalent to zero in the following contexts.

1. As operands of relational operators. For example, the logical expression x>=0.0 evaluates to true if x is zero, regardless of whether it is a positive or negative zero.

2. As actual arguments to intrinsic procedures, except where the intrinsic procedure explicitly specifies that negative zero is distinguished. The SIGN intrinsic function

is one that specifies special treatment of negative zero; it provides a means to distinguish positive from negative zeros for those cases where that is desired.

3. As the expression in an arithmetic if statement. Both positive and negative zero values result in taking the zero branch.

In other contexts, it is processor dependent whether positive and negative zeros are treated differently. For example, if x has the value negative zero, it is processor-dependent whether the expression 2.0*x yields a positive or negative zero as a result.

4.3.2.3 Operators

The intrinsic binary arithmetic operators for the real type are: +, −, *, /, and **. The intrinsic unary arithmetic operators are: + and −. The relational operators are: <, <=, ==, /=, >=, and >. The result of an intrinsic arithmetic operation on real operands is a real value. If one of the operands of an arithmetic operation is an integer, the result is still a real value. The result of an intrinsic relational operation is a default logical value.

4.3.2.4 Forms for Constants

A real constant is distinguished from an integer constant by containing either a decimal point, an exponent, or both. Forms for a real literal constant (R417) are:

```
digit-string exponent-letter exponent [ _ kind-parameter ]
whole-part . [ fraction-part ] [ exponent-letter exponent ] [ _ kind-parameter ]
. fraction-part [ exponent-letter exponent ] [ _ kind-parameter ]
```

where the exponent letter (R419) is E or D, the whole part and fraction part are digit strings (R409), and an exponent (R420) is a signed digit string (R408). If both a kind parameter and an exponent letter are present, the exponent letter must be E. If a kind parameter is specified, the real constant is of that kind; if a D exponent letter is specified, the constant is of type double precision real; otherwise, the constant is of type default real. A real constant may have more decimal digits than are significant for reals of its kind. Examples of real literal constants are:

```
2.1
0.45_LOW
.123
3E4
2.718281828459045D0
```

where LOW is a named integer constant.

A signed real literal constant (R416) is a real literal constant preceded by an optional sign. As with signed integer literal constants, signed real literal constants are used in only a few places in the language. Examples of signed real literal constants are:

```
-14.78
+1.6E3
1111111111.1111111
-16.E4_HIGH
```

where HIGH is a named integer constant.

4.3.3 Complex Type

4.3.3.1 Name, Type Parameters, and Type Specifier

The name of the complex type is COMPLEX. It has a single kind type parameter named KIND. The supported kind values for complex type are required to be the same as those for real. The amount of storage occupied by a complex must be twice the amount of storage occupied by a real of the same kind.

The form of the complex type specifier is:

```
COMPLEX [ ( [ KIND = ] kind-value ) ]
```

If the kind value is omitted, the kind value for default real is implied.

Examples of type declaration statements using complex type specifiers are:

```
COMPLEX CC, DD
COMPLEX (KIND = QUAD), POINTER :: CTEMP (:)
```

where QUAD is a named integer constant.

4.3.3.2 Values

The complex type has values that approximate the mathematical complex numbers. A complex value is represented as a pair of real values; the first is called the real part and the second is called the imaginary part. Each approximation method used to represent data entities of type real is available for entities of type complex with the same kind parameter values. Therefore, there are at least two approximation methods for complex, one of which corresponds to default real and one of which corresponds to double precision real. There is no double precision complex keyword; a double precision complex can be specified only by using the appropriate kind parameter value in the complex type specifier. The intrinsic functions KIND, PRECISION, and RANGE can be used with complex arguments and have the same interpretation as when they are applied to real arguments. Because the kind values and representations for complex correspond to those of real, the SELECTED_REAL_KIND intrinsic function may be used in a declaration of a complex object. For example:

```
COMPLEX (SELECTED_REAL_KIND (8, 70)) CX
```

CX must have at least eight decimal digits of precision and a range that includes values between 10^{-70} and 10^{70} in magnitude for the real and imaginary parts.

4.3.3.3 Operators

The intrinsic binary arithmetic operators for the complex type are: +, –, *, /, and **. The intrinsic unary arithmetic operators are: + and –. The intrinsic relational operators are: == and /=. The arithmetic operators specify complex arithmetic; the relational operators compare operands to produce default logical results. The result of an intrinsic arithmetic operation on complex operands is a complex entity. If one of the operands is an integer or real entity, the result is still a complex entity.

4.3.3.4 Form for Constants

A complex constant is written as two constants that are real or integer, separated by a comma, and enclosed in parentheses. The form for a complex literal constant (R421) is:

(real-part , imaginary-part)

where the real part and imaginary part each may be either a signed integer literal constant (R405) or a signed real literal constant (R416), or a named constant of type real or integer.

Examples are:

```
(3.0, -3.0)
(6, -7.6E9)
(3.0_HIGH, 1.6E9_LOW)
(x_offset, y_offset)
```

where HIGH and LOW are named integer constants, and x_offset and y_offset are named real or integer constants.

The types and kinds of the two parts of a complex literal constant need not be the same. If both parts are real, the complex constant has the kind of one of the parts; it is the part with greater precision unless the parts have the same precision, in which case the choice of part is processor-dependent. If one part is real and the other integer, the complex constant has the kind of the real part. If both parts are integer, the complex constant has the kind of default real. In any case, the complex value is formed by converting each part to a real of the same kind as the complex constant.

This form is only for complex constants. It is not a general constructor for complex values that are not constants. If this form were allowed as such a general constructor, there are contexts where it would cause syntactic ambiguity. The CMPLX intrinsic function serves the purpose of a general complex constructor which can be used with variables.

4.3.4 Logical Type

4.3.4.1 Name, Type Parameters, and Type Specifier

The name of the logical type is LOGICAL. It has a single kind type parameter named KIND. Only one kind of logical, referred to as default logical, is required by the standard. The default logical type must occupy the same amount of storage as default integer.

The form of the logical type specifier is:

LOGICAL [([KIND =] kind-value)]

If the kind value is omitted, default logical kind is implied.

Examples of type declaration statements using logical type specifiers are:

```
LOGICAL IR, XT
LOGICAL (KIND = BIT), SAVE :: XMASK (3000)
```

where BIT is a named integer constant.

4.3.4.2 Values

The logical type has two values that represent true and false. A processor is required to provide one logical kind, but may provide other kinds to allow the packing of logical values; for example, one value per bit or one per byte. The intrinsic function KIND may be used to determine the kind of its logical argument. There is no intrinsic function to select a logical kind analogous to the functions SELECTED_INT_KIND, SELECTED_ REAL_KIND, and SELECTED_CHAR_KIND; the only way to determine the logical kinds supported by a compiler are from the compiler documentation or by experimentation.

4.3.4.3 Operators

The intrinsic binary operators for the logical type are: conjunction (.AND.), inclusive disjunction (.OR.), logical equivalence (.EQV.), and logical nonequivalence (or exclusive disjunction) (.NEQV.). The intrinsic unary operation is negation (.NOT.).

4.3.4.4 Form for Constants

There are only two logical literal constants. Optionally, they may include a trailing underscore and a kind parameter. The forms for logical literal constants (R428) are:

```
.TRUE. [ _ kind-parameter ]
.FALSE. [ _ kind-parameter ]
```

If a kind parameter is specified, the constant is of that kind; otherwise, it is of type default logical.

Examples are:

```
.FALSE.
.TRUE._BIT
```

4.3.5 Character Type

4.3.5.1 Name, Type Parameters, and Type Specifier

The name of the character type is CHARACTER. It has a single kind type parameter named KIND and a single length type parameter named LEN (4.2.2). Only one kind of character, referred to as default character, is required by the standard. The amount of storage occupied by a default character is referred to as a character storage unit (13.6.1, 16.2.3.1) and is not necessarily (or usually) the same as a numeric storage unit.

The form of a character type specifier is more complicated than that of the specifiers for the other intrinsic types. This is partly because of the multiple type parameters and partly because of the need to support historical forms as well as newer, more flexible forms. The complete form is:

```
CHARACTER [ character-selector ]
```

where a character selector (R424) has one of the forms:

```
( length-value [ , [ KIND= ] kind-value ] )
( LEN= length-value [ , KIND= kind-value ] )
( KIND= kind-value [ , LEN= length-value ] )
* character-length [ , ]
```

and a character length (R426) has one of the forms:

```
( length-value )
integer-literal-constant
```

Kind value and length value are described in 4.2.3. If the kind value is omitted, the kind value for default character is implied. If the length is not explicitly specified, a length of one is implied. If a length value is negative, it specifies a length of zero.

The * character-length form is obsolescent in a type specifier (but the similar form in a component declaration or entity declaration is not).

Rules and restrictions:

1. The optional comma after * character-length is permitted only in a type declaration statement that has no double colon separator.
2. The integer literal constant that specifies a character length must not include a kind parameter.
3. A character length may optionally be specified in an entity declaration of a character type declaration statement (5.1) or in a component declaration of a character component definition statement. If so, the particular entity or component has that length, overriding the length specified by the type specifier.
4. A length of * may be used only in the following ways:
   a. It may be used to declare a dummy argument of a procedure, in which case the dummy argument assumes the length of the associated actual argument when the procedure is invoked.
   b. It may be used to declare a named constant, in which case the length is that of the constant value.
   c. It may be used in the type-spec of an ALLOCATE statement in which each allocate-object is a dummy argument of type character with an assumed character length, in which case the length is that of the associated actual argument.
   d. It may be used to declare the result variable for an external function. Any scoping unit that invokes the function must declare the function with a length other than *, or it must access such a declaration by host or use association. When the function is invoked, the length of the result is the value specified in the program unit referencing the function. This use is obsolescent.

Note that an implication of rule 4 is that a length of * must not appear in an IMPLICIT statement.

5. A function name may be declared with a length of * only if the function is an external or dummy function; it must not be an internal or module function. The function must not be pure or recursive. The function result must not be an array or a pointer.

6. The length of a character-valued statement function or statement function dummy argument of type character must be an initialization expression.

   Examples of type declaration statements using character type specifiers are:

```
CHARACTER answer
CHARACTER (80) LINE
CHARACTER (KIND=ASCII, LEN=20) GREETING
CHARACTER (LEN=30, KIND=CYRILLIC), DIMENSION(10) :: C1
character (len=*), parameter :: title="Fortran 2003 Handbook"
character (len=:), allocatable :: job_title
```

where ASCII and CYRILLIC are named integer constants, ASCII possibly having been defined using the SELECTED_CHAR_KIND intrinsic function.

4.3.5.2 Values

The character type has a set of values composed of character strings. A character string is a sequence of characters, numbered from left to right 1, 2, ..., n, where n is the length of (number of characters in) the string. A character string may have length 0. The maximum length permitted for character strings is processor-dependent.

A standard-conforming processor must support one character kind and may support more. The intrinsic function KIND may be used to determine the kind of its character argument. The intrinsic function SELECTED_CHAR_KIND returns a character kind value based on the name of a character type. The standard defines names for the default, ASCII, and ISO_10646 character sets, but requires that only the default be supported.

Each character kind must contain a character designated as a blank that can be used as a padding character in character operations and input/output data transfer. The characters in all processor-supported character kinds are considered to be representable characters. The default character kind must include the characters that make up the Fortran character set (3.1.1).

4.3.5.3 Collating Sequence

Each character kind has a collating sequence, which is used in the definition of comparison operators. A collating sequence assigns a unique nonnegative integer to each character in the character set. The intrinsic functions CHAR and ICHAR provide conversions between the characters and these integers.

The standard specifies a partial collating sequence for the default character type so that some character relational operations will be portable across different processors.

The standard specifies properties of the collating sequence that are consistent with all processor character sets in common use. Thus it tells the programmer what properties can be counted on, while allowing most processors to use their native character sets. The blank must precede both the alphabetic and numeric characters in the collating sequence. The alphabetic characters, whether uppercase or lowercase, must be in the normal alphabetic sequence. The numeric characters must be in the normal numeric sequence, 0, 1, ..., 9. Numeric characters and alphabetic characters of each case must not be interspersed. Other than blank, there are no constraints on the position of the special characters and the underscore, nor is there any specified relationship between the uppercase and lowercase alphabetic letters.

If the processor supports ASCII or ISO_10646 character kinds, those kinds are required to have the collating sequences specified by the corresponding character set standards. The intrinsic functions ACHAR and IACHAR convert between characters of any kind and positions in the ASCII collating sequence, provided that a corresponding ASCII character exists. The intrinsic functions LGT, LGE, LLE, and LLT provide comparisons between default character strings based on the ASCII collating sequence, whereas the relational operators, such as < and >, use the processor's collating sequence, which might not be the ASCII sequence.

4.3.5.4 Operators

The binary operation concatenation (//) is the only intrinsic operation on character operands that has a character value as a result. A number of intrinsic functions are provided that perform character operations. These are described in 13 and A. The intrinsic relational operators on objects of type character are <, <=, ==, /=, >=, and >. The relational operations may be used to compare character operands, but, because of possible processor-dependent collating sequences, the intrinsic functions LGT, LGE, LLE, and LLT provide more portable results. The relational operators and relational intrinsic functions have default logical results.

4.3.5.5 Form for Constants

A character literal constant is written as a sequence of characters, enclosed either by apostrophes or quotation marks. Forms for character literal constants (R427) are:

```
[ kind-parameter _ ] ' [ representable-character ] ... '
[ kind-parameter _ ] " [ representable-character ] ... "
```

where a representable character is any character in that character set kind that the processor can represent. The use of control characters in character literal constants may be restricted by the processor. Note that, unlike the other intrinsic types, the kind parameter for the character literal constant precedes the constant. If a kind is not specified, the type of the constant is default character. If the string delimiter character (either an apostrophe or quotation mark) is required as part of the constant, two consecutive such characters with no intervening blanks serve to represent a single such character in the string.

Examples are:

```
GREEK_"πβϕ"
GERMAN_"gemütlichkeit"
"DON'T"
'DON''T'
```

The last two both have the value DON'T. A zero-length character constant is written as two consecutive single or double quotes.

4.4 Derived Types

Unlike the intrinsic types that are defined by the language, derived types must be defined by the programmer by means of a derived-type definition. It is intended that these types have the same utility as the intrinsic types. For example, variables of these types may be declared, passed as procedure arguments, and returned as function results.

Like an intrinsic type, a derived type has a name, a set of type parameters, a set of values, a set of operations, and a means to represent constants of the type. Unlike with intrinsic types, with derived types there are considerations of accessibility, components, default initialization, procedure type bindings, type equivalence, and type extension.

4.4.1 A Simple Example of a Derived-Type Definition

The simplest derived-type definitions specify just a type name and some components. For example, the following is a definition of type COLOR:

```
TYPE COLOR
   INTEGER :: HUE, SHADE, INTENSITY
   CHARACTER(LEN=30) :: NAME
END TYPE COLOR
```

The type has four components, integer components named HUE, SHADE, and INTENSITY, and a character component of length 30 named NAME. A variable of this type could be declared with a type declaration statement such as

```
type(color) :: background
```

where color is the type specifier. This variable could be assigned a value with an assignment statement such as

```
background = color(0, 0, 0, "black")
```

The four components of background can be individually referred to as background%hue, background%shade, background%intensity, and background%name.

Note that the initial statement of a type definition and the statement used to declare objects of derived type both begin with the keyword TYPE. The initial statement of a type definition is called a derived-type statement, and the statement used to declare objects of derived type is called a type declaration statement. The type name in a derived-type statement is not enclosed in parentheses, whereas the type name in a type declaration statement is.

4.4.2 Derived-Type Definition Overview

The general form of a type definition (R429) is:

```
TYPE [ [ , type-attribute-list ] : : ] type-name [ ( type-parameter-name-list ) ]
   [ type-parameter-definition-statement ] ...
   [ private-or-sequence-statement ] ...
   [ component-definition-statement ] ...
   [ procedure-binding-part ]
END TYPE [ type-name ]
```

where the first statement in the definition is called the derived-type statement.

A type attribute (R431) is one of:

```
access-spec
EXTENDS ( parent-type-name )
ABSTRACT
BIND (C)
```

where an access specification is either PRIVATE or PUBLIC and a private-sequence statement is PRIVATE or SEQUENCE. Accessibility (PRIVATE and PUBLIC) and SEQUENCE are discussed in 4.4.5 and 4.4.10.

The same type attribute must not appear more than once in a given derived-type statement. The same private or sequence statement must not appear more than once in a given type definition.

The name of the type is type-name. If the END TYPE statement has a type name, it must be the same as the one in the derived-type statement.

Type parameters are declared by the type parameter name list and type parameter definition statements; components are defined by the component definition statements; and procedure bindings are defined by the procedure binding part. The EXTENDS and ABSTRACT type attributes relate to type extension and extended types (4.4.12). The BIND attribute declares a derived type to be interoperable. Interoperable types are subject to additional restrictions described in 15.

Contrary to some expectations, the order of the component declaration statements does not imply a storage order, except in the cases of sequence and bind types (4.4.10).

4.4.3 Type Parameters

A derived type is said to be parameterized if it has any type parameters. The parameters of a derived type are specified by the type parameter name list in the derived-type statement. For an extended type (4.4.12), the type parameters are those of the parent type, followed by those specified by the type parameter name list of the extended type.

Additional information about a type parameter is specified by a type parameter definition statement (R435), which has the form:

INTEGER [([KIND=] kind-value)] , kind-or-len :: type-parameter-declaration-list

where kind-or-len is KIND or LEN, specifying whether the type parameters named in the list are kind or length parameters. A type parameter declaration (R436) has the form:

type-parameter-name [= scalar-integer-initialization-expression]

Each type parameter name in a type parameter declaration must be one of the type parameter names specified in the derived-type statement. Each type parameter name in the derived-type statement must appear in exactly one type parameter declaration in the type definition.

Each type parameter is of type integer and therefore has a kind type parameter, which is specified by the kind value in the INTEGER type specifier of the type parameter definition statement. If no kind value appears, the type parameter is of type default integer. Discussion of the type parameter of a type parameter can be confusing; fortunately, such discussion is not often needed.

If a type parameter declaration has a scalar integer initialization expression, the expression specifies a default value for the type parameter.

An example of a derived-type definition with some simple type parameter definition statements is:

```
type :: some_type(KIND, M, N)
  integer, len :: N, M
  integer, kind :: KIND
     . . .
end type some_type
```

This example has a kind parameter named KIND, which is likely to be a common style, and also has two length parameters named M and N. A more complicated example is:

```
type :: matrix(M, N, KIND, K)
  integer, kind :: KIND = kind(0.0), K=kind(0)
  integer(K), len :: M, N
  real(KIND) :: body(M, N)
    . . .
end type matrix
```

In this example, KIND is a kind parameter with a default that is the kind value for default real. K is a kind parameter with a default that is the kind value for default integer. M and N are length parameters and have kind K. The REAL statement in this example is a component declaration statement (4.4.6). To help clarity of exposition, the above examples use a convention that type parameter names are all upper case; attributes and function names are all lower case.

4.4.4 Type Specifier

The form of a derived-type specifier (R455) is a generalization of the forms of the intrinsic type specifiers. Its description is more complicated because it is expressed in general terms which cover having an arbitrary number of type parameters. The form is similar to that of an actual argument list. The general form is:

type-name [(type-parameter-spec-list)]

where a type-parameter-spec (R456) is:

[type-parameter-name =] type-parameter-value

Type parameter values are described in 4.2.3; the following rules and restrictions apply in addition to those of that section.

Rules and restrictions:

1. The type-name must be an accessible name of a derived type.
2. Each type parameter name must be the name of a type parameter declared in the type definition.
3. Each type parameter of the derived type may have no more than one corresponding type parameter specification.
4. Each type parameter of the derived type must have a corresponding type parameter specification unless that type parameter has a default value, as specified in its type parameter definition statement.
5. If a type parameter specification specifies a type parameter name, all subsequent type parameter specifications in the list must also specify a type parameter name.
6. A type parameter value for a kind type parameter must be a kind value (4.2.3); a type parameter value for a length type parameter must be a length value (4.2.3).
7. If the type name is the name of an abstract type, the derived-type specifier can appear only in a CLASS declaration.

The correspondence between type parameters and type parameter specifications is established as follows: a type parameter specification with a type parameter name corresponds to the type parameter with that name. The type parameter specifications without type parameter names correspond to the type parameters in type parameter order. The type parameter order of a nonextended type is the order of the type parameters in the derived-type statement for the type. The type parameter order of an extended type is the type parameter order of the parent type, followed by the type parameters in the derived-type statement for the type in order.

If necessary, each specified type parameter value is converted to the kind of the corresponding type parameter. A type parameter that is not in a particular type parameter specification list takes its default value.

The following examples of type definition statements use type specifiers for the derived-type matrix defined in 4.4.3.

```
type (matrix(n=10,m=10)) :: x
type (matrix(10,10,kind(0.0),kind(0)) :: y
type (matrix(20,10,kind(1.0D0)) :: z
```

In this example. x%body and y%body are default real with shape [10, 10], while z%body is double precision with shape [20, 10].

4.4.5 Accessibility

For a derived type declared in the specification part of a module, it is possible to specify the accessibility of the derived-type name, its component names, and its procedure bindings. The accessibility is either public or private. If an identifier has public accessibility, then it is available for use outside of the module via USE statements. If an identifier has private accessibility, then it may be used only inside of the module where it is defined. This definition of accessibility is inherently related to modules; it would be meaningless to specify accessibility of something not declared in the specification part of a module; therefore, that is not allowed.

It is important to understand that accessibility applies to identifiers (most commonly names, but also things such as operators, which do not have the form of names). If, for example, a type name has private accessibility, that means only that the type name may not be accessed via USE statements. It does not prevent objects of that type from being used outside of the module, as long as that use does not involve the type name.

The accessibility of a type name, its component names, and its procedure bindings are orthogonal; all combinations are allowed and meaningful.

The accessibility of a type name may be specified either by an access specification in the derived-type statement or by a separate accessibility statement (5.8.1). If the accessibility of the type name is not individually specified, then that type name has the default accessibility for the module (5.8.1).

Having a private type name substantially restricts the allowable uses of the type outside of the module. In particular, objects of the type cannot be declared outside of the module. However, public objects of the type may be accessed.

The accessibility of a component name is specified by an access specification in the component definition statement. If a component definition statement has no access specification, then the accessibility of that component name is the default component accessibility for the type. The default component accessibility for a type is private if the type definition has a PRIVATE statement preceding the component definition statements; otherwise, it is public. A default of public can be specified only by omission of the PRIVATE statement; there is no explicit PUBLIC statement in a derived-type definition. The accessibility of a component is not influenced by the accessibility of the type name or by the default accessibility of the module.

The accessibility of a procedure binding (4.4.11) is specified by an access specification in the procedure binding statement. If a procedure binding statement has no access specification, then the accessibility of that binding is the default binding accessibility for the type. The default binding accessibility for a type is private if there is a PRIVATE statement in the procedure binding part of the type definition; otherwise, it is public. The accessibility of a binding is not influenced by the accessibility of components, the accessibility of the type name, or the default accessibility of the module. Accessibility is not relevant for final bindings (4.4.11.3) because they do not have identifiers.

Accessibility cannot be specified for type parameters. Effectively they are always public.

Example:

```
type, public :: some_type
  private
  real, public :: x
  integer :: i,j
end type some_type
```

The type name some_type is public, as is the component name x. The component names i and j are private because of the PRIVATE statement. A scoping unit that uses this module could have a declaration like

```
type(some_type) :: z
```

and could then refer to z%x. The only way to access the i and j components of z would be by means of some procedure defined in the same module that defines the type.

4.4.6 Data Component Definition

A component definition statement is either a data component definition statement or a procedure component definition statement. The form of a data component definition statement (R440) is

> declaration-type-spec [[, component-attribute-spec-list] : :] &
> component-declaration-list

where a component attribute specification is one of

> POINTER
> ALLOCATABLE
> DIMENSION (component-array-spec)
> access-spec

A component declaration is

> component-name [(component-array-spec)] [* character-length] [initialization]

and a component array specification is one of

> explicit-shape-spec-list
> deferred-shape-spec-list

The form of a data component definition statement is similar to that of a type declaration statement (5.1); both forms use several of the same terms, the definitions of which are not duplicated here. Declaration type specification is defined in 5.1, explicit-shape specification and deferred-shape specification are defined in 5.4.1, and initialization is defined in 5.7.2. There are several other attributes that are allowed in a type declaration statement, but not in a data component definition statement.

Rules and restrictions:

1. The declaration type specification must not specify the type being defined or a derived type defined later in the same scoping unit unless the POINTER attribute is specified.

2. A particular attribute specification may appear at most once in a given component attribute specification list.

3. A component must not have both the POINTER and ALLOCATABLE attributes.

4. If CLASS appears in the declaration type specification (5.1), either the POINTER or ALLOCATABLE attribute must be specified.

5. If either the POINTER or ALLOCATABLE attribute is specified in a data component definition statement, then each component array specification in that statement must be a deferred-shape specification list.

6. If neither the POINTER nor ALLOCATABLE attribute is specified in a data component definition statement, then each component array specification in that statement must be an explicit-shape specification list.

7. Each bound in an explicit-shape specification must either be an initialization expression or be a specification expression that contains neither variables nor references to specification functions. Type parameters are not variables; they are allowed in such specification expressions and are the only way for a component to have nonconstant explicit bounds.

8. Each type parameter value in a component definition statement must be either a colon, an initialization expression, or a specification expression that contains neither variables nor references to specification functions.

9. A * character-length is allowed only if the type specified is character.

10. If initialization appears, the double colon separator must appear.

11. If initialization appears, the ALLOCATABLE attribute must not appear.

A data component is an array if there is a component array specification in its component declaration or in a DIMENSION component attribute specification in its component definition statement. If it is specified in both places, the specification in the component declaration overrides the one in the DIMENSION component attribute specification. In the example

```
type :: some_arrays
  real, dimension(2) :: x, y(10,10)
  integer, allocatable :: p(:)
end type some_arrays
```

the x component is a rank 1 array with dimension (2), the y component is a rank 2 array with dimension (10,10), and the p component is an allocatable array of rank 1.

An example of a derived type with a pointer component is

```
type summary
  character(len=50) :: title
  integer :: no_of_pages
  character(len=:), pointer :: text
end type summary
```

The space for the target of the TEXT component may be allocated (6.7.1.2) during execution, or the pointer may be assigned (7.5.5) to point to existing space.

4.4.7 Procedure Component Definition

The form of a procedure component declaration statement (R445) is

```
PROCEDURE ( [ interface-spec ] ) , &
   procedure-component-attribute-spec-list :: procedure-component-declaration-list
```

where a procedure component attribute specification is one of

```
POINTER
PASS [ ( argument-name ) ]
NOPASS
access-spec
```

and a procedure component declaration is

```
procedure-component-name [ initialization ]
```

The form and interpretation of an interface specification is defined in 5.11. The form of a procedure component declaration statement is similar to that of a procedure declaration statement except for differences in the allowed attributes. The PASS and NOPASS attributes are unique to derived types and are discussed in 4.4.8.

Rules and restrictions:

1. The POINTER attribute must always be specified.
2. A particular attribute specification may appear at most once in a given procedure component attribute specification list.

Example:

```
module proc_component_example
  type t
    real :: a
    procedure(print_me), pointer, nopass :: proc
  end_type t
contains
  subroutine print_me (arg, lun)
    type(t), intent(in) :: arg
    integer, intent(in) :: lun
    write (lun,*) arg%a
  end subroutine print_me
  subroutine print_my_square (arg, lun)
    type(t), intent(in) :: arg
    integer, intent(in) :: lun
    write (lun,*) arg%a**2
  end subroutine print_my_square
end module proc_component_example
program main
  use proc_component_example
  use iso_fortran_env, only :: output_unit
  type(t) :: x
  x%a = 2.71828
  x%proc => print_me
  call x%proc(x, output_unit)
  x%proc => print_my_square
  call x%proc(x, output_unit)
end program main
```

The proc component of type t is declared to be a procedure pointer to a procedure with the same interface as print_me. Note that this does not imply that the target procedure actually is print_me—just that it has the same interface as print_me; one could have alternatively written an abstract interface (12.5.5) to use in defining the procedure component, but it is simpler to just give the name of a procedure with the desired interface if such a procedure is handy.

The main program assigns a value to the x%a component and then invokes both print_me and print_my_square using the procedure component. In this simple case, the same thing could have been achieved without procedure pointers at all. In more realistic situations, there could be multiple variables of type t, different variables having different targets for their procedure pointer component, and the procedure invocations could be far removed from the pointer assignments.

For a description of the ISO_FORTRAN_ENV intrinsic module and its output_unit constant, the details of which are peripheral to this example, see 13.6.1.

4.4.8 The Passed-Object Dummy Argument

A procedure component or a procedure binding (4.4.11) may optionally be declared to have a passed-object dummy argument. A passed-object dummy argument is associated with a special actual argument, which is not explicitly written in the actual argument list. In effect, the compiler automatically adds the appropriate actual argument to the argument list.

The appropriate actual argument is inherent in the form of reference to the procedure. A reference to a procedure component or procedure binding always involves the general form x%p, where x is an object of derived type, and p is the name of a procedure component or binding of the type. The object x is the actual argument associated with the passed-object dummy argument.

The determination of the passed-object dummy argument depends on the PASS and NOPASS attributes specified and on the interface of the procedure component or procedure binding, as described below.

Rules and restrictions:

1. PASS and NOPASS must not both be specified for the same procedure component or binding.
2. NOPASS must be specified if the procedure component or binding has an implicit interface.
3. If NOPASS is specified, there is no passed object dummy argument.
4. If PASS (argument-name) is specified, then the dummy argument named argument-name is the passed-object dummy argument; there must be such a dummy argument.
5. If PASS is specified without an argument name, or if neither PASS nor NOPASS is specified, the first dummy argument is the passed-object dummy argument. There must be at least one dummy argument in these cases.
6. The passed object dummy argument must be a scalar, nonpointer, nonallocatable dummy data object. Its declared type must be the type in which the component or binding appears. All of its length type parameters must be assumed. It must be polymorphic (5.2) if and only if the type is extensible (4.4.12).

The following example illustrates the use of a passed object dummy argument. Except for the passed object dummy argument, this example is the same as the procedure component example above.

```
module passed_object_example
  type t
    real :: a
    procedure(print_me), pointer, pass(arg) :: proc
  end_type t
contains
```

```
  subroutine print_me (arg, lun)
    type(t), intent(in) :: arg
    integer, intent(in) :: lun
    write (lun,*) arg%a
  end subroutine print_me
  subroutine print_my_square (arg, lun)
    type(t), intent(in) :: arg
    integer, intent(in) :: lun
    write (lun,*) arg%a**2
  end subroutine print_my_square
end module passed_object_example
program main
  use passed_object_example
  use iso_fortran_env, only :: output_unit
  type(t) :: x
  x%a = 2.71828
  x%proc => print_me
  call x%proc(output_unit)
  x%proc => print_my_square
  call x%proc(output_unit)
end program main
```

Other than the module name, the only difference between this module and the one in the previous example is that the proc component is given the PASS attribute instead of NOPASS. The attribute specification could have been omitted from the example because PASS is the default. The pass(arg) makes it explicit which argument is the passed one, although again, the specification just emphasizes what would have been the default (the first argument).

In the main program, x is not included in the actual argument list; instead, it is passed implicitly. It would be an error to put x in the actual argument list; that would count as another actual argument rather than a confirmation of the automatically passed one.

For all of its somewhat intimidating terminology, the effect of a passed object dummy argument is just to remove a redundancy in some procedure references. Without the passed object dummy argument, it is necessary to specify x both in the form x%proc and as an actual argument. That happens to be a very common type of redundancy for procedure components. The redundancy is more of an issue if the object name is longer than the short x of this example, or particularly if the object is some expression rather than a simple name.

Note that if print_me is referenced directly as in

```
call print_me(x, output_unit)
```

instead of being referenced via the procedure pointer component, the x actual argument must be provided explicitly. The passed object property affects only invocation via the procedure pointer component; it does not change anything about the procedure or unrelated invocations of it. The same procedure may even have different passed-object dummy arguments in different contexts.

4.4.9 Default Initialization

If a component declaration includes initialization (5.7.2), that component of the type is said to have default initialization. For a pointer component, the only default initialization allowed is to a pointer association status of disassociated (nullified), which is specified by a reference to the intrinsic function NULL. For an allocatable component, user-specified default initialization is not allowed, but the normal behavior of an allocatable component is essentially equivalent to default initialization to an unallocated status. For a nonpointer, nonallocatable data component, the initialization follows the same rules as intrinsic assignment (7.5.2); the type and type parameters of the initialization expression must be compatible with intrinsic assignment to the component, and the expression must either be scalar or have the same shape as the component.

Default initialization for a type applies whenever any object of the type is created. This may be when program execution begins, when the object is allocated, or when a procedure is invoked.

Example. A derived type may have a pointer component that is of the type being defined. This is useful in creating linked lists and trees.

```
TYPE LINK
   REAL VALUE
   TYPE(LINK), POINTER :: PREVIOUS => NULL()
   TYPE(LINK), POINTER :: NEXT => NULL()
END TYPE LINK

TYPE (LINK), POINTER :: A_LINK
ALLOCATE (A_LINK)
```

When A_LINK is allocated, its PREVIOUS and NEXT pointers are nullified; its VALUE component is undefined.

If initialization is specified at multiple levels, the highest level specification overrides. That is, explicit initialization of a variable overrides any default initialization specified for the type of the variable; default initialization specified for a component of a type overrides any default initialization specified for the type of the component.

Example:

```
TYPE TEMPERATURES
   REAL :: LOW = 0.0, HIGH = 100.0
END TYPE TEMPERATURES

TYPE (TEMPERATURES) :: WATER, &
   HEAVY_WATER = TEMPERATURES(3.82,101.2)
```

WATER is not initialized explicitly, so the default initialization specified in the type definition for TEMPERATURES applies; the LOW component of WATER is initialized to 0 and the HIGH component is initialized to 100. HEAVY_WATER (water with deuterium) is explicitly initialized with a structure constructor (4.4.15) to have a LOW component of 3.82 and a HIGH component of 101.42; that explicit initialization overrides the default initialization.

4.4.10 Sequence Types and Type Equivalence

The question of whether two entities are of the same type arises in many contexts, such as the association of actual and dummy arguments. There are two ways for entities to be declared to be of the same type. The simplest way is for them to be declared with reference to the same derived-type definition. If the two objects are in different scoping units, the only ways to declare them with reference to the same derived-type definition are by using host association (16.2.1.3) or use association (16.2.1.2). Except in some special cases of sequence and bind types, each derived-type definition defines a distinct type; if two entities are declared with reference to two distinct derived-type definitions, those entities are of different type, even if the derived-type definitions are textually identical.

Example:

```
MODULE SHOP
  TYPE COMPONENT
    CHARACTER(LEN=20) NAME
    INTEGER CATALOG_NO
    REAL WEIGHT
  END TYPE COMPONENT
  TYPE(COMPONENT)  PARTS(100)
CONTAINS
SUBROUTINE GET_PART(PART, NAME)
  TYPE(COMPONENT) PART
  CHARACTER(LEN=*) NAME
  DO I=1,100
    IF(NAME == PARTS(I)%NAME) THEN
      PART = PARTS(I)
      RETURN
    ENDIF
  ENDDO
  PRINT *, "Part not available"
  PART%NAME = "none"
  PART%CATALOG_NO = 0
  PART%WEIGHT = 0.0
  END SUBROUTINE GET_PART
    ...
END MODULE SHOP

PROGRAM BUILD_MACHINE
  USE SHOP
  TYPE(COMPONENT) MOTOR(20)
  TOTAL_WEIGHT = 0.0
  CALL GET_PART(MOTOR(1), "VALVE")
  TOTAL_WEIGHT = TOTAL_WEIGHT + MOTOR(1)%WEIGHT
    ...
END PROGRAM BUILD_MACHINE
```

Module procedure GET_PART has access to the type COMPONENT because the type definition appears in its host. Program BUILD_MACHINE has access to the same type because it uses module SHOP. This allows a variable of the type, such as MOTOR(1), to be passed as an actual argument.

The other way to declare entities to be of the same derived type involves sequence and bind types. A sequence type is a derived type whose type definition has a SEQUENCE statement. Details of bind types are in 15.4.4. Bind types share many of the features of sequence types and might have been more clearly categorized as a special case of sequence types, but the standard does not categorize them that way; as a result, there are several places in the standard where material about sequence and bind types is nearly identical.

Rules and restrictions:

1. A sequence type must not have the EXTENDS, ABSTRACT, or BIND attributes.
2. A sequence type is not extensible.
3. The type definition for a sequence type must not have a procedure binding part.
4. Each data component of a sequence type must be declared to be of an intrinsic or sequence type.

Entities declared with reference to two distinct derived-type definitions are of the same type if both type definitions specify SEQUENCE or both specify BIND; they specify the same type name; they have no PRIVATE components; and they have type parameters and components that agree in order, name, and attributes. The example for program BUILD_MACHINE above is restated to illustrate the differences between the two ways:

```
PROGRAM BUILD_MACHINE
  TYPE COMPONENT
    SEQUENCE
    CHARACTER(LEN=20) NAME
    INTEGER CATALOG_NO
    REAL WEIGHT
  END TYPE COMPONENT
  TYPE(COMPONENT) PARTS, MOTOR(20)
  COMMON /WAREHOUSE/ PARTS(100)
  TOTAL_WEIGHT=0.0
  CALL GET_PART(MOTOR(1), "VALVE")
  TOTAL_WEIGHT = TOTAL_WEIGH + MOTOR(1)%WEIGHT
    ...
END PROGRAM BUILD_MACHINE
```

```
SUBROUTINE GET_PART(PART, NAME)
  TYPE COMPONENT
    SEQUENCE
    CHARACTER(LEN=20) NAME
    INTEGER CATALOG_NO
    REAL WEIGHT
  END TYPE COMPONENT
  TYPE(COMPONENT) PART, PARTS
  CHARACTER(LEN=*) NAME
  COMMON /WAREHOUSE/ PARTS(100)
  DO I = 1, 100
    IF (NAME .EQ. PARTS(I)%NAME) THEN
      PART = PARTS(I)
      RETURN
    END IF
  END DO
  PART%NAME = "none"
  PART%CATALOG_NO = 0
  PART%WEIGHT = 0.0
  PRINT *, "Part not available"
END SUBROUTINE GET_PART
  ...
```

In this example, type COMPONENT in program BUILD_MACHINE and type COMPONENT in subroutine GET_PART are the same because they are sequence types with the same name; have no private components; and have type parameters and components that agree in order, name, and attributes. This example is less concise, particularly if there are more procedures that need access to the type definition. The necessity to replicate the type definition also introduces extra chances for errors, which might not be caught by the compiler.

In addition to their role in type equivalence, sequence types also play a role in storage association; a derived-type object in COMMON or EQUIVALENCE must be of a sequence type. Additional forms of COMMON and EQUIVALENCE association are allowed if the sequence type meets the extra conditions required to be a numeric sequence type or character sequence type, as follows:

1. A numeric or character sequence type must not have type parameters. It must not have allocatable or pointer components.

2. Each component of a numeric sequence type must be of type default integer, default real, double precision real, default complex, default logical, or a numeric sequence type.

3. Each component of a character sequence type must be of type default character or a character sequence type.

There is no way to explicitly declare that something is a numeric or character sequence type; the terms just categorize sequence types that meet the extra conditions.

Storage sequences for sequence types are described in 16.2.3.1. For numeric and character sequence types, the allowed storage associations essentially require that the components of objects of the type be stored in the specified order with no padding. For other sequence types, although the standard does specify a sequence of storage units for the components, this specification has no practical effect because it cannot be detected by a standard-conforming program; therefore, the compiler is free in practice to rearrange the internal storage of such types as long as it is done consistently so that the rules of type equivalence still work. For nonsequence types, no internal storage order is even implied by the standard.

4.4.11 Procedure Type Bindings

A procedure type binding connects a derived type and a procedure. A procedure that has a binding to a type is often referred to as a type-bound procedure. This term is somewhat misleading in that being type-bound is not a property of the procedure. A given procedure may be bound to multiple types and may also be invoked in ways having no connection with type binding. There is nothing about the type binding in the code of the procedure. Type bindings are specified in the type binding part of the derived-type definition.

The form of the procedure binding part (R448) of a derived-type definition is

```
CONTAINS
 [ PRIVATE ]
 procedure-binding-statement
 [ procedure-binding-statement ] ...
```

The optional PRIVATE statement is discussed in 4.4.5. Note that at least one procedure binding statement is required in a procedure binding part, although the procedure binding part as a whole is optional. The form of a procedure binding statement (R450) is one of

```
specific-binding
generic-binding
final-binding
```

In the scope of the type definition, the procedure is identified by a **binding name**. It may be the binding name of a specific binding or the generic name for a generic binding. A final binding, or a generic binding whose generic specification is not a name, has no binding name.

4.4.11.1 Specific Bindings

A specific binding is either deferred or nondeferred. The form of a nondeferred specific binding is

```
PROCEDURE [ [ , NON_OVERRIDABLE ] [ , binding-attribute-list ] : : ] &
    binding-name [ => procedure-name ]
```

The form of a deferred specific binding is

PROCEDURE (interface-name) , DEFERRED [, binding-attribute-list] : : binding-name

A binding attribute is one of

PASS [(argument-name)]
NOPASS
access-spec

Although these forms show the NON_OVERRIDABLE and DEFERRED attributes separately and preceding the other binding attributes above, this ordering is not required. The PASS and NOPASS attributes are discussed in 4.4.8; access specifications are discussed in 4.4.5; the NON_OVERRIDABLE and DEFERRED attributes relate to inheritance, which is further discussed in 4.4.12.

Rules and restrictions:

1. The same binding attribute must not be specified more than once in a given binding attribute list.
2. If the procedure name is omitted from a nondeferred specific binding, it is as though it were specified to be the same as the binding name.
3. If the procedure name explicitly appears in a nondeferred specific binding, the double colon separator must appear.
4. The procedure name in a nondeferred specific binding must be the name of an accessible module procedure or external procedure with an explicit interface.

Other rules about procedure type bindings apply only in the context of type inheritance and are discussed in 4.4.12.

A specific procedure type binding has similarities, in both syntax and function, to a procedure pointer component. Each involves referencing a procedure and the syntax of each reference is identical. The difference is in how the particular procedure to be referenced is determined. For a procedure pointer component with a particular component name, every object of the type has a separate pointer; these pointers can, in general, point to different procedures just like the data components of different objects of the type can have different values. For a procedure type binding with a particular binding name, all objects of the same type use the same procedure. Procedure type bindings are thus more restricted than procedure pointer components. In applications where the restriction fits, they are consequently less verbose and less error prone. However, procedure type bindings are so restricted that, unless type inheritance is involved, they are little more than an alternative syntax for an ordinary call to a procedure using its name.

Example:

```
module procedure_binding_example
  type t
    real :: a
```

```
    contains
      procedure, pass(arg) :: print_me
      procedure, pass(arg) :: print_my_square
    end_type t
  contains
    subroutine print_me (arg, lun)
      type(t), intent(in) :: arg
      integer, intent(in) :: lun
      write (lun,*) arg%a
    end subroutine print_me
    subroutine print_my_square (arg, lun)
      type(t), intent(in) :: arg
      integer, intent(in) :: lun
      write (lun,*) arg%a**2
    end subroutine print_my_square
  end module procedure_binding_example
  program main
    use procedure_binding_example
    use iso_fortran_env, only :: output_unit
    type(t) :: x
    x%a = 2.71828
    call x%print_me(output_unit)
    call x%print_my_square(output_unit)
  end program main
```

This example directly parallels our prior examples of procedure components and passed object dummy arguments. It illustrates the similarities and differences. The actual subroutines in question are identical; only the means of invoking them differ. Type bindings are defined for both print_me and print_my_square. The bindings are defined in the type definition and cannot be changed subsequently, so there are two different binding names for the two different procedures. For this example, the binding names are the same as the subroutine names. The syntax difference between

```
call x%print_me(output_unit)
```

and

```
call print_me(x, output_unit)
```

seems like a pretty trivial result to be worth a special feature in the language. The power of the feature is only evident in conjunction with type extension and inheritance, and in particular with polymorphism (5.2).

4.4.11.2 Generic Bindings

The form of a generic binding is

GENERIC [, access-spec] :: generic-spec => binding-name-list

Generic specifications are defined in 12.5.4.

Rules and restrictions:

1. Each binding name must be the name of a specific binding of the type.

2. If the generic specification is for an operator, an assignment, or derived-type input/output, each binding must have a passed-object dummy argument. The interface of the binding must be as specified in 12.5.4.2, 12.5.4.3, or 9.5.1.4, respectively.

3. All generic bindings with the same generic specification in the same derived-type definition must have the same accessibility.

4. The set of specific bindings for a particular generic specification must satisfy the requirements of 12.5.4, which allow generic resolution. This set of specific bindings includes any inherited ones for the same generic specification.

As an example of generic binding, the following could be procedure binding statements in the definition of the matrix type shown in 4.4.3.

```
procedure :: invert_single
procedure :: invert_double
procedure :: invert_huge_single
procedure :: invert_huge_double
generic :: operator(.invert.) => invert_single, invert_double, &
   & invert_huge_single, invert_huge_double
```

The first four statements define specific bindings using procedures that are not shown here. The last statement defines a generic binding built from the specific ones.

4.4.11.3 Final Bindings

The form of a final binding is

FINAL [: :] final-subroutine-name-list

Rules and restrictions:

1. Each final subroutine name must be the name of a module procedure that has exactly one dummy argument, which must be of the type being defined. That dummy argument must not be optional, pointer, allocatable, polymorphic, or INTENT (OUT). Any length type parameters of the dummy argument must be assumed.

2. Any two final subroutines for a type must differ in the rank or kind type parameters of the dummy argument. The same final subroutine may not be specified twice.

A final binding is comparable to a destructor in some other languages.

A type is **finalizable** if it has any final subroutines or if it has any nonpointer, nonallocatable components of a finalizable type. A data entity is finalizable if it is of a finalizable type and is not a pointer.

Finalization is the process of executing the appropriate final subroutines for a data entity and its components. It applies only to finalizable data entities. Finalization of an entity consists of three steps in the following order:

1. The final subroutine for the data entity is called with the data entity as the actual argument. The final subroutine for the data entity is one that is bound to the dynamic type of the data entity and is compatible with being called with the data entity as an actual argument. This implies that the final subroutine's dummy argument has the same kind type parameter values as the data entity. It also implies that either the dummy argument has the same rank as the data entity or that the subroutine is elemental. If there are both elemental and nonelemental compatible final subroutines, the nonelemental one is called. If there are no compatible final subroutines, nothing is called for this step.

2. Each finalizable component specified in the type definition is finalized. If the data entity is an array, then this component finalization is done separately for each element of the data entity. The order of the component finalizations is processor-dependent.

3. If the data entity is of an extended type and the parent type is finalizable, then the parent component is finalized.

Conceptually, finalization occurs when a finalizable data entity goes out of existence. Specifically, it occurs for finalizable data entities in the following situations:

1. When a pointer is deallocated, its target is finalized. When an allocatable entity is deallocated, it is finalized.

2. When an entity becomes undefined due to completion of execution of an instance of a procedure (16.3.3(3)), it is finalized.

3. A function result or structure constructor referenced in an executable construct is finalized after execution of the innermost executable construct containing the reference.

4. A function result or structure constructor referenced in a specification expression is finalized before execution of the first executable statement in the scoping unit.

5. When a procedure is invoked, any nonpointer, nonallocatable actual argument associated with an INTENT (OUT) dummy is finalized.

6. When an intrinsic assignment statement is executed, the variable is finalized immediately before it is defined, after the expression is evaluated.

7. If a target allocated through a pointer becomes unreachable by any pointer, it may be finalized at any subsequent time, at the processor's option.

If multiple entities are finalized as a consequence of a single event, the order of their finalization is processor-dependent. A final subroutine must not reference or define an object that has already been finalized; this restriction effectively prohibits any dependence between finalizations that are triggered by the same event.

As an example of finalization, consider the following type definition and final procedure.

```
module linked_list_module
   type linked_list_node_type
      real, allocatable :: data(:)
      type(linked_list_node_type), pointer :: next
   contains
      final :: finalize_node
   end type
contains
   recursive subroutine finalize_node (node)
      type(linked_list_node_type) :: node
      if (associated(node%next)) deallocate(node%next)
   end subroutine
      . . .
end module
```

If a node of this type is deallocated, the finalize_node procedure is automatically invoked as a precursor to the actual deallocation. That procedure deallocates the next node, if one exists. Deallocation of the next node causes it to be finalized, which results in a recursive invocation of finalize_node. The procedure must be declared recursive even though the recursive invocation is a result of finalization instead of an explicit call. With this final procedure, deallocation of a single node causes recursive deallocation of the remainder of the list beginning with that node. Note that the allocatable data component of a node is automatically deallocated when the node is deallocated (6.7.3.1); the final procedure does not have to do this deallocation explicitly.

4.4.12 Type Extension and Inheritance

Type extension is a means of defining a new type by building on a previously defined type. The new type starts with the previously defined one and can add type parameters, components, and procedure bindings; the new type can also override procedure bindings.

Any derived type that is neither a sequence nor a bind type is extensible. A type definition with the EXTENDS type attribute defines an extended type. The parent type specified in the EXTENDS attribute must be an extensible type. An extended type is also extensible and may in turn be a parent of other extended types in a tree-like structure. An extensible type is an extension of itself and of any type below it in the tree (if we think of parent types as being lower and the tree branching out upwards).

An extended type inherits all of the type parameters and components from its parent type. It also inherits those specific and generic bindings that are not overridden. These inherited entities retain all of the attributes that they had in the parent type. The extended type has the type parameters and components specified in its type definition in addition to the inherited ones. The procedure bindings specified in the definition of the extended type can either be additional bindings or overriding ones.

It is allowed for the number of type parameters, components, and bindings in a type definition to be zero; although that is not the norm, there are cases where it can be useful, particularly with type extension.

An extended type additionally has a special implicitly declared component called the parent component. The parent component is scalar, nonpointer, and nonallocatable, with the type and type parameters of the parent type. The name of the parent component is the parent type name and it has the same accessibility as the parent type name.

The parent component provides a second way to refer to the inherited components. If an object x has an inherited component named y, that component can be referred to with the syntax x%y. If the parent type is named p, then the same inherited component can also be referred to as x%p%y. In isolation, the longer form might seem pointless. Its main benefit is in the ability to refer to all of the inherited components as the single object x%p.

Type extension is most useful as an enabler for polymorphism (5.2). Type extension can be used without polymorphism, but then it is little more than a syntax convenience, allowing the shorter form of reference to the inherited components.

4.4.12.1 Type Extension Versus its Alternatives

In order to illustrate the role of type extension, consider a base type and some alternative ways to define new types based on it, but with additional components. The base type for this illustration is

```
type :: base_type
   real :: a
   integer :: i
end type
```

It is desired to define two new types: one adding a character component, and the other adding a logical component.

One way to define the new types is to declare them from scratch, taking no explicit advantage of the previous declaration of the base type. The appropriate components are added to each type definition, as in

```
type :: new_type_from_scratch_c
   real :: a
   integer :: i
   character  :: c
end type

type :: new_type_from_scratch_d
   real :: a
   integer :: i
   logical :: d
end type
```

Then with variables declared as

```
type(base_type) :: x
type(new_type_from_scratch_c) :: y
type(new_type_from_scratch_d) :: z
```

the components of y are y%a, y%i, and y%c.

This approach has maintenance problems in keeping the three type definitions synchronized if there are future changes to the base type. Also, because the three type definitions are independent, there is no simple way to write one procedure that will work on the common components of x, y, or z. One either has to write three procedures or copy the common components into a temporary and back.

Another approach is sometimes referred to as type embedding. In this approach the new types are defined with an embedded component of type base_type as in

```
type :: new_type_with_embedding_c
  type(base_type) :: base
  character  :: c
end type

type :: new_type_with_embedding_d
  type(base_type) :: base
  character  :: d
end type
```

Then with variables declared as

```
type(base_type) :: x
type(new_type_with_embedding_c) :: y
type(new_type_with_embedding_d) :: z
```

the components of y are y%c and y%base, with subcomponents y%base%a and y%base%i.

With this approach, the maintenance problems are ameliorated because the common components are defined only once. Also, a single procedure can operate on x, y%base, or z%base.

However, the syntax is a bit awkward with the "extra" %base thrown in. As multiple levels of extension are added, the awkwardness gets worse. One is likely to be driven to make temporary pointers or use the ASSOCIATE construct (8.2) solely to simplify notation.

Using type extension for the same situation is illustrated by

```
type, extends(base_type) :: new_type_c
   character :: c
end type

type, extends(base_type) :: new_type_d
   logical :: d
end type
```

Then with variables declared as

```
type(base_type) :: x
type(new_type_c) :: y
type(new_type_d) :: z
```

the components of y are y%a, y%i, and y%c, much like with new types defined from scratch. But y also has the parent component y%base_type, with its subcomponents y%base_type%a and y%base_type%i. The parent component does not have physically separate storage; it is just another view of the y%a and y%i components. The y%base_type%a component is inheritance associated with y%a, and the y%base_type%i component is inheritance associated with y%i.

As with type embedding, a single procedure can operate on x, y%base_type, or z%base_type. Thus type inheritance has the maintenance benefits of type embedding combined with the notational simplicity of new types declared from scratch.

Although this example illustrates maintenance and notational benefits of type inheritance, these benefits alone would probably not be sufficient to justify the inclusion of type inheritance in the language. Type extension is most useful as an enabler for polymorphism (5.2).

4.4.12.2 Overriding Procedure Bindings

If a specific binding in an extended type has the same binding name as a binding from the parent type, then the binding in the extended type overrides that from the parent type. This override blocks the inheritance of the binding from the parent.

When overriding bindings are used in conjunction with polymorphism, the compiler might not be able to determine at compile time which specific procedure is invoked by a particular reference in the code; the same reference might invoke different procedures for different executions of it. Therefore, the parent binding and the overriding one must be similar enough that the same invoking code makes sense for both bindings. The rules and restrictions to ensure that are:

1. Either both or neither must have passed-object dummy arguments, which must correspond in name and position.

2. They must have the same number of dummy arguments. Dummy arguments that correspond by position must have the same names and characteristics, except that the passed-object dummy will differ in type.

3. Either both must be subroutines or both must be functions.

4. Either both or neither must be elemental.

5. If the parent binding is pure, then the overriding binding must be pure. (But the overriding binding may be pure even when the parent is not.)

6. If the parent binding is public, then the overriding binding must be public. (But the overriding binding may be public even when the parent is not.)

An additional restriction is that a parent binding with the NON_OVERRIDABLE attribute may not be overridden; this restriction is, in fact, the only effect of the NON_OVERRIDABLE attribute.

Generic bindings are never overridden. Instead, a generic binding extends any inherited generic binding with the same generic specification.

Final bindings are neither inherited nor overridden. Instead, if an extended type and its parent both have final bindings, then the subroutines specified by both bindings will be called as described in 4.4.11.3.

4.4.12.3 Abstract Types and Deferred Procedure Bindings

An abstract type is one that has the ABSTRACT attribute. An abstract type can be used to define a parent for type extension, but the restrictions on derived-type specifiers make it impossible to create an object whose dynamic type (5.2) is abstract. One could think of an abstract type as a placeholder in the type extension tree.

Similarly, a deferred procedure binding can be thought of as a placeholder for a binding that can never be invoked, but may be overridden in a type extension.

Abstract types and deferred procedure bindings are never strictly necessary; one could alternatively use nonabstract types and nondeferred bindings, but that would require writing stub procedures to bind to, even though the procedures would never be invoked. The limitations on abstract types and deferred bindings allow compile-time verification that a deferred binding will never be invoked, thus obviating the need for a stub procedure.

Rules and restrictions:

1. An abstract type must be extensible.
2. A deferred binding, whether declared or inherited, is allowed only in an abstract type. Thus, if an abstract type with a deferred binding is extended, and the extended type is not also abstract, then the extended type must override the deferred binding with a nondeferred one.
3. A nondeferred binding may not be overridden with a deferred one.

4.4.12.4 Example of Inheriting and Overriding Bindings

The following example illustrates several of the features relating to inheritance of procedure type bindings. The first part of the example is a module that defines an abstract type for a time history file (a file with time series data). A simple rewind procedure is probably adequate for many possible formats and is provided. A general seek procedure that uses the rewind and next_frame bindings is also provided. A general next_frame procedure is not provided because that depends too strongly on the specific file format details. Therefore, the next_frame procedure is deferred. Next_frame is specified to have the same interface as the rewind procedure; if a different interface were appropriate, an abstract interface block could be used to define the necessary interface. Because there is a deferred binding, the type must be abstract. In order to create an object in this class, it is necessary to extend the type and provide a nondeferred binding in that extension. Other bindings and details are elided from the example.

```
module th_file_module

   private
   type, public, abstract :: th_file_type
      integer :: lun = 0
      logical :: frame_ok = .false.
      real :: frame_time
         . . .
   contains
      procedure :: rewind
      procedure(rewind), deferred :: next_frame
      procedure :: seek
         . . .
   end type

contains
   subroutine rewind (file)
      class(th_file_type) :: file

      rewind(file%lun)
      file%frame_ok = .false.
      return
   end subroutine rewind

   subroutine seek (file, time)
      class(th_file_type) :: file
      real, intent(in) :: time

      call file%rewind
      seek_loop: do
         call file%next_frame
         if (.not. file%frame_ok) return
         if (file%frame_time >= time) exit seek_loop
      end do seek_loop
         . . .
      return
   end subroutine seek
      . . .
end module th_file_module
```

The next part of the example extends the type to support a simple file format (the details of which are elided in the example). A nondeferred binding is provided for next_frame, but the inherited bindings are used for rewind and seek.

```
module simple_1_file_module

   use th_file_module

   type, extends(th_file_type) :: simple_1_file_type
   contains
     procedure :: next_frame
        . . .
   end type

contains
  subroutine next_frame (file)
     type(simple_1_file_type) :: file
        . . .
  end subroutine next_frame
     . . .
end module simple_1_file_module
```

The final part of the example extends the type to support a different file format. In this case, the file has an index to support positioning by a more efficient means that rewinding and reading sequentially through the file. Therefore, the rewind and seek bindings are overridden by ones tailored to this particular file format. The extended type probably also has extra components needed to maintain the index information.

```
module indexed_1_file_module

   use th_file_module

   type, extends(th_file_type) :: indexed_1_file_type
      . . .
   contains
      procedure :: rewind
      procedure :: next_frame
      procedure :: seek
         . . .
   end type

contains
   subroutine rewind (file)
      type(indexed_1_file_type) :: file
         . . .
   end subroutine rewind
   subroutine next_frame (file)
      type(indexed_file_type) :: file
         . . .
   end subroutine next_frame
```

```
      subroutine seek (file, time)
         type(indexed_file_type) :: file
         real, intent(in) :: time
            . . .
      end subroutine next_frame
         . . .
   end module indexed_1_file_module
```

See 5.2 for an example of using this module.

4.4.13 Values

The set of values of a derived type consists of all combinations of the possibilities for component values that are consistent with the components specified in the type definition.

The component value of a nonpointer, nonallocatable component is the value of the component. The component value of a pointer component is its pointer association. The pointer association includes the association status. If the pointer is associated, the association also includes any array bounds of the pointer and the identification of the target. The dynamic type and type parameters of a pointer are implicit in the identification of its target. The value of the pointer target is not part of the component value. The component value of an allocatable component is its allocation status, its dynamic type and type parameters, its bounds, and its value.

4.4.14 Operators

Any operation on derived-type entities must be defined explicitly by a function with an OPERATOR interface. Assignment, other than the intrinsic assignment provided for entities of the same derived type, must be defined by a subroutine with an ASSIGNMENT interface. These are described in 12.5.4.3.

A simple example is provided. Suppose it is desirable to determine the number of words and lines in a section of text. The information is available for each paragraph. A type named PARAGRAPH is defined as follows:

```
TYPE PARAGRAPH
   INTEGER NO_OF_WORDS, NO_OF_LINES
   CHARACTER (LEN = 30) SUBJECT
END TYPE PARAGRAPH
```

It is now desirable to define an operator for adding the paragraphs. An OPERATOR interface is required for the function that defines the addition operation for objects of type PARAGRAPH.

```
INTERFACE OPERATOR (+)
   MODULE PROCEDURE ADDP
END INTERFACE
```

This definition of addition for objects of type PARAGRAPH adds the words and lines, and concatenates the trimmed subjects inserting " and ".

```
TYPE (PARAGRAPH) FUNCTION ADDP (P1, P2)
   TYPE (PARAGRAPH), INTENT (IN) :: P1, P2
   ADDP % NO_OF_WORDS =  &
         P1 % NO_OF_WORDS + P2 % NO_OF_WORDS
   ADDP % NO_OF_LINES =  &
         P1 % NO_OF_LINES + P2 % NO_OF_LINES
   ADDP%SUBJECT = trim(p1%subject) // " and " // trim(p2%subject)
END FUNCTION ADDP
```

If the following variables were declared:

```
TYPE (PARAGRAPH) BIRDS, BEES
```

the expression BIRDS + BEES would be defined and could be evaluated in the module subprograms as well as any program unit accessing the module.

4.4.15 Structure Constructor

A **structure constructor** is used to construct a value of the type from component values. When a derived type is defined, a structure constructor for that type is defined automatically. Although there is no special form for derived-type constants, a structure constructor whose component values are all initialization expressions serves as a constant.

For a simple example, a value of type COLOR (defined in 4.4.1) may be constructed with the following structure constructor:

```
COLOR (I, J, K, "MAGENTA")
```

The form for a structure constructor (R457) is:

derived-type-spec ([component-spec-list])

where a component-spec (R458) is

[component-name =] component-source

and a component source (R459) is one of

expression
data-target
procedure-target

Expression, data-targets, and procedure-targets are described in 7.1.2, 7.5.5.1, and 7.5.5.2, respectively.

Rules and restrictions:

1. The component name may be omitted from a component specification only if the component name is also omitted from each preceding component specification in the list. The component sources without explicit component names are assigned to

the components of the type in component order. The component order of a type is the component order of the parent type, followed by the order of the component declarations in the type definition.

2. For each component of the type, a value must be specified, either explicitly in the constructor or implicitly by default initialization. Specifying the value of the parent component in an extended type is equivalent to specifying the values of all the inherited components; either is allowed.

3. No more than one value may be explicitly specified for a component.

4. The type name and each component name that is explicitly specified in a component specification must be accessible in the scoping unit where the structure constructor appears.

5. A component source corresponding to a pointer component must be allowable as a target for such a pointer in a pointer assignment statement (7.5.5). The component value is the result of such a pointer assignment.

6. A component source corresponding to a nonpointer component must be allowable as an expression in an intrinsic assignment statement for the component (7.5.2), except that the component source for an allocatable component may be unallocated or may be a reference to the intrinsic function NULL with no arguments. The component value is the result of such an intrinsic assignment or, in the exceptional case, is an unallocated allocatable.

7. A component that has no corresponding component source is defined as specified by the default initialization for the type.

8. A structure constructor for a type must not appear before the type is defined.

One consequence of rule 2 above is that if a type has private components that do not have default initialization, it is not possible to write a constructor for the type outside of the module where the type is defined.

If all of the values in a structure constructor are initialization expressions, the structure constructor may be used to define a named constant, for example, using types defined earlier in this chapter:

```
PARAMETER ( TEAL = COLOR (14, 7, 3, "TEAL") )
PARAMETER ( NO_PART = COMPONENT ("none", 0, 0.0) )
```

The form of a structure constructor has much in common with the form of a function reference (12.2.3). If there is an accessible generic function with the same name as a derived type, any consequent ambiguity is resolved by first applying generic function resolution as described in 12.8. Only if the generic resolution fails is a form considered for interpretation as a structure constructor. This essentially allows a user to override the interpretation of a structure constructor

Following are several examples of structure constructors.

Example 1. A structure constructor for a type that has a derived type as a component must provide a value for each of the components. A component may be of derived type, in which case a structure constructor is required for the component. In the example below, type RING has a component of type STONE. This example also illustrates the use of default initialization and keywords in a constructor; the insurer in this example is the default Lloyds.

```
TYPE STONE
   REAL              CARETS
   INTEGER           SHAPE
   CHARACTER (30)    NAME
END TYPE STONE

TYPE RING
   REAL              EST_VALUE
   CHARACTER (30)    INSURER = "Lloyds"
   TYPE (STONE)      JEWEL
END TYPE RING
```

If OVAL is an integer, a structure constructor for a value of type RING is:

```
RING (5000.00, jewel = STONE (2.5, OVAL, "emerald") )
```

Example 2. If a type is specified with an array component, the value that corresponds to the array component in the expression list of the structure constructor must conform with the specified shape. For example, type ORCHARD has an array component:

```
TYPE ORCHARD
    INTEGER                 AGE, NO_OF_TREES
    CHARACTER (LEN = 20)    VARIETIES (10)
END TYPE
```

Given the declaration:

```
CHARACTER (LEN = 20) CATALOG (16, 12)
```

a structure constructor for a value of type ORCHARD is:

```
ORCHARD (5, ROWS * NO_PER_ROW, CATALOG (LEMON, 1:10) )
```

Example 3. If a component of the type is a pointer, the corresponding structure constructor expression must evaluate to an entity that would be an allowable target for such a pointer in a pointer assignment statement (7.5.5). If the variable SYNOPSIS is declared:

```
CHARACTER(4000), TARGET :: SYNOPSIS
```

a value of the type SUMMARY (defined in 4.4.6) may be constructed:

```
SUMMARY("War and Peace", 1025, SYNOPSIS)
```

Example 4. For an extended type, the component order includes the components of the parent type, but does not include the parent component. This implies that the keyword form must be used if the parent type is to be specified. Using the type definitions in 4.4.12.1, a constructor for new_type_c can be written in positional form using the components of the parent type, as:

```
new_type_c(1.2, 7, 'Z')
```

But if bt is an object of type base_type, using it in a constructor for new_type_c requires the keyword form as in:

```
new_type_c(base_type=bt, c='Z')
```

Example 5. The following constructor uses the type matrix defined in 4.4.3 and illustrates a constructor with type parameters.

```
matrix(2, 2)(reshape([7.0, 0.0, 0.0, 7.0], [2, 2])
```

The type parameters kind and k keep their default values in this example. The RESHAPE intrinsic is used to construct the needed rank-2 array.

Example 6. A constructor for a type with allocatable components such as:

```
type item
  integer :: code
  character(:), allocatable :: description
end type
```

can have the allocatable components allocated or not, as in:

```
item(0, null())

item(1, 'Firewire 400 cable, 1 meter A-B')
```

4.5 Array Constructors

An **array constructor** is a mechanism that is used to construct a rank-one array from a sequence of values. Syntactically, it is a sequence of scalar values, arrays, and implied-do specifications enclosed in either square brackets or in parentheses and slashes. For example:

```
REAL VECTOR_X(3), VECTOR_Y(2), RESULT(100)
  . . .
RESULT (1:8) = [ 1.3, 5.6, VECTOR_X, 2.35, VECTOR_Y ]
```

The value of the first eight elements of RESULT is constructed from the values of VECTOR_X and VECTOR_Y and three real constants in the specified order. If an array appears in the value list, the values of its elements are taken in array element order. If it is necessary to construct an array of rank greater than one, the RESHAPE intrinsic function may be applied to an array constructor.

The form for an array constructor (R465) is one of:

```
[ [ type-spec :: ] [ ac-value-list ] ]
(/ [ type-spec :: ] [ ac-value-list ] /)
```

The outermost square brackets [ ] in the first syntax form are literal square brackets rather than indications of optionality.

An ac-value is one of

```
expression
ac-implied-do
```

where the expression can be either a scalar or an array.

The form for an ac-implied-do (R470) is:

```
( ac-value-list , ac-do-variable = scalar-integer-expression ,   &
   scalar-integer-expression [ , scalar-integer-expression ] )
```

Rules and restrictions:

1. The type specifier and the ac-value list may not both be omitted.
2. If the type specification is omitted, each ac-value expression in the array constructor must have the same type and type parameters, including length parameters; the type and type parameters of the constructor are those of the expressions.
3. If the type specification is included, each ac-value expression must be of a type and type parameters compatible with intrinsic assignment to the specified type. The constructor has the specified type and type parameters.
4. An ac-do-variable must be a scalar integer named variable. This variable has the scope of this ac-implied-do.
5. If an ac-implied-do is contained within another ac-implied-do, they must not have the same ac-do-variable.

If an ac-value is an array expression, the values of the elements of the expression in array element order (6.6.6) become the values of the array constructor.

If an ac-value is an implied-do specification, it is expanded to form a sequence of values under control of the ac-do-variable as in the DO construct (8.7.2.1).

If every expression in an array constructor is an initialization expression, the array constructor is an initialization expression as in the example above. Such an array constructor may be used to give a value to a named constant, for example:

```
REAL X(3), BIGGER_X(4)
PARAMETER (X = [ (I, I = 2, 6, 2) ] )
PARAMETER (BIGGER_X = [ 0.0, X ] )
```

Following are several examples of array constructors.

Example 1. To create a value for an array of rank greater than one, the RESHAPE intrinsic function must be used. With this function, a one-dimensional array may be reshaped into any allowable array shape.

```
Y = RESHAPE (SOURCE = [ 2.0, [ 4.5, 4.0 ], Z ], SHAPE = [ 3, 2 ] )
```

If Z has the value [1.2 3.5 1.1], Y is a 3×2 array with the elements:

```
2.0   1.2
4.5   3.5
4.0   1.1
```

Example 2. It may be necessary to construct an array value of derived type.

```
TYPE PERSON
   INTEGER AGE
   CHARACTER (LEN = 40) NAME
END TYPE PERSON

TYPE (PERSON) CAR_POOL (3)

CAR_POOL = [ PERSON (35, "SCHMITT"),  &
      PERSON (57, "LOPEZ"), PERSON (26, "YUNG") ]
```

Example 3. A type specifier in an array constructor can be used to coerce all elements to the same type and type parameters, as in:

```
[real:: 42, 1.234, 57]
[character(16):: 'Tom', 'Dick', 'Harry']
```

It also provides a simple way to write a zero-sized constructor as in:

```
[integer:: ]
```

Without the type specifier, this form would be invalid because there would be no specification of the type of the array.

4.6 Enumerations

An enumeration is a set of named integer constants, each of which is called an enumerator. Enumerations are designed primarily to facilitate C interoperability. Although an enumeration can be used independently of C, it then provides little utility that could not be achieved using other syntax. The main functionality of enumerations is in their automatic selection of the appropriate integer kinds to be compatible with corresponding C enums.

The form of an enumeration is

```
ENUM, BIND(C)
 enumerator-definition-statement
 [ enumerator-definition-statement ] ...
END ENUM
```

The BIND(C) is mandatory. The form of an enumerator definition statement is

```
ENUMERATOR [ :: ] enumerator-list
```

where the form of an enumerator is

```
named-constant [ = scalar-integer-initialization-expression ]
```

If any enumerator in a list includes the optional initialization expression, then the double colon in that enumerator definition statement is required.

The enumeration declares all of its enumerators to be integer named constants with values as described below. The kind of the named constants is automatically selected so that they are interoperable with a C enumeration type that specified the same values in the same order. Note that an enumeration does not define a distinct type; it just facilitates the selection of an appropriate integer kind. The effect of an enumeration is identical to that of a corresponding set of integer parameter declarations.

The value of an enumerator is determined as follows:

1. If the enumerator definition has an initialization expression, that expression gives the value.

2. If the enumerator definition does not have an initialization expression, the value is 1 greater than the value of the previous enumerator, or is 0 if it is the first enumerator.

Example:

```
enum, bind(c)
   enumerator :: red=4, blue=9
   enumerator yellow
end enum
```

In this example, the named constant red will have the value 4, blue will be 9, and yellow will be 10. Note that the effect is the same whether the enumerators of an enumeration are declared all in a single statement or in multiple ones.

Although the main functionality of an enumeration is to automatically select the appropriate integer kind, the syntax provides no way to directly find what kind was selected. The only way to find that is to use the KIND intrinsic on one of the resulting named constants. For example, the following is a way to declare an integer variable of the kind selected by the above enumeration.

```
integer(kind(red)) :: x
```

5 Declarations

- **Declarations** are used to specify the attributes and relationships of the entities in a program.
- The declared **Type** of a variable, function, or named constant is specified explicitly by a type declaration or implicitly by the first letter of the entity's name. An IMPLICIT statement associates a type with specific letters or disables implicit typing.
- A **Polymorphic** entity is one whose dynamic type can change during program execution. For a polymorphic entity, the dynamic type may be different from the declared type. A nonpolymorphic entity also has a dynamic type, but it is always the same as the declared type.
- The **DIMENSION attribute** specifies an array. The array may have explicit shape (with all bounds specified), deferred shape (if it also has the ALLOCATABLE or POINTER attribute), or assumed shape/size (if it is a dummy argument).
- The **ALLOCATABLE** or **POINTER attribute** specifies an entity that may be dynamically allocated during program execution. Alternatively, a pointer variable may be associated with an existing target. The TARGET attribute specifies that a variable may be the target of a pointer.
- **Initialization** of a variable may be specified in a type declaration or in a DATA statement. Pointers may be initially disassociated.
- The **EXTERNAL** or **INTRINSIC attribute** specifies the nature of a procedure.
- The **INTENT**, **VALUE**, or **OPTIONAL attribute** specifies properties of a dummy argument.
- The **PARAMETER attribute** specifies a named constant.
- The **PUBLIC**, **PRIVATE**, or **PROTECTED attribute** allows a programmer to control the accessibility and use of entities specified in modules.
- The **BIND(C) attribute** facilitates interoperation with C data and functions.
- The **ASYNCHRONOUS** or **VOLATILE attribute** specifies that a variable's value might be referenced or redefined outside of the normal flow of program execution.

Declarations are used to specify the type and other attributes of program entities. The attributes that an entity possesses determine how the entity may be used in a program. Every variable and function has a type, which is the most important of the attributes; type is discussed in 4. However, type is only one of a number of attributes that an entity may possess. Attributes may be specified in type declaration or procedure declara-

J.C. Adams et al., *The Fortran 2003 Handbook*,
DOI: 10.1007/978-1-84628-746-6_5, © Springer-Verlag London Limited 2009

tion statements (entity-oriented form), in separate attribute declaration statements (attribute-oriented form), or in a mix of these forms. Attributes of a procedure may be specified in an interface body, which is discussed in 12.5.2. Some entities, such as subroutines and namelist groups, do not have a type but may possess other attributes.

In addition, there are relationships among objects that can be specified by EQUIVALENCE, COMMON, and NAMELIST statements. A NAMELIST statement specifies a name for a list of objects that may be referenced in an input/output statement. The EQUIVALENCE statement indicates that some variables share storage. The COMMON statement specifies a name for a block of storage and the names of objects in the block; this block can then be shared among different program units. COMMON and EQUIVALENCE are provided primarily for compatibility with older versions of the language; they are seldom needed in new programs, where modules can provide replacement functionality in a more structured way.

In general, Fortran keywords are used to declare the attributes for an entity. The following list summarizes these keywords:

Type	INTEGER REAL (and DOUBLE PRECISION) COMPLEX LOGICAL CHARACTER TYPE (*user-defined name*)
Array properties	DIMENSION
Allocatable property	ALLOCATABLE
Pointer properties	POINTER TARGET
Value definition properties	DATA PARAMETER SAVE ASYNCHRONOUS VOLATILE
Module entity properties	PUBLIC PRIVATE PROTECTED BIND
Dummy argument properties	INTENT OPTIONAL VALUE
Procedure properties	EXTERNAL INTRINSIC

In earlier versions of the language, it was necessary to use a different statement for each attribute given to a variable or a collection of variables, for example:

```
INTEGER A, B, C
SAVE    A, B, C
```

In later versions, for objects that have a type, the other attributes may be included in the type declaration statement. For example:

```
INTEGER, SAVE :: A, B, C
```

Collecting the attributes into a single statement is sometimes more convenient for readers of programs. It eliminates searching through many declaration statements to locate all the attributes of a particular object. Emphasis can be placed on an object and its attributes (entity-oriented declaration) or on an attribute and the objects that possess the attribute (attribute-oriented declaration), whichever is preferred by a programmer.

It is also allowed to use a mixed form, specifying some attributes for an entity together and some separately. The terms "entity-oriented" and "attribute-oriented" are used here for expository purposes, but are not actually distinctions made by the standard. The attributes of an entity are collected from those specified for it by all forms of specification. Unfortunately, there is no way within the language for the programmer to specify enforcement of an entity-oriented form.

The same attribute must not be specified explicitly for an entity more than once, regardless of the form of specification.

The following are examples of entity-oriented and attribute-oriented forms:

- entity-oriented declarations

  ```
  REAL, DIMENSION(20), SAVE :: X
  ```

  or

  ```
  REAL, SAVE :: X(20)
  ```

- attribute-oriented declarations

  ```
  REAL X
  DIMENSION X(20)
  SAVE X
  ```

  or

  ```
  REAL X (20)
  SAVE X
  ```

Although most attributes are determined statically at compilation time, some attributes can be specified to vary during program execution. A variable that has such a varying attribute is called dynamic. The attributes that can be dynamic are the type, length type parameters, and array bounds. There are four categories of dynamic variables: automatic, allocatable, pointer, and polymorphic. An **automatic variable** is one whose dynamic attributes are automatically determined on entry to a procedure, but which is not a dummy argument or function result. The exception for dummy arguments and function results is largely historical in that they predate dynamic allocation and can be implemented without it, so the standard does not refer to them as dynamic. Length type parameters and array bounds are the attributes that can be automatic. Allocatable, pointer, and polymorphic variables can have their dynamic attributes specified by executable statements and can be dummy arguments, function results, or other variables.

5.1 Type Declaration Statements

A type declaration statement begins with a type specifier, optionally lists other attributes, then ends with a list of entities that possess these attributes. In addition, a type declaration statement may include an initial value for a variable or association status for a pointer. The form of a type declaration statement (R501) is:

```
declaration-type-spec [ [ , attribute-spec ] ... :: ] entity-declaration-list
```

where a declaration type specification (R502) is one of:

```
intrinsic-type-spec
TYPE ( derived-type-spec )
CLASS ( derived-type-spec )
CLASS ( * )
```

and where an attribute specification (R503) is one of:

```
ALLOCATABLE
ASYNCHRONOUS
BIND( C [ , NAME= scalar-character-initialization-expression ] )
DIMENSION ( array-spec )
EXTERNAL
INTENT ( intent-spec )
INTRINSIC
OPTIONAL
PARAMETER
POINTER
PRIVATE
PROTECTED
PUBLIC
SAVE
TARGET
VALUE
VOLATILE
```

where an entity declaration (R504) has one of the forms:

```
object-name [ ( array-spec ) ] [ * character-length ] [ initialization ]
function-name [ * character-length ]
```

Rules and restrictions:

1. Each expression used as a length type parameter value in a declaration type specification must be a specification expression. The same restriction applies to an expression used as a character length in an entity declaration, which is just an alternate syntax for the same thing.
2. If an expression used as a length type parameter value in a declaration type specification is not an initialization expression, the declaration type specification must be in the scoping unit of a subprogram or interface body. The same restriction applies to an expression used as a character length in an entity declaration.
3. A derived type specified with the CLASS keyword in a declaration type specification must be an extensible type (4.4.12).
4. A derived type specified with the TYPE keyword in a declaration type specification must not be an abstract type (4.4.12.3).
5. A type declaration statement with the TYPE keyword must not specify a derived type that is defined later in the same scoping unit.
6. The same attribute specification must not appear more than once in a given type declaration statement.
7. An entity must not be given the same attribute explicitly more than once in a scoping unit.
8. The character length option may appear only if the type specified is CHARACTER.
9. If initialization appears in any entity declaration of the statement, the double colon separator before the entity declaration list must be used.
10. A function name must be the name of an external function, an intrinsic function, a function dummy procedure, a procedure pointer, or a statement function.

There are other rules and restrictions that pertain to particular attributes; these are covered in the sections describing those attributes. Some attributes are incompatible with others; a table of these incompatibilities is in 5.12.

Item 5 above is a bit inconsistent in that it does not apply to type declaration statements with the CLASS keyword and is more stringent than similar restrictions on component declarations.

The form of entity declaration that has a function name applies only to declarations outside of the function or interface body. Within a function or an interface body for it, the function result variable is a data object, which can be declared using the form with an object name. Elsewhere, the only cases where a function can be declared with a type declaration statement are where it has an implicit interface or is intrinsic.

If an expression used as a length type parameter value in a declaration type specification is not an initialization expression, the expression is evaluated and establishes the value of the length type parameter on each entry to the procedure in which it appears. The value is not affected by any changes to the values of variables in the expression during execution of the procedure. A data object declared using such an expression is an automatic variable unless it is a dummy argument or function result. This also applies to an expression used as a character length in an entity declaration. These issues do not apply to a specification that is an initialization expression; this is because the value of an initialization expression does not depend on anything that can change during execution.

Some example type declaration statements are:

```
REAL A(10)
LOGICAL, DIMENSION(5, 5) :: MASK_1, MASK_2
COMPLEX :: CUBE_ROOT = (-0.5, 0.867)
INTEGER, PARAMETER :: SHORT = SELECTED_INT_KIND(4)
INTEGER(SHORT) K              ! Range of -9999 to 9999
REAL, ALLOCATABLE :: A1(:, :), A2(:, :, :)
TYPE(PERSON) CHAIRMAN
TYPE(NODE), POINTER :: HEAD_OF_CHAIN => NULL ( )
REAL, INTENT(IN) :: ARG1
REAL, INTRINSIC :: SIN
```

5.2 Polymorphism

A polymorphic entity is one whose type can change during program execution. The term refers to having many (poly) forms (morph). The CLASS keyword is used to declare polymorphic entities.

With polymorphism comes a distinction between the declared and dynamic type of an entity. The declared type of an entity is the type that it is declared to have, either explicitly or implicitly. The dynamic type is the type that it has at a particular time during program execution. In general, when the type of an entity is used without qualification in the standard or this book, it refers to the dynamic type.

A nonpolymorphic entity also has both a declared type and a dynamic type, but they are always the same. The dynamic type of a polymorphic object that is not allocated or associated is the same as its declared type. The dynamic type of an allocated or associated polymorphic object is the type that it was allocated with or the dynamic type of the entity that it is associated with.

An object declared with CLASS(*) is an unlimited polymorphic object; it has no declared type. It is not considered to have the same declared type as any other entity, even another unlimited polymorphic entity.

The concept of type compatibility is used in several contexts; it is defined as follows. A nonpolymorphic entity is type compatible with entities of the same declared type. A polymorphic entity that has a declared type is type compatible with entities whose declared type is the same type or any extension of it. An unlimited polymorphic entity is type compatible with all entities. An entity is type compatible with a type if it is type compatible with entities declared to have that type.

Note that type compatibility is not symmetric. Two entities are type incompatible if neither is type compatible with the other.

Rules and restrictions:

1. An entity declared with the CLASS keyword must be a dummy argument, a pointer, or an allocatable variable.

2. An allocatable object may be allocated only with a type with which it is type compatible.

3. A pointer or dummy argument may be associated only with a target or actual argument with which it is type compatible.

The following simple example of polymorphism uses the example modules from 4.4.12.4 and assumes that they have additional type bindings for open and close procedures.

```
program read_file
  use simple_1_file_module
  use indexed_1_file_module
  class(th_file_type), allocatable :: th_file
  character :: file_name*128, file_type*16
  real :: start_time

  read(*,*) file_name, file_type, start_time
  select case(file_type)
  case ('simple_1')
    allocate(simple_1_file_type:: th_file)
  case ('indexed_1')
    allocate(indexed_1_file_type:: th_file)
  case default
    write (*,*) 'Unrecognized file type: ', file_type
    stop
  end select

  call th_file%open(file_name)
  call th_file%seek(start_time)
  do while (th_file%frame_ok)
    write (*,*) th_file%frame_time
    call th_file%next_frame
  end do
  call th_file%close
end program read_file
```

Based on the value that is read for the string file_type, the program allocates the polymorphic variable th_file to be one of the two possible types. The corresponding type-bound versions of the open, seek, next_frame, and close procedures are invoked depending on the dynamic type given to th_file by the ALLOCATE statement.

5.3 Implicit Typing

Implicit typing is a method of inferring a type and type parameter values based on the first letter of the name of an entity. Implicit typing applies to each named variable, named constant, or nonintrinsic specific function whose type is not otherwise specified.

In each scoping unit, there is a mapping of each of the letters A, B, ..., Z to a particular type and type parameter values or to no type. The mapping does not distinguish between upper and lower case. The type and type parameters of an implicitly typed entity are those of the mapping for the first letter of the entity name. If an entity would be implicitly typed, but the applicable mapping is to no type, then the program is invalid.

If a scoping unit has no IMPLICIT statements, then its default mapping applies. The default mapping for an internal or module procedure is the mapping from its host. For other scoping units, the default mapping is as shown in Figure 5-1. That is, each applicable entity whose name begins with any of the letters I, J, K, L, M, or N is of type default integer and all others are of type default real. Note that although interface bodies have a host scoping unit, their default mapping does not come from the host.

Figure 5-1 Default implicit mapping for a program unit

IMPLICIT statements can be used to specify a mapping different from the default. The IMPLICIT statement (R549) has two forms:

```
IMPLICIT implicit-spec-list
IMPLICIT NONE
```

where an implicit specification (R550) is:

```
declaration-type-spec ( letter-spec-list )
```

and a letter specification (R551) is:

```
letter [ - letter ]
```

An IMPLICIT NONE statement specifies that the mappings for all letters are to no type, with the consequence that implicit typing cannot be used in that scoping unit—type declaration statements must be used for all appropriate entities. The other form of IMPLICIT statement specifies mappings from the specified letters to the type and type parameter values specified by the declaration type specifiers; the letter–letter form indicates a range of letters from the first to the last. If the mapping for a letter is not specified by any IMPLICIT statements in a scoping unit, then the mapping for that letter remains the same as in the default mapping for the scoping unit.

Rules and restrictions:

1. If IMPLICIT NONE appears, it must precede any PARAMETER statements and there must be no other IMPLICIT statements in the scoping unit.

2. If the letter–letter form appears in a letter specification, the second letter must not precede the first alphabetically.

3. The same letter must not appear as a single letter or be included in a range of letters more than once in all of the IMPLICIT statements in a scoping unit.

   For example, the statement

```
IMPLICIT COMPLEX (A-C, Z)
```

indicates that all implicitly typed entities whose names begin with the letters A, B, C, or Z are of type default complex. If this is the only IMPLICIT statement, implicitly typed entities whose names begin with I-N are of type default integer; implicitly typed entities whose names begin with D-H and O-Y are of type default real.

The statement

```
IMPLICIT NONE
```

indicates that there is no implicit typing in the scoping unit and that each named variable, named constant, and nonintrinsic specific function used in the scoping unit and not accessed by use or host association must be declared explicitly in a type declaration statement.

It is generally recommended to use IMPLICIT NONE in new code. This facilitates compiler detection of some common coding errors. With IMPLICIT NONE, inadvertent misspellings are usually detected during compilation. Without IMPLICIT NONE, such misspellings are often interpreted as implicitly declared variables, with the consequence that the program compiles, but does not work as intended. An IMPLICIT statement may specify a user-defined type. However, it is usually recommended to avoid this; the error-proneness of implicit typing is exacerbated when applied to derived types. This is because a derived-type name has local scope; the same name can mean something completely different in a different scope. The following example shows one resulting oddity.

```
program main
  implicit type(t) (a-z)
  type t
    . . .
  end type t
  call sub
contains
  subroutine sub
    integer :: t
      . . .
  end subroutine sub
end program main
```

The IMPLICIT statement in this example causes all implicitly typed entities to have type t. The INTEGER statement in the subroutine blocks host association of the type name t. Thus nothing can be explicitly declared to be of type t in the subroutine. But the implicit mapping to type t still holds. Thus variables in the subroutine can get type t implicitly, but not explicitly. Fortunately, this odd state of affairs seems unlikely to arise in real code.

The complexity that implicit typing causes in determining the scope of an undeclared variable in a nested scope is explained in 16.2.1.3.

Some examples of IMPLICIT statements are:

```
IMPLICIT INTEGER (A-G), LOGICAL (KIND = BIT) (M)
IMPLICIT CHARACTER *10 (P, Q)
IMPLICIT TYPE (COLOR) (X-Z)
IMPLICIT REAL (QUAD) (H-J, U-W, R)
```

5.4 Array Properties

An object with the dimension attribute is an array. An array specification specifies an array's rank and information about the bounds. The rank must be known at compile time. The bounds may be dynamic in one of several ways.

5.4.1 Array Specifications

There are four forms that an array specification (R510) may take:

explicit-shape-spec-list
assumed-shape-spec-list
deferred-shape-spec-list
assumed-size-spec

The specified rank is the number of comma-separated items in the array specification, which is one more than the number of commas. The maximum rank of an array is 7.

5.4.1.1 Explicit-Shape Arrays

An explicit-shape array is one whose bounds are entirely determined from its array specification, which is an explicit-shape specification list. Each dimension has an explicit-shape specification (R514), which has the form:

[lower-bound :] upper-bound

where the lower bound, if present, and the upper bound are specification expressions (7.4.2).

Rules and restrictions:

1. If the lower bound is omitted, the default value is 1.

2. If any of the bound expressions is not an initialization expression, then the array specification must be in the scoping unit of a subprogram or interface body. The expression is evaluated and establishes the bound on each entry to the procedure. The bound is not affected by any changes to the values of variables in the expression during execution of the procedure. An array declared using such an expression is an automatic variable unless it is a dummy argument or function result.

The subscript range of the array in a given dimension is the set of integer values between and including the lower and upper bounds, provided the upper bound is not less than the lower bound. If the upper bound is less than the lower bound, the range is empty, the extent in that dimension is 0, and the size of the array is 0.

Examples of explicit-shape arrays:

```
REAL Q (-10:10, -10:10, 2)
```

or in a subroutine

```
SUBROUTINE EX1 (Z, I, J)
   REAL, DIMENSION (2:I + 1, J) :: Z
      . . .
```

5.4.1.2 Assumed-Shape Arrays

An assumed-shape array is a dummy argument that takes the shape of the actual argument passed to it. Each dimension has an assumed-shape specification (R514), which has the form:

[lower-bound] :

Rules and restrictions:

1. An assumed-shape array must be a dummy argument.
2. If the lower bound is omitted, the default value is 1. Note that the lower bound is not assumed from the actual argument; only the shape (extents) is assumed.
3. If the lower bound is specified, it must be a specification expression and it is evaluated on entry to the procedure. The bound is not affected by any changes to the values of variables in the expression during execution of the procedure.
4. The upper bound is the extent of the corresponding dimension of the associated array plus the lower bound minus 1.
5. An assumed-shape array cannot have the POINTER or ALLOCATABLE attribute, but this is more a matter of definition than a restriction. Pointer or allocatable dummy arrays are deferred shape, which has a similar syntax as described below.

Examples of assumed-shape arrays:

```
SUBROUTINE EX2 (A, B, X)
   REAL, DIMENSION (2:, :) :: X
   REAL, INTENT(IN) :: A(:), B(0:)
```

Suppose EX2 is called by the statement

```
CALL EX2 ( U, V, W (4:9, 2:6))
```

For the duration of the execution of subroutine EX2, the dummy argument X is an array with bounds (2:7, 1:5). The lower bound of the first dimension is 2 because X is declared to have a lower bound of 2. The upper bound is 7 because the dummy argument takes its shape from the actual argument W.

5.4.1.3 Deferred-Shape Arrays

A deferred-shape array is one whose bounds may change at times other than entry to a procedure. Each dimension has a deferred-shape specification (R515), which has the form:

```
:
```

Rules and restrictions:

1. The array must have either the ALLOCATABLE or POINTER attribute. This is the only form of array specification allowed for allocatables or pointers.

2. The size, bounds, and shape of a disassociated array pointer or unallocated allocatable array are undefined.

A deferred-shape array may or may not be a dummy argument. If a deferred-shape array is a dummy argument, then on entry to the procedure the definition status and values of the dummy argument's bounds are those of the actual argument; this applies to both the lower and upper bounds and is thus different from assumed-shape arrays, which do not assume the lower bounds. Another important difference between assumed-shape and deferred-shape dummy arguments is that a deferred-shape dummy argument can change bounds during execution of the procedure it is in; such changes propagate back to the actual argument on termination of the procedure. The bounds of an assumed-shape dummy argument cannot change during execution of the procedure. Details of argument association for deferred-shape dummy arrays are discussed in 12.6.5 and 12.6.7.

The bounds of an allocatable array are determined when it is allocated. Allocation of allocatable variables is discussed in 6.7.1.1. The bounds of a pointer array are determined when it is associated with a target, through allocation or other means, as discussed in 6.7.1.2, 7.5.5.1, and 12.6.5.

Examples of deferred-shape arrays:

```
REAL, POINTER :: D (:,:), P (:) ! pointer arrays
REAL, ALLOCATABLE :: E (:)      ! allocatable array
```

5.4.1.4 Assumed-Size Arrays

An assumed-size array is a dummy argument array whose size is assumed from that of the associated actual argument. Only the size is assumed—the rank and bounds (except for the upper bound and extent in the last dimension) are determined from its assumed-size array specification (R516) which has the form:

[explicit-shape-spec-list ,] [lower-bound :] *

The form and interpretation of the array specification for an assumed-size dummy array is identical to that of an explicit-shape array except for the replacement of the last upper bound by an asterisk.

Rules and restrictions:

1. An assumed-size array must be a dummy argument.
2. If any lower bound is omitted, the default value is 1.
3. Each bound expression must be a specification expression and it is evaluated on entry to the procedure. The bound is not affected by any changes to the values of variables in the expression during execution of the procedure.
4. If an assumed-size array has the INTENT (OUT) attribute, the array must not be of a type that has default initialization.
5. The name of an assumed-size array must not be used as a whole-array reference except as an actual argument in a procedure reference for which the array's shape is not required.
6. The upper bound and extent of the last dimension of an assumed-size array are not defined.
7. In an array section (6.6.4) of an assumed-size array, the second subscript (upper limit) must not be omitted from a subscript triplet in the last dimension.

Conceptually, the requirements on assumed-size arrays derive from the presumption that the compiler will not necessarily "know" the actual array size. The array has a size, as defined below, but this serves only as a definition of the limits that the programmer is required to adhere to. An assumed-size array cannot be used in a context where the compiler would need knowledge of the size. For example, a statement such as

```
write (*,*) x
```

requires that the compiler know the size of x in order to write out the appropriate number of values; therefore, this statement is disallowed if x is an assumed-size array.

Similarly, because the extent and upper bound of the last dimension of an assumed-size array are not defined, an array slice such as x(3,:) is disallowed. However, a slice such as x(:,3) is allowed.

The size of an assumed-size array is defined as follows:

1. If the actual argument associated with the assumed-size dummy argument is an array of any type other than default character, the size is that of the actual array.

2. If the actual argument associated with the assumed-size dummy array is an array element of any type other than default character with a subscript order value (6.6.6) of v in an array of size x, the size of the dummy argument is $x - v + 1$.

3. If the actual argument is a default character array, default character array element, or a default character array element substring (6.4), and if it begins at character storage unit t of an array with c character storage units, the size of the dummy array is

   MAX (INT $((c - t + 1) / e)$, 0)

   where e is the length of an element in the dummy character array.

This complicated-sounding definition can be stated simply in informal terms: the assumed-size array is big enough to fill to the end of the array in the actual argument; the filling is done by characters if the type is default character, and by elements otherwise.

Some implementations track the actual size of assumed-size arrays in order to facilitate debugging, but the standard is written so as not to require this.

Examples of assumed-size arrays:

```
SUBROUTINE EX3 (N, S, Y)
   REAL, DIMENSION (N, *) :: S
   REAL Y (10, 5, *)
      . . .
```

5.4.1.5 Limitations on Whole Arrays

There are some limitations on appearances in a program of whole arrays declared with each of the four forms of array specification. Table 5-1 gives a partial summary of the allowable appearances.

Table 5-1 Partial summary of allowable appearances of whole arrays declared in each of the four ways

	An array declared with			
May appear as a	**Explicit shape**	**Assumed shape**	**Deferred shape**	**Assumed size**
Primary in an expression	Yes	Yes	Yes	No
Vector subscript	Yes	Yes	Yes	No
Dummy argument	Yes	Yes	Yes	Yes

Table 5-1 *(Continued)* Partial summary of allowable appearances of whole arrays declared in each of the four ways

Actual argument	Yes	Yes	Yes	Yes
Equivalence object	Yes	No	No	No
Common object	Yes	No	Yes	No
Namelist object	Yes	Yes	Yes	No
Saved object	Yes	No	Yes	No
Data initialized object	Yes	No	No	No
I/O list item	Yes	Yes	Yes	No
Format	Yes	Yes	Yes	No
Internal file	Yes	Yes	Yes	No
Allocate object	No	No	Yes	No
Pointer object in pointer assignment statement	No	No	Yes	No
Target object in pointer assignment statement	Yes	Yes	Yes	No

5.4.2 The DIMENSION Attribute

An array specification can appear in several contexts. It can be in a DIMENSION attribute specification or a DIMENSION statement. Alternatively, it can be in an entity declaration in a type declaration statement or in parentheses following the variable name in an ALLOCATABLE, POINTER, TARGET, or COMMON statement. As with other attributes, the dimension attribute must not be specified more than once for a given array. However, an array specification may appear in an entity declaration in a type declaration statement that also has a DIMENSION attribute specification. In this case, the array specification in the DIMENSION attribute specification does not apply to that particular name; the array specification in the entity declaration applies instead.

The form of a DIMENSION statement (R526) is:

```
DIMENSION [ :: ] array-name ( array-spec ) &
   [ , array-name ( array-spec ) ] ...
```

Examples of specifying the DIMENSION attribute are:

- entity-oriented

```
INTEGER, DIMENSION (10), TARGET, SAVE :: INDICES
INTEGER, ALLOCATABLE, TARGET :: LG (:, :, :)
```

- attribute-oriented

```
INTEGER INDICES, LG (:, :, :)
DIMENSION INDICES (10)
TARGET INDICES, LG
ALLOCATABLE LG
SAVE INDICES
```

- with the array specification in other statements

```
INTEGER INDICES, LG
TARGET INDICES (10), LG
ALLOCATABLE LG (:, :, :)
SAVE INDICES
```

- with the array specification in a COMMON statement

```
COMMON / UNIVERSAL / TIME (80), SPACE (20, 20, 20, 20)
```

5.5 The ALLOCATABLE Attribute

The ALLOCATABLE attribute signifies a variable whose space is allocated by executable statements, in particular ALLOCATE or assignment statements. The details of allocation and deallocation of allocatable variables are discussed in 6.7.

In some cases, either an automatic or allocatable variable could be used; the following differences between automatic and allocatable variables are relevant to making the choice. The syntax for using an automatic variable is simpler in that it involves only the declaration, while an allocatable variable requires the declaration plus a separate executable statement to do the allocation. The benefit of the allocatable alternative is increased flexibility. The allocation of an allocatable variable is not restricted to being at the beginning of the procedure; it may thus use values that are computed during execution of the procedure. It may even be allocated and deallocated multiple times during execution of the procedure, or the allocation from one execution of the procedure may be saved for subsequent executions. Also, ALLOCATE statements allow for user-specified error handling for problems such as insufficient resources to successfully allocate the requested space.

In previous standards the ALLOCATABLE attribute was restricted to arrays. It is now also allowed for scalars. Polymorphism and dynamic length type parameters (particularly for character type) are two examples of situations where allocatable scalar variables can have dynamic attributes. It is also allowed for a scalar with no dynamic attributes to be allocatable; although less common, this can conceivably be useful in managing memory use if the scalar is of a derived type with large array components.

The ALLOCATABLE attribute can be specified in a type declaration statement or in an ALLOCATABLE statement. The form of an ALLOCATABLE statement (R520) is:

```
ALLOCATABLE [ :: ] object-name [ ( deferred-shape-spec-list ) ] &
               [ , object-name [ ( deferred-shape-spec-list ) ] ] ...
```

An allocatable array must be deferred shape. Note that there is no similar restriction on length type parameters; a declaration can specify a fixed length for an allocatable scalar string, but cannot specify a fixed size for an allocatable array of characters.

Examples of specifying the ALLOCATABLE attribute are:

- entity-oriented

```
REAL, ALLOCATABLE :: A (:, :)
LOGICAL, ALLOCATABLE, DIMENSION (:) :: MASK1
CHARACTER(LEN=:), ALLOCATABLE :: STRING, STRINGS(:)
```

- attribute-oriented

```
REAL A (:, :)
LOGICAL MASK1
DIMENSION MASK1 (:)
ALLOCATABLE A, MASK1
```

5.6 Pointer Properties

An entity with the POINTER attribute is referred to as a **pointer**. A pointer is either a data pointer or a procedure pointer. A data pointer may be either a scalar or an array.

A pointer can be thought of as a dynamic alias. It does not stand on its own, but is an additional name for some other data object or procedure. That other data object or procedure is called the **target** of the pointer. At different times, a pointer may point to different targets or possibly to no target.

The standard's technical terminology refers to a pointer as being associated with a target; this means the same thing as the phrase "pointing to a target", which is widely used in less formal contexts. When a pointer has no target, the pointer is disassociated.

It is also possible for the association status of a pointer to be undefined, which is quite different from being disassociated. When a pointer is disassociated, it is "known" to have no target. When the association status is undefined, it is not "known," possibly even to the compiler. There is no way to inquire whether a pointer's association status is undefined or not; it is not a testable state. It is illegal to use the ASSOCIATED intrinsic on a pointer with undefined association status; doing so might cause the program to abort or return a misleading value. The ASSOCIATED intrinsic can distinguish associated from disassociated, but cannot be used to detect undefined association status. Undefined status is the standard's way of describing situations where it is the programmer's responsibility to avoid using the pointer.

When a pointer is associated with a target, you can use the pointer like another name for the target. There is no special syntax needed to indicate that you are referring to the target instead of the pointer; the notion of a dynamic alias may be a more helpful mental model than direct analogy to pointers in some other languages. In many ways, a pointer is like a dummy argument. When a pointer is not associated with a target, you are not allowed to reference or define the pointer.

A pointer can be initialized (5.7.2) to be disassociated; otherwise, its initial association status is undefined.

A pointer can become associated with an existing target in several different ways, including pointer assignment statements, intrinsic assignment statements for derived types, argument association, and the MOVE_ALLOC intrinsic procedure. All of these ways share the general nature of pointing at existing data or procedures. Alternatively, a data pointer can be associated with a newly allocated target using the ALLOCATE statement.

In all cases, it is important to understand that a pointer and its target are distinct entities; their association is temporary. A pointer does not uniquely "own" its target. There may be multiple pointers associated with the same target. If any one of those associations is severed, that does not cause the target to stop existing; the other pointers would still be associated with the target. This is often a source of confusion when the ALLOCATE statement is used to allocate a new target. In that case, the newly allocated target does not have its own name, but it still has its own existence. The effect of the ALLOCATE statement is to create an anonymous target and then to associate the pointer with that target. Even though the pointer is specified in the ALLOCATE statement that creates such an anonymous target, the association between that pointer and that target has no special standing. Other pointers may subsequently become associated with that target; that pointer may subsequently become associated with other targets.

Another way of thinking about a pointer is as a **descriptor** with space to contain information about the type, type parameters, rank, extents, and location of a target. Thus, a pointer to a scalar object of type real would be quite different from a pointer to an array of user-defined type. In fact, each of these pointers is considered to occupy a different amount of storage. When an object with the POINTER attribute is declared to be in a common block, it is likely to be the descriptor that occupies the storage. This is why every declaration of a common block that contains a pointer must specify the same sequence of storage units.

5.6.1 The POINTER Attribute

The POINTER attribute can be specified in a type declaration statement or in a POINTER statement. The POINTER attribute for a procedure pointer can alternatively be specified in a procedure declaration statement (5.11).

The form of a POINTER statement (R540) is:

```
POINTER [ :: ] pointer-declaration-list
```

where a pointer declaration (R541) is one of

```
object-name [ ( deferred-shape-spec-list ) ]
procedure-pointer-name
```

Rules and restrictions:

1. A pointer array must be deferred shape.
2. A procedure pointer must have the EXTERNAL attribute explicitly declared.

3. A pointer must not be referenced or defined unless it is associated with a target that may be referenced or defined.

   Examples of specifying the POINTER attribute are:

- entity-oriented

```
TYPE (NODE), POINTER :: CURRENT
REAL, POINTER :: X (:, :), Y (:)
PROCEDURE (), POINTER :: HANDLER
```

- attribute-oriented

```
TYPE (NODE) CURRENT
REAL X (:, :), Y (:)
PROCEDURE () :: HANDLER
POINTER CURRENT, X, Y, HANDLER
```

5.6.2 The TARGET Attribute

An object with the TARGET attribute may become the target of a pointer during execution of a program. The main purpose of the TARGET attribute is to provide aid to a compiler in the production of efficient code. If an object does not have the TARGET attribute, no part of it can be accessed via a pointer. If an object has the TARGET attribute, then so do all of its subobjects.

The TARGET attribute can be specified in a type declaration statement or in a TARGET statement.

The form of the TARGET statement (R546) is:

```
TARGET [ :: ] object-name [ ( array-spec ) ] &
   [ , object-name [ ( array-spec ) ] ] ...
```

Examples of specifying the TARGET attribute are:

- entity-oriented

```
TYPE (NODE), TARGET :: HEAD_OF_LIST
REAL, TARGET, DIMENSION (100, 100) :: V, W (100)
```

- attribute-oriented

```
TYPE (NODE) HEAD_OF_LIST
REAL V, W (100)
DIMENSION V (100, 100)
TARGET HEAD_OF_LIST, V, W
```

5.7 Value Definition Properties

Several properties relate to the definition of data values. These include specification of initial values and of the circumstances in which variable values can change.

Named constant definition is specified in a type declaration statement or a PARAMETER statement. It applies to the particular entities specified and can be used with any type. Named constants are defined only once and cannot be redefined.

Default initialization is covered in 4.4.9. It applies only to derived types and is specified in the derived-type definition. It applies to all objects of the type; it occurs whenever such an object comes into existence.

Explicit initialization is specified in a type declaration statement or a DATA statement. It applies to the particular entities specified and can be used with any type. Explicit initialization occurs exactly once for each explicitly initialized entity.

The SAVE attribute specifies that a variable's value will be saved between invocations of a procedure. The ASYNCHRONOUS and VOLATILE attributes specify that a variable's value might be referenced or redefined outside of the normal flow of program execution.

5.7.1 The PARAMETER Attribute

The PARAMETER attribute indicates a named constant. A named constant is defined exactly once; it cannot be redefined. The value of a named constant is known at compile time, which allows it to be used in several contexts where a variable is not allowed. A named constant may be of any type.

A named constant is sometimes informally referred to as a parameter, after the name of the attribute. However, the English word "parameter" has meanings that could apply broadly to many things in a Fortran program. Technical uses of the term in other programming languages often refer to something different, such as what are called procedure arguments in Fortran. Therefore, it is important to clarify exactly what is meant when someone uses the term informally in reference to Fortran.

The PARAMETER attribute can be specified in a type declaration statement or in a PARAMETER statement. The form of a PARAMETER statement (R538) is:

```
PARAMETER ( named-constant=initialization-expression &
    [ , named-constant=initialization-expression ] ... )
```

Rules and restrictions:

1. The PARAMETER attribute must not be specified for a dummy argument, function, or objects in a common block.
2. A named constant must have a corresponding initialization expression, specified either in a type declaration statement or a PARAMETER statement.
3. The value of the initialization expression is converted, using the rules of intrinsic assignment, to the type, type parameters, and shape of the named constant. The value must be compatible with such a conversion.
4. A named constant defined by a PARAMETER statement may appear in a subsequent type declaration statement only if that declaration confirms the implicit type.
5. A named array constant must have its array properties established previously or in the same statement as its initialization.

6. A named constant must not be referenced prior to its definition.

   Examples of named constant declarations:

- entity-oriented

```
INTEGER,PARAMETER::STATES=50
INTEGER,PARAMETER::M=MOD(28,3),&
                   NUMBER_OF_SENATORS=2*STATES
character(1), parameter :: digits(0:9) = &
        ['0', '1', '2', '3', '4', '5', '6', '7', '8', '9']
```

- attribute-oriented

```
INTEGER STATES, M, NUMBER_OF_SENATORS
PARAMETER (STATES=50)
PARAMETER (M=MOD(28,3), NUMBER_OF_SENATORS=2*STATES)
character*1 :: digits
dimension digits(0:9)
parameter (digits=['0', '1', '2', '3', '4', '5', '6', '7', '8', '9'])
```

5.7.2 Explicit Initialization

Explicit initialization specifies that a nonpointer variable has an initial value or that a pointer is initially disassociated. The initialization is done exactly once. In most implementations, the initialization happens when the program is loaded; this is probably the simplest way to think of it. An alternative implementation is to perform the initialization on first entry to a scoping unit where the variable appears. If it is necessary to reinitialize a variable on every entry, this can be accomplished with assignment statements at the beginning of the executable code for the procedure.

Explicit initialization can be specified in a type declaration statement by including initialization in an entity declaration. Initialization has one of the forms:

= initialization-expression

=> null-initialization

where null initialization is a reference to the intrinsic function NULL with no arguments. In almost all cases, null initialization will have the form NULL(), but it is possible for the intrinsic to be referenced by a different name via a USE statement with renaming.

Explicit initialization of a variable in a type declaration statement overrides any default initialization that would otherwise apply to the variable based on its type.

The same form for initialization is used for explicit initialization in a type declaration statement, parameter definition in a type declaration statement, and default initialization in a component definition statement in a derived-type definition. The following rules and restrictions apply to all these uses:

1. The form with an initialization expression is allowed only for nonpointers.
2. The form with null initialization is allowed only for pointers.

3. The value of the initialization expression is converted, using the rules of intrinsic assignment, to the type, type parameters, and shape of the entity being initialized. The value must be compatible with such a conversion.

The PARAMETER attribute may be used with initialization, but in this case the object is a named constant instead of an explicitly initialized variable; named constants are covered separately.

Explicit initialization also can be specified in a DATA statement. The form of a DATA statement is complicated enough, and the issues involved are different enough, that it is covered in a separate section.

The following rules apply to explicit initialization, whether specified in a type declaration statement or a DATA statement.

1. An object, or the same part of an object, must not be explicitly initialized more than once in a program.
2. None of the following may be explicitly initialized:
   a. a dummy argument
   b. an object made accessible by use or host association
   c. a function result
   d. an automatic object
   e. an allocatable object
   f. an object in a named common block, unless the initialization is in a block data program unit
   g. an object in a blank common block
3. If a variable is explicitly initialized, in whole or in part, that variable implicitly has the SAVE attribute unless it is in common. The implicit SAVE attribute may be confirmed by explicit declaration.

   The following are examples of explicit initialization in type definition statements:

```
CHARACTER(LEN=10)::NAME="John Doe"
INTEGER,DIMENSION(0:9)::METERS=0
TYPE (LINK), POINTER :: START => NULL( )
TYPE(PERSON)::ME=PERSON(21,"John Smith"),&
YOU=PERSON(35,"Fred Brown")

REAL::SKEW(100,100)=RESHAPE([((1.0,K=1,J-1),&
(0.0,K=J,100),J=1,100)],[100,100])
```

In these examples, the character variable NAME is initialized with the value JOHN DOE with padding on the right because the length of the constant is less than the length of the variable. All ten elements of the integer array METERS are initialized to 0. The pointer START is initially nullified. ME and YOU are structures declared using the user-defined type PERSON defined in 4.5. The two-dimensional array SKEW is initialized so that the lower triangle is 0 and the strict upper triangle is 1.

5.7.3 The DATA Statement

The DATA statement is the attribute-oriented statement for specifying explicit initialization. Unlike most of the attribute-oriented specification statements, the DATA statement provides some extra functionality that is not available in the entity-oriented form of declaration. In particular, the DATA statement allows explicit initialization of parts of an object; explicit initialization in a type declaration statement is always for the entire named object. Also, the repeat factor in the DATA statement allows some initializations to be written in substantially more compact form than that needed for the same initialization in a type declaration statement.

The full description of the DATA statement form requires several special-case rules, making it fairly complicated. The repeat factor is the root of most of the special-case issues in that, for example, 2*3 gets interpreted as 2 repetitions of the value 3 instead of a single expression with the value 6. The form of a DATA statement (R524) is:

```
DATA data-statement-object-list / data-value-list /  &
   [ [ , ] data-statement-object-list / data-value-list / ] ...
```

where a data statement object (R526) is one of:

```
variable
data-implied-do
```

and a data value (R530) is:

```
[ repeat-factor * ] data-constant
```

where a repeat factor (R531) is a scalar integer constant or a scalar integer constant subobject, and a data constant (R532) is one of:

```
scalar-constant
scalar-constant-subobject
signed-integer-literal-constant
signed-real-literal-constant
null-initialization
structure-constructor
```

Two aspects of the use of scalar-constant in this form are worth special comment. First, this is one of the very few contexts where a constant is allowed to be a BOZ literal constant (4.3.1.4). The general definition of constant includes that form, but its use is prohibited in almost all contexts. Second, the separate itemization of the signed integer and real literal constants above is needed because those forms are not included in the syntax term constant; in most contexts, this oddity of definition is invisible because a sign before a numeric literal constant is allowed in an expression. Expressions are not allowed here, so the signed cases are itemized separately.

The form of a data-implied do (R527) is:

```
( data-implied-do-object-list , named-scalar-integer-variable = &
   scalar-integer-expression , scalar-integer-expression &
   [ , scalar-integer-expression ] )
```

where a data-implied-do object (R528) is one of:

```
array-element
scalar-structure-component
data-implied-do
```

Rules and restrictions:

1. A data constant of null-initialization is allowed only for pointers.
2. Data constants other than null-initialization are allowed only for nonpointers.
3. A data constant, other than a BOZ literal constant, corresponding to a nonpointer object must be one that could be assigned to the object using an intrinsic assignment statement.
4. A BOZ literal constant used as a data constant in a DATA statement must correspond to an integer variable. The BOZ literal constant is treated as if it were an integer constant of the kind with the largest range supported by the processor.
5. An nonpointer object of derived type with default initialization must not be initialized in a DATA statement. This is because of the complications relating to the possibility of partial initialization; initialization of such an object is allowed in a type declaration statement, where partial initialization cannot happen.
6. A variable that appears in a DATA statement may appear in a subsequent type declaration statement only if that declaration confirms the implicit declaration. An array name, array section, or array element appearing in a DATA statement must have had its array properties established previously.
7. An array element or structure component that is a data-implied-do object must be a variable.
8. A DATA statement repeat factor must be positive or zero.
9. A structure constructor used as a data constant must be an initialization expression.
10. For a variable used as a data statement object in a DATA statement, each subscript, section subscript, substring starting point, or substring ending point must be an initialization expression.
11. For an array element or array structure component used as a data implied do object, each subscript, section subscript, substring starting point, or substring ending point must be an expression whose primaries are constants, subobjects of constants, or implied-do variables; each operation must be intrinsic.

12. A scalar integer expression in an implied-do must contain as operands only constants, subobjects of constants, or DO variables; each operation in it must be an intrinsic operation.

The data statement object list is expanded to form a sequence of scalar variables. An array or array section is equivalent to the sequence of its array elements in array element order. A data-implied-do is expanded to form a sequence of array elements, under the control of the implied-do variable, as in the DO construct. A zero-sized array or an implied-do with an iteration count of zero contributes no variables to the expanded list, but a scalar character variable declared to have zero length does contribute a variable to the list.

The data value list is expanded to form a sequence of scalar constant values. A DATA statement repeat factor indicates the number of times the data constant after it is to be included in the sequence. If the repeat factor is zero, the following data constant is not included in the sequence.

Scalar variables and data constants of the expanded sequence are placed in one-to-one correspondence. Each data constant specifies the initial value or status for the corresponding variable. The lengths of the two expanded sequences must be the same. Each value is converted to the type and type parameters of the corresponding object; it is the initial value for the object.

The following are examples of explicit initialization with DATA statements:

```
CHARACTER (LEN = 10) NAME
INTEGER METERS
DIMENSION METERS (0:9)
TYPE (LINK) START
POINTER START
DATA START / NULL( ) /
DATA NAME / "JOHN DOE" /, METERS / 10*0 /
TYPE (PERSON) ME, YOU
DATA ME / PERSON (21, "JOHN SMITH") /
DATA YOU % AGE, YOU % NAME / 35, "FRED BROWN" /

REAL SKEW (100, 100)
DATA ((SKEW (K, J), K = 1, J-1), J = 1, 100) / 4950 * 1.0 /
DATA ((SKEW (K,J), K=J,100), J=1,100) / 5050*0.0 /
```

The effect of these examples is identical to that of the previous examples of explicit initialization in type definition statements. The following is an example of a nonzero-sized array of zero-length characters:

```
character(len=0) :: empty_strings(3)
data empty_strings/ 3*""/
```

5.7.4 The SAVE Attribute

The SAVE attribute applies to local variables of a subprogram, module variables, and common blocks. An entity with the SAVE attribute is often referred to as being saved.

A local variable (2.4) with the SAVE attribute retains its value and its definition, association, and allocation status after the subprogram in which it is declared completes execution. When the subprogram is next invoked, that variable will have the same value and status. A local variable without the SAVE attribute becomes undefined when the subprogram completes; if it is a pointer, its association status becomes undefined; if it is allocatable, it is deallocated.

For a local variable in a recursive subprogram, the SAVE attribute has the additional effect of causing all instances of the procedure to share the same variable. A local variable without the SAVE attribute has a separate instance for each instance of the procedure.

A module variable with the SAVE attribute likewise retains its value and status in a situation where a module variable without the SAVE attribute does not. Modules are not directly executed, so the situation in question is not the completion of execution of the module. Instead, the situation is the completion of execution of all subprograms that use the module. Note that a local variable in a module procedure is not a module variable; the rules for local variables apply to it.

The SAVE attribute for a common block follows rules like those for a module variable with two distinctions. First, having an instance of a common block accessible in a subprogram plays the same role as using a module. Second, a common block is given the SAVE attribute as a whole; the SAVE attribute may not be given to an individual variable in the common block.

The SAVE attribute is most commonly implemented by allocation of static storage such that the variables in question remain in memory throughout program execution. However, such an implementation is not required by the standard. There have in the past been implementations where the saved data was stored on disk and subsequently reloaded. The difference between these implementation choices cannot be distinguished by a standard-conforming program.

Some implementations allocate most variables statically, with the result that they act as though they were saved (except for variables that are allocatable or are local to a recursive subprogram). However, code that assumes such behavior is nonstandard and nonportable, particularly with newer compilers. For portability, the SAVE attribute should always be specified where its behavior is needed, even if it appears to make no difference on some implementations.

If a variable is explicitly initialized, in whole or in part, that variable implicitly has the SAVE attribute unless it is in common. The implicit SAVE attribute may be confirmed by explicit declaration. This implied SAVE attribute is triggered only by explicit initialization—not by default initialization. If a variable or common block has the BIND attribute, it implicitly has the SAVE attribute, which may be confirmed by explicit declaration.

The SAVE attribute can be specified in a type declaration statement or in a SAVE statement. The SAVE statement is the only form for specifying the SAVE attribute for a common block. It has the form (R524):

```
SAVE [ [ : : ] saved-entity-list ]
```

where a saved entity (R544) is one of:

```
object-name
procedure-pointer-name
/ common-block-name /
```

Rules and restrictions:

1. A SAVE statement without a saved entity list is treated as though it specified all items that could be saved in the scoping unit. No other explicit SAVE statements or attributes may appear in the scoping unit.
2. The SAVE attribute is allowed in a main program, but it has no effect.
3. Specifying a common block in a main program has the same effect as saving that common block. Using a module in a main program has the same effect as saving every allowable variable of the module.
4. The following data objects must not be saved:
   a. function result
   b. a dummy argument
   c. an automatic data object
   d. an object in a common block
5. If a common block is saved in one scoping unit of a program, it must be saved in every scoping unit of the program in which it is defined (other than the main program).

   The following are examples of SAVE specifications:

- entity-oriented

```
CHARACTER(LEN=12), SAVE :: NAME
```

- attribute-oriented

```
CHARACTER(LEN=12) NAME
SAVE NAME
```

- saving objects and common blocks

```
SAVE A, B, /BLOCKA/, C, /BLOCKB/
```

5.7.5 The ASYNCHRONOUS Attribute

The ASYNCHRONOUS attribute specifies that a variable might be involved in asynchronous input/output (9.4.3, 9.5.1.3). This information facilitates compiler optimizations. If a variable is involved in asynchronous input/output, its value can be changed or referenced by an input/output process executing at the same time as the normal flow of control. Some classic optimizations cannot be safely applied to such variables.

The number of variables involved in asynchronous input/output is typically small. By identifying those variables, we allow the compiler more freedom in applying optimizations to other variables.

A variable that is accessible in multiple scoping units may have the ASYNCHRONOUS attribute in some scoping units, while not necessarily having it in others. This is because the attribute is not fundamental to the variable in isolation, but is about the relationship of the variable to currently executing code (the input/output). It is possible for there to be a some scoping units that can be in execution during the asynchronous input/output, and other scoping units that cannot be. The ASYNCHRONOUS attribute is one of the few attributes that can vary among scoping units for the same variable.

If an object has the ASYNCHRONOUS attribute, then so do all of its subobjects.

The ASYNCHRONOUS attribute can be specified in a type declaration statement or in an ASYNCHRONOUS statement. The form of an ASYNCHRONOUS statement (R521) is:

```
ASYNCHRONOUS [ [ : : ] variable-name-list ]
```

Rules and restrictions:

1. A variable must have the ASYNCHRONOUS attribute in a scoping unit if both of the following conditions hold.
    a. The variable appears in any executable statement or specification expression in the scoping unit. This condition could be informally described as the variable being used in the scoping unit, where mere declaration of the variable does not count as usage.
    b. Any statement in the scoping unit is executed while the variable is involved in asynchronous input/output.
2. Using a variable in an asynchronous input/output statement in a scoping unit implicitly confers the ASYNCHRONOUS attribute. This is a case which is evident to the compiler without the help of explicit declaration, but such a confirming explicit declaration is allowed.

    Examples of specifying the ASYNCHRONOUS attribute are:

- entity-oriented

```
REAL, ASYNCHRONOUS :: BUFFER(2048)
```

- attribute-oriented

```
ASYNCHRONOUS :: INPUT_BUFFER, OUTPUT_BUFFER
```

5.7.6 The VOLATILE Attribute

The VOLATILE attribute specifies that a variable might be used or modified by means not specified in the program. It is similar to the ASYNCHRONOUS attribute in that it identifies variables that might be involved in processes not immediately evident. However, the VOLATILE attribute is not directly related to any other Fortran language feature; it facilitates interaction with unspecified processes outside the scope of the Fortran language.

If a pointer is volatile, then the possible modifications include the pointer association status and array bounds in addition to the value of its target.

If an object has the VOLATILE attribute, then so do all of its subobjects.

The VOLATILE attribute can be specified in a type declaration statement or in a VOLATILE statement. The form of a VOLATILE statement (R548) is:

```
VOLATILE [ [ :: ] variable-name-list ]
```

Examples of specifying the VOLATILE attribute are:

- entity-oriented

```
REAL, VOLATILE :: SHARED_MEMORY_REGION(2048)
```

- attribute-oriented

```
VOLATILE :: SEMAPHORE
```

A typical application for the VOLATILE attribute is to identify a variable whose memory is shared by a separate program or by a memory-mapped hardware device. Establishing such shared memory areas is outside the scope of standard Fortran, but the VOLATILE attribute provides a standard syntax for accommodating their existence.

If a variable is volatile, the processor is expected to fetch the value from memory every time that the variable is referenced, even if a value was previously fetched and there is no evident way for the value to have changed in the interim. A simple example is:

```
subroutine wait_for_value(i)
   integer, intent(out), volatile :: i
   i = 0
   do
      if (i /= 0) return
   end do
end subroutine wait_for_value
```

Without the VOLATILE attribute, this subroutine would clearly never return and the compiler would be justified in assuming as much. With the attribute, there is the possibility that some independent process might cause the variable i to become nonzero.

Similarly, the processor is expected to store the value in memory every time that the variable is defined, even if there is no evident reference to the value or if the same value was previously stored. Independent processes might be monitoring or modifying the same memory location.

5.8 Module Entity Properties

Several attributes pertain specifically to entities in modules. The PUBLIC and PRIVATE attributes control the accessibility of an identifier in a module. The PROTECTED attribute controls how a module object may be used. The BIND attribute specifies interoperability with C. The BIND attribute for common blocks is not specific to modules, but it is covered here because it shares syntax with the BIND attribute for variables, which is specific to modules.

5.8.1 PUBLIC and PRIVATE Accessibility

The PUBLIC and PRIVATE attributes are collectively referred to as accessibility attributes. They control whether or not identifiers are accessible via use association. They apply to identifiers known in the scoping unit of a module. An identifier with the PUBLIC attribute is available outside the module by use association. An identifier with the PRIVATE attribute in a module cannot be accessed from that module by use association, but can still be accessed within the module. The identifiers are most commonly names, but can also include generic specifications, which do not have the same form as names.

Accessibility applies only to a particular identifier rather than to the entity identified. There are several ways that an entity can be known via multiple identifiers. Declaring an identifier of an entity to be PRIVATE does not inherently preclude access to the same entity via some other identifier.

The accessibility attribute of an identifier can be specified in one of several ways. It can be specified in a type declaration statement or a procedure declaration statement (5.11). The accessibility of a derived-type name can be specified in the derived-type statement (4.4.2). Accessibility of derived types involves several additional issues covered in 4.4.5. The PUBLIC and PRIVATE statements, collectively referred to as accessibility statements, can specify the accessibility of some entities that cannot be specified in any other way because they do not have a type or do not have a name; these are subroutines, generic specifiers, and namelist groups. Additionally, the accessibility statements can be used to specify the default accessibility for identifiers in a module. Forms for accessibility statements (R518) are:

```
PUBLIC [ [ :: ] access-id-list ]
PRIVATE [ [ :: ] access-id-list ]
```

where an access-id (R519) is one of:

```
use-name
generic-spec
```

A generic specification (R1207) is one of:

```
generic-name
OPERATOR ( defined-operator )
ASSIGNMENT ( = )
dtio-generic-specification
```

Generic specifications are explained in 12.5.4. Examples of accessibility statements that might be used with generic specifications are:

```
PUBLIC HYPERBOLIC_COS, HYPERBOLIC_SIN ! generic names
PRIVATE HY_COS_RAT, HY_SIN_RAT        ! specific names
PRIVATE HY_COS_INF_PREC               ! specific name
PUBLIC :: OPERATOR(.MYOP.), OPERATOR(+), ASSIGNMENT(=)
PUBLIC :: read(formatted), write(formatted)
```

Rules and restrictions:

1. Accessibility attributes may be specified only in the specification part of a module (the part above the CONTAINS).
2. A use name may be the name of a variable, procedure, derived type, named constant, or namelist group.
3. Only one accessibility statement without an access-id list is permitted in the scoping unit of a module.
4. A module may specify an accessibility attribute for an identifier that is accessed from some other module via use association. This is an exception to the general prohibition against respecifying attributes of identifiers accessed via use association (11.3.8). To understand this exception, consider accessibility not so much as a property of any entity, but more as controlling whether a particular module exports that entity. An entity might be accessible via some modules that use it, but not via others; this does not change anything about the entity itself, which had to have been PUBLIC in the module where it was declared in order for the situation to arise.

If the accessibility of a particular module entity is not explicitly specified, the module's default accessibility applies to that entity. The module's default accessibility is specified by the accessibility statement without an access-id list; if there is no such statement, the default accessibility is PUBLIC.

The following are examples of accessibility specifications:

- entity-oriented

```
REAL, PUBLIC :: GLOBAL_X
type, private :: local_data
   logical :: flag
   real, dimension (100) :: density
end type local_data
```

- attribute-oriented

```
REAL GLOBAL_X
PUBLIC GLOBAL_X
```

```
TYPE LOCAL_DATA
   LOGICAL FLAG
   REAL DENSITY (100)
END TYPE LOCAL_DATA
PRIVATE LOCAL_DATA
```

- changing the default accessibility

```
MODULE M
   PRIVATE
   REAL R, K, TEMP (100)  ! R, K, TEMP are private.
   REAL, PUBLIC :: A(100), B(100) ! A, B are public.
      . . .
END MODULE M
```

- accessibility via different names

```
module t
   private
   interface sqrt
      module procedure sqrt_for_my_type
   end interface sqrt
   interface assignment(=)
      module procedure assignment_for_my_type
   end interface assignment(=)
   public :: sqrt, assignment(=)
contains
   . . .
end module t
```

In this example, the names of the specific procedures sqrt_for_my_type and assignment_for_my_type are private. However, the procedures can be accessed outside of the module by the generic name sqrt or by assignment.

5.8.2 The PROTECTED Attribute

The PROTECTED attribute limits the ways in which a module variable may be modified. If a module variable has the PROTECTED attribute, that variable is not definable outside of that module. One could think of PROTECTED as allowing read access, but not writing. For a pointer, it is the pointer association that must not be modified.

Although the PROTECTED attribute has a surface similarity to the PRIVATE attribute, there are some fundamental differences. The obvious difference is that PROTECTED distinguishes between reading and modification. In some ways a more fundamental difference is that the PRIVATE attribute applies only to an identifier, while the PROTECTED attribute applies to the underlying entity. That is, if an entity has the PROTECTED attribute, modification of that entity outside of the module is disallowed regardless of how it is done.

A related difference is that, as with most attributes, the PROTECTED attribute for an entity may be specified only in the module where it is declared; you cannot access an entity from one module via use association in a second module and then give it the

PROTECTED attribute in the second module. The PRIVATE attribute is an exception to this general rule, but the PROTECTED attribute is not. This is because the PROTECTED attribute is considered an attribute of the underlying entity.

The PROTECTED attribute can be specified in a type declaration statement or in a PROTECTED statement. The form of a PROTECTED statement (R542) is:

```
PROTECTED [ :: ] entity-name-list
```

Rules and restrictions:

1. The PROTECTED attribute may be specified only in the specification part of a module.

2. The PROTECTED attribute is allowed only for a procedure pointer or a variable.

3. The PROTECTED attribute is not allowed for entities in a common block.

The following are examples of PROTECTED specifications:

- entity-oriented

```
REAL,PUBLIC,PROTECTED::GLOBAL_X
```

- attribute-oriented

```
REAL GLOBAL_X
PUBLIC GLOBAL_X
PROTECTED GLOBAL_X
PROCEDURE(), POINTER :: proc_ptr
PROTECTED :: proc_ptr
```

The following example illustrates the limitations established by the PROTECTED attribute.

```
module m
  integer, protected :: i
  integer, pointer, protected :: ip
contains
  subroutine set_i (value)
    integer, intent(in) :: value
    i = value
    return
  end subroutine set_i
  subroutine set_arg (arg, value)
    integer, intent(out) :: arg
    integer, intent(in) :: value
    arg = value
    return
  end subroutine set_arg
```

```
    subroutine point_ip (target)
      integer, target :: target
      ip => target
      return
    end subroutine set_arg
  end module m

  program main
    use m
    integer, target :: ip_target
    i = 1                       !-- Invalid
    call set_i(2)
    call set_arg(i, 3)          !-- Invalid
    ip => ip_target             !-- Invalid
    call point_ip(ip_target)
    ip = 4
  end program main
```

The i=1 statement is invalid because it defines i outside of the module. The call to set_i shows a valid way to define i. The call to set_arg is invalid because the actual argument i is not definable, even though the assignment statement in set_arg is in the module. Similarly, the ip=>ip_target statement is invalid pointer assignment. The call to point_ip shows a valid way to achieve that effect. The ip=4 is allowed because the PROTECTED attribute for a pointer restricts modification of the pointer association, not its value.

The PROTECTED attribute has several useful applications, although they are not illustrated by this simple example. It can be used to enforce validation and consistency of assigned values, to protect against accidental changes, and for such things as counting modifications.

5.8.3 The BIND Attribute

The BIND attribute pertains to interoperability with the C language. The BIND attribute applies to variables, common blocks, types, procedures, and procedure interfaces. The syntax described in this section applies only to variables and common blocks. The syntax for the BIND attribute for derived types is described in 4.4.2; the syntax for the BIND attribute for procedures and interfaces is described in 5.11, 12.1.1, 12.2.1 and 12.4.5. Chapter 15 covers what the BIND attribute means and how to use it.

The BIND attribute for a variable is restricted to module variables. There is no such restriction for the BIND attribute for common blocks, but the same statement can be used to specify the BIND attribute for variables and common blocks, so we describe the cases together.

The BIND attribute for a variable can be specified in a type declaration statement or in a BIND statement. The BIND statement is the only form for specifying the SAVE attribute for a common block. It has the form:

BIND (C [, NAME= scalar-character-initialization-expression]) [::] bind-entity-list where a bind entity (R523) is one of

```
variable-name
/ common-block-name /
```

1. The scalar character initialization expression in the NAME specifier of a BIND attribute specification must be of default kind.

2. All leading and trailing blanks in the value of the expression in the NAME specifier are ignored. After discarding them, the result must either have zero length or be valid as an identifier for the C processor.

3. If an entity is given the BIND attribute by a type declaration statement, the entity must be an interoperable variable. A variable named in a BIND statement must be interoperable.

4. The BIND attribute for a variable may be specified only in the specification part of a module.

5. If a common block has the BIND attribute, each variable in the COMMON block must be interoperable.

6. If a common block has the bind attribute in one scoping unit, it must have the bind attribute in every scoping unit where it is declared. The binding label (15.2) must be the same in all the scoping units.

7. If a variable or common block has the BIND attribute, it implicitly has the SAVE attribute. The implicit SAVE attribute may be confirmed by explicit declaration.

8. If a BIND attribute specification in a type declaration statement or BIND statement has a NAME specifier, the entity declaration list or bind entity list must have exactly one item. This is related to the restriction against multiple entities being associated with the same C variable with external linkage (15.2).

   The following are examples of BIND specifications:

- entity-oriented

```
REAL, BIND(C) :: X
INTEGER, BIND(C, NAME='MixedCase') :: mono_case
```

- attribute-oriented

```
BIND(C) :: X, /COM/
```

5.9 Dummy Argument Properties

The INTENT, VALUE, and OPTIONAL attributes specify properties particular to dummy arguments.

5.9.1 The INTENT Attribute

The INTENT attribute specifies the intended use of a dummy argument. If specified, it can help detect errors, provide information for readers of the program, and give the compiler information that can be used to make the code more efficient.

Some dummy arguments only provide input data for a subprogram; some are only for output from the subprogram; others may be used for both input and output. INTENT has three explicit forms: IN, OUT, and INOUT which correspond respectively to the above three situations. A fourth case, where INTENT is not explicitly specified, is a bit more complicated.

If the intent of an argument is IN, the argument must not be modified during the execution of the subprogram. For a pointer dummy argument, it is the argument's pointer association that must not be modified; for a nonpointer dummy argument, it is the argument's value. These restrictions on INTENT (IN) arguments apply broadly, having implications both at compile time and at run time. The run-time implication is that the modification is prohibited even if it happens in some other lower-level procedure; this is not in general detectable at compile time. The compile-time implication is that an INTENT (IN) dummy argument must not appear in the subprogram in a context that would cause the argument to be modified; this applies regardless of whether or not the statement that it appears in would actually get executed for any particular invocation. For nonpointers, the forbidden contexts are called variable definition contexts and are described in 16.3.1. For pointers, the forbidden contexts are:

1. A pointer object in a nullify statement.
2. The left-hand side of a pointer assignment statement.
3. An allocate object in an allocate or deallocate statement.
4. An actual argument corresponding to an INTENT (OUT) or INTENT (INOUT) pointer dummy argument.

If the intent of an argument is OUT, the argument becomes undefined on invocation of the procedure. If the argument is a pointer, its association status becomes undefined. If the argument is a nonpointer and is of a type that has default initialization, the default initialization is applied. The actual argument associated with an INTENT (OUT) dummy must be definable.

For an INTENT (OUT) dummy, any previous value of the actual argument is irrelevant. If there is any situation in which you want to leave the previous value of the actual argument unchanged or reference that value in any way, then INTENT (OUT) is the wrong choice. Even if no executable statement in the subroutine refers to the dummy argument, the actual argument does not retain its previous value; the actual argument will become undefined except for components that get default initialization. Being undefined means that any program that references the value is nonstandard. Actual implementations might realistically leave the value unchanged, change it to random garbage, or detect an error.

If the intent is INOUT, the argument may be used to communicate information to the subprogram and return information. As with INTENT (OUT), the corresponding actual argument is required to be definable. The difference between INTENT (OUT) and INTENT (INOUT) is that an INTENT (INOUT) dummy acquires its starting definition status and value from that of the actual argument; it does not become undefined or have default initialization applied.

An unspecified intent is similar to INTENT (INOUT), with just one subtle distinction. For INTENT (INOUT), the actual argument is required to be definable. For unspecified intent, the actual argument is required to be definable if execution of the procedure causes definition or undefinition of the dummy. This is a run-time requirement that applies independently to each invocation of the procedure. There can be some invocations that trigger the requirement and other invocations of the same procedure that do not. The following illustrates a trivial case of this:

```
program illustrate_unspecified
   real :: x

   call maybe_set(.false., 4.567)
   call maybe_set(.true., x)
end program

subroutine maybe_set (set, x)
   logical, intent(in) :: set
   real :: x

   if (set) x = 1.23
   print *, x
end subroutine
```

This is valid because the first call to the subroutine does not cause definition or undefinition of the dummy x. The second call does cause such definition, but the actual argument for that call is definable, so it is ok. Using unspecified intent makes it difficult for the compiler to diagnose some kinds of problems; it is usually recommended to specify intent explicitly in new code and to regard unspecified intent as primarily a compatibility feature for old codes.

The INTENT attribute can be specified in a type declaration statement or in a INTENT statement. The form of a INTENT statement (R536) is:

```
INTENT ( intent-spec ) [ :: ] dummy-argument-name-list
```

where an intent specification is IN, OUT, or INOUT.

Rules and restrictions:

1. The INTENT attribute may be specified only for a dummy argument.

2. An intent must not be specified for a procedure unless it is a procedure pointer. This is because the concepts of definition and modification do not apply to procedures other than procedure pointers.

   The following are examples of INTENT specifications:

- entity-oriented

```
SUBROUTINE MOVE (FROM, TO)
   USE PERSON_MODULE
   TYPE (PERSON), INTENT (IN) :: FROM
   TYPE (PERSON), INTENT (OUT) :: TO
      . . .
SUBROUTINE SUB (X, Y)
   INTEGER, INTENT (INOUT) :: X, Y
      . . .
```

- attribute-oriented

```
SUBROUTINE MOVE (FROM, TO)
   USE PERSON_MODULE
   TYPE (PERSON) FROM, TO
   INTENT (IN) FROM
   INTENT (OUT) TO
      . . .
SUBROUTINE SUB (X, Y)
   INTEGER X, Y
   INTENT (INOUT) X, Y
      . . .
```

5.9.2 The VALUE Attribute

The VALUE attribute specifies a form of argument association for a dummy argument. The dummy argument is not associated with the actual argument itself, but rather with an anonymous temporary variable. The initial value of this temporary variable is taken from the value of the actual argument. The dummy argument's value may be modified during execution of the procedure (unless the dummy also has the INTENT (IN) attribute), but such modifications affect only the temporary variable—not the actual argument.

Although C interoperability was a major motivation for the VALUE attribute, it has utility independent of C in situations where there is a need to modify the dummy argument's value without having such modifications change the actual argument's value, or even where it would not be allowed to change the actual argument's value. Without the VALUE attribute, such situations would require that the programmer explicitly copy the dummy argument to a temporary variable. The VALUE attribute makes such a copy automatic and transparent.

The VALUE attribute can be specified in a type declaration statement or in a VALUE statement. The form of a VALUE statement (R547) is:

VALUE [::] dummy-argument-name-list

Rules and restrictions:

1. The VALUE attribute may be specified only for a dummy data argument.

2. For a variable with the VALUE attribute, any length type parameter values must either be specified by initialization expressions or be omitted (in which case they would take their default values).

   The following are examples of VALUE specifications:

- entity-oriented

```
subroutine sub(x)
   real, value :: x
      . . .
```

- attribute-oriented

```
subroutine sub(x)
   real :: x
   value :: x
      . . .
```

5.9.3 The OPTIONAL Attribute

The OPTIONAL attribute for a dummy argument specifies that a procedure reference may omit the corresponding actual argument. The PRESENT intrinsic function can be used to test whether the actual argument was or was not present in a particular invocation of the procedure.

The syntax for referencing a procedure with omitted optional arguments is presented in 12.6.2.

The OPTIONAL attribute can be specified in a type declaration statement or in an OPTIONAL statement. The form of an OPTIONAL statement (R537) is:

`OPTIONAL [ :: ]` dummy-argument-name-list

Rules and restrictions:

1. The OPTIONAL attribute may be specified only for dummy arguments.

   The following are examples of OPTIONAL specifications:

- entity-oriented

```
INTEGER, INTENT (IN), OPTIONAL :: SIZEX
LOGICAL, INTENT (IN), OPTIONAL :: FAST
```

- attribute-oriented

```
OPTIONAL SIZEX, FAST
```

Argument optionality is useful in several situations. An argument might be irrelevant to some invocations of a procedure. A procedure might have several output arguments, some of which are not needed from a particular invocation. Although there is no direct mechanism for specifying a default value for an omitted argument, the effect of a default value can be achieved by using a local variable in the procedure, as in the following example.

```
subroutine do_something(...other arguments..., tolerance)
   real, optional, intent(in) :: tolerance
      . . .
   real :: tolerance_local
      . . .
   if (present(tolerance)) then
      tolerance_local = tolerance
   else
      tolerance_local = 0.001
   end if
      . . .
```

Such default values allow a procedure to accommodate common simple situations with simple references to the procedure, while still allowing detailed specification when needed.

The presence of an optional argument can also be used like a logical input variable to select an option in the code. This is most natural when applied to an argument that has data needed for one option, but irrelevant to the other as illustrated in the following example:

```
subroutine minimize(tolerance)
   real, optional, intent(in) :: tolerance
   if (present(tolerance)) then
      call full_method(tolerance)
   else
      call simple_method
   end if
      . . .
```

If an optional argument is not present for a particular invocation of a procedure, that argument is subject to the restrictions detailed in 12.6.2.

5.10 Procedure Properties

The EXTERNAL and INTRINSIC attributes are particular to procedures. Procedures can have other attributes; for example, a function can have a dimension.

5.10.1 The EXTERNAL Attribute

The EXTERNAL attribute specifies that an entity is an external procedure, dummy procedure, procedure pointer, or block data subprogram.

The terminology is historical and unfortunately misleading in that entities other than external procedures also have the EXTERNAL attribute. For example, just because a dummy procedure has the external attribute, that does not imply that it is an external procedure or that the corresponding actual argument has to be one; the actual argument could be an intrinsic or module procedure as well.

The simplest use for declaration of the EXTERNAL attribute is to distinguish an EXTERNAL procedure from any possible intrinsic procedure of the same name. This is the use where the terminology for the attribute makes most sense. If you attempt to reference a procedure without declaring the EXTERNAL attribute, but there is an intrinsic procedure of the same name, the reference will be to the intrinsic procedure instead. This can happen even for vendor-defined intrinsic procedures—not just the standard ones. It can also happen if new versions of the standard add new intrinsic procedures. Thus, referencing an external procedure without declaring the attribute is a potential portability problem.

Declaring the EXTERNAL attribute for an external procedure or dummy procedure allows it to be used in contexts where it would not otherwise be evident that it was a procedure instead of a data object. Declaring the EXTERNAL attribute for a procedure pointer is always required.

Declaring the EXTERNAL attribute for a block data subprogram does not directly affect the interpretation of a Fortran program, but can be useful as a hint to the system that the block data subprogram should be included as part of the program. There is no other mechanism within the language to specify this because block data subprograms are never referenced. However, the process of building a program is outside the scope of the standard, so a processor is not required to make use of such a hint. In building a program that makes use of block data, it is prudent to verify that one understands how to ensure that the particular processor includes the block data in the program.

The EXTERNAL attribute can be declared by several means. It can be declared by a procedure declaration statement; that option can be used in all cases except for block data, which is not a procedure. It can be declared by an EXTERNAL attribute in a type declaration statement; that option can be used only with functions because subroutines and block data subprograms do not have types. The form does not define an explicit interface (12.5.1) and so cannot be used in situations where an explicit interface is required (12.5.1.2). The EXTERNAL statement (R1210) provides an attribute-oriented form for declaring the attribute. It can be used for subroutines and block data program units as well as functions. It also does not provide an explicit interface. It has the form:

```
EXTERNAL [ :: ] external-name-list
```

Rules and restrictions:

1. If a dummy argument has the EXTERNAL attribute, it is a dummy procedure (and possibly also a procedure pointer).

2. If a pointer has the EXTERNAL attribute, it is a procedure pointer (and possibly also a dummy procedure).

3. If a procedure that is neither a dummy argument nor a pointer has the EXTERNAL attribute, it is an external procedure or a block data subprogram.

4. If an external procedure or dummy procedure is used as an actual argument or as a target in a procedure pointer assignment, then it must be explicitly declared to have the EXTERNAL attribute.

In some cases, explicit declaration is needed to establish the interpretation of the code, but the requirements mandate explicit declaration even in some cases where there is no ambiguity. For example, the main program

```
program one
   external :: s
   call t(s)
end
```

would have a different meaning without the EXTERNAL statement; s would be an implicitly declared real variable instead of a procedure. But in the program

```
program two
   external :: s
   call s
   call t(s)
end
```

the first call statement makes it unambiguous that s must be a subroutine; nonetheless, the declaration is still required.

The rules for resolving procedure references, including the effects of the EXTERNAL attribute, are in 12.8.

The following are examples of EXTERNAL specifications using type declaration statements and EXTERNAL statements:

- entity-oriented

```
SUBROUTINE SUB (FOCUS)
   INTEGER, EXTERNAL :: FOCUS
   LOGICAL, EXTERNAL :: SIN
```

- attribute-oriented

```
SUBROUTINE SUB (FOCUS)
   INTEGER FOCUS
   LOGICAL SIN
   EXTERNAL FOCUS, SIN
```

FOCUS is declared to be a dummy procedure. SIN is declared to be an external procedure. Both are functions. The intrinsic function SIN is no longer available by that name in subroutine SUB.

5.10.2 The INTRINSIC Attribute

The INTRINSIC attribute specifies that a name is the name of an intrinsic function. It may be either a standard intrinsic or a vendor-defined intrinsic. Of course, a program that uses a vendor-defined intrinsic might not be portable to other vendor's compilers.

Specifying the INTRINSIC attribute is required in order to use an intrinsic procedure as an actual argument. Other specifications of the attribute are largely for documentation or for confirmation that a particular intrinsic exists.

The INTRINSIC attribute can be declared in a type declaration statement; that option can be used only with functions because subroutines do not have types.

The INTRINSIC statement (R1216) provides an attribute-oriented form for specifying the attribute. It can be used for both subroutines and functions. Its form is:

```
INTRINSIC [ :: ] intrinsic-procedure-name-list
```

Rules and restrictions:

1. Each intrinsic procedure name must be the name of an intrinsic procedure.
2. If an intrinsic procedure is used as an actual argument, it must be specified to have the INTRINSIC attribute. This is allowed only for the specific intrinsic names listed in 13.4.
3. Specifying a type for a generic intrinsic function is allowed, but has no effect. It does not remove the generic properties of the function name. The function can still be referenced generically with any other types for which it is defined.
4. Specifying a type for a specific intrinsic function that has a name different from the generic name is allowed as long as it confirms the type that the specific function has anyway; it is never required.

The rules for resolving procedure references, including the effects of the INTRINSIC attribute, are in 12.8.

The following are examples of INTRINSIC specifications:

- entity-oriented

```
REAL, INTRINSIC :: SIN, COS
```

- attribute-oriented

```
REAL SIN, COS
INTRINSIC SIN, COS
```

5.11 The Procedure Declaration Statement

The procedure declaration statement is an entity-oriented form for declaring procedures. It can be used for procedure pointers, dummy procedures, and external procedures. It is particularly convenient for declaring multiple procedures that have the same abstract interface because it avoids the need to replicate the interface body.

The form of a procedure declaration statement (R1211) is:

```
PROCEDURE ( [ interface-spec ] ) [ [ , procedure-attribute-spec ] ... :: ] &
   procedure-declaration-list
```

where an interface specification is one of:

```
interface-name
declaration-type-spec
```

a procedure attribute specification (R1213) is one of:

```
BIND ( C [ , NAME = scalar-character-initialization-expression ] )
INTENT ( intent-spec )
OPTIONAL
POINTER
PRIVATE
PUBLIC
SAVE
```

and a procedure declaration (R1214) has the form:

```
procedure-entity-name [ initialization ]
```

The interface of the declared procedures is specified by the interface specification. There are three possibilities

1. If the interface specification is omitted, the procedures are declared to have implicit interfaces; the statement does not specify whether they are subroutines or functions.

2. If the interface specification is a declaration type specification, the procedures are declared to be functions with implicit interfaces and with the specified type and type parameters.

3. If the interface specification is an interface name, the procedures are declared to have the specified explicit interface. The interface name must either be the name of an abstract interface or of a procedure that has an explicit interface. If it is the name of a procedure, the abstract interface of that procedure is used.

Except for the BIND attribute, the meanings of and restrictions on the procedure attribute specifications are the same as those for attribute specifications in a type declaration statement or attribute specification statement. The meaning of the BIND attribute for procedures is discussed in 15.6.1.

Rules and restrictions:

1. A procedure name used as an interface name must not be declared in a subsequent procedure declaration statement. This restriction avoids the circularity of declaring a procedure x to have the same interface as y, while also declaring y to have the same interface as x.

2. An intrinsic procedure name used as an interface name must be one of the specific names not marked with an asterisk in 13.4.

3. An elemental explicit interface may be specified only for external procedures; procedure pointers and dummy procedures must not be elemental.

4. If a procedure entity has initialization or the INTENT or SAVE attribute, it must be a pointer.

5. The scalar character initialization expression in the NAME specifier of a BIND attribute specification must be of default kind.

6. All leading and trailing blanks in the value of the expression in the NAME specifier are ignored. After discarding them, the result must either have zero length or be valid as an identifier for the C processor.

7. If the BIND attribute is specified, the procedure must have an interoperable explicit interface.

8. If there is a BIND attribute with a NAME specifier, the procedure declaration list must consist of a single external procedure name.

The following example illustrates procedure declaration statements.

```
subroutine sub(arg)
  abstract interface
    function real_func (x)
      real, intent(in) :: x
      real :: real_func
    end function real_func
  end interface

  procedure(real_func) :: arg
  procedure(arg), pointer : p, q
  procedure(real) :: ext, sqrt
```

The dummy procedure arg is declared to have abstract interface real_func. The procedure pointers p and q are declared to have the same abstract interface as arg. The external procedures ext and sqrt are declared to be implicit interface functions returning reals. The intrinsic function sqrt is no longer available by that name in this scope.

5.12 Attribute Compatibility

No single entity can possess all of the attributes because some attributes are incompatible with others. For example, OPTIONAL is an attribute that can be applied only to dummy arguments, and dummy arguments must not have the SAVE attribute. Table 5-2 shows which attributes may be used together to specify an entity.

Many of the incompatibilities indicated in the table are for attributes that apply only to limited categories of entities. For example, the INTENT, OPTIONAL, and VALUE attributes apply only to dummy arguments, while the PRIVATE, PROTECTED, and PUBLIC attributes apply only to module entities. A dummy argument cannot be a module entity, so those two sets of attributes are incompatible.

Table 5-2 Attribute compatibility

	init	alloc	async	bind	dim	extern	intent	intrin	opt	param
init		x	ok	ok	ok	ok	x	x	x	ok
allocatable	x		ok	x	ok	ok	ok	x	ok	x
asynchronous	ok	ok		ok	ok	x	ok	x	ok	x
bind	ok	x	ok		ok	ok	x	x	x	x
dimension	ok	ok	ok	ok		ok	ok	x	ok	ok
external	ok	ok	x	ok	ok		ok	x	ok	x
intent	x	ok	ok	x	ok	ok		x	ok	x
intrinsic	x	x	x	x	x	x	x		x	x
optional	x	ok	ok	x	ok	ok	ok	x		x
parameter	ok	x	x	x	ok	x	x	x	x	
pointer	ok	x	ok	x	ok	ok	ok	x	ok	x
private	ok	ok	ok	ok	ok	ok	x	ok	x	ok
protected	ok	ok	ok	ok	ok	ok	x	x	x	x
public	ok	ok	ok	ok	ok	ok	x	ok	x	ok
save	ok	ok	ok	ok	ok	ok	x	x	x	x
target	ok	ok	ok	ok	ok	x	ok	x	ok	x
value	x	x	ok	x	ok	x	ok	x	ok	x
volatile	ok	ok	ok	ok	ok	ok	ok	x	ok	x

	ptr	priv	prot	public	save	target	value	volat
init	ok	ok	ok	ok	ok	ok	x	ok
allocatable	x	ok	ok	ok	ok	ok	x	ok
asynchronous	ok	ok	ok	ok	ok	ok	ok	ok
bind	x	ok	ok	ok	ok	ok	x	ok
dimension	ok	ok	ok	ok	ok	ok	ok	ok
external	ok	ok	ok	ok	ok	x	x	ok
intent	ok	x	x	x	x	ok	ok	ok
intrinsic	x	ok	x	ok	x	x	x	x
optional	ok	x	x	x	x	ok	ok	ok
parameter	x	ok	x	ok	x	x	x	x

Table 5-2 Attribute compatibility

	init	alloc	async	bind	dim	extern	intent	intrin	opt	param
pointer		ok	ok	ok	ok	x	x	ok		
private	ok		ok	x	ok	ok	x	ok		
protected	ok	ok		ok	ok	ok	x	ok		
public	ok	x	ok		ok	ok	x	ok		
save	ok	ok	ok	ok		ok	x	ok		
target	x	ok	ok	ok	ok		ok	ok		
value	x	x	x	x	x	ok		x		
volatile	ok	ok	ok	ok	ok	ok	x			

The table shows attributes as incompatible only if there is no circumstance where they can be used together. In some cases where attributes are shown as compatible, there are limitations on the compatibility. This is particularly so for the EXTERNAL and INTENT attributes because those two attributes have multiple possible meanings.

The EXTERNAL attribute can apply to an external procedure, a dummy procedure, a procedure pointer, or block data. The initialization, PROTECTED, SAVE, and VOLATILE attributes are compatible with a procedure pointer, but not with the other possible meanings of the EXTERNAL attribute. An additional subtlety of the EXTERNAL attribute relates to the fact that there are several ways to specify it; two of those ways are with an interface body or a procedure declaration statement, neither of which use the EXTERNAL keyword. The ALLOCATABLE and DIMENSION attributes are compatible with the EXTERNAL attribute, but not with the EXTERNAL keyword because they require explicit interfaces.

The INTENT attribute has the forms INTENT (IN), INTENT (OUT), and INTENT (INOUT). The VALUE attribute is compatible with INTENT (IN), but not with the other two forms. Conversely, the VOLATILE attribute is compatible with INTENT (OUT) and INTENT (INOUT), but not with INTENT (IN).

5.13 The NAMELIST Statement

A NAMELIST statement establishes the name for a collection of objects that can then be referenced by the group name in input/output statements (9.4.2). The form of the NAMELIST statement (R552) is:

```
NAMELIST / namelist-group-name / variable-name-list &
    [ [ , ] / namelist-group-name / variable-name-list ] ...
```

Rules and restrictions:

1. A variable in the variable name list must not be an assumed-size array.

2. If a namelist group name has the PUBLIC attribute, no item in the namelist group object list may have the PRIVATE attribute.

3. The order in which the variables are specified in the NAMELIST statement determines the order in which the values appear on output. Multiple specifications of the same variable are allowed, in which case its value will appear multiple times.

4. A namelist group name may occur in more than one NAMELIST statement in a scoping unit. The variable list following each successive appearance of the same namelist group name in a scoping unit is treated as a continuation of the list for that namelist group name.

5. A variable may be a member of more than one namelist group.

6. A variable that is not accessed by use or host association must have its type, type parameters, and shape specified previously in the same scoping unit, or must be determined by implicit typing rules. If a variable is typed by the implicit typing rules, its appearance in any subsequent type declaration statement must confirm this implicit type and type parameters.

   Examples of NAMELIST statements are:

```
NAMELIST / N_LIST / A, B, C
NAMELIST / S_LIST / A, V, W, X, Y, Z
```

5.14 Storage Association

In general, the physical storage and storage order for data objects are not specified. However, the COMMON, EQUIVALENCE, and SEQUENCE statements and the BIND attribute provide sufficient control over the order and layout of storage units to permit data to share storage units.

Prior to Fortran 90, storage association was a fundamental feature for sharing data and managing storage. Almost all large programs, as well as many small ones, made extensive use of storage association.

In modern Fortran, modules and dynamic allocation provide tools for sharing data and managing storage. These tools are often more effective and have fewer subtleties and complications than storage association. However there remain situations where storage association is still a useful concept.

The concept of storage association involves storage units and storage sequence. These concepts are used to explain how the COMMON and EQUIVALENCE mechanisms work. This description does not imply that any particular memory allocation scheme is required by a Fortran system, but the system must function as though storage were actually managed according to these descriptions.

Storage association is based on sequences of storage units. These concepts are discussed in 16.2.3.

5.14.1 The EQUIVALENCE Statement

To indicate that two or more variables are to share storage, they may be placed in an equivalence set in an EQUIVALENCE statement. If the objects in an equivalence set have different types or type parameters, no conversion or mathematical relationship is implied. If a scalar and an array are equivalenced, the scalar does not have array properties and the array does not have the properties of a scalar. The form of the EQUIVALENCE statement (R554) is:

```
EQUIVALENCE equivalence-set-list
```

where an equivalence set (R555) is:

```
( equivalence-object , equivalence-object-list )
```

and an equivalence object (R556) is one of:

```
variable-name
array-element
substring
```

Rules and restrictions:

1. An equivalence object must not be a designator with a base object (6.2) that is:

   a dummy argument
   a pointer
   an allocatable variable
   a nonsequence structure
   a structure with an allocatable ultimate component
   a structure with a pointer at any level
   an automatic object
   a function name, result name, or entry name
   a variable with the BIND attribute
   a variable in a common block with the BIND attribute
   a named constant
   accessed by use association

2. An equivalence object must not be a designator with more than one part reference.

3. Any subscript or substring range must be an integer initialization expression.

4. If an equivalence object is of type default integer, default real, double precision real, default complex, default logical, or numeric sequence type, all of the objects in the set must be of one of these types.

5. If an equivalence object is of default character or character sequence type, all of the objects in the set must be of these types. The lengths do not need to be the same.

6. If an equivalence object is of sequence type other than numeric or character sequence type, all of the objects in the set must be of the same type with the same type parameter values.

7. If an equivalence object is of intrinsic type other than default integer, default real, double precision real, default complex, default logical, or default character, all of the objects in the set must be of the same type with the same kind type parameter value.

8. If an equivalence object has the PROTECTED attribute, all of the objects in the set must have the PROTECTED attribute.

9. An EQUIVALENCE statement must not specify that the same storage unit is to occur more than once in a storage sequence or that consecutive storage units are to be nonconsecutive. For example, the following is illegal because it would indicate that storage for X(2) and X(3) is shared.

```
EQUIVALENCE (A, X (2)), (A, X (3))
```

10. An equivalence object must not have the TARGET attribute.

11. A substring must not be zero length.

An EQUIVALENCE statement specifies that the storage sequences of the data objects in an equivalence set have storage sequences (16.2.3.1) with the same initial point. This causes storage association of the objects in the set and may cause storage association of other data objects.

Note that the effect of equivalencing a zero-sized array with two nonzero-sized objects is to equivalence the two nonzero-sized objects. For example,

```
INTEGER A(5), B(0), X
EQUIVALENCE (B, A(2)), (B, X)
```

causes X and A(2) to share the same storage unit.

The restriction in item 1 against nonsequence structures seems anomalous in that it disallows structures with the BIND attribute. As discussed in 4.4.10, sequence and BIND types share many of the same properties. Most contexts that allow sequence types also allow BIND types. This is one of the rare exceptions. This exception is particularly anomalous because BIND types are allowed in COMMON. COMMON and EQUIVALENCE are closely related in that both establish sequence association. It is possible to use COMMON in a roundabout way to get the effect of equivalencing BIND types.

As an example of equivalence:

```
CHARACTER (LEN = 4) :: A, B
CHARACTER (LEN = 3) :: C (2)
EQUIVALENCE (A, C (1)), (B, C (2))
```

causes the alignment illustrated below:

A(1:1)	A(2:2)	A(3:3)	A(4:4)			
			B(1:1)	B(2:2)	B(3:3)	B(4:4)
C(1)(1:1)	C(1)(2:2)	C(1)(3:3)	C(2)(1:1)	C(2)(2:2)	C(2)(3:3)	

As a result, the fourth character of A, the first character of B, and the first character of C(2) all share the same character storage unit.

```
REAL, DIMENSION (6) :: X, Y
EQUIVALENCE (X (5), Y(3))
```

causes the alignment illustrated below:

X(1)	X(2)	X(3)	X(4)	X(5)	X(6)		
		Y(1)	Y(2)	Y(3)	Y(4)	Y(5)	Y(6)

The statements

```
character :: string*8
character :: array(8)
equivalence (string,array)
```

illustrate equivalencing a character string and a character array.

For rules on the interaction of equivalence and default initialization, see 16.2.3.2.

5.14.2 The COMMON Statement

The COMMON statement establishes blocks of storage called **common blocks** and specifies objects that are contained in the blocks. Two or more program units may share this space and thus share the values of variables stored in the space. Thus, the COMMON statement provides a global data facility based on storage association. A common block may be named, in which case it is called a **named common block**, or may be unnamed, in which case it is called **blank common**.

A common block may contain a mixture of storage units and may contain unspecified storage units; however, if a common block contains a mixture of storage units, every declaration of the common block in the program must contain the same sequence of storage units. The form of the COMMON statement (R557) is:

```
COMMON [ / [ common-block-name ] / ] common-block-object-list &
    [ [ , ] / [ common-block-name ] / common-block-object-list ] ...
```

where a common block object (R558) is one of:

```
variable-name [ ( explicit-shape-spec-list ) ]
procedure-pointer-name
```

Rules and restrictions:

1. A common block object must not be:

   a dummy argument
   an allocatable variable
   a structure with an allocatable ultimate component
   an automatic object
   a function name, result name, or entry name
   a variable with the BIND attribute
   accessed by use association

2. If a common block object is of derived type, the type must either be a sequence type or a type with the BIND attribute. In either case, the type must have no default initialization.

3. The object list following a common block name declares objects in the common block of that name. The object list following two slashes with no common block name between them or an object list with no preceding slashes declares objects in blank common.

4. A common block name or an indication of blank common may appear more than once in one or more COMMON statements in the same scoping unit. The object list following each successive block name or blank common indication is treated as a continuation of the previous object list.

5. An object may appear in only one common block within a scoping unit.

6. The DIMENSION attribute for an array can be declared by an explicit-shape specification list in a COMMON statement. Alternatively, the DIMENSION attribute for a variable in common may be specified by other statements as described in 5.4.2. Pointer arrays are allowed in common, but because they must have a deferred shape, their DIMENSION attribute must be specified by other statements. (The reason for this restriction is not clear; it might be accidental.) Because automatic objects are not allowed in common, each bound in the explicit-shape specification list must be an initialization expression.

7. A nonpointer object of type default integer, default real, double precision real, default complex, default logical, or numeric sequence type must become associated only with nonpointer objects of these types.

8. A nonpointer object of type default character or character sequence must become associated only with nonpointer objects of these types.

9. A nonpointer object of a type and kind not listed in the previous two rules must become associated only with nonpointer objects of the same type and type parameter values.

10. A pointer must become storage associated only with pointers of the same type, type parameters, and rank.

11. An object with the TARGET attribute must become storage associated only with objects that have the TARGET attribute and the same type and type parameters.

For each common block, a common block storage sequence is formed. It consists of the sequence of storage units of all the variables listed for the common block in the order of their appearance in the common block list. The storage sequence may be extended on the end to include the storage units of any variable equivalenced to a variable in the common block. Similar extension at the beginning is not allowed. Data objects that are storage associated with a variable in a common block are considered to be in that common block.

The following examples illustrate the distinction between extension at the end and the beginning.

```
COMMON A(5)
REAL B(5)
EQUIVALENCE (A(2), B(1))
```

is legal and results in the following alignment::

A(1)	A(2)	A(3)	A(4)	A(5)	
	B(1)	B(2)	B(3)	B(4)	B(5)

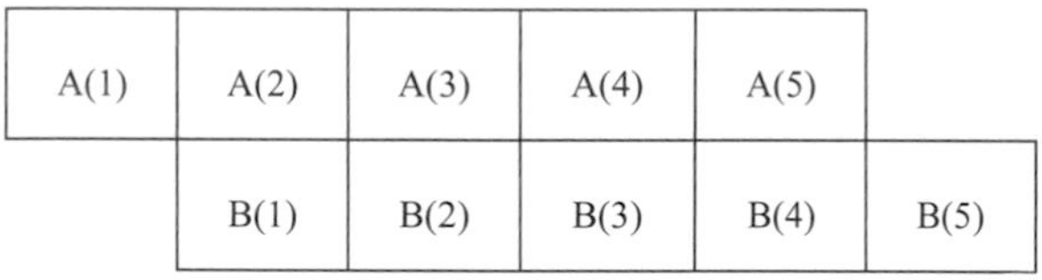

On the other hand, the following is not legal:

```
EQUIVALENCE (A(1), B(2))
```

because it would place B(1) ahead of A(1) as in the following alignment:

	A(1)	A(2)	A(3)	A(4)	A(5)
B(1)	B(2)	B(3)	B(4)	B(5)	

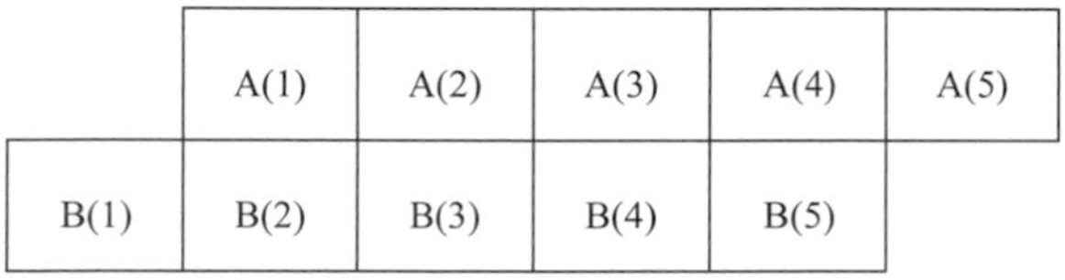

and a common block must not be extended from the beginning of the block.

Equivalence association must not cause two different common blocks to become associated.

The size of a common block is the size of its storage sequence including any extensions of the sequence resulting from equivalence association.

Zero-sized common blocks are permitted. Frequently a program is written with array extents and character lengths specified by named constants. When there is a need for a different-sized data configuration, the values of the named constants can be changed and the program recompiled. Allowing extents and lengths to be specified to have the value zero, and thus possibly specifying zero-length common blocks, permits the maximum generality.

Within a program, all named common blocks with the same name must have the same size. If that size is zero, the common blocks with that name are associated with one another. If that size is nonzero, the storage sequences of the common blocks with that name all have the same first storage unit. Note that the storage sequence association does not depend on the variable names.

Corresponding rules apply to blank common blocks except that they may be of different sizes. Thus, it is possible for a zero-sized blank common block in one scoping unit to be associated with the first storage unit of a nonzero-sized blank common block in another scoping unit.

A blank common block has the same properties as a named common block except for the following:

1. Variables with explicit initialization are allowed in named common in a block data program unit. Variables with explicit initialization are never allowed in blank common.

2. Blank common is, in effect, always saved, even though it cannot be specified in a SAVE statement and the standard does not use that terminology. A named common block is not saved unless it is mentioned in a SAVE statement.

3. Named common blocks of the same name must be the same size in all scoping units of a program. Blank common blocks may be of different sizes.

The following is an example of common block usage:

```
SUBROUTINE FIRST

   INTEGER, PARAMETER :: SHORT = 2
   REAL B(2)
   COMPLEX C
   LOGICAL FLAG
   TYPE COORDINATES
      SEQUENCE
      REAL X, Y
      LOGICAL Z_O
   END TYPE COORDINATES
   TYPE (COORDINATES) P
   COMMON / REUSE / B, C, FLAG, P
```

```
      REAL MY_VALUES (100)
      CHARACTER (LEN = 20) EXPLANATION
      COMMON / SHARE / MY_VALUES, EXPLANATION
      SAVE / SHARE /

      REAL, POINTER :: W (:, :)
      REAL, TARGET, DIMENSION (100, 100) :: EITHER, OR
      INTEGER (SHORT) :: M (2000)
      COMMON / MIXED / W, EITHER, OR, M
         . . .

   SUBROUTINE SECOND

      INTEGER, PARAMETER :: SHORT = 2
      INTEGER I(8)
      COMMON / REUSE / I

      REAL MY_VALUES (100)
      CHARACTER (LEN = 20) EXPLANATION
      COMMON / SHARE / MY_VALUES, EXPLANATION
      SAVE / SHARE /

      REAL, POINTER :: V (:)
      REAL, TARGET, DIMENSION (10000) :: ONE, ANOTHER
      INTEGER (SHORT) :: M (2000)
      COMMON / MIXED / V, ONE, ANOTHER, M   ! ILLEGAL
           . . .
```

Common block REUSE has a storage sequence of 8 numeric storage units. It is being used to conserve storage. The storage referenced in subroutine FIRST is associated with the storage referenced in subroutine SECOND as shown below:

B(1)	B(2)	C		FLAG	X	Y	Z_O
I(1)	I(2)	I(3)	I(4)	I(5)	I(6)	I(7)	I(8)

There is no guarantee that the storage is actually retained and reused because, in the absence of a SAVE attribute for REUSE, some processors may release the storage when either of the subroutines completes execution.

Common block SHARE contains both numeric and character storage units and is being used to share data between subroutines FIRST and SECOND.

The declaration of common block MIXED in subroutine SECOND is illegal because it does not have the same sequence of storage units as the declaration of MIXED in subroutine FIRST. The array pointer in FIRST has two dimensions; the array pointer in SECOND has only one. Pointers must match in type, kind and rank in order to have the same storage units.

6 Using Data

- A **Data Object** may be a variable, a constant, or a subobject of a constant. It may be a scalar or an array. A subobject of a data object is also a data object.
- The value and properties of a **Variable** may change during program execution.
- A **Constant** has a specified value and cannot be changed. It may be a literal constant or a named constant.
- A **Scalar** has a rank of zero. A scalar object that is defined has a single value from the set of values permitted for its type.
- A **Structure** is an object of derived type. Although it consists of parts specified by the components of the type definition, a defined scalar structure has a single value made up of its component values.
- An **Array** is a set of scalar elements of the same type and type parameters. The rank is the number of dimensions and may be between one and seven.
- A **Designator** is used to identify a data object. It may identify a whole object or a part of an object, such as a substring, structure component, array element, or array section.
- A **Substring** is a contiguous portion of a character string that has a starting point and an ending point within the string.
- A **Structure Component** is part of an object of derived type.
- An **Array Element** is one of the scalar elements that make up an array. The element is selected by a subscript list.
- An **Array Section** is a selected set of elements of an array. The elements are selected by a section subscript list that consists of subscripts, triplet subscripts, or vector subscripts. The subscript list must contain one or more triplet or vector subscripts.
- The **ALLOCATE Statement** may be used to create space for allocatable variables and pointer targets.
- The **NULLIFY Statement** or the **NULL Intrinsic Function** may be used to disassociate a pointer from any target.
- The **DEALLOCATE Statement** releases the space allocated for an allocatable variable or a pointer target and nullifies the pointer.

Chapter 5 explains how data objects and their attributes are specified. Chapter 6 goes further and explains how these objects can be used. A designator is used to identify a data object. The appearance of the designator where its value is required is a **reference**

J.C. Adams et al., *The Fortran 2003 Handbook*,
DOI: 10.1007/978-1-84628-746-6_6, © Springer-Verlag London Limited 2009

to the object. When an object is referenced, it must be **defined**; that is, it must have a value. The reference makes use of the value. For example:

```
A = 1.0
B = A + 4.0
```

In the first statement, the constant value 1.0 is assigned to the variable A. It does not matter whether A was previously defined with a value or not; it now has a value and can be referenced in an executable statement. In the second statement, A is referenced; its value is obtained and added to the value 4.0 to obtain a value that is then assigned to the variable B. The appearances of A in the first statement and B in the second statement are not references because their values are not required. The appearance of A in the second statement is a reference.

A data object may be a constant or a variable. If it is a constant, either a literal or a named constant, its value cannot change. If it is a variable, it may take on different values as program execution proceeds. Variables and constants may be scalar objects (with a single value) or arrays (with a number of values, all of the same type).

Objects may have type parameters. The value of a type parameter of an object of intrinsic or user-defined type is returned by a type parameter inquiry.

Variables generally have storage space set aside for them by the compiler. If, however, the variable is a pointer or an allocatable object, the compiler does not set aside any space for its value. The programmer must allocate space or, in the case of a pointer, the programmer might assign existing space.

Variables are dynamic if their size, parameters, or location, may change. The declared rank of an array variable may not change, but the extents of its dimensions may. Automatic variables are discussed in 2.2.4; they are created on entry to a procedure and their size and location are determined at that time. The dynamic properties of allocatable and pointer objects may change with each allocation or pointer assignment.

Sometimes it is desirable to reallocate an array to a different size while retaining some of its values. An intrinsic function, MOVE_ALLOC, and assignment (7.5) aid in resizing allocatable arrays.

If a variable or constant is a portion of another object, it is called a **subobject**. A subobject that may be identified by a designator is one of:

a substring
a structure component
an array element
an array section

The real and imaginary parts of a complex object are also subobjects; they may be accessed by intrinsic functions. Each subobject has a **parent** and is a portion of that parent. Each of the subobjects in the list above is described in this chapter; first, substrings and structure components, and then array subobjects (array elements and array sections) along with the use of subscripts, subscript triplets, and vector subscripts. A number of additional aspects of arrays are covered: array terminology, use of whole arrays, and array element order.

Finally, this chapter explains how pointers and allocatable objects can be allocated and deallocated. It describes the ALLOCATE and DEALLOCATE statements and how pointers can be disassociated from any target object by using the NULLIFY statement or pointer assignment with the intrinsic function NULL (13.3.3.1).

6.1 Constants and Variables

A constant has a value that cannot change. A reference to a constant is always permitted, but a constant cannot be redefined.

A constant (R305) has one of the forms:

```
literal-constant
named-constant
```

As explained in 4, each of the intrinsic types has a form that specifies the type, type parameters, and value of a literal constant of the type. For user-defined types, there is a structure constructor to specify values of the type. Array constructors are used to form array values of any intrinsic or user-defined type.

Variables may be of any type, but there are contexts in which a variable must be of a certain type. In most of these cases, terms such as logical-variable, character-variable, or default-character-variable, provide precise limitations.

A subobject with a constant parent is not a variable and might not be a constant. It is classified as a subobject of a constant.

6.2 Designators

A data object may have a designator such as A or B(I). A designator (R603) is one of the following:

```
object-name
array-element
array-section
structure-component
substring
```

A single object of any type is a scalar. A set of scalar objects, all of the same type and type parameters, may be arranged in a pattern involving columns, rows, planes, and higher-dimensioned configurations to form arrays. An array has a rank between one and seven. A scalar has rank zero. An array element is one of the elements of an array and is a scalar. An array section is a selected subset of the elements and is itself an array. A structure component is one of the components of an object of derived type; it may be a scalar or an array. A substring is a contiguous portion of a character string; it is a scalar.

The form of a designator (R612) is:

```
part [ % part ]... [ ( substring-range ) ]
```

where a part (R613) has the form

```
part-name [ ( section-subscript-list ) ]
```

and a substring range (R611) is:

```
[ starting-position ] : [ ending-position ]
```

The starting and ending positions must be scalar integer expressions.

A section subscript (R619) is one of:

```
subscript
subscript-triplet
vector-subscript
```

The simplest form of a section subscript list is a subscript list.

Rules and restrictions:

1. If the designator ends with a substring range, the rightmost part name must be of type character.
2. If a substring range appears and the rightmost part name has the dimension attribute, a section subscript list must be present in the rightmost part.
3. In a part containing a section subscript list, the number of section subscripts must equal the rank of the part name.

If an object designator contains more than one part, the **base object** is the data object specified by the leftmost part.

There are rules for determining whether a particular object designator identifies a character string, character substring, structure component, scalar, array, array element, or array section. This is important in the case of array sections and character substrings because they have a similar syntax; however, in general, these classifications are perhaps of more interest to compiler writers than to users of the language, but knowing how an object designator is classified makes it clearer which rules and restrictions apply to the object and easier to understand some of the explanations for the formation of expressions.

To determine the classification of a valid designator, two aspects must be considered: the syntax of the designator and the type and attributes of the part names, particularly the rightmost part name.

1. A designator with one part may specify a scalar or array of any type, a substring, an array element, or an array section.
2. A designator specifies a substring if a substring range appears and no part has a rank greater than zero.
3. A designator specifies an array element if no part other than the rightmost has a rank greater than zero and a subscript list appears in the rightmost part.

4. A designator specifies an array section if any part other than the rightmost has a rank greater than zero or the rightmost part contains a section subscript list with at least one subscript triplet or vector subscript.

5. A designator specifies a structure component if there is more than one part and none of the previous situations is true. The component may be a scalar or a whole array.

   For example, given the declarations:

```
TYPE PERSON
   INTEGER AGE
   CHARACTER (LEN = 40) NAME
END TYPE PERSON

TYPE(PERSON) CHIEF, FIREMEN(50)
CHARACTER (20) DISTRICT, STATIONS(10)
```

the following designators are classified as indicated by the comments on each line.

```
DISTRICT(1:6)          ! substring
STATIONS               ! array of character strings
STATIONS(1)            ! array element (character string)
STATIONS(1:4)          ! array section of character strings
STATIONS(1)(15:16)     ! substring of an array element
STATIONS(:)(15:16)     ! array section (of substrings)
CHIEF%AGE              ! structure component (integer scalar)
FIREMEN(I) % AGE       ! structure component (integer scalar)
FIREMEN%AGE            ! array section (of integers)
```

A subobject may have a constant parent, for example:

```
CHARACTER(*), PARAMETER :: MY_DISTRICT = "DISTRICT A7"
CHARACTER (2) DISTRICT_NUMBER
DISTRICT_NUMBER = MY_DISTRICT (10:11)
```

DISTRICT_NUMBER has the value A7, a character string of length 2.

6.3 Type Parameter Inquiry

An object of either intrinsic or user-defined type may have type parameters. The intrinsic types have the type parameters kind and length (4.4.3); user-defined types may have user-named parameters (4.4.3); for example, SIZE or NUM. A type parameter inquiry returns the value of a type parameter of a data object. It has the form

   designator % type-parameter-name

Although a type parameter inquiry has a syntax similar to a structure component reference, it does not have the same semantics. It is not a variable and thus can never be defined in an assignment statement; it may appear only as a primary in an expression or as an actual argument. It is scalar even if the designator is an array. The designator need not be defined. The intrinsic functions KIND and LEN also may be used to inquire about some type parameters.

Rules and restrictions:

1. The name of the type parameter being queried must be that of a type parameter of the type of the specified designator.

2. A deferred type parameter of a pointer that is not associated or an allocatable object that is not allocated must not be queried.

   Given the declarations:

```
INTEGER, PARAMETER :: DOUBLE = KIND (0.0D0)
REAL (DOUBLE) :: X, TEMP (20)
CHARACTER (10) :: TITLE

TYPE PROPERTIES (NUM)
     INTEGER, LEN :: NUM
     REAL :: ACREAGE (NUM)
END TYPE

TYPE (PROPERTIES(:)), ALLOCATABLE :: LIST
```

and the executable statement:

```
ALLOCATE (TYPE (PROPERTIES (NUM = 1000)) :: LIST (100) )
```

the following are examples of type parameter inquiries:

```
X % KIND            ! Same value as KIND (X)
TITLE % LEN         ! Same value as LEN (TITLE)
TEMP(10) % KIND     ! Same value as KIND (TEMP)
LIST % NUM         ! 1000 because LIST has been allocated with NUM = 1000
```

6.4 Substrings

A **character string** consists of zero or more characters. Even though it is made up of individual characters, a character string is a scalar. This is a significant difference between Fortran and other languages, such as C, where a "character string" is an array of single characters. As with any data type, it is possible to declare an array of character strings, with all elements of the same length.

A **substring** is a contiguous portion of a parent string that has a starting point and an ending point within the parent string.

A parent string (R610) is one of:

scalar-variable-name
array-element
scalar-structure-component
scalar-constant

Rules and restrictions:

1. The parent string of a substring must be of type character. The substring is of type character.
2. A substring is the contiguous sequence of characters within the string beginning with the character at the starting position and ending at the ending position. If the starting position is omitted, the default is 1; if the ending position is omitted, the default is the length of the character string.
3. The length of a character string or substring may be 0, but not negative. Zero-length strings result when the starting position is greater than the ending position. The formula for calculating the length of a string is:

   MAX (ending-position – starting-position + 1, 0)

4. The first character of a parent string is at position 1 and the last character is at position n where n is the length of the string. The starting position of a substring must be greater than or equal to 1 and the ending position must be less than or equal to the length n, unless the length of the substring is 0. If the parent string is of length 0, the substring must be of length 0.

   In the following example,

```
CHARACTER (14) NAME
NAME = "John Q. Public"
NAME(1:4) = "Jane"
PRINT *, NAME (9:14)
```

NAME is a scalar character variable, a string of 14 characters, that is assigned the value "John Q. Public" by the first assignment statement. NAME(1:4) is a substring of four characters that is reassigned the value "Jane" by the second assignment statement, leaving the remainder of the string NAME unchanged; the string NAME then becomes "Jane Q. Public". The PRINT statement prints the characters in positions 9 through 14, in this case, the surname, "Public".

Given the definition and declarations:

```
TYPE PERSON
   INTEGER AGE
   CHARACTER (LEN = 40) NAME
END TYPE PERSON

TYPE(PERSON) CHIEF, FIREMEN(50)
CHARACTER (20) DISTRICT, STATIONS(10)
```

the following are all substrings:

```
STATIONS (1) (1:5)    ! array element parent string
CHIEF%NAME(4:9)       ! structure component parent string
```

```
DISTRICT(7:14)          ! scalar variable parent string
'0123456789'(N:N+1)     ! character constant parent string
```

The last example is a substring where the parent is a constant and the starting and ending positions are variable. This substring is an expression that is neither a constant nor a variable, but is a primary; it is called a subobject of a constant (6.1).

Whenever an array is constructed of character strings and any part of it (other than the whole object) is selected, an array section subscript must appear before the substring range specification, if any. Otherwise, the substring range specification will be treated as an array section specification because the two have the same form. STATIONS (1:5) is an array section designator that specifies the entire character strings of the first five elements of STATIONS. The designator STATIONS (:) (1:5) is permitted. It specifies an array with elements that are substrings. Even though all elements of the array are selected, this designator is an array section. STATIONS (1:5) (1:5) is also permitted. It specifies an array section with elements that are substrings. Array sections are described in 6.6.4.

If a character string is declared with a deferred length parameter, it is a variable length string. The value of the deferred length parameter is determined by successful execution of an ALLOCATE statement (6.7.1), intrinsic assignment statement (7.5.2), pointer assignment statement (7.5.5.1), or by argument association (12.6)

6.5 Structure Components

A structure is an object of derived type. It is an aggregate of zero or more components of intrinsic or derived types. The types, type parameters, and attributes of the components are specified in the type definition; they may be scalars or arrays. Each structure usually has at least one component; however, the base type of an extensible derived type often has no components (see 4.4.12 for extensible types). There may be arrays of structures. In the example given above, CHIEF is a structure; FIREMEN is an array of structures of type PERSON.

A component of a structure may be specified by placing the name of the component after the name of the parent structure, separated by a percent sign (%). For example, CHIEF % NAME specifies the character string component of the variable CHIEF of type PERSON.

Rules and restrictions:

1. In a structure component designator, the leftmost part name must be the name of a data object, and each part name except the leftmost must be the name of a component of the derived-type definition of the type of the preceding part name.
2. The type, as well as the type parameters if any, of an object of derived type are those of the rightmost part name.
3. If the rightmost part name is of abstract type (4.4.12.3), the data object must be polymorphic (5.2).

4. A structure may have nested array components, but a designator must not contain more than one part with nonzero rank. (See the example at the end of this section.)

5. In a structure component designator, a part name to the right of a part with nonzero rank must not have the ALLOCATABLE or POINTER attribute.

A structure component is a pointer or allocatable object only if the rightmost part name has the POINTER or ALLOCATABLE attribute. It is possible to declare an array of structures that have a pointer or allocatable array as a component, but it is not possible to treat such an object as an array. This ensures that structure component arrays have a regular structure in memory, simplifying implementation.

The rank of a part reference consisting of just a part name is the rank of the part name. The rank of a part reference of the form

part-name (section-subscript-list)

is the number of subscript triplets and vector subscripts in the list. The rank is less than the rank of the part name if any of the section subscripts are subscripts other than subscript triplets or vector subscripts. The shape of an object of derived type is the shape of the part with nonzero rank, if any; otherwise, the object is a scalar and has rank zero.

Given the type definition and structure declarations:

```
TYPE PERSON
   INTEGER AGE
   CHARACTER (LEN = 40) NAME
END TYPE PERSON

TYPE(PERSON) CHIEF, FIREMEN(50)
```

examples of designators are:

```
CHIEF % AGE                 ! scalar component of scalar parent
FIREMEN(J) % NAME           ! component of array element parent
FIREMEN(1:N) % AGE          ! array section of integers
FIREMEN % NAME              ! array section of character strings
FIREMEN (1:N) % NAME        ! array section of character strings
FIREMEN % NAME (39:40)      ! array section of character strings,
                            !       each two characters long
```

If a derived-type definition contains a component that is of derived type, then a designator can contain more than two part references. Given the type definitions and declarations:

```
TYPE REPAIR_BILL
   REAL PARTS
   REAL LABOR
END TYPE REPAIR_BILL
```

```
TYPE VEHICLE
   CHARACTER (LEN = 40) OWNER
   INTEGER MILEAGE
   TYPE(REPAIR_BILL) COST
END TYPE VEHICLE

TYPE (VEHICLE) BLACK_FORD, RED_FERRARI
```

examples of designators are:

```
BLACK_FORD % COST % PARTS
RED_FERRARI % COST
RED_FERRARI % OWNER
```

Given the following type definition and declaration, the designators X % A(1) and X(1) % A may appear, but the designator X % A must not.

```
Type T
   Real, Dimension(100) :: A
End Type T

Type(T) :: X(9)
```

6.6 Arrays

An **array** is a collection of scalar elements of any intrinsic or derived type. All of the elements of an array have the same type and type parameters. An object that is specified to have the DIMENSION attribute is an array. The value returned by a function may be an array. The appearance of an array designator has no implications for the order in which the individual elements are processed unless array element ordering is specifically required, such as for input/output statements.

6.6.1 Array Terminology

An array consists of elements that extend in one or more dimensions to represent columns, rows, planes, etc. There may be up to seven dimensions in an array declaration. The number of dimensions in an array is called the **rank** of the array. The number of elements in a dimension is called the **extent** of the array in that dimension. Limits on the size of extents are not specified in the Fortran standard.

The **shape** of an array is determined from the rank and the extents; to be precise, the shape is a vector where each element of the vector is the extent in the corresponding dimension. The **size** of an array is the product of the extents; that is, it is the total number of elements in the array. Note that an array of size one is not a scalar.

For example, given the declaration

```
REAL X (0:9, 2)
```

the rank of X is 2; X has two dimensions. The extent of the first dimension is 10; the extent of the second dimension is 2. The shape of X is 10 by 2, that is, a vector of two values, [10 2]. The size is 20, the product of the extents.

An object is given the DIMENSION attribute in a type declaration statement or in one of several other declaration statements. The following are some ways of declaring that A has rank 3 and shape [10 15 3]:

```
DIMENSION A(10, 15, 3)
REAL, DIMENSION(10, 15, 3) :: A
REAL A(10, 15, 3)
COMMON A(10, 15, 3)
TARGET A(10, 15, 3)
```

Arrays have a lower and upper bound along each dimension. For arrays of nonzero size, the **lower bound** is the smallest subscript value along a dimension; the **upper bound** is the largest subscript value along that dimension. The default lower bound is 1 if the lower bound is omitted in the declaration. Array bounds may be positive, zero, or negative. In the example:

```
REAL Z(-3:10, 12)
```

the first dimension of Z ranges from –3 to 10, that is, –3, –2, –1, 0, 1, 2, ..., 9, 10. The lower bound is –3; the upper bound is 10. In the second dimension, the lower bound is 1; the upper bound is 12. The bounds for array expressions and zero-sized arrays are described in 7.2.4.

6.6.2 Whole Arrays

The name of an array object or array component without a section subscript list specifies all the elements of the array except when the name appears in an equivalence set (5.14.1). The name may be that of a variable or a constant. Designators for a single element of an array or a section of an array are permitted. In general, most attributes for the whole array also apply to an element or section of an array. An element or section of an array never has the ALLOCATABLE or POINTER attribute. An element never has the DIMENSION attribute, but a section does.

6.6.3 Array Elements

An **array element** is one of the scalar elements that make up an array. A subscript list is used to indicate which element is selected. If A is declared to be a one-dimensional array:

```
REAL, DIMENSION (10) :: A
```

then A(1) selects the first element, A(2) to the second, and so on. The number in the parentheses is the subscript that indicates which scalar element is selected. If B is declared to be a seven-dimensional array:

```
REAL B (5, 5, 5, 5, 4, 7, 5)
```

then B (2, 3, 5, 1, 3, 7, 2) selects one scalar element of B, by specifying a subscript in each dimension. The set of numbers that specify the position along each dimension in turn (in this case, 2, 3, 5, 1, 3, 7, 2) is called a **subscript list**.

Rules and restrictions:

1. In an array element designator, a subscript must be present for each dimension of the array.
2. For a structure component designator to be classified as an array element designator, every part must have zero rank and the last part must have a subscript.

6.6.4 Array Sections

Sometimes only a portion of an array is needed for a calculation. It is possible to designate a selected portion of an array as an array; this portion is called an **array section**. A **parent array** is the array from which the portion that forms the array section is selected.

The designator for an array section is the array variable name followed by a section subscript list that consists of subscripts, triplet subscripts, or vector subscripts. At least one subscript must be a triplet or vector subscript; otherwise, the designator indicates an array element, not an array. The following example uses a section subscript to specify an array section:

```
REAL A (10)
   . . .
A (2:5) = 1.0
```

The parent array A has 10 elements. The array section consists of the elements A(2), A(3), A(4), and A(5) of the parent array. The section A(2:5) is an array itself and the value 1.0 is assigned to all four of its elements.

A section subscript (R619) can be any of:

subscript
subscript-triplet
vector-subscript

where a subscript triplet (R620) is:

[subscript] : [subscript] [: stride]

Subscripts and strides must be scalar integer expressions and a vector subscript (R622) must be an integer array expression of rank one. The rank of a section is the number of subscript triplets or vector subscripts that appear in the designator.

Rules and restrictions:

1. For a designator to be classified as an array section designator, exactly one part must have nonzero rank. Either the final or only part must have a section subscript list and nonzero rank or another part must have nonzero rank.

2. A section subscript must be present for each dimension of an array. If any section subscript is simply a subscript, the section will have a lesser rank than its parent.

3. In an array section of an assumed-size array, the second subscript must not be omitted from a subscript triplet in the last dimension.

6.6.4.1 Subscripts

A subscript, other than one in a subscript triplet, must be within the bounds for that dimension. A subscript may appear in an array section designator. Whenever this occurs, it decreases the rank of the section by one.

6.6.4.2 Subscript Triplets

If the first subscript in a subscript triplet is omitted, the lower bound for the array in that dimension is used. If the second subscript is omitted, the upper bound is used. The stride is the increment between successive subscripts in the sequence and must be nonzero. If it is omitted, it is assumed to be one. If the subscripts and stride are omitted and only the colon (:) appears, the extent for the dimension is used. For an assumed-size array, the second subscript in the last dimension must not be omitted.

When the stride is positive, an increasing sequence of integer values is specified from the first subscript in increments of the stride, up to the last value that is not greater than the second subscript. The sequence is empty if the first subscript is greater than the second. If any subscript sequence is empty, the array section is a zero-sized array, because the size of the array is the product of its extents. For example, given the array declared A(5, 4, 3) and the section A(3:5, 2, 1:2), the array section is of rank 2 with shape [3 2] and size 6. The elements are:

A(3, 2, 1) A(3, 2, 2)
A(4, 2, 1) A(4, 2, 2)
A(5, 2, 1) A(5, 2, 2)

The stride must not be 0.

When the stride is negative, a decreasing sequence of integer values is specified from the first subscript, in increments of the stride, down to the last value that is not less than the second subscript. The sequence is empty if the second subscript is greater than the first, and the array section is a zero-sized array. For example, given the array declared B(10) and the section B(9:4:–2), the array section is of rank 1 with shape [3] and size 3. The elements are:

B(9)
B(7)
B(5)

However, the array sections B(9:4) and B(4:9:–1) are zero-sized arrays.

A subscript in a subscript triplet is not required to be within the bounds for the dimension as long as all subscript values selected by the triplet are within the bounds. For example, the section B(3:11:7) is permitted. It has rank 1 with shape [2] and size 2. The elements are:

B(3)
B(10)

B(99:98) is a zero-sized array.

6.6.4.3 Vector Subscripts

While subscript triplets specify values in increasing or decreasing order with a specified stride to form a regular pattern, vector subscripts specify values in arbitrary order. The values must be within the bounds for the dimension. A **vector subscript** is a rank-one array of integer values used as a section subscript to select elements from a parent array. For example:

```
INTEGER K(3)
REAL A(30)
    . . .
K = [ 8, 4, 7 ]
A(K) = 1.0
```

The last assignment statement assigns the value 1.0 to A(4), A(7), and A(8) but not necessarily in that order. The section A(K) is a rank-one array with shape [3] and size 3.

If K were assigned [4 7 4] instead, the element A(4) would be accessed in two ways: as A(K(1)) and as A(K(3)). Such an array section is called a many-one array section. A many-one section must not appear on the left of the equal sign in an assignment statement or as an input item in a READ statement. The reason is that the result will depend on the order of evaluation of the subscripts, which is not specified by the language. The results would not be predictable and the program containing such a statement would not be portable.

Array sections with vector subscripts must not appear:

1. as internal files
2. as pointer targets
3. as actual arguments for dummy arguments that become defined

If IV is declared:

```
INTEGER, DIMENSION(3) :: IV=[4,5,4]
```

then the section B(8:9, 5:4, IV) is a zero-sized array of rank 3 and the section B(8:9, 5, IV) is a 2 by 3 array consisting of the six elements:

B(8, 5, 4) B(8, 5, 5) B(8, 5, 4)
B(9, 5, 4) B(9, 5, 5) B(9, 5, 4)

6.6.5 Examples of Array Elements and Array Sections

The following designators are classified as indicated by the comments on each line.

```
ARRAY_A(1,2)                              ! array element
ARRAY_A(1:N:2,M)                          ! rank-one array section
ARRAY_B(:,:,:)(2:3)                       ! array section whose elements
                                          !    are substrings of length 2
SCALAR_A % SCALAR_B                       ! scalar structure component
SCALAR_A % ARRAY_C                           ! array structure component
SCALAR_A%ARRAY_C(L)                          ! array element
SCALAR_A%ARRAY_C(1:L)                        ! array section
SCALAR_C%ARRAY_D%SCALAR_D                    ! array section
ARRAY_E(1:N:2)%ARRAY_F(I,J)%STRING(K)(:) ! array section
```

If a part of a designator other than the last part has nonzero rank or the last part has nonzero rank and also contains a section subscript list, the designator identifies an array section. There may be at most one part with rank greater than zero. This is a somewhat arbitrary restriction imposed for the sake of simplicity.

In the last example above, each component of the type definition is an array and the object ARRAY_E(1:N:2) is an array. The designator is valid; each part except the first is scalar. The substring range is not needed because it specifies the entire string; however, it serves as a reminder that the last component is of type character.

The following examples demonstrate the allowable combinations of scalar and array parents with scalar and array components.

```
TYPE REPAIR_BILL
   REAL PARTS(20)
   REAL LABOR
END TYPE REPAIR_BILL

TYPE(REPAIR_BILL) FIRST
TYPE(REPAIR_BILL) FOR_2003(6)
```

Scalar parent

```
1. FIRST%LABOR              ! structure component (scalar)
2. FIRST%PARTS(I)           ! array element
3. FIRST%PARTS              ! structure component (array)
4. FIRST%PARTS(I:J)         ! array section
5. FOR_2003(K)%LABOR        ! structure component (scalar)
6. FOR_2003(K)%PARTS(I)     ! array element
7. FOR_2003(K)%PARTS        ! structure component (array)
8. FOR_2003(K)%PARTS(I:J)   ! array section
```

Array parent

```
9.  FOR_2003%LABOR            ! array section
10. FOR_2003%PARTS(I)         ! array section
11. FOR_2003%PARTS            ! ILLEGAL
12. FOR_2003%PARTS(I:J)       ! ILLEGAL
13. FOR_2003(K:L)%LABOR       ! array section
14. FOR_2003(K:L)%PARTS(I)    ! array section
```

```
15. FOR_2003(K:L)%PARTS         ! ILLEGAL
16. FOR_2003(K:L)%PARTS(I:J)    ! ILLEGAL
```

Examples 11, 12, 15, and 16 are illegal because only one component may be of rank greater than zero. Examples 3 and 7 are compact (contiguous) array objects and are classified as whole arrays. Examples 9, 10, 13, and 14 are noncontiguous array objects and are classified as array sections.

6.6.6 Array Element Order

The elements of an array form a sequence whose ordering is called **array element order**. This is the sequence that occurs when the subscripts along the first dimension vary most rapidly, and the subscripts along the last dimension vary most slowly. Thus, for an array declared as:

```
REAL A(3, 2)
```

the elements in array element order are: A(1, 1), A(2, 1), A(3, 1), A(1, 2), A (2, 2), A(3, 2).

The position of an array element in this sequence is its **subscript order value**. Element A(1, 1) has a subscript order value of 1. Element A(1, 2) has a subscript order value of 4. Table 6-1 shows how to compute the subscript order value for any element in arrays of rank 1 through 7.

Table 6-1 Computation of subscript order value

Rank	Subscript bounds	Subscript list	Subscript order value
1	$j_1:k_1$	s_1	$1+(s_1-j_1)$
2	$j_1:k_1,j_2:k_2$	s_1,s_2	$1+(s_1-j_1)$ $+(s_2-j_2)\times d_1$
3	$j_1:k_1,j_2:k_2,j_3:k_3$	s_1,s_2,s_3	$1+(s_1-j_1)$ $+(s_2-j_2)\times d_1$ $+(s_3-j_3)\times d_2\times d_1$
.	.	.	.
.	.	.	.
.	.	.	.
7	$j_1:k_1,\ldots,j_7:k_7$	$s_1,\ldots,s_7$	$1+(s_1-j_1)$ $+(s_2-j_2)\times d_1$ $+(s_3-j_3)\times d_2\times d_1$ $+\ldots$ $+(s_7-j_7)\times d_6\times d_5\times\ldots\times d_1$

1. $d_i = \max(k_i - j_i + 1, 0)$ is the size of the ith dimension.
2. If the size of the array is nonzero, $j_i \le s_i \le k_i$ for all $i = 1, 2, \ldots, 7$.

This ordering determines the effects of the input and output of arrays; it is needed for features that depend on storage association such as EQUIVALENCE; and it determines the result of certain intrinsic functions such as MAXLOC. When arrays are used as operands in expressions, the indicated operation is performed on corresponding elements, but no order is implied for these elemental operations; they may be executed in any order or simultaneously.

The subscript order of the elements of an array section is that of the array object that the section represents. That is, given the array A(10) and the section A(2:9:2) consisting of the elements A(2), A(4), A(6), and A(8), the subscript order value of A(2) in the array section A(2:9:2) is 1; the subscript order value of A(4) in the section is 2 and A(8) is 4.

Given the section A(9:4:–2), consisting of the elements A(9), A(7), and A(5), the subscript order values of A(9), A(7), and A(5) are 1, 2, and 3, respectively.

6.7 Pointers and Allocatable Variables

There are several categories of dynamic data objects. Automatic objects are discussed in 2.2.4. In addition, there are two data attributes that can be used to specify dynamic data objects: ALLOCATABLE and POINTER. Arrays and scalars of any type may have the ALLOCATABLE or POINTER attribute. Chapter 5 describes how such objects are declared. This section describes how space is created for these objects, how it may be released, and how pointers can be disassociated from any target.

The ALLOCATE statement is not the only means by which allocation may occur. In Fortran 2003, because of changes to assignment, the ALLOCATE statement is no longer needed in some cases. Assignment to an allocatable variable causes allocation if the variable is unallocated or if the expression being assigned is an array of different shape or any of the corresponding length type parameters of the expression and the variable differ. This provides a shortcut for the Fortran programmer. See the examples at the ends of 6.7.1 and 6.7.1.1.

The association status of a pointer is either defined or undefined; initially (unless the pointer is initialized), it is undefined. An undefined pointer may be used in very limited ways (16.2.2.1). The association status of any pointer becomes defined by nullification, allocation, or pointer assignment. If the status is defined, the pointer is either associated with a target or disassociated from any target. A disassociated pointer has a defined status and can be used as an argument to the ASSOCIATED intrinsic function, but a pointer with undefined status must not.

When a pointer designator appears in an expression, the pointer must have both a defined association status and its target must be defined with a value. Figure 6-1 shows the various states that a pointer may assume.

At the top left, an uninitialized pointer P is declared; it has an undefined association status. Its association status becomes defined if it is nullified (lower left) or if space is allocated for it (upper right). Its target may be undefined (upper right) or defined when a value is assigned to it (lower right).

Section 7.5.5.1 describes how pointers can be associated with existing space and how dynamic objects can acquire values.

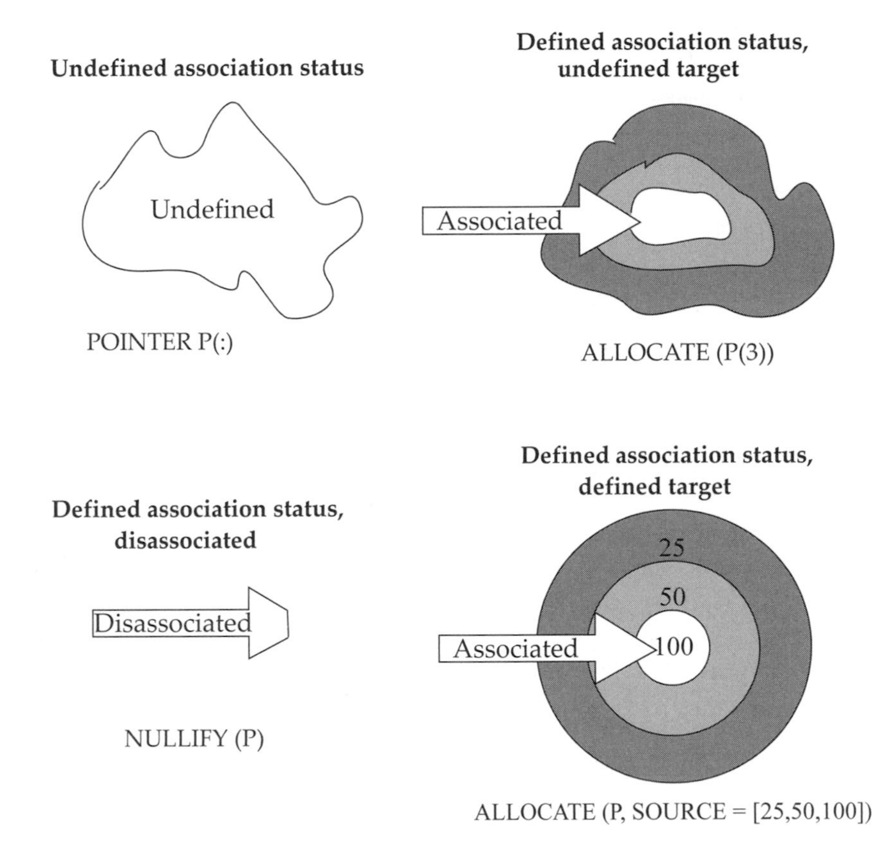

Figure 6-1 States of a pointer

6.7.1 ALLOCATE Statement

The ALLOCATE statement creates space for variables with the ALLOCATABLE or POINTER attribute. If the variable is a pointer, it becomes associated with the newly created space.

The form of the ALLOCATE statement (R623) is:

```
ALLOCATE ( [ type-specifier :: ] allocation-list [ , allocate-option-list ] )
```

Type specifiers are described in 4. They may specify intrinsic or derived types.

An allocation (R628) is:

```
allocate-object [ ( allocate-shape-specification-list ) ]
```

An allocate object (R629) is one of:

```
variable-name
structure-component
```

and an allocate shape specification (R630) is:

```
[ allocate-lower-bound : ] allocate-upper-bound
```

An allocate-option (R624) is one of:

```
STAT = scalar-integer-variable
ERRMSG = scalar-default-character-variable
SOURCE = source-expr
```

Rules and restrictions:

1. The allocate lower bound and allocate upper bound must be scalar integer expressions.
2. Each allocate object must be a data pointer or an allocatable variable.
3. An allocate-shape-specification-list may appear if and only if the allocate object is an array. The number of allocate shape specifications must agree with the declared rank of the array.
4. If an allocate object in a statement has a deferred type parameter, a type specifier or a SOURCE option must appear in the statement.
5. If a type specifier appears, it must specify a type that is compatible with each allocate object (5.2).
6. Either a type specifier or a SOURCE option must appear if any allocate object in a statement is unlimited polymorphic or is of abstract type (4.4.12.3).
7. A type parameter value in a type specifier must be an asterisk if and only if each allocate object is a dummy argument for which the corresponding type parameter is assumed.
8. If a type specifier appears, the kind type parameter values of each allocate object must be the same as the corresponding kind type parameters in the type specifier; length type parameter values may differ.
9. If the SOURCE option appears, type-specifier must not appear, and the allocation list must contain only one allocate object, which must be type compatible (5.2) with source-expr.
10. The rank of source-expr must be either zero or the same as that of the single allocate object. Corresponding kind type parameters must have the same values.
11. Neither the STAT variable, source-expr, nor the ERRMSG variable may be allocated in the ALLOCATE statement in which they appear; nor may they depend on the value, bounds, length type parameters, allocation status, or association status of any allocate object in the same statement.

12. An allocate object or a bound or type parameter must not depend on the value of the STAT variable, the value of the ERRMSG variable, or on the value, bounds, length type parameters, allocation status, or association status of any allocate object in the same ALLOCATE statement.

If a type-specifier or a source-expr appears, it determines the dynamic type and type parameters of the allocate object(s). If neither appears, allocation of a polymorphic object creates an object with a dynamic type and type parameters that are the same as its declared type.

If source-expr appears, it must be conformable with the allocate object. If the allocation is successful, the value of the allocate object becomes that of source-expr.

If a STAT variable appears, it is set to zero if the allocation is successful and is set to a processor dependent positive value if there is an error condition. Each allocate object that was successfully allocated will have an allocation status of allocated or a pointer association status of associated; each allocate object that was not successfully allocated will retain its previous allocation status or pointer association status. If there is no STAT variable, the program terminates when an error condition occurs.

An error condition occurs if:

1. there is insufficient memory for the requested allocations or some other anomaly is detected by the processor,
2. an allocate object in an ALLOCATE statement has an allocation status of allocated,
3. the value specified for a type parameter in a type specification differs from a corresponding nondeferred value specified in the declaration of any of the allocate objects, or
4. the value of a type parameter in source-expr is different from the value of a nondeferred length type parameter of the allocate object.

If the ERRMSG option appears and an error condition occurs during execution of an ALLOCATE statement, the processor will assign an explanatory message to the errmsg character variable. Otherwise, the processor will not change the value of the errmsg variable.

An example of an allocate statement is:

```
ALLOCATE (pressure (i), mat (-1 : total, 0:50), STAT = alloc_err)
```

When an ALLOCATE statement is executed for an array, the values of the lower and upper bound expressions determine the shape of the array. If an entity in one of these expressions is subsequently redefined, the shape of the allocated array is not changed. If the lower bound is omitted, the default is 1. If the upper bound is less than the lower bound, the extent in that dimension is 0 and the array has zero size, in which case no memory is allocated for the array.

An allocate object may be of type character. If it has a character length of zero, no memory is allocated.

An example of an ALLOCATE statement in which the value and dynamic type are determined by reference to another object is:

```
CLASS (*), ALLOCATABLE :: ANY
CLASS (*), POINTER :: PICK
   . . .
PICK => . . .
ALLOCATE (ANY, SOURCE = PICK)   ! Allocate ANY with the value and
                                !    dynamic type of PICK
```

An example of an (unnecessary) ALLOCATE statement with a type specifier is:

```
TYPE BOOK
   CHARACTER (LEN = : ), ALLOCATABLE :: TITLE
END TYPE BOOK

TYPE (BOOK) :: BOOKLIST (100)
CHARACTER (LEN = 1000) :: HOLDER

DO I = 1, 100
   READ *, HOLDER                     ! Get title
   J = LEN_TRIM (HOLDER)              ! Get length of title
   IF (J <= 1) EXIT
   ALLOCATE (CHARACTER (LEN = J) :: BOOKLIST(I) % TITLE)
   BOOKLIST(I) % TITLE = HOLDER(1:J)
END DO
```

The ALLOCATE statement can be omitted because allocation is accomplished as a part of the assignment in the statement following the ALLOCATE statement.

6.7.1.1 Allocation of Allocatable Variables

An allocatable variable has an allocation status of allocated or unallocated at any time during the execution of a program. Unlike pointers, there is no undefined allocation status. At the beginning of execution of a program, an allocatable variable has a status of unallocated. Its status changes to allocated if it appears in a successfully executed ALLOCATE statement, if it is allocated during assignment (7.5.2), or if it is given that status by the allocation transfer intrinsic MOVE_ALLOC (13.3.3.1). An allocatable variable with this status may be referenced, defined, or deallocated. The intrinsic function ALLOCATED (13.3.1.4) returns true for such a variable.

The status of an allocatable variable becomes unallocated if it is successfully deallocated (6.7.3.1) or if it is given that status by the allocation transfer intrinsic. An allocatable variable with this status must not be referenced, defined, or supplied as an actual argument corresponding to a nonallocatable dummy argument, except to certain intrinsic inquiry functions. The intrinsic function ALLOCATED returns false for such a variable.

When the allocation status of an allocatable variable changes, the allocation status of any associated allocatable variable changes accordingly. Allocation of an allocatable variable establishes values for the deferred type parameters of all associated allocatable variables.

An example of using the intrinsic function ALLOCATED to query the allocation status of an allocatable variable is:

```
REAL, ALLOCATABLE :: X(:,:,:)
   . . .
IF(.NOT. ALLOCATED(X)) ALLOCATE (X(-6:2,10,3))
```

The array X cannot be referenced until it has been allocated and assigned a value; it can be used as an argument to the ASSOCIATED intrinsic, as that is not a reference (2.4). X must be declared with a deferred-shape array specification and the ALLOCATABLE attribute.

An unsaved allocatable object that is a local variable of a procedure has a status of unallocated at the beginning of each invocation of the procedure. The status may change during execution of the procedure. An unsaved allocatable object that is a local variable of a module has an initial status of unallocated. The status may change during execution of the program. When an object of derived type is created by an ALLOCATE statement without a SOURCE option, any allocatable ultimate components have an allocation status of unallocated.

In the following example, allocation occurs when an array constructor is assigned to the allocatable array GAMEBOARD. In the first ALLOCATE statement, the value and dynamic type and properties of STORE are determined by reference to GAMEBOARD. In the second ALLOCATE statement, they are determined by reference to CELLS.

```
TYPE POSITION
   INTEGER :: COLOR, PIECE
   LOGICAL :: FILLED = .FALSE.
END TYPE POSITION
TYPE (POSITION), ALLOCATABLE :: GAMEBOARD (:,:)
CLASS (*), ALLOCATABLE :: TEMP_STORE (:,:)
REAL, POINTER :: CELLS (:,:)
   . . .
READ *, SIZE
GAMEBOARD = RESHAPE (   &
      [ ( ( [ POSITION (COLOR = I, PIECE = J), &
                        I = 1, SIZE), J = 1, SIZE) ] , &
             SHAPE = [SIZE, SIZE] )
   . . .
ALLOCATE (TEMP_STORE, SOURCE = GAMEBOARD)
   . . .
CELLS => . . .
IF (.NOT.ALLOCATED(TEMP_STORE)) ALLOCATE (TEMP_STORE, SOURCE = CELLS)
```

6.7.1.2 Allocation of Pointers

When an object with the POINTER attribute is allocated, space is created, and the pointer is associated with that space, which becomes the pointer target. Such an allocated pointer target implicitly has the target attribute which allows additional pointers to point to that target or part of that target. Additional pointers may become associated with the same target by pointer assignment (7.5.5.1). A pointer target may be a variable with the ALLOCATABLE attribute if the variable also has the TARGET attribute.

It is not an error to allocate a pointer that is already associated with a target. In this case, a new pointer target is created as required by the attributes of the pointer and any array bounds, type, and type parameters specified by the ALLOCATE statement. The previous association of the pointer is lost. If there was no other way to access the previous target, it becomes inaccessible. This is sometimes referred to as a "memory leak".

The ASSOCIATED intrinsic function may be used to query the association status of a pointer only if the association status of the pointer is defined. There is no means to determine whether a pointer with defined association status was associated by an ALLOCATE statement; the ALLOCATED intrinsic function cannot have a pointer argument. The ASSOCIATED function also may be used to inquire whether a pointer is associated with a particular target or whether two pointers are associated with the same target.

At the beginning of execution of a function with a pointer result, the association status of the result pointer is undefined. Before such a function returns, it must associate a target with this pointer or cause the association status of the pointer to become disassociated.

Pointers can be used in many ways; an important usage is the creation of linked lists. For example:

```
TYPE NODE
   INTEGER :: VALUE
   TYPE (NODE), POINTER :: NEXT => NULL( )
END TYPE NODE

TYPE(NODE), POINTER :: LIST
    . . .

ALLOCATE (LIST)
LIST % VALUE = 17
ALLOCATE (LIST % NEXT)
```

The first two executable statements create a node pointed to by LIST and put the value 17 in the VALUE component of the node. The next statement creates a second node pointed to by the NEXT component of the first node. The NEXT component of the second node is disassociated because of default initialization specified for the derived type NODE. Its VALUE component is undefined.

6.7.2 NULLIFY Statement

The NULLIFY statement causes a pointer to be disassociated from any target. Unless initialized, pointers have an initial association status that is undefined. One way to give a pointer a defined association status of disassociated (pointing to no target) is to execute a NULLIFY statement for the pointer. Another way is to execute a pointer assignment statement to the intrinsic function NULL. The intrinsic function NULL can be used to initialize a pointer as well.

The form of the NULLIFY statement (R633) is:

```
NULLIFY ( pointer-object-list )
```

where a pointer object (R634) is one of:

```
variable-name
structure-component
procedure-pointer-name
```

Rules and restrictions:

1. Each pointer object must have the POINTER attribute.
2. A pointer object must not depend on the value, bounds, or association status of another pointer object in the same NULLIFY statement.
3. When a NULLIFY statement is applied to a polymorphic pointer (5.2), its dynamic type becomes the declared type.

6.7.3 DEALLOCATE Statement

The DEALLOCATE statement releases the space allocated for an allocatable variable or a pointer target and nullifies the pointer. After a pointer or an allocatable variable has been deallocated, it cannot be referenced or defined until it is allocated or assigned again.

The form of the DEALLOCATE statement (R635) is:

```
DEALLOCATE ( allocate-object-list [ , deallocate-option-list ] )
```

where an allocate object is (R629) one of:

```
variable-name
structure-component
```

and a deallocate-option (R636) is one of:

```
STAT = scalar-integer-variable
ERRMSG = scalar-default-character-variable
```

Rules and restrictions:

1. Each allocate object must be a data pointer or an allocatable variable.

2. Neither the STAT variable nor the ERRMSG variable may be deallocated in the same DEALLOCATE statement; nor may they depend on the value, bounds, allocation status, or association status of any allocate object in the same DEALLOCATE statement.

3. An allocate object must not depend on the value, bounds, allocation status, or association status of another allocate-object in the same DEALLOCATE statement; it also must not depend on the value of the STAT variable or the ERRMSG variable in that statement.

The STAT variable is set to zero if the deallocation is successful and is set to a processor-dependent positive value if there is an error condition. If an error occurs, each allocate object that was successfully deallocated will have an allocation status of unallocated or a pointer association status of disassociated. Each allocate object that was not successfully deallocated will retain its previous allocation status or pointer association status. The status of the allocate objects can be individually checked with the ALLOCATED or ASSOCIATED intrinsic functions. If there is no STAT variable, the program terminates when an error condition occurs. An error condition occurs if an allocate object has a status of unallocated.

If the ERRMSG option appears and an error condition occurs during the execution of a DEALLOCATE statement, the processor will assign an explanatory message to the errmsg character variable; otherwise, the processor will not change the value of the errmsg variable.

An example of a DEALLOCATE statement is:

```
DEALLOCATE (PRESSURE, MAT, ERRMSG = MSG, STAT = DERR)
```

An example of the allocation and deallocation of an allocatable array is:

```
REAL, ALLOCATABLE :: X(:,:)
   . . .
ALLOCATE (X(10,2), STAT=IERR)
IF (IERR > 0) CALL HANDLER
X = 0.0
   . . .
DEALLOCATE (X)
   . . .
ALLOCATE (X(-10:10,5), STAT=JERR)
```

X is declared to be a deferred-shape, two-dimensional, real array with the ALLOCATABLE attribute. Space is allocated for it and it is given bounds, extents, shape, and size and then initialized to have zero values in all elements. Later X is deallocated, and still later, it is again allocated with different bounds, extents, shape, and size, but its rank remains as declared.

6.7.3.1 Deallocation of Allocatable Variables

An allocatable variable may have the TARGET attribute. If such a variable is deallocated, the association status of any pointer associated with the variable will become undefined. Such a variable can be deallocated only by the appearance of its name in a DEALLOCATE statement. It must not be deallocated by the appearance of the pointer name in a DEALLOCATE statement.

When a RETURN or END statement is executed in a procedure, an allocatable variable that is a named local variable of the procedure retains its allocation and definition status if it has the SAVE attribute or is a function result variable or a subobject thereof; otherwise, it is deallocated.

In the example

```
SUBROUTINE TASK
   REAL, ALLOCATABLE :: WORK
   REAL, ALLOCATABLE, SAVE :: VALUES
      . . .
END SUBROUTINE TASK
```

on return from subroutine TASK, the allocation status of VALUES is preserved because VALUES has the SAVE attribute. WORK does not have the SAVE attribute, so it will be deallocated. On the next invocation of TASK, WORK will have an allocation status of unallocated.

If an allocatable variable declared in a module is allocated and, on the execution of a RETURN or END statement, no active scoping unit has access to the module, the allocation status of the variable is processor dependent. The allocation status can be tested and the variable can be reallocated.

When a variable of derived type is deallocated, any allocated allocatable subobject is deallocated.

A function may have a result that is allocatable or is a structure with a subobject that is allocatable. If such a function appears in a specification expression that is executed, the result or any allocated subobject of the result is deallocated before execution of the executable constructs in the scoping unit. If such a function appears in an executable construct that is executed, the result or any allocated subobject of the result is deallocated after execution of the innermost executable construct containing the function reference.

When a procedure is invoked, any allocated allocatable object that is an actual argument associated with an INTENT (OUT) allocatable dummy argument is deallocated; any allocated allocatable subobject of the actual argument is also deallocated.

Deallocation may occur when intrinsic assignment takes place (7.5.2).

If an allocatable component is a subobject of a finalizable object (4.4.11.3), that object is finalized before the component is automatically deallocated.

The effect of automatic deallocation is the same as that of a DEALLOCATE statement without a deallocate-option-list.

6.7.3.2 Deallocation of Pointers

A pointer may be deallocated only if it has a defined association status. Deallocating a pointer that is disassociated causes an error condition in the DEALLOCATE statement. A pointer associated with an allocatable variable must not be deallocated. (Of course, the variable itself may be deallocated which would cause the association status of any associated pointers to become undefined.)

It is possible (by pointer assignment) to associate a pointer with a portion of an object such as an array section, an array element, or a substring. A pointer associated with only a portion of an object must not be deallocated. If more than one pointer is associated with an object, deallocating one of the pointers causes the association status of the others to become undefined. There are other events that cause the association status of a pointer to become undefined (16.2.2.1.3). When its status is undefined, a pointer can no longer be referenced, defined, deallocated, or be an argument to the ASSOCIATED intrinsic function. It may be allocated, nullified, or pointer assigned to a new target.

An example of the allocation of a pointer is:

```
REAL, POINTER :: X (:, :)
    . . .
ALLOCATE (X (10, 2), STAT = IERR)
IF (IERR .GT. 0) CALL HANDLER
X = 0.0
    . . .
ALLOCATE (X(-10:10, 5), STAT = JERR)
```

X is declared to be a deferred-shape, two-dimensional, real array with the POINTER attribute. Space is allocated for it and it is given bounds, extents, shape, and size and then initialized to have zero values in all elements. Later X is allocated with different bounds, extents, shape, and size. This example is quite similar to the previous example for allocatable arrays, except that, in the case of pointers, it is not necessary that X be deallocated before it is reallocated.

7 Expressions and Assignment

- An **Expression** (made up of primaries, operators, and parentheses) usually produces a value as a result of evaluation. An expression has a type, type parameters, and shape. Primaries and results may be scalars or arrays. If a primary is a pointer, the value of the target is used in most cases. If the result is a pointer, it might not have a value or any other particular attribute.
- An **Initialization Expression** is an expression that can be evaluated at compile time. It is used whenever a value is needed at compile time, such as for kind type parameters, named constants, or to initialize variables.
- A **Specification Expression** is a scalar expression of type integer that can be evaluated on entry to a subprogram. Specification expressions may be used to specify array bounds and character lengths.
- **Assignment** is a process that gives a variable a value which is the result of evaluating an expression. Assignment is provided for all types. Assignment can cause changes in the variable's dynamic type, length type parameters, and bounds.
- **Defined Assignment** is provided by a user-supplied subroutine with an assignment interface.
- **Pointer Assignment** associates a pointer with a target, disassociates the pointer, or makes its association status undefined. A target is either a variable, a procedure, or a function that returns a pointer.
- **Masked Array Assignment** assigns values to array elements selected by a mask. This is accomplished with a WHERE statement or construct.
- **Indexed Array Assignment** assigns values to array elements selected by index values and an optional mask. This is accomplished with a FORALL statement or construct.

In Fortran, calculations are specified by writing expressions. **Expressions** look much like algebraic formulas in mathematics, particularly if the expressions involve calculations on numerical values. In fact, the attempt to give the programmer a programming language that reflects, as much as possible, ordinary mathematical notation is what inspired the name Fortran (**For**mula **tran**slation).

An expression represents a computation that results in a value or a pointer. This chapter describes how expressions are formed, how they are interpreted, and how they are evaluated. Almost anywhere a value is needed, the value can be provided by a general expression rather than just a simple variable or constant. The result of an expression also has a type, type parameters, and shape. Context sometimes limits the allowable expressions; for example, subscripts must be scalar integer expressions.

J.C. Adams et al., *The Fortran 2003 Handbook*,
DOI: 10.1007/978-1-84628-746-6_7,

The result value is a scalar or an array. A complex value or a structure value is a scalar, even though it may consist of more than one part (for example, a complex value consists of two parts).

Expressions are described in terms of the following three sets of rules:

- The rules for forming expressions (syntax) (7.1)
- The rules for interpreting expressions (semantics) (7.2)
- The rules for evaluating expressions (optimization of the computation) (7.3)

The syntax rules indicate which forms of expressions are valid. The semantics indicate how each expression is to be interpreted. Once an expression has been given an interpretation, a compiler may evaluate another completely different expression, provided the expression evaluated is mathematically equivalent to the original.

One of the major uses of expressions is in assignment statements where the value of an expression is assigned to a variable. Assignment may be

- intrinsic assignment
- defined assignment
- masked array assignment
- indexed parallel array assignment
- pointer assignment

Intrinsic assignment evaluates an expression and uses the result to define a variable. Defined assignment evaluates an expression and invokes a user-provided subroutine. For masked array assignment, multiple scalar values are involved and the mask determines which computations and assignments are performed. For indexed parallel array assignment, a set of parallel assignments is specified by an index set and an optional masked scalar expression. For pointer assignment, a pointer, the object on the left side, is associated with (points to) the target indicated by the right side. The forms of assignment are described in detail in this chapter.

7.1 Formation of Expressions

An expression is formed from primaries, operators, and parentheses. The simplest form of an expression is a constant or a variable. Some examples are:

3.1416	A real constant
X	A scalar variable

Slightly more complicated expressions consist of a designator, an array constructor, a structure constructor, a function reference, a type parameter inquiry, or a type parameter name. Examples are:

[X, Y, X]	An array constructor
Y (2:10:2)	A variable that is a section of array Y
M % N	Either a variable that is a component of a structure M or a type parameter inquiry
FX (Y + Z)	A function reference

These simple objects may be combined with operators to form more complicated expressions. Examples are:

A + B	An expression using intrinsic +
A + B .X. C	An expression using intrinsic + and defined operator .X.

Finally, any expression enclosed in parentheses is an expression; this recursion allows the formation of arbitrarily complicated expressions.

7.1.1 Operators and Operations

There are two classes of operations within an expression: intrinsic operations and nonintrinsic operations. The latter class is often called defined operations or user-defined operations. A nonintrinsic operation can be defined as a new operator, such as .MatrixDivide., or as an extension of an existing intrinsic operator, such as .NOT. or +. Extending an existing operator is often called overloading.

7.1.1.1 Operators

An **intrinsic operator** is built into the Fortran language. Table 7-1 lists the intrinsic operators and their allowed operand types. A **defined operator** (12.5.4.2) is defined by the programmer using a function subprogram. There are two sorts of defined operators: extensions of intrinsic operators and new operators.

An operator is either unary or binary; a unary operator requires one operand and a binary operator requires two operands.

Note that the operators + and – are both unary and binary. The only intrinsic unary operators are +, –, and .NOT. The operator .NOT. is the only intrinsic operator that is a unary operator but not a binary operator.

There is a precedence ordering among the operators. This precedence is used to determine the interpretation of expressions containing more than one operator. The precedences of all operators are described in more detail in 7.2.1.

Table 7-1 Intrinsic operators and the allowed types of their operands

Operator category	Intrinsic operator	Operand types
Arithmetic	**, *, /, +, −, unary +, unary −	Of any numeric type and any kind type parameters
Character	//	Both of type character of any length with the same kind type parameter
Relational	.EQ., .NE., ==, /=	Either both of any numeric type and any kind type parameters, or both of type character of any length and with the same kind type parameter
Relational	.GT., .GE., .LT., .LE., >, >=, <, <=	Both of any numeric type except complex and any kind type parameter, or both of type character of any length and with the same kind type parameter
Logical	.NOT., .AND., .OR., .EQV., .NEQV.	Of type logical with any kind type parameters

Note: The relational operator symbols ==, /=, >, >=, <, and <= are synonyms for the operators .EQ., .NE., .GT., .GE., .LT., and .LE., respectively.

7.1.1.2 Operations

The term **operation** refers both to the syntax forms of an operator and its operands and also to the action performed.

An operation is unary or binary, depending on whether its operator is unary or binary.

A unary operation has one operand as in:

operator x_1

Examples are:

```
- C
+ J
.NOT. L
```

A binary operation has two operands as in:

x_1 operator x_2

Examples are

```
A + B
2 * C
```

An operation is intrinsic or defined. Intrinsic operations are those defined by the language. For an operation to be intrinsic, an intrinsic operator symbol must be used and the operands must be of the intrinsic types specified in Table 7-1.

A **defined operation** is any nonintrinsic operation that is interpreted and evaluated by a function subprogram. The defined operation uses a defined operator, either an extension of an intrinsic operator or a new operator. Defined operations are described in 12.5.4.2. The forms of a defined operation are:

intrinsic-unary-operator x_2
nonintrinsic-unary-operator x_2
x_1 intrinsic-binary-operator x_2
x_1 nonintrinsic-binary-operator x_2

where x_1 and x_2 are operands. Intrinsic operations cannot be redefined. Therefore, when either an intrinsic unary or binary operator symbol is used in a defined operation, the operand types must not be the same as the types of the operands specified in Table 7-1 for the particular intrinsic operator symbol. Examples of each of the forms are:

```
- Person
.PLUS. A
Matrix_A / Matrix_B
A .HIGHER. B
```

7.1.2 Rules for Forming Expressions

The set of syntax rules defines an expression in terms of operators and operands which may themselves be expressions. As a result, the formal set of rules is recursive. The basic or lowest level of an expression is a primary. The rules for forming expressions are described from the lowest or most primitive level to the highest or most complex level; that is, the rules are stated from a primary up to an expression.

Primary. A primary has one of the following forms (R701):

constant
designator
array constructor
structure constructor
function reference
type parameter inquiry
type parameter name
(expression)

Examples of primaries are:

3.2	A real constant
A	A designator
[1, J, 7]	An array constructor
RATIONAL (I, J)	A structure constructor
STRING % LEN	A type parameter inquiry

NROWS	A type parameter name
FCN (A)	A function reference
(A 1 B)	A parenthesized expression

A designator (6.2, R603) is a very general term and includes named objects, array elements and sections, structure components, and substrings.

A general expression is built up by combining operators with their operands, which can be primaries or other forms of expressions. Complicated expressions usually enclose some of the operations in parentheses either to control operation precedence or associativity or to make the expression more readable.

The grammar that follows is a simpler, easier to understand, version compared to the one given by rules R310, R311, R312, and R701 through R723 of the Fortran 2003 standard. The simplification has lost some of the suggested precedence and associativity information of the standard grammar. The grammar needs to be read in conjunction with the restrictions and explanations below and also the discussion of operator precedence (7.2.1).

defined-unary-expr	is	[nonintrinsic-unary-op] primary
power-expr	is	[defined-unary-expr **] ... defined-unary-expr
mult-expr	is	power-expr [mult-like-op power-expr] ...
add-expr	is	[unary-add-like-op] mult-expr [add-like-op mult-expr] ...
concat-expr	is	add-expr [// add-expr] ...
comparison-expr	is	concat-expr relational-op concat-expr
not-expr	is	[.NOT.] comparison-expr
and-expr	is	not-expr [.AND. not-expr] ...
or-expr	is	and-expr [.OR. and-expr] ...
equiv-expr	is	or-expr [equiv-op or-expr] ...
expr	is	equiv-expr [nonintrinsic-binary-op equiv-expr] ...
mult-like-op	is	*
	or	/
unary-add-like-op	is	add-like-op
add-like-op	is	+
	or	-

```
relational-op        is  ==
                     or  .EQ.
                     or  /=
                     or  .NE.
                     or  <
                     or  .LE.
                     or  <=
                     or  .LE.
                     or  >
                     or  .GT.
                     or  >=
                     or  .GE.

equiv-op             is  .EQV.
                     or  .NEQV.
```

Note that power-expr has a different form which suggests a right-to-left order of evaluation of exponents. The form of the other operations suggests a left-to-right order of evaluation.

Each operand must have an appropriate type as described in Table 7-1 (for intrinsic operators) or specified by the function that defines the operation (for defined operators).

The syntax rules for expressions have some consequences that may not be obvious.

- A unary plus or minus followed by a constant is not a constant; it is an expression. As a consequence, parentheses must be used for some common formulas; there is little other effect on the language. Except for expression syntax, there are few cases where it matters that 6, for example, is a constant whereas –6 is an expression. The major one is that on a 2s complement 16-bit machine –32768 is representable as an integer, but this expression is likely to cause an integer overflow as written.

- Except for some cases involving unary operators, two operators cannot be adjacent. Thus, for example

  ```
  X ** -Y           ! Invalid syntax
  (A+B) * -2        ! Invalid syntax
  .not. .not. OK    ! Invalid syntax
  ```

  are illegal. Parentheses can be used to express the intention of the above illegal expressions

  ```
  X ** (-Y)
  (A+B) * (-2)
  .not. (.not. OK)
  ```

On the other hand, the following are legal:

```
X > -Y

X > - Y  .OR.  -X > 0

A .and. .not. B

C + .MatrixInverse. B

.not. -A  ! for an appropriate extension of - or .NOT.
```

- The relational operators (e.g., < or .NE.) cannot occur in a series. That is, expressions such as

```
A > B > C   ! Invalid syntax
```

are illegal.

7.2 Interpretation of Expressions

Interpretation of an expression determines the meaning of the expression. As with the rules for forming an expression, the rules for interpreting an expression are described from the bottom up, from the interpretation of constants, variables, constructors, and functions to the interpretation of each subexpression to the interpretation of the entire expression.

For the purpose of evaluation of expressions, it is required that each referenced operand (2.4) be defined, including all of its parts. If the operand is a subobject (part of an array, structure, or string), only the selected part is required to be defined. If the operand is a pointer, it must be associated with a target that is defined. Note that function references are not in themselves references. Inquiry functions, such as SIZE, do not require that their arguments be defined. Whether or not a function requires that its arguments be defined depends on the particular function.

For the numeric intrinsic operations, the operands must have values for which the operation is well-defined on the processor being used. For example, on some processors, the result of any of the numeric operations must be within the exponent range for the result data type. Most processors support some form of IEEE arithmetic (14) and can process values that are out of range.

When an expression is interpreted, the meaning of the simplest primaries, such as constants and variables, is determined. Once these are determined, the operations for which they are operands are interpreted in precedence order, and a meaning for the operation is determined by the interpretation rules for each operator. This repeats recursively until the entire expression is interpreted and a meaning is determined.

The interpretation rules for operations are either rules for the intrinsic operations (intrinsic operators with operands of the intrinsic types specified by Table 7-1) or rules for the defined operations (provided by function subprograms). Except for integer division, the intrinsic operations are interpreted in the usual mathematical way, subject to representation limitations (for example, a finite range of integers, or finite precision of real numbers). The defined operations are interpreted by a function program that is specified in an interface block with a generic specifier of the form OPERATOR (defined-operator).

The interpretation rules for an intrinsic or a defined operation are independent of the context in which the expression occurs, except for the NULL intrinsic function. That is, the type, type parameters, and interpretation of any expression do not depend on any part of a larger expression in which it occurs. This statement is often misunderstood. It does not mean that in all cases the results of individual operations with the same operands must be the same in all contexts. The reason is that the actual results of the intrinsic operations for real and complex operands are not specified precisely. For example, the following code fragment

```
REAL :: A, B, X
X = A+B
PRINT *, A+B .EQ. X
```

may print the value false because the result of A + B is required to be only an approximation of the mathematical result of adding A to B, and different numerical approximations are allowed in different contexts. This allows an implementation the freedom to optimize the evaluation of expressions. Many processors keep intermediate values of an expression in registers and these values may have higher precision than values stored in memory. When a value in a register is compared with a value that is fetched from memory, the comparison may give surprising results because of the precision differences. Because of the approximate nature of floating-point computations, programmers should program defensively if small differences are important.

This section covers the precedence of operators, which determines how the operations in an expression are grouped; then it covers the data type and type parameters of an expression, the shape and bounds of an expression, and the meaning of an expression.

These properties are determined inside-out in the sense that they are determined first for the primaries. These properties then are determined repeatedly for the operations in precedence order, resulting eventually in the properties for the expression.

For example, consider the expression A + B ∗ C, where A, B, and C are of numeric type. First, the data types, type parameter values, and shapes of the three variables A, B, and C are determined. Because ∗ has a higher precedence than +, the type, type parameters, and shape of the expression B ∗ C are determined next, and then these properties for the entire expression are determined from those of A and B ∗ C.

7.2.1 Precedence of Operators

It is the precedence rules, not the formation rules, that determine how an expression is interpreted. Table 7-2 summarizes the relative precedence of operators, including the precedence when operators of equal precedence are adjacent. An entry "N/A" in the column titled "In context of equal precedence" indicates that the operator cannot appear in such contexts. Note that these operators are not intrinsic operators unless the types of the operands are those specified in Table 7-1.

Table 7-2 Categories of operations and relative precedences

Category of operator	Operator	Precedence	In context of equal precedence
Defined	Defined unary operator	Highest	N/A
Numeric	**	.	Right-to-left
Numeric	* or /	.	Left-to-right
Numeric	Unary + or –	.	N/A
Numeric	Binary + or –	.	Left-to-right
Character	//	.	Left-to-right
Relational	.EQ., .NE., .LT., .LE., .GT., .GE. ==, /=, <, <=, >, >=	.	N/A
Logical	.NOT.	.	N/A
Logical	.AND.	.	Left-to-right
Logical	.OR.	.	Left-to-right
Logical	.EQV. or .NEQV.	.	Left-to-right
Defined	Defined binary operator	Lowest	Left-to-right

For example, in the expression

```
A .AND. B .AND. C .OR. D
```

Table 7-2 indicates that the .AND. operator is of higher precedence than the .OR. operator, and the .AND. operators are combined left-to-right when in contexts of equal precedence; thus, A and B are combined by the .AND. operator, the result A .AND. B is combined with C using the .AND. operator, and that result is combined with D using the .OR. operator. This expression is thus interpreted the same way as the following fully parenthesized expression

```
((A .AND. B) .AND. C) .OR. D
```

Exponentiation is right associative; all of the other binary operators are left associative (except for the relational operators, which cannot appear in a series). Thus

```
A ** B ** C
```

is interpreted as

```
A ** ( B ** C )
```

while

```
A * B / C
```

is interpreted as

```
( A * B ) / C
```

Note that all the defined operators have fixed precedences; defined unary operators have the highest precedence of all operators and are all of equal precedence; defined binary operators have the lowest precedence, are all of equal precedence, and are combined left-to-right when in contexts of equal precedence. Both kinds of defined operators may be generic.

Recall that new defined unary and new defined binary operators have the form **.**letter [letter...]**.** (3.2.4). It is also possible to give additional meanings to the intrinsic operators (7.2.7.2 or 12.5.4.2); they are called extended intrinsic operators or, sometimes, overloaded operators. They have the same precedence as the intrinsic operator. An interesting consequence is that if the **.**NOT**.** operator is extended, it is a defined operator and a unary operator. However, it is not a defined unary operator; it is an extended intrinsic operator. It has the same precedence as the intrinsic **.**NOT**.** operator.

As a consequence of the expression formation rules, unary operators in the same category cannot appear in a context of equal precedence; parentheses must be used. There is thus no left-to-right or right-to-left rule for any unary operators. Similarly, the relational operators cannot appear in a context of equal precedence; consequently, there is no left-to-right or right-to-left rule for the relational operators.

7.2.2 Data Type and Type Parameters of an Expression

Once the interpretation of an expression is complete, the data type and type parameters of the expression are determined recursively from the primaries and operations that make up the expression.

Expressions have both a declared type and a dynamic type. The declared type of an expression is determined by the following rules using the declared types of the entities and defined operators. The declared type can be determined at compile-time. The dynamic type of an expression is also determined by the same rules; however, the dynamic types of the entities and operands are used as the expression is evaluated at runtime. If none of the entities or operators are polymorphic, the declared type will be the same as the dynamic type. If the expression is a polymorphic primary or a defined operation with a polymorphic result, the declared and dynamic types might be different.

7.2.2.1 Data Type and Type Parameters of a Primary

The type and type parameters of a literal constant are determined by the form of the constant (4.2.6) and not from the context. For example, the form of the constant

```
0.3333333333333333333333333333333
```

indicates that it is of type default real. Neither the large number of digits in the constant, nor usage in a context where double precision would be appropriate makes the constant anything but a default real constant. Similarly

```
(1.3_LONG, 2.9_LONG)
```

indicates that it is of type complex and of kind LONG, regardless of where it appears.

The type of a named constant is determined by its declaration.

The type and type parameters of a variable are determined by its declaration and possibly partly when it is allocated or pointer assigned. For example,

```
TYPE :: D3(N)
   INTEGER, LEN :: N
   REAL, DIMENSION(N) :: A, B, C
END TYPE D3
TYPE(D3(N=:)), ALLOCATABLE :: QQ
```

indicates that QQ is of type D3, but until it is allocated, the length parameter N is not established.

The type and type parameters of a structure constructor are described in 4.4.15. The type and type parameters of a structure component are those given by the declaration of that component.

The type and type parameters of an array constructor are described in 4.5. The type and type parameters of an array element or array section are those of the array.

The type of a substring is character, the kind type parameter is that of the string, and the length parameter is the length of the substring.

The type and type parameters of the result of an intrinsic function are described in A. The type and type parameters of a user-defined function are determined by the function subprogram; however, see 4.3.5.1 for an obsolescent character exception. If the function is generic, the type and type parameters are determined from the specific function referenced and the actual arguments.

A type parameter name (which can be used only within a type definition or a type parameter inquiry) is a scalar integer with the same kind as the type parameter.

The type and type parameters of an expression in parentheses are the same as those of the expression.

The type and type parameters of the intrinsic function NULL are context dependent and are described in A. This is an exception to the general rule about context not determining type.

If a pointer appears as a primary in a defined or intrinsic operation, in parenthesis as an expression, or as a single primary to the right of the equals sign in an intrinsic assignment statement, the reference is to the target. The type and type parameters are those of the target. If a pointer is not associated with a target, it may appear as the target in a pointer assignment statement or as a primary only as an actual argument associated with a dummy argument that is also a pointer. It may also appear as the data target in a pointer assignment statement (7.5.5), but, in that case, it is not a primary in an expression.

7.2.2.2 Data Type and Type Parameters of an Operation

The type of the result of an intrinsic operation is determined by the type of the operands and the intrinsic operation and is specified by Table 7-3.

For nonnumeric intrinsic operations, the type parameters of the result of an operation are determined as follows.

Table 7-3 Type of operands, x_1 and x_2, and result for intrinsic operations

Intrinsic operator	Type of x_1	Type of x_2	Type of result
Unary +, –		I, R, Z	I, R, Z
Binary +, –, *, /, **	I	I, R, Z	I, R, Z
	R	I, R, Z	R, R, Z
	Z	I, R, Z	Z, Z, Z
//	C	C	C
.EQ., .NE.	I	I, R, Z	L, L, L
==, /=	R	I, R, Z	L, L, L
	Z	I, R, Z	L, L, L
	C	C	L
.GT., .GE., .LT., .LE.	I	I, R	L, L
>, >=, <, <=	R	I, R	L, L
	C	C	L
.NOT.		L	L
.AND., .OR., .EQV., .NEQV.	L	L	L

Note: The symbols I, R, Z, C, and L stand for the types integer, real, complex, character, and logical, respectively. Where more than one type for x_2 is given, the type of the result of the operation is given in the same relative position in the next column. For the intrinsic operations with operands of type character, the kind type parameters of the operands must be the same.

- For the relational operations, it is that of the default logical type.
- For the logical operations, it is that of the operands if the operands have the same kind type parameter; otherwise, it is processor dependent but must be that of one of the operands.
- For the unary **.NOT.** operation, it is that of the operand.
- For the character operation //, the operands must have the same kind type parameter and the result has that kind type parameter. The length type parameter value is the sum of the length type parameters of the operands.

For numeric intrinsic operations, the kind type parameter value of the result is determined as follows:

- For unary operations, it is that of the operand.
- For binary operations, if one operand is of type integer and the other is of type real or complex (for example, 1 + 2.0), it is the kind type parameter of the real or complex operand.

- For binary operations, if the operands are of the same type and kind type parameters or one is real and one is complex with the same kind parameters, it is the kind type parameter of the operands.
- For binary operations, if the operands are both of type integer but with different kind type parameters, it is the kind type parameter of the operand with the larger decimal exponent range. If the decimal exponent ranges of the two kinds are the same, it is processor dependent, but must be that of one of the operands.
- For binary operations, if the operands are both of type real or complex but with different kind type parameters, it is the kind type parameter of the operand with the larger decimal precision. If the decimal precisions are the same, the kind type parameter is processor dependent, but must be that of one of the operands.

For numeric intrinsic operations, an easy way to remember the result type and type parameter rules is to consider that the three numeric types—integer, real, and complex—are ordered by the increasing generality of numbers: integers are contained in the set of real numbers and real numbers are contained in the set of complex numbers. Within the integer type, the kinds are ordered by increasing decimal exponent ranges. Within the real and complex types, the kinds for each type are ordered by increasing decimal precision. If there is more than one kind of integer with the same decimal exponent range, the ordering is processor dependent; a similar processor-dependent ordering is selected for the real and complex types if there is more than one kind with the same decimal precision. Because the result precision is that of the higher precision operation, operations between double precision real and single precision complex produce a result of double precision complex. Most processors do not support integers or reals that have different kinds but have the same exponent range or precision; the processor dependent exceptions for kind rarely occur in practice.

The type and type parameter values of a defined operation are determined from the interface block (or blocks) for the referenced operation and are the type and type parameters of the name of the function specified by the interface block. Note that the operator may be generic and therefore the type and type parameters may be determined by the operands. For example, consider the interface:

```
INTERFACE OPERATOR (.PLUS.)

   TYPE (SET) FUNCTION FCN_SET_PLUS (X, Y)
      TYPE (SET), INTENT (IN) :: X, Y
   END FUNCTION FCN_SET_PLUS

   TYPE (RATIONAL) FUNCTION FCN_RAT_PLUS (X, Y)
      TYPE (RATIONAL), INTENT(IN) :: X, Y
   END FUNCTION FCN_RAT_PLUS

END INTERFACE
```

The operation A .PLUS. B where A and B are of type RATIONAL is an expression of type RATIONAL with no type parameters. The operation C .PLUS. D where C and D are of type SET is an expression of type SET with no type parameters.

7.2.3 Shape of an Expression

The shape of an expression is determined by the shape of each operand in the expression in the same recursive manner as for the type and type parameters for an expression. That is, the shape of an expression is the shape of the result of the last operation.

However, the shape rules are simplified considerably by the requirement that the operands of binary intrinsic operations must be in shape conformance. Two operands are in **shape conformance** if both are arrays of the same shape, or one or both operands are scalars. When one operand is an array and the other is a scalar, the operation behaves as if the scalar operand were broadcast to an array of the result shape and the operation performed. Broadcasting a scalar to an array means creating an array of elements all equal to the scalar. This broadcast need not actually occur if the operation can be performed without it. The operands of a defined operation have no such requirement:

- they must match the shape of the corresponding dummy arguments of the defining function, or
- they must be in shape conformance with each other, the dummy arguments of the defining function must be scalar, and the defining function must be elemental.

For primaries that are constants, variables, constructors, or functions, the shape is that of the constant, variable, constructor, or function name. Type parameter inquiries and type parameter names are scalars. If the primary is a reference to the intrinsic function NULL, the shape of the result is not relevant; the type, type parameters, and rank are determined by the pointer that becomes associated with the result (A). Recall that structure constructors are always scalar, and array constructors are always rank-one arrays of size equal to the number of elements in the constructor. For unary intrinsic operations, the shape of the result is that of the operand. For binary intrinsic operations, the shape is that of the array operand if there is one and is scalar otherwise. For defined operations, the shape is that of the function name specifying the operation if the operands match the shapes of the dummy arguments or is the shape of an array operand if the defining function is elemental.

For example, consider the intrinsic operation A + B where A and B are of type default integer and default real respectively; assume A is a scalar and B is an array of shape [3 5]. Then, the result is of type default real with shape [3 5].

As a second example, consider the expression A // B as a defined operation where A is a scalar of type character with kind type parameter value 1 and of length 25, and B is an array of type character with kind type parameter value 2, of length 30, and of shape [10]. This is permitted because there is no intrinsic concatenation between character operands of different kinds. Suppose further there is the following interface for the // operator:

```
INTERFACE OPERATOR (//)

   FUNCTION FCN_CONCAT (X, Y)
      CHARACTER (*, KIND=1), INTENT (IN) :: X
      CHARACTER (*, KIND=2), INTENT (IN) :: Y (:)
      CHARACTER (LEN (X) + LEN (Y), KIND=2) :: FCN_CONCAT (SIZE (Y))
   END FUNCTION FCN_CONCAT

END INTERFACE
```

The type declaration for FCN_CONCAT specifies that the result of the expression A // B is of type character with kind type parameter 2. In addition, the length of the result is the sum of the lengths of the operands A and B, that is, of length 55. The shape is specified to be of rank one and of size equal to the size of the actual argument B corresponding to the dummy argument Y, that is, of shape [10].

7.2.4 Bounds of an Expression

For most contexts, the lower and upper bounds of the dimensions of an array expression are not needed; only the sizes of each dimension are needed to satisfy array conformance requirements for expressions. The bounds of an array expression can be found by using the LBOUND and UBOUND intrinsic functions.

The bounds of the dimensions of whole arrays and whole array components are described in 5.4.1. If the array is anything but a whole array or whole array component, the lower bound in each dimension is one and the upper bound is the number of elements in that dimension, which might be zero.

Note that the LBOUND and UBOUND functions distinguish between whole arrays, including whole array structure components, and arrays that either have section subscripts, are assumed-shape arrays, or are other expressions. It may seem strange to distinguish between a simple array name and an array name with a section subscript. For example, it seems obvious that the lower bound of ARRAY(2:4) should be 2 rather than 1. Problems arise with more complicated forms, ARRAY(2:6:2) has only 3 elements; it would be odd to say the upper bound is either 3 or 6 if the lower bound were 2. Similarly, an expression such as ARRAY(2:4) + ARRAY(9:7:–1) has no obvious natural bounds. Rather than try to distinguish between "simple" sections and "complicated" sections, Fortran treats all sections as if they were complicated and returns 1 for the lower bound and the actual extent for the upper bound.

As a practical matter, the LBOUND and UBOUND functions are usually applied to pointer targets or dummy argument arrays that are assumed shape; the bounds for other arrays are usually obvious from the declarations. With dummy arguments, the compiler has no information at all about the associated actual argument and whether or not it has "simple" section subscripts.

7.2.5 Elemental Operations and Functions

For both the unary and binary intrinsic operators, the operation is interpreted element-by-element; that is, the scalar operation is performed on each element of the operand or operands. Similarly, if the operation is an elemental function reference, the function is invoked for each element of the array arguments. For example, if A and B are arrays of the same shape, the expression A * B is interpreted by taking each element of A and the corresponding element of B and multiplying them together using the scalar intrinsic operation * to determine the corresponding element of the result. Note that this is not the same as matrix multiplication. As a second example, the expression SQRT(A) is interpreted by taking each element of A and invoking the square root function to determine the corresponding element of the result.

Note that there is no order specified for the interpretation of these array operations. Indeed, a processor is allowed to perform them in any order, including all at once (possible for vector and parallel processors). A processor also has the option to invoke elemental functions on an element-by-element basis or to invoke a version of the function that accepts array operands and evaluates the results in an optimized way. The rules for elemental functions (12.7.2) allow either method.

7.2.6 Value of a Primary

The value of a primary that is a constant, designator, array or structure constructor, or type parameter name or inquiry is the obvious value of the entity.

The value of a primary that is a function reference is the value returned by the function.

The value of a primary that is an expression in parentheses is that of the expression.

7.2.7 Value of an Operation

The value of the result of an operation depends on the operator and the values of the operands.

7.2.7.1 Value of Intrinsic Operations

When the operands of the intrinsic operators satisfy the requirements of Table 7-1, the operations are intrinsic and are interpreted in the usual mathematical way as described in Table 7-4, except for integer division. For example, the binary operator * is interpreted as the mathematical operation multiplication and the unary operator – is interpreted as numeric negation. Intrinsic operations reference their operands for their value; therefore, the operands must be defined and allocated or associated as appropriate.

7.2.7.1.1 Value of Numeric Intrinsic Operations

Except for exponentiation to an integer power, when an operand for a numeric intrinsic operation does not have the same type, type parameters, or shape as the result of the operation, the operand is converted to the type, type parameter, and shape of the result and the operation is then performed. For exponentiation to an integer power, the operation may be performed without the conversion of the integer power, say, by developing binary powers of the first operand and multiplying them together to obtain an efficient computation of the result.

For integer division, when both operands are of type integer, the result is of type integer, but the mathematical quotient is, in general, not an integer. In this case, the re-

Table 7-4 Interpretation of the intrinsic operations

Use of operator			Interpretation
x_1	**	x_2	x_1 raised to the power x_2
x_1	/	x_2	x_1 divided by x_2
x_1	*	x_2	x_1 multiplied by x_2
x_1	−	x_2	x_2 subtracted from x_1
	−	x_2	x_2 negated
x_1	+	x_2	x_1 added to x_2
	+	x_2	Same as x_2
x_1	//	x_2	x_1 concatenated with x_2
x_1	.LT.	x_2	True if x_1 less than x_2
x_1	<	x_2	True if x_1 less than x_2
x_1	.LE.	x_2	True if x_1 less than or equal to x_2
x_1	<=	x_2	True if x_1 less than or equal to x_2
x_1	.GT.	x_2	True if x_1 greater than x_2
x_1	>	x_2	True if x_1 greater than x_2
x_1	.GE.	x_2	True if x_1 greater than or equal to x_2
x_1	>=	x_2	True if x_1 greater than or equal to x_2
x_1	.EQ.	x_2	True if x_1 equal to x_2
x_1	==	x_2	True if x_1 equal to x_2
x_1	.NE.	x_2	True if x_1 not equal to x_2
x_1	/=	x_2	True if x_1 not equal to x_2
	.NOT.	x_2	True if x_2 is false
x_1	.AND.	x_2	True if x_1 and x_2 are both true
x_1	.OR.	x_2	True if x_1 or x_2 or both are true
x_1	.NEQV.	x_2	True if either x_1 or x_2 is true, but not both
x_1	.EQV.	x_2	True if both x_1 and x_2 are true or both are false

sult is specified to be the integer value closest to the quotient and between zero and the quotient inclusively.

For exponentiation, there are four special cases that need to be described further.

- When both operands are of type integer, the result is of type integer; when x_2 is negative, the operation x_1 ** x_2 is interpreted as the quotient 1/(x_1 ** ($-x_2$)). Note that it is subject to the rules for integer division and in most cases is zero. For example, 4 ** (–2) is 0.
- The second case occurs when x_1 is a negative value of type integer or real and x_2 is of type real; this is not permitted.
- The third case occurs when x_2 is of type real or of type complex. In this case, the result returned is the principal value of the mathematical power function $x_1^{x_2}$. If x_1 is integer or real, it must not be negative.
- The standard does not specify what zero raised to the zero power is, nor even if it is a valid operation.

7.2.7.1.2 Value of Nonnumeric Intrinsic Operations

The intrinsic character operation performs the usual concatenation operation. For this operation, the operands must be of type character with the same kind type parameters. The length parameter values may be different. The result is of type character with the kind type parameter of its operands and a length type parameter value equal to the sum of the lengths of the operands. The result consists of the characters of the first operand in order followed by those of the second operand in order. For example, 'Fortran*b*' // '*b*2003' yields the result Fortran*bb*2003.

The intrinsic relational operations perform the usual comparison operations for character and most numeric operands. For these operations, the operands must both be of numeric type or both be of character type. The kind type parameter values of the operands of the numeric types may be different but must be the same for operands of type character. However, the lengths of the character operands may be different. Complex operands may be compared only for equality and inequality; the reason is that complex numbers are not totally ordered. The result in all cases is of type default logical.

When the operands of an intrinsic relational operation are both numeric, but of different types or type parameters, each operand is converted to the type and type parameters of the sum of the two operands. Then, the operands are compared according to the usual mathematical interpretation of the particular relational operator.

When the operands are both of type character, the shorter one is padded on the right with blank characters until the operands are of equal length. Then, the operands are compared one character at a time in order, starting from the leftmost character of each operand until the corresponding characters differ. The first operand is less than or greater than the second operand according to whether the characters in the first position where they differ are less than or greater than in the collating sequence (4.3.5.3). The operands are equal if both are of zero length or all corresponding characters are equal, including the padding characters. Note that the padding character is the Fortran blank when the operands are of default, ASCII, or ISO_10646 character kind and is a processor specified character for nondefault character kinds (3.1.2). Also, all comparisons, except equality (.EQ. or ==) and inequality (.NE. or /=), are processor dependent as they depend on the processor-dependent collating sequence. However, the collating sequences for the default kind is partially specified (4.3.5.3) and the important sorting cases work as expected. The collating sequences for ASCII and ISO 10646 characters are specified by the appropriate standards.

There is no ordering defined for logical values. However, logical values may be compared for equality and inequality by using the logical equivalence and not equivalence operators .EQV. and .NEQV. That is, L1 .EQV. L2 is true when L1 and L2 are equal and is false otherwise; L1 .NEQV. L2 is true if L1 and L2 are not equal and is false otherwise.

For logical operations, the operands must both be of logical type but may have different kind type parameters. When the kind type parameters are the same, the kind parameter value of the result is that value; if different, the kind parameter value of the result is processor dependent, but is that of either L1 or L2. The values of the result in all cases are specified in Table 7-5.

Table 7-5 The values of operations involving logical intrinsic operators

x_1	x_2	.NOT. x_2	x_1 .AND. x_2	x_1 .OR. x_2	x_1 .EQV. x_2	x_1 .NEQV. x_2
true	true	false	true	true	true	false
true	false	true	false	true	false	true
false	true	false	false	true	false	true
false	false	true	false	false	true	false

7.2.7.2 Value of Defined Operations

The interpretation of a defined operation is provided by a function subprogram with an OPERATOR interface (12.5.4). When there is more than one function with the same OPERATOR interface, the function giving the interpretation of the operation is the one whose dummy arguments match the operands in argument order, type, kind type parameters, and rank (12.8). In the following example, for the operation A .PLUS. B, where A and B are structures of the derived type RATIONAL, the generic interface specifies that the function RATIONAL_PLUS provides the interpretation of this operation.

```
TYPE( RATIONAL )
   INTEGER :: N, D
END TYPE

INTERFACE OPERATOR (.PLUS.)

   FUNCTION RATIONAL_PLUS (L, R)
      IMPORT RATIONAL
      TYPE (RATIONAL), INTENT (IN) :: L, R
      TYPE (RATIONAL)              :: RATIONAL_PLUS
   END FUNCTION RATIONAL_PLUS
```

```
      FUNCTION LOGICAL_PLUS (L, R)
         LOGICAL, INTENT (IN) :: L, R
         LOGICAL              :: LOGICAL_PLUS
      END FUNCTION LOGICAL_PLUS

   END INTERFACE
```

The result of A .PLUS. B is the same as RATIONAL_PLUS (A, B).

As with the intrinsic operations, the type, type parameters, and interpretation of a defined operation are independent of the context of the larger expression in which the defined operation appears. The interpretation of the same defined operation in different contexts is the same; however, the results may be different because the results of the procedure being invoked may depend on values that are not operands and that are different for each invocation.

The relational operators ==, /=, >, >=, <, and <= are synonyms for the operators .EQ., .NE., .GT., .GE., .LT., and .LE., even when they are defined operators. It is invalid, therefore, to have an interface block for both == and .EQ., for example, for which the order, types, type parameters, and rank of the dummy arguments of two functions are the same.

Defined operations are either unary or binary. An intrinsic unary operator cannot be defined as a new binary operator unless it is also an intrinsic binary operator. Note that this applies only to the .NOT. operator. Similarly, an intrinsic binary operator cannot be defined as a new unary operator unless it is also an intrinsic unary operator. However, a nonintrinsic defined operator, .PLUS. say, (that is, one that is not the same as an intrinsic operator) can be defined as both a unary and binary operator.

7.3 Evaluation of Expressions

The form of the expression, the precedence rules, and the meaning of the operations establish the interpretation. Once the interpretation is established, the compiler is free to evaluate the expression in any way that provides the same interpretation with one exception: parentheses specify an order of evaluation that cannot be modified. This applies to both intrinsic operations and defined operations. For defined operations, it is more difficult to determine whether an alternative evaluation scheme provides the same interpretation.

Another way to state this is to say that, except for the presence of parentheses, the compiler may evaluate any expression that is equivalent to the one written.

7.3.1 Equivalent Expressions

Two expressions are **equivalent** if they have the same value for all possible values of the operands in the expressions. For example, because addition is commutative and associative, the expressions A + B + C and C + A + B are equivalent.

This freedom for the compiler to use alternative equivalent evaluations permits the compiler to produce code that is more optimal in some sense (for example, fewer operations, array operations rather than scalar operations, or a reduction in the use of registers or work space), and thereby produce more efficient executable code.

For the numeric intrinsic operations, two expressions are equivalent if they are mathematically equivalent, not computationally equivalent. Mathematical equivalence assumes exact arithmetic (no rounding errors and infinite exponent range) and thus assumes the rules of commutativity, associativity, and distributivity as well as other rules that can be used to determine equivalence (except that the grouping of operations specified by parentheses must be honored). A + B + C and C + A + B are thus mathematically equivalent, but are not necessarily numerically equivalent because of possibly different rounding errors. On the other hand, K / 2 and 0.5 * K (where K is an integer) is a mathematical difference because of the special Fortran definition of integer division.

Parentheses within the expression must be honored. This is particularly important for computations involving numeric values where rounding errors or range errors may occur or for computations involving functions with side effects. Of course, if there is no computational difference between two evaluation schemes where parentheses are provided, the compiler can violate the parentheses integrity because no one can tell the difference. For example, the expression (1.0/3.0)*3.0 must be evaluated by performing the division first because of the explicit parentheses. Evaluating the expression as 1.0 would be valid if the value obtained by performing the division first and then the multiplication produced a result that is equal to 1.0 despite rounding errors. Although this sort of rearrangement might be possible in theory, it is not a practical option in general, unless all of the operands are constants as in the above example.

Table 7-6 gives examples of equivalent expressions where A, B, and C are operands of type real or complex, and X, Y, and Z are of any numeric type, I and J are of type integer, L1, L2, and L3 are of type logical, and C1, C2, and C3 are of type character with the length of C1 greater than or equal to the length of C3. If the expression in the left-hand column is written in a Fortran program, the compiler may evaluate the expression as if it were written as the equivalent expression in the right-hand column. All of the variables are assumed to be defined and have values that make all of the operations in this table well-defined.

Table 7-7 provides examples of invalid alternative expression evaluations. In the first three examples, the expressions are not mathematically equivalent; recall that when both operands of the division operator are of type integer, a Fortran integer division truncates the result toward zero to obtain the nearest integer quotient. The last three are not allowed because of parentheses in the expression written.

There are three concerns raised by alternative evaluation of an expression. They are:

- The rearrangement of an expression may cause the resulting computation to yield a different computational result. For example, evaluating the equivalent expressions A – B – C and A – (B + C) might produce significantly different results due to roundoff error for real or complex results.
- An unevaluated portion of an expression may reference a function with a side effect and so the side effect may or may not take place if the function is not invoked.

Table 7-6 Valid alternative expression evaluation

Expression	Equivalent evaluation
X + Y	Y + X
– X + Y	Y – X
X – Y + Z	X – (Y – Z)
X * A / Z	X * (A / Z)
X * Y – X * Z	X * (Y – Z)
A / B / C	A / (B * C)
A / 5.0	0.2 * A
I > J	(I–J) > 0
L1 .AND. L2 .OR. L1 .AND. L3	L1 .AND. (L2 .OR. L3)
L1 .AND. L1	L1
L1 .OR. F(X)	L1 ! if L1 is true
F(X) .AND. L1	L1 ! if L1 is false
C3 = C1 // C2	C3 = C1 ! LEN(C1) >= LEN(C3)

Table 7-7 Invalid alternative expression evaluation

Expression	Prohibited evaluations
I / 2	0.5 * I
X * I / J	X * (I / J)
I / J / A	I / (J * A)
(X + Y) + Z	X + (Y + Z)
(X * Y) – (X * Z)	X * (Y – Z)
X * (Y – Z)	X * Y – X * Z

- The rearrangement of an expression may result in an error that the programmer thought would be avoided by a particular order of evaluation. For example, in the logical expression that represents the condition for the IF test

```
if (present(x) .and. x > 0) then
```

the condition x > 0 may be evaluated first, resulting in an error if x is not present.

7.3.2 Side Effects and Partial Evaluation

With some exceptions described below, functions are allowed to have side effects; that is, they are allowed to modify the state of the program so that the state is different after the function is invoked than before it is invoked. This possibility potentially affects the results of a program when an equivalent expression is evaluated.

The first exception is pure procedures (12.7.1), which, in effect, are not allowed to have any side effects.

Some side effects are prohibited in all procedures: a function (or defined operation) within a statement must not affect nor be affected by a change in any entity in the same statement. Exceptions are those statements that have statements within them—for example, an IF statement or a WHERE statement. In these cases, the evaluation of functions in the logical expressions in parentheses after the IF or WHERE keyword or within the subscripts and stride in a FORALL statement are allowed to affect objects in the statement following the closing right parenthesis. For example, if F and G below are functions that change their actual argument I, the statements

```
IF (F (I))  A = I
WHERE (G (I))  B = I
```

are valid, even though I is changed when the functions are evaluated. Examples of invalid statements are:

```
A (I) = F (I)    ! Invalid code
Y = G (I) + I    ! Invalid code
```

because F and G change I, which is used elsewhere in the same statement.

The rules for equivalent evaluation schemes allow the compiler to elide evaluating any part of an expression that has no effect on the resulting value of the expression. Consider the expression X * F(Y), where F is a function and X has the value 0. The result will be the same regardless of the value of F(Y); therefore, it need not be evaluated. This shortened evaluation is allowed in all cases, even if F(Y) has side effects. In this case every data object that F could affect is undefined after the expression is evaluated—that is, it does not have a predictable value.

This normally applies to functions in logical expressions where expression evaluation is often "short-circuited". Some processors evaluate every term in a logical expression, others use run-time tests and skip further evaluation once the result is clear. Consider

```
PRESENT( A )  .AND.  A > 0  .AND.  LOG( A ) < 3.5
```

where A is an optional argument. If A is not present, the processor is allowed to evaluate the A > 0 term, and the program is invalid. Similarly, if A is present and has a negative value, the processor is allowed to evaluate LOG(A) and the program is again invalid.

The conclusion to be drawn from all of this is that the result of a program using a function with side effects is not predictable and hence not portable. To be completely safe and portable, a subroutine should be used in place of a function when a procedure is needed with a side effect. However, in practice, the side effect will occur as expected in most cases.

The execution of an array element, an array section, or a character substring reference requires, in most cases, the evaluation of the expressions that are the subscripts, strides, or substring ranges. It is not necessary for these expressions to be evaluated if, for example, the array section can be shown to be zero-sized or the substring can be shown to be of a zero-length by other means. For example, in the expression

A (1:0) + B ($expr_1$:$expr_2$)

$expr_1$ and $expr_2$ need not be evaluated as the conformance rules for intrinsic operations require that the section of B be zero sized.

7.4 Special Expressions

Expressions may appear in many places. In many contexts, expressions are restricted in some way. There are two particularly important special categories of expressions. Expressions that need to be evaluated at compile time are called **initialization expressions**—they can be used for variable initialization or kind values. Expressions that need to be evaluated on entry to a subprogram at the time of execution are called **specification expressions**; they can be used as array bounds and character lengths in specification statements, for example. Figure 7-1 shows the relationship between these categories of expressions.

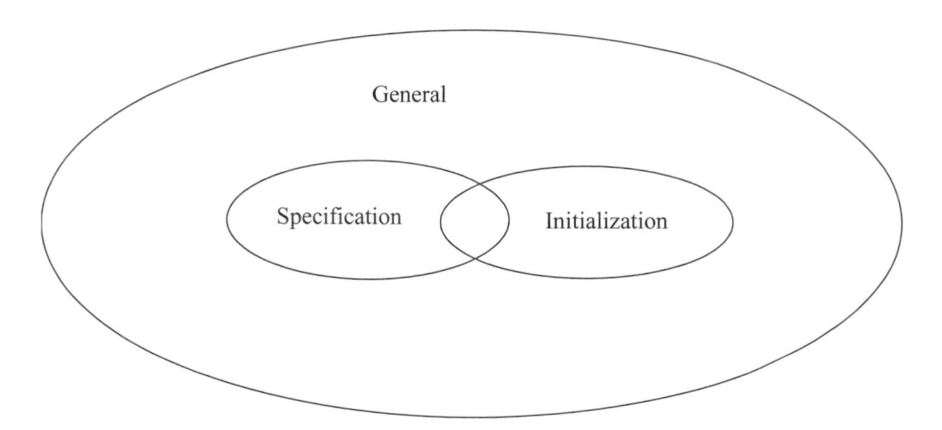

Figure 7-1 Diagram describing relationships between the kinds of expressions

7.4.1 Initialization Expressions

An initialization expression is built up from constants or attributes which are constant. In an initialization expression

1. Each primary is one of the following:
   a. a literal or named constant, or a subobject of a constant
   b. an array constructor

c. an implied-DO variable within an array constructor

d. a structure constructor

e. a reference to a standard elemental intrinsic function

f. a reference to a standard transformational intrinsic function other than NULL

g. a reference to NULL where any type parameter for its argument that is assumed or defined is an initialization expression

h. a reference to IEEE_SELECTED_REAL_KIND from the intrinsic module IEEE_ARITHMETIC (14.3)

i. a specification inquiry (7.4.2.1)

j. a kind type parameter of the derived type being defined

k. an initialization expression enclosed in parentheses

2. Each operation is intrinsic.

3. Each subscript, section subscript, starting and ending point of a substring range, type parameter value, and argument of an intrinsic function (except for a specification inquiry) must be an initialization expression.

4. Each component of a structure constructor must be an initialization expression, except that one corresponding to an allocatable component must be a reference to the intrinsic function NULL.

5. Each element of an array constructor must be an initialization expression.

6. Each expression specifying the initial, final, or stride value in an implied-DO in an array constructor must be an initialization expression.

7. If a specification inquiry designator or function argument is not an initialization expression, it must be a variable whose properties being inquired about are not assumed, not deferred, and not defined by an expression that is not an initialization expression.

8. If a specification inquiry depends on a type parameter or array bound, it must be specified prior to the specification inquiry, but not in the same entity declaration (5.1). See the examples below.

Examples of initialization expressions follow.

3.0E+01	A real literal constant
[7, (I, I = 1, 10)]	An array constructor
RATIONAL (1, 2+J)	A structure constructor where RATIONAL is a derived type and J is a named integer constant

LBOUND (A,1)+3	A reference to an inquiry intrinsic function where A is an explicit-shape array
LOG (2.0)	An intrinsic function
INT (N, 2)	An intrinsic function where N is a named constant
KIND (X)	An intrinsic function where X is a real variable with a known type parameter
I/3.3 + J**3.3	A numeric expression where I and J are named integer constants
SUM (A)	A reference to a transformational intrinsic function where A is a named integer array constant
KIND (0.0D0)	An inquiry function with a constant argument
SELECTED_REAL_KIND (6, 30)	An inquiry function with constant arguments
4.0 * ATAN(1.0)	A reference to an intrinsic function to compute an approximation to p
ceiling(DIGS/log10(radix(0.0)))	References to intrinsic functions to compute the number of model digits equivalent to a given number of decimal digits DIGS

7.4.2 Specification Expressions

A specification expression is restricted to using constants and variables whose values can be determined on entry to a scoping unit before any executable statement is executed. For example, variables that are dummy arguments, are in a common block, are in a host program unit, or are in an accessible module can appear in a specification expression. Specification expressions are used as bounds for arrays and length parameter values in type declarations, attribute specifications, dimension declarations, and other specification statements. Usually specification expressions are evaluated at run-time as a subprogram begins execution; however, simple forms are often evaluated at compile-time.

In order to describe specification expressions, a slightly more general sort of expression, a general specification expression, is defined first. A **specification expression** is then defined to be a general specification expression with a scalar integer value; the general category is used to describe what can occur within a specification expression, for example, as a function argument. In the statement

```
Real  ::   X( INT( TAN( 3.14 ) ) )
```

TAN(3.14) is a general specification expression and INT (TAN(3.14)) is a specification expression. A general specification expression is called a **restricted expression** in the Fortran standard.

A **general specification expression** is an expression that has the following limitations.

1. Each primary is one of the following:
   a. a literal or named constant, or a subobject of a constant
   b. an object designator with a base object (6.2) that is a dummy argument with neither the OPTIONAL nor the INTENT (OUT) attribute.
   c. an object designator with a base object that is in a common block
   d. an object designator with a base object that is made accessible from a module or the host
   e. an array constructor
   f. an implied-DO variable within an array constructor
   g. a structure constructor
   h. a specification inquiry (7.4.2.1)
   i. a reference to a standard intrinsic function that is not a specification inquiry
   j. a reference to a specification function (7.4.2.2)
   k. a type parameter of the derived type being defined
   l. a general specification expression enclosed in parentheses
2. Each operation is intrinsic.
3. Each subscript, section subscript, starting and ending point of a substring range, type parameter value, and argument of a function (except for a specification inquiry) must be a general specification expression.
4. Each element of an array constructor must be a general specification expression or an implied-DO variable.
5. Each expression specifying the initial, final, or stride value in an implied-DO must be a general specification expression.
6. Each component of a structure constructor must be a general specification expression.
7. Each designator and function argument in a specification inquiry must be a general specification expression or it must be a variable whose properties being inquired about are not assumed, not deferred, and not defined by an expression that is not a general specification expression.
8. Each final subroutine invoked must be pure.

Rules and restrictions:

1. When a specification expression is evaluated, it must not directly or indirectly invoke any procedure defined by the subprogram in which it appears. Neither a recursive reference to the subprogram in which the specification expression appears nor a reference to any contained internal procedures or statement functions is allowed.

2. A specification expression that is not also an initialization expression may appear only within the specification part of a subprogram or the type specification of a FUNCTION statement, but not in the main program. For example, the variable N in the program segment:

   ```
   INTEGER N
   COMMON N
   REAL A(N)
   ```

   is providing a value that determines the size of the array A. This program segment must not appear in a main program but may appear in the specification part of a subprogram.

Specification expressions are often used to declare the dimensions of dummy arguments or temporary arrays as in the following example:

```
Subroutine Example (A, B, C, N)

   Use values, only: J, K
   Real  ::  A ( N, J+2, 2*K-1)
   Real  ::  B ( Size(A) )
   Integer  ::  C ( Dot_Product ( [ J, K, N ], [ J, K, N] )
   Real :: Temp ( N, N )
```

A specification function (7.4.2.2) may be used in the declaration of bounds. The following example illustrates the declaration and use of specification functions.

```
MODULE SPEC_FNS
   IMPLICIT NONE
   PUBLIC :: N_ROWS, N_COLS
CONTAINS
   PURE FUNCTION N_ROWS(X)
      INTEGER :: N_ROWS
      REAL, INTENT(IN) :: X(:,:)
      N_ROWS = SIZE(X,DIM=1)
   END FUNCTION N_ROWS
```

```
      PURE FUNCTION N_COLS(X)
         INTEGER :: N_COLS
         REAL, INTENT(IN) :: X(:,:)
         N_COLS = SIZE(X,DIM=2)
      END FUNCTION N_COLS
   END MODULE SPEC_FNS

   SUBROUTINE S(A)
      USE SPEC_FNS
      IMPLICIT NONE
      REAL :: A(:,:)
      REAL :: TEMP(N_ROWS(A), N_COLS(A))
      REAL :: TEMP_TRANSPOSE(N_COLS(A), N_ROWS(A))
         . . .
   END SUBROUTINE S
```

7.4.2.1 Specification Inquiry

A **specification inquiry** is one of the following:

1. an array inquiry function (13.3.1.4)
2. the bit inquiry function BIT_SIZE
3. the character inquiry functions LEN or NEW_LINE
4. the kind inquiry function KIND
5. a numeric inquiry function (13.3.1.3)
6. a type parameter inquiry (6.3)
7. an IEEE inquiry function (Table 14-4, Table 14-9, B)

7.4.2.2 Specification Functions

A function is a **specification function** if it

1. is pure,
2. is not a standard intrinsic function,
3. is not an internal function,
4. does not have a dummy argument that is a procedure,
5. is not a statement function, and
6. has an explicit interface.

7.4.3 Differences Between Specification and Initialization Expressions

Initialization expressions are not a subset of specification expressions because the result of an initialization expression can be of any type, whereas the result of a specification expression must be of type integer and scalar. Also, specification expressions are not a subset of initialization expressions because specification expressions allow certain variables (such as dummy arguments and variables in common blocks) to be primaries, whereas initialization expressions do not allow such variables. Table 7-8 summarizes the allowed properties of the two forms of expressions.

Table 7-8 Differences and similarities between initialization and specification expressions

	Kind of expression	
Property	**Initialization**	**Specification**
Integer result	Yes	Yes
Any noninteger result	Yes	No
Scalar result	Yes	Yes
Array or structure result	Yes	No
Variables as primaries (limited to dummy arguments, common objects, host objects, module objects)	No	Yes
Intrinsic functions of any type	Yes[1]	Yes
Specification functions	No	Yes
Specification inquiry	Yes	Yes
Only constants as primaries	Yes	No
Only constant subscripts, strides, character lengths	Yes	No

[1]With restrictions on NULL

There is a good deal of commonality between the forms of general, specification, and initialization expressions; context determines which is required. For example:

```
Subroutine EXAMPLE ( A )
   Integer, Parameter    ::  Two = 2
   Real, Dimension(2)    ::  A
   Print *, 2
```

In the second line, the 2 is an initialization expression. In the next line, it is a specification expression and in the last line it is a general expression.

7.4.4 Uses of Specification and Initialization Expressions

An expression used in the following contexts must be a specification expression (7.4.2):

1. as an array bound in a type declaration specification
2. as a length type parameter value in a type declaration specification or character specification

Note that in some cases, such as a declarations for an item in a common block, the specification expression must also meet the requirements for an initialization expression.

An expression used in the following contexts must be an initialization expression (7.4.1).

1. as the value following the equal sign in a PARAMETER statement and in a type declaration statement with the PARAMETER attribute
2. as a subscript or substring range expression of a data object in a DATA statement
3. as a value in a DATA statement value list
4. as a kind type parameter value in a type specifier, type declaration statement, or constant
5. as the default value for a type parameter in a derived-type definition
6. as the default value for a component in a derived-type definition
7. as a value in an enumerator
8. as an initial value in a type declaration statement
9. as the name in a BIND attribute
10. as an ASYNCHRONOUS specifier in an input/output control list
11. as a length specifier for a character statement function or one of its dummy arguments
12. as a bound in an explicit-shape array dimension in some contexts
13. as a length type parameter value in a type declaration specification or character specification in some contexts
14. as a length specifier for an entity with the VALUE attribute
15. as an actual argument for a KIND dummy argument of an intrinsic function
16. as a case value in the CASE statement
17. as a subscript or substring range expression of an equivalence object in an EQUIVALENCE statement
18. as a part of any of the above items

The following rules and restrictions apply to the use of initialization and specification expressions.

Rules and restrictions:

1. If an entity is implicitly typed and then is explicitly declared in a subsequent type declaration statement, it must confirm the implicit type and type parameters.

```
PARAMETER  ( K = 2,  X = 3.0 )
REAL   X                              ! Valid
REAL   K                              ! Invalid
```

2. If an element of an array is referenced in one of these expressions, the array bounds must be specified in a prior specification.

3. If a specification or initialization expression depends on an attribute or value of an entity defined in the same specification part, the attribute or value must have been completely specified in a prior specification.

A prior specification in the above cases may be in the same specification statement, but to the left of the reference and not in the part of the statement specifying the entity. For example, the following declarations are valid:

```
INTEGER, DIMENSION(4), PARAMETER :: A = [4,3,2,1]
REAL, DIMENSION(A(2)) :: B, C(SIZE(B) + 1)
```

B and C are of size 3 and 4 respectively. But the following declaration is invalid because SIZE (E) precedes E:

```
REAL, DIMENSION(2) :: D(SIZE(E)), E      ! Invalid
```

The following also is invalid because the size of X is specified in the same entity declaration (5.1) as the occurrence of SIZE(X):

```
REAL :: X(9) = SIZE(X)                  ! Invalid
```

The following is an example using specification expressions to declare a working array with the same shape as the dummy argument array.

```
SUBROUTINE S(A)
REAL :: A(:,:)
REAL :: WORK(SIZE(A,DIM=1), SIZE(A,DIM=2))
   . . .
END SUBROUTINE S
```

7.5 Assignment

A common use of the result of an expression is to give a value to a variable. This is done with an assignment statement. For example,

```
RUG = BROWN + 2.34 / TINT
```

Assignment involves the evaluation of an expression and the use of the result to establish the value of a variable. In the assignment statement example above, the expression on the right of the assignment symbol (=) is evaluated and assigned to the variable RUG.

There are five forms of assignment: intrinsic, defined, pointer, masked array, and indexed parallel array.

Examples of the five forms of assignment are:

X = X + 1	Intrinsic assignment for reals
POLAR_0 = (0.0, 0.0)	Defined assignment for a derived type
PTR => X	Pointer assignment
WHERE (Z /= 0.0) A = B / Z	Masked array assignment
FORALL (I=1:N) A(I) = 1.0/I	Indexed parallel assignment

7.5.1 The Assignment Statement

The **assignment statement** is used for intrinsic assignment and defined assignment; its form is

```
variable = expression
```

The assignment statement is used to assign a value to a nonpointer variable of any type or to the target associated with a pointer variable. It defines or redefines the value of the variable or the target, as appropriate. In general, the value is determined from the result of evaluation of the expression on the right-hand side of the equal sign.

Rules and restrictions:

1. The variable must not be a whole assumed-size array (5.4.1.4); however, it may be an element or section of an assumed-size array.

7.5.2 Intrinsic Assignment

An assignment statement is an intrinsic assignment if it does not meet the requirements (12.5.4.3) of a defined assignment statement.

Rules and restrictions:

1. Assignment of an array to a scalar is not allowed, even if the size of the array is 1.

2. The variable must not be polymorphic.

3. The types and kind parameters of the variable and expression in an intrinsic assignment statement must be as given in Table 7-9.

4. If the variable and expression are of type character with different kinds, each must be either default, ASCII, or ISO 10646 kind.

Table 7-9 Type and type parameter requirements for the variable and expression in an intrinsic assignment

Variable	Expression
Integer, real, or complex	Integer, real, or complex
ISO 10646, ASCII, or default character kind	ISO 10646, ASCII, or default character kind
Other character kind	Character with the same kind as the variable
Logical	Logical
Derived type	Same derived type and same kind parameters as the variable; each length parameter must be the same as that of the variable unless the variable is allocatable and its corresponding length parameter is deferred

5. If the expression is an array, it must have the same rank as the variable. If the variable is not allocatable, the shapes of the variable and the expression must conform.

6. If the variable is a pointer, it must be associated with a target and the target must satisfy all of the conditions required of the variable.

Before the assignment begins, any necessary type conversions are completed if the variable has a different numeric type or type parameter from the expression. For numerical and logical types, the conversion is the same as that performed by the conversion intrinsic functions INT, REAL, CMPLX, and LOGICAL, as specified in Table 7-10; for the character type, see below.

Table 7-10 Conversion performed on an expression before assignment

Type of the variable	Value assigned
Integer	INT (expression, KIND (variable))
Real	REAL (expression, KIND (variable))
Complex	CMPLX (expression, KIND (variable))
Logical	LOGICAL (expression, KIND (variable))

For character assignment, if the variable and expression have different character kinds, the value of each character *c* in the expression is converted to the character kind of the variable by applying ACHAR (IACHAR(*c*), KIND(*variable*)). If a character in the expression is not representable in the character kind of variable, the result is processor dependent. If the variable and expression have different lengths and the expression length is greater than the length of the variable, characters are truncated from the right of the expression. If the variable length is greater than the expression length, blank characters are appended on the right of the expression. Except for default, ASCII, and ISO 10646 character kinds, the blank padding character is processor dependent.

The evaluation of subscript and section subscript expressions that are part of the expression or part of the variable and the complete expression on the right-hand side of the equal sign is performed before any portion of the assignment is performed. This may require temporary storage to hold values before they can be stored into the variable, however, the as-if rule often allows the compiler to optimize temporary storage. For example, in evaluating a character string expression on the right-hand side of an assignment, the values in the variable on the left-hand side may be used, as in

```
DATE (2:5) = DATE (1:4)
```

This is not the same as the similar appearing DO loop

```
DO  I = 1, 4
   DATE( I+1 : I+1 )  =  DATE (I : I )    ! not the same thing
END DO
```

It is, however, the same as

```
DO  I = 4, 1, -1
   DATE( I+1 : I+1 )  =  DATE ( I : I )
END DO
```

Compilers are free to evaluate the expressions and perform the assignments in an order that avoids use of temporary storage.

Similarly, all subscripts and substrings are established before any values are assigned.

```
A = [ 1, 2 ]
A(A) = A(A) + 1 ! Increments A(1) and A(2) by 1
```

Array assignment is element-by-element but the order is not specified. If A and B are real arrays of size 10, in the whole array assignment:

```
A = B
```

the first element of B would be assigned to the first element of A, the second element of B would be assigned to the second element of A, and this would continue element-by-element for 10 elements. The assignment of elements, however, may be performed in any order, as long as the effect is as if all elements were assigned simultaneously.

When a scalar is assigned to an array, the assignment behaves as if the scalar is broadcast to an array of the shape of the variable; it is then in shape conformance with the variable. In the example:

```
REAL A(10)
A = 1.0
```

all ten elements of the array A are assigned the value 1.0.

The evaluation of expressions in the variable on the left-hand side, such as subscript expressions, must have no affect on, nor be affected by, the evaluation of the expression on the right-hand side. The right-hand side expression is evaluated completely before any assignment is made to a variable on the left-hand side. (As usual, this requirement that the expression on the right be evaluated first is specifying the semantics of the statement and does not imply that an implementation must perform the computation in this way if there is an equivalent order that computes the same result.)

Consider the case where the variable is allocatable. If it is allocated, but if the expression has a different shape or any of the corresponding length parameters differ, the variable is deallocated. Whether originally unallocated or deallocated as described in the previous sentence, it is allocated with each deferred type parameter equal to the corresponding type parameter of the expression, with the same shape as the expression, and with each lower bound equal to the corresponding lower bound of the expression. Regardless of the original state of the variable, it will be allocated with shape and length parameters to match the expression. The assignment then proceeds as an ordinary array assignment as described above.

Note that even if A is an allocatable array, A(:) is not an allocatable array; it is an array subsection.

```
integer, dimension(:), allocatable :: a
a = [ 1, 2, 3 ]     ! Size of a becomes 3
a = [ 1, 2, 3, 4 ] ! Size of a becomes 4
a(:) = [ 1, 2, 3 ] ! Illegal because a(:) is size 4 and not allocatable
```

If the variable is not allocatable and the variable and expression are of character type with different lengths, the assignment occurs as follows: if the length of the variable is less than that of the expression, the value of the expression is truncated from the right; if the length of the variable is greater than the expression, the value of the expression is filled with blanks on the right. The character used as the blank character for default, ASCII, or ISO 10646 character kind is the Fortran blank character and otherwise is a blank padding character specified by the processor (3.1.2).

If an allocatable variable of type character has deferred length, the variable assumes the length of the expression. For example

```
character(len=:), allocatable :: c
c = "Brahms"
print *, len(c)
c(:) = "Beethoven"
print *, len(c)
```

prints the value 6 in both cases. In the first assignment, c is an allocatable variable and it assumes the length of the expression, 6. In the second assignment, c(:) is not a variable, it is a character substring, and, therefore, does not have the allocatable attribute. Ordinary assignment takes place, the length remains 6, and the new value is Beetho. Deferred length characters provides much of the functionality of what is commonly called "variable length character strings". Similar functionality can be provided by derived types with a length parameter.

Derived-type intrinsic assignment is performed as if the assignment were expanded, component-by-component with corresponding elements from the variable and the expression, into separate assignment statements. Each assignment is processed as follows:

1. If the component is a pointer, pointer assignment (7.5.5) is used.
2. If the component is not allocatable, is not a pointer, and there is a type-bound assignment available, that defined assignment is used.
3. If the component is not allocatable, is not a pointer, and there is no type-bound assignment available, intrinsic assignment, following the rules given above, is used. This is true even if a non-type-bound defined assignment is available.
4. If the component is allocatable
   a. if it is allocated, it is deallocated.
   b. if the corresponding component of the expression is allocated, the variable component is allocated with the same dynamic type and type parameters and, if it is an array, with the same bounds. The value of the expression component is then assigned to the variable component using defined assignment if there is a consistent type-bound assignment available; otherwise, intrinsic assignment is used.

If the variable is a subobject, the assignment does not affect any of the parts of the object not designated.

7.5.3 Defined Assignment

Defined assignment is an assignment operation provided by a subroutine with the generic specifier ASSIGNMENT (=) (12.5.4.3). When the variable and expression in the assignment statement are of intrinsic types and do not satisfy the type matching rules in Table 7-9 or are of different derived types, a defined assignment operation will be used. Defined assignment also may replace the intrinsic assignment operation for derived types.

The effect of the defined assignment on variables in the program is determined by the referenced subroutine.

Rules and restrictions:

1. There must be an accessible generic interface for the subroutine with the generic specifier of the form ASSIGNMENT (=). This can be either in an interface block or a type definition (12.5.4.3, 4.4.7)
2. The types and kind type parameters of the variable and expression in the assignment statement must be compatible with the dynamic types of those of the dummy arguments.

3. For a nonelemental subroutine, the rank of the variable and the expression in the assignment must match the ranks of the corresponding dummy arguments of the subroutine. For an elemental subroutine, the variable must be an array and the expression must be conformable with the variable, or both the variable and expression must be scalar. If the variable and expression match both the interface to a nonelemental and elemental subroutine, the nonelemental subroutine defines the assignment operation.

Example:

```
INTERFACE  ASSIGNMENT (=)

   ELEMENTAL SUBROUTINE  RATIONAL_TO_REAL (L, R)
      IMPORT  RATIONAL
      TYPE (RATIONAL), INTENT (IN) :: R
      REAL, INTENT(OUT)            :: L
   END SUBROUTINE  RATIONAL_TO_REAL

   ELEMENTAL SUBROUTINE  REAL_TO_RATIONAL (L, R)
      IMPORT  RATIONAL
      REAL, INTENT(IN)              :: R
      TYPE (RATIONAL), INTENT (OUT) :: L
   END SUBROUTINE  REAL_TO_RATIONAL

END INTERFACE
```

The above interface block specifies two defined assignments for two assignment operations in terms of two external subroutines, one for assignment of objects of type RATIONAL to objects of type default real and another for assignment of objects of type default real to objects of type RATIONAL. With this interface block, the following assignment statements are defined:

```
REAL  R_VALUE
TYPE (RATIONAL)  RAT_ARRAY(10)

R_VALUE = RATIONAL (1, 2)
RAT_ARRAY = 3.7
```

The second example is equivalent to

```
CALL REAL_TO_RATIONAL ( RAT_ARRAY, (3.7) )
```

7.5.4 Polymorphic Assignment

There is the restriction that the variable on the left of an intrinsic assignment statement must not be polymorphic. However, it is possible to assign to a polymorphic variable using a defined assignment. In the following example, the two assignment statements in the main program assign a value of two different types to the variable X.

```
module types
   integer, parameter, public :: &
      red = 1, blue = 2, green = 3
   type, public :: line_type
      real :: x, y
   end type line_type
   type, public, extends(line_type) :: painted_line_type
      integer :: color
   end type painted_line_type
   type, public, extends(line_type) :: labeled_line_type
      character(len=99) :: label
   end type labeled_line_type
end module types

module poly_assign_mod
   use types
   interface assignment(=)
      module procedure poly_assign_sub
   end interface
   private :: poly_assign_sub
   public :: assignment (=)
contains
subroutine poly_assign_sub(v, e)
   class(line_type), intent(in) :: e
   class(line_type), intent(out), allocatable :: v
   allocate (v, source = e)
end subroutine poly_assign_sub
end module poly_assign_mod

program p
use types
use poly_assign_mod
class(line_type), allocatable :: line

line = painted_line_type(1.1, 2.2, blue)
   . . .
line = labeled_line_type(4.4, 6.6, "long")
   . . .
end program p
```

Note that in subroutine poly_assign_sub, v is an INTENT (OUT) variable and therefore deallocated on entry to poly_assign_sub. The subroutine then allocates v with the dynamic type and value of e.

7.5.5 Pointer Assignment

A pointer is a variable that points to another object. The term **pointer association** is used for the concept of "pointing to" and the term **target** is used for the object associated with a pointer.

A pointer assignment associates a pointer with a target, and terminates any previous association for that pointer. If the target is a disassociated or undefined pointer, the pointer becomes disassociated or undefined, respectively.

There are two forms of pointer assignment, data pointer assignment and procedure pointer assignment.

There is no pointer analog to defined assignment; all pointer assignments are intrinsic.

7.5.5.1 Data Pointer Assignment

The forms of a data pointer assignment are (R735):

```
data-pointer-object [ ( bounds-specification-list ) ] => data-target
data-pointer-object ( bounds-remapping-list ) => data-target
```

where a data pointer object (R736) has one of the forms:

```
variable-name
structure-component
```

a data target (R739) has one of the forms:

```
variable
expression
```

a bounds specification is of the form (R737):

```
lower-bound :
```

and a bounds remapping is of the form (R738):

```
lower-bound : upper-bound
```

If the variable on the right of => has the TARGET attribute, the pointer object on the left of => becomes associated with this target.

If the variable on the right of => has the POINTER attribute and is associated, the pointer object on the left of => points to the same data that the target points to after the pointer assignment statement is executed. If the variable on the right of => has the POINTER attribute and is disassociated or if the expression on the right is a reference to the intrinsic function NULL, the data pointer object on the left of => becomes disassociated.

If the variable on the right of => has the POINTER attribute and has an undefined association status, the association status of the data pointer object on the left of => becomes undefined.

Data pointer assignment associates the pointer with the new target. If the pointer was previously associated with allocated memory, the assignment does not deallocate the old memory. This can cause memory leaks.

Rules and restrictions:

1. If the pointer object is a variable name, the name must have the POINTER attribute. If the pointer object is a structure component, the component must have the POINTER attribute.

2. If the target is not unlimited polymorphic, the pointer object must be type compatible with the target and the corresponding kind type parameters must have the same value.

3. If the target is unlimited polymorphic, the pointer object must be unlimited polymorphic, of a sequence derived type, or of a type with the BIND attribute.

4. If the target is a variable, it must have the TARGET or POINTER attribute.

5. The target expression must be a pointer. The only form of expression which satisfies this restriction is a function whose result is a pointer. This can a defined operation, a user written function, or the intrinsic function NULL.

6. The target must not have a vector subscript.

7. If the target is allocatable, it must be allocated.

8. If there is a bounds list, the number of bounds must be the same as the rank of the pointer object.

9. If there is a bounds remapping list, the number of bounds remappings must be the same as the rank of the pointer object.

10. If there is a bounds remapping list, the target must have rank one; otherwise, the ranks of the pointer object and the target must be the same.

11. If there is a bounds remapping list, the target must not be a disassociated or undefined pointer. If s is the size of the target, s must be greater than or equal to the size of the pointer object. The first s elements of the target, in array element order, become the target of the pointer object.

12. If the pointer object is polymorphic, it assumes the dynamic type of the target.

13. If the pointer object is of a type that has the BIND attribute or is of a sequence type, the dynamic type of the target must be the same type.

14. If the pointer object is not polymorphic and the target is polymorphic with a dynamic type that differs from its declared type, the assignment will be to the ancestor component of the target that has the same type as the pointer object; otherwise, the assignment is to the target.

15. If the target is a disassociated pointer, all nondeferred type parameters of the declared type of the pointer object must be the same as the corresponding type parameters of the target. Otherwise, all nondeferred type parameters of the declared

type of the pointer object that correspond to nondeferred type parameters of the target must have the same value as the corresponding type parameters of the target.

16. If the pointer object has nondeferred type parameters that correspond to deferred type parameters of the target, the target must not be an undefined pointer.

17. The target must not be a whole assumed-size array. If it is an array section of an assumed-size array, it must have a subscript or a triplet section with the upper bound specified in the last dimension.

18. If the target of a pointer must not be referenced or defined, the pointer must not be referenced or defined while it is an alias of that target.

Note that, when a pointer appears on the right side of => in a pointer assignment, the pointer on the left side of => is defined or redefined to be associated with the target of the pointer on the right side of the =>. To put it another way, the pointer on the right does not become the target of the pointer on the left; this does not create "a pointer to a pointer".

Examples:

```
MONTH => DAYS(1:30)
PTR => X(:, 5)
NUMBER => JONES % SOCSEC
HEAD_OF_CHAIN => NULL( )
```

An example where a target is another pointer is:

```
REAL, POINTER :: PTR, P
REAL, TARGET :: A
REAL  B
A = 1.0
P => A
PTR => P
B = PTR + 2.0
```

This program segment defines A with the value 1.0, associates P with A; then PTR is associated with A, not with P. The value assigned to B in the regular assignment statement is 3.0, because the reference to PTR in the expression yields the value of the target A which is the value 1.0.

If the pointer object is an array, the pointer assignment statement establishes the extents for each dimension of the array. If bounds remapping is specified, the extents and lower and upper bounds are specified by the remapping. If no bounds remapping is specified, the extents are those of the target. If a bounds specifier is present, it specifies the lower bounds; otherwise the lower bounds for each dimension are the same as the result of the LBOUND (7.2.4, A) function applied to that dimension. For example, if the following statements have been processed:

```
INTEGER, TARGET :: T(11:20)
INTEGER, POINTER :: P1(:), P2(:)
P1 => T
P2 => T(:)
```

the extents of P1 are those of T, namely 11 and 20, but those of P2 are 1 and 10, because T(:) has a section subscript list.

Bounds specifications and remapping may be used to define the subscript extents in the pointer object.

```
REAL, TARGET  :: DATA(1000)
REAL, POINTER :: DP(:), DQ(:), DR(:)
   . . .
DP(FIRST:LAST) => DATA(FIRST:LAST)
DQ             => DATA(FIRST:LAST)
DR(0:)         => DATA(FIRST:LAST)
```

In the first case, DP is assigned with a simple form of bounds remapping and will have lower and upper bounds of FIRST and LAST, respectively. In the second case, no subscripts are specified for DQ and it will have 1 and LAST-FIRST+1 as its lower and upper bounds, respectively. In the last case, a bounds specification is used and DR will have bounds of 0 and LAST-FIRST, respectively.

Bounds remapping may also be used to give multi-dimensional views of a rank one array.

```
REAL, DIMENSION(1000*1000), TARGET :: LOTSA_DATA
REAL, DIMENSION(:, :), POINTER :: SQUARE, SMALL_SQUARE
REAL, DIMENSION(:), POINTER :: DIAGONAL
SQUARE(1:1000, 1:1000) => LOTSA_DATA
```

SQUARE is a two-dimensional representation of the data

A target array may have triplets for subscripts. With the definitions above and

```
DIAGONAL => LOTSA_DATA(1 : : 1001)
SMALL_SQUARE => SQUARE (1 : 10, 1 : 10)
```

DIAGONAL is an alias of the diagonal of SQUARE and SMALL_SQUARE is an alias for the upper left corner of SQUARE.

Pointers may become associated using the ALLOCATE (6.7.1) statement instead of a pointer assignment statement. Pointers may become disassociated using the DEALLOCATE (6.7.3) or NULLIFY (6.7.2) statements, as well as with the pointer assignment statement.

A pointer may be used in an expression (see 7.2.2.1 for the details). Briefly, any reference to a pointer in an expression, other than in a pointer assignment statement or in certain procedure references, yields the value of the target associated with the pointer. When a pointer appears as an actual argument corresponding to a dummy argument that has the POINTER attribute, the reference is to the pointer and not the value. Note that a procedure must have an explicit interface (12.5.1) if it has a dummy argument with a POINTER attribute.

7.5.5.2 Procedure Pointer Assignment

Procedure pointer assignment is similar to data pointer assignment, except that the pointer must be a procedure pointer and the target must be a procedure.

The form of a procedure pointer assignment is (R735):

```
procedure-pointer-object => procedure-target
```

where a procedure pointer object has one of the forms (R740):

```
procedure-pointer-name
structure-component
```

and a procedure target has one of the forms (R742):

```
procedure-name
procedure-component-reference
expression
```

If the procedure target is not a pointer, the procedure pointer object is pointer associated with the target. If the procedure target is a pointer, the procedure pointer object assumes the definition status of the pointer target and, if the pointer target is associated, the procedure pointer becomes associated with the same target.

Procedure pointers are declared with the PROCEDURE statement (5.11).

Rules and restrictions:

1. If the pointer object is a structure component, the component must be a procedure pointer.

2. The target expression must be a pointer. The only form of expression which satisfies this restriction is a function whose result is a pointer. This can be a defined operation, a user written function, or the intrinsic function NULL.

3. If the target is a procedure name, it must be the name of an external procedure, module procedure, dummy procedure, a specific intrinsic function (not marked with a asterisk in Table 13-1), or a procedure pointer. However, it must not be a nonintrinsic elemental procedure.

4. If the pointer object has an explicit interface, it must have the same characteristics (12.5.1.1) as the target, except that the pointer need not be pure or elemental, even if the target is.

5. If the characteristics of either the pointer object or the target require an explicit interface, both must have an explicit interface.

6. If the pointer object has an implicit interface and is typed or referenced as a function, the target must be a function; if it is referenced as a subroutine, the target must be a subroutine.

7. If both the pointer and target are functions, they must have the same type; corresponding type parameters must either have the same value or must both be deferred.

8. If the target is the name of both a specific and generic intrinsic procedure, only the specific procedure is associated with the pointer.

Examples:

```
PROCEDURE, POINTER :: PP
PROCEDURE (REAL) :: BESSEL
   . . .
PP => BESSEL

TYPE :: T
   REAL :: X
   PROCEDURE(REAL), POINTER :: TPP
END TYPE T

INTRINSIC :: SORT

TYPE(T) :: TP
   . . .
TP % TPP => SQRT ! TPP becomes associated with the specific SQRT

ABSTRACT INTERFACE
   FUNCTION EXT_FCN(X)
      REAL :: X
      REAL :: EXT_FCN
   END FUNCTION
END INTERFACE
PROCEDURE (EXT_FCN), POINTER :: P
PROCEDURE (EXT_FCN)          :: GAMMA
P => GAMMA
```

In this example, the abstract interface EXT_FCN declares functions that have one real argument and return one real result. After execution of the pointer assignment statement, the pointer P points to the GAMMA function.

7.5.6 Masked Array Assignment—WHERE

Sometimes, it is desirable to assign only certain elements of one array to another array. The masked array assignment often is used for such selective assignment, as the following example illustrates:

```
REAL, DIMENSION(10,10) :: A, RECIP_A
  ...

WHERE( A /= 0.0 )
    RECIP_A = 1.0 / A    ! Assign only where the
                         !    elements are nonzero.
```

```
ELSEWHERE
    RECIP_A = 1.0          ! Use the value 1.0 for
                           !    the zero elements.
END WHERE
```

The first array assignment statement is executed for only those elements where the mask A /= 0.0 is true. Next, the second assignment statement (after the ELSEWHERE statement) is executed for only those elements where the same mask is false. If the values of RECIP_A where A is 0 are never used, this example can be simply written using the WHERE statement rather than the WHERE construct as follows:

```
WHERE( A /= 0.0 ) RECIP_A = 1.0 / A
```

A **masked array assignment** is an intrinsic assignment statement in a WHERE block, an ELSEWHERE block, or a WHERE statement for which the variable being assigned is an array. The WHERE statement and WHERE construct appear to have the characteristics of a control statement or construct such as the IF statement and IF construct. But there is a major difference; every assignment statement in a WHERE construct is executed, whereas at most one block in the IF construct is executed. Similarly, the assignment statement following a WHERE statement is always executed. For this reason, WHERE statements and constructs are discussed here under assignment rather than under control constructs.

In a masked array assignment, the assignment is made to certain elements of an array based on the value of a logical array expression serving as a mask for picking out the array elements. The logical array expression acts as an array-valued condition on the following:

- elemental intrinsic operations
- elemental intrinsic function references
- elemental user-defined operations
- elemental user-defined function references
- intrinsic assignment
- elemental user-defined assignment

for each array assignment statement in the WHERE statement or WHERE construct.

7.5.6.1 Form of the WHERE Construct

The form of the WHERE construct (R744) is:

```
[ where-construct-name : ] WHERE ( logical-expression )
    [ where-body-construct ] ...
[ ELSEWHERE ( logical-expression ) [ where-construct-name ]
    [ where-body-construct ] ... ]
      ...
```

[ELSEWHERE [where-construct-name]
  [where-body-construct] ...]
END WHERE [where-construct-name]

and a where body construct (R746) is one of:

assignment-statement
where-construct
where-statement

Note that a FORALL is not allowed in a WHERE construct, although a WHERE may appear in a FORALL.

The logical expression that appears on the initial WHERE statement forms a mask that controls the evaluation of expressions and assignment of values in array assignment statements that appear in the WHERE body constructs. If a logical expression appears on an ELSEWHERE statement, that statement is referred to as a masked ELSEWHERE statement. That logical expression further restricts the mask, as described below in 7.5.6.2, that would otherwise apply to the WHERE body constructs following the ELSEWHERE statement.

Rules and restrictions:

1. In each assignment statement in a WHERE construct, the variable being defined must have the same shape as the mask. If a WHERE construct contains a masked ELSEWHERE statement or if one of the WHERE body constructs is a WHERE statement or another WHERE construct, each mask expression must have the same shape.
2. Each statement and construct in a WHERE construct is executed in sequence as it appears in the construct. Subsequent masks may use the assigned values.
3. Each mask is evaluated only once. Subsequent changes to the values of entities in the logical expression that defines the mask have no effect on the value of the control mask.
4. A defined assignment (12.5.4.3) in a WHERE construct must be defined by an elemental subroutine (12.7.2).
5. In a WHERE construct, only the WHERE statement may be a branch target.

7.5.6.2 Execution of a WHERE Construct

Except as described below, an elemental operation or function within the expression or variable of an assignment statement in the construct is evaluated only for the elements corresponding to true values in the control mask. For example:

```
REAL, DIMENSION(10, 20) :: A, SQRT_A
   . . .
```

```
WHERE (A>0.0)
   SQRT_A = SQRT(A)
END WHERE
```

Square roots are calculated only for positive elements of A.

If an array constructor appears in a logical expression or an assignment statement in the construct, the array constructor is evaluated completely without any mask control.

Nonelemental function references in a logical expression or the variable or expression of an assignment statement in the construct are completely evaluated even though all elements of the resulting array may not be used. For example:

```
REAL A(2,3), B(3,10), C(2,10), D(2,10)
INTRINSIC MATMUL
   . . .
WHERE (D<0.0)
   C = MATMUL(A,B)
END WHERE
```

The matrix product A × B is performed, yielding all elements of the product. The only elements of C assigned a value are those corresponding to elements of D that are negative.

When a WHERE construct is executed, both a control mask and a pending control mask are established. It is the control mask that governs the execution of the following block of statements. If the WHERE construct is not a nested WHERE construct, the control mask, *mask*, has the value of the logical expression. The pending control mask has the value .NOT. *mask*. The calculation of the mask and the pending mask for subsequent blocks in a WHERE construct can be illustrated with the following example.

```
WHERE (C1)              ! Statement 1
   . . .                      ! Block 1
ELSEWHERE (C2)          ! Statement 2
   . . .                      ! Block 2
ELSEWHERE               ! Statement 3
   . . .                      ! Block 3
END WHERE
```

Following execution of statement 1, the control mask has the value C1 and the pending control mask has the value .NOT. C1. Following execution of statement 2, the control mask has the value (.NOT. C1) .AND. C2 and the pending control mask has the value (.NOT. C1) .AND. (.NOT. C2). Following execution of statement 3, the control mask has the value of the pending control mask (.NOT. C1) .AND. (.NOT. C2). This complicated looking formulation has a simple effect: it guarantees that each corresponding location in the various conformable arrays will only be processed once. The expression for the pending control mask is equivalent to .NOT. (C1 .OR. C2 .OR. ...). Once an element in the control mask becomes true, the corresponding element in the pending control mask will become false. No subsequent ELSEWHERE (logical expression) block will process that element because the pending control mask is ORd with the expression mask for each subsequent block. The process acts as if there were a pending control mask with all true values for the WHERE logical expression. The final ELSEWHERE block (if there is one) will process all of the elements that were not processed by any previous blocks (the ones that still have a true in the pending control mask), and only those elements.

If the WHERE construct in the example above is a nested WHERE construct, it appears in a block that is governed by a control mask *outer-mask*. The control mask for Block 1 of the nested construct is then *outer-mask* .AND. C1 and the pending control mask is .NOT. (*outer-mask* .AND. C1). The following control masks and pending control masks are calculated from these initial masks as above. Only elements selected by *outer-mask* can be processed in the inner nested WHERE construct. On execution of the inner END WHERE statement, the control mask reverts to *outer-mask*. This is also the case for a nested WHERE statement.

Consider:

```
INTEGER :: N(9) = [1,2,3,4,5,6,7,8,9]
WHERE ( MOD (N,2) == 0 )
   N = 2                          ! N is now [1 2 3 2 5 2 7 2 9]
ELSEWHERE ( MOD (N,3) == 0 )
   N = 3                          ! N is now [1 2 3 2 5 2 7 2 3]
ELSEWHERE ( MOD (N,5) == 0)
   N = 5                          ! N is now [1 2 3 2 5 2 7 2 3]
ELSEWHERE
   N = 0                          ! N is now [0 2 3 2 5 2 0 2 3]
ENDWHERE
```

The masks for the various blocks are shown in Table 7-11.

Table 7-11 Masks for various WHERE blocks

Statement	Mask expression value	Control mask	Pending control mask
before block (as if)			[T T T T T T T T T]
where block	[F T F T F T F T F]	[F T F T F T F T F]	[T F T F T F T F T]
first elsewhere block	[F F T F F T F F T]	[F F T F F F F F T]	[T F F F T F T F F]
second elsewhere block	[F F F F T F F F F]	[F F F F T F F F F]	[T F F F F F T F F]
final elsewhere block		[T F F F F F T F F]	

7.5.6.3 WHERE Statement

The form of the WHERE statement (R743) is:

```
WHERE ( logical-expression ) array-assignment-statement
```

It is equivalent to the WHERE construct

```
WHERE ( logical-expression )
   array-assignment-statement
END WHERE
```

Examples:

```
WHERE( TEMPERATURES > 90.0 )  HOT_TEMPS = TEMPERATURES
WHERE( TEMPERATURES < 32.0 ) COLD_TEMPS = TEMPERATURES
```

7.5.7 Indexed Parallel Array Assignment—FORALL

The FORALL statement and construct provide a mechanism to specify an indexed parallel assignment of values to an array for the following sorts of formulas often found in mathematical treatises:

$$a_{ij} = i + j, \text{ for } i = 1 \text{ to } n, j = 1 \text{ to } m$$

or

$$a_{ii} = b_i, \text{ for } i = 1 \text{ to } n$$

The first formula above can be translated into nested DO loops:

```
DO J = 1, M
   DO I = 1, N
      A(I,J) = I + J
   END DO
END DO
```

But this formulation does not allow for the optimization that can be achieved on some computers when array notation is used.

FORALL statements and constructs provide a notational convenience, but also, because of the rules that govern their execution, they express data parallel computations that can be optimized on certain machine architectures. However, they sometimes are less efficient than a corresponding nest of DO loops on other machine architectures. One of the rules that is imposed is that any procedures referenced in the FORALL body or the mask expression must be pure (12.7.1). A pure procedure is one that is virtually free of side effects.

The Fortran array assignment statement requires that the expression on the right-hand side be conformable with the array on the left. The first formula above can be expressed in Fortran with an array assignment that makes use of the SPREAD intrinsic function on the right side to create a conformable array:

```
A = SPREAD ( (/ (I, I=1,N) /), DIM=2, NCOPIES=M) + &
         SPREAD ( (/ (I, I=1,M) /), DIM=1, NCOPIES=N)
```

It is not obvious at a glance that this assignment statement has the same effect as the first formula. A FORALL statement is provided that makes use of array element and section references to express such calculations more naturally and at the same time indicate computations that may be executed in parallel.

```
FORALL (I=1:N, J=1:M) A(I,J) = I+J
```

The second formula above cannot be expressed with array section notation, but a FORALL statement can be used to assign the elements of the array B of rank one to the diagonal of array A:

```
FORALL (I=1:N) A(I,I) = B(I)
```

The information in parentheses following the FORALL keyword is called the FORALL header. The header exerts some control over the following statement or block of statements. If there is a need to control more than one statement in this way, a FORALL construct can be used, for example:

```
FORALL (I=2:N-1, J=2:N-1)
   A(I,J) = (A(I+1,J) + A(I-1,J) + A(I,J+1) + A(I,J-1))/4.0
   B(I,J) = 1.0/A(I+1,J+1)
END FORALL
```

The statements and constructs that appear between the FORALL statement and END FORALL statement make up the FORALL body. The following are permitted in a FORALL body:

1. assignment statements
2. pointer assignment statements
3. WHERE constructs and statements
4. FORALL constructs and statements

Each construct or statement in a FORALL body is completely evaluated in statement order for all selected index values before any evaluation is performed on the next one. For an assignment statement, such as one of those in the previous example, all expressions on the right hand side are evaluated for all selected index values and these evaluations may occur in any order of the selected index values. After all of these evaluations have been performed for a particular statement, the assignments for this statement may occur in any order. Thus in the first assignment statement in the construct above, it is always the original values of the elements in array A that participate in the calculation. In the second assignment statement, it is the new values of the elements of array A that determine the values of the elements of array B.

The FORALL statement resembles a loop construct, but its evaluation rules really treat the statements within the construct as indexed parallel operations, in which a particular statement is executed for all selected index values before the next statement in the FORALL body is executed. As such, it is not a control construct, but a special kind of parallel assignment statement. On the other hand, a DO construct executes each statement in its range in order for a particular index value and then returns to the first statement in the range to repeat the computations for the next index value.

Sometimes it is desirable to exclude some elements from taking part in a calculation. Thus an optional mask expression may appear in a FORALL header. For example,

```
FORALL (I=1:N, J=1:M, A(I)<9.0 .AND. B(J)<9.0) C(I,J) = A(I) + B(J)
```

7.5.7.1 Form of the FORALL Construct

The form of the FORALL construct (R752) is:

```
[ forall-construct-name : ] FORALL ( forall-triplet-specification-list &
        [ , scalar-logical-expression ] )
    [ forall-body-construct ] ...
END FORALL [ forall-construct-name ]
```

where a forall triplet specification (R755) is:

```
index-name = scalar-integer-expression : &
        scalar-integer-expression [ : scalar-integer-expression ]
```

and a forall body construct (R756) is one of:

```
assignment-statement
pointer-assignment-statement
where-construct
where-statement
forall-construct
forall-statement
```

Rules and restrictions:

1. The index name is the name of a scalar integer variable. The name has the scope of the FORALL construct itself. It has the type (which must be integer) and type parameters it would have if it were the name of a variable in the scope that contains the FORALL construct, but it has no other attributes. For example:

   ```
   SUBROUTINE CALC (II, A)
   INTEGER :: A(:)
   INTEGER, INTENT(IN) :: II
      . . .
   FORALL (II = 1:SIZE(A)) !OK even though II is intent IN
      A(II) = II
   END FORALL
      . . .
   END SUBROUTINE CALC
   ```

   After execution of the FORALL construct, A has the value (1, 2, 3, ...) and II retains the value it had on entry to the subroutine. The definition of II in the FORALL construct does not violate the intent specification of IN for II. However, this is such a confusing style that it is never recommended.

2. An expression that appears in a triplet specification must not contain a reference to any index name from the list in which the expression appears. Thus, the following FORALL statement is invalid:

   ```
   FORALL (I = 1:J, J = 1:N) A(I,J) = 0.0
   ```

but can be rewritten as:

```
FORALL (I = 1:N, J = 1:N, I<=J) A(I,J) = 0.0
```

3. Any procedure referenced in the scalar logical expression that defines the mask or in any FORALL body construct (including one referenced by a defined operation or assignment) must be a pure procedure.

4. An index name must not be assigned a value within the FORALL body constructs.

5. A nested FORALL construct or statement must not use as an index name one of the index names of an outer construct. The value of an inner construct index name, however, may depend on the values of outer index variables.

6. A many-one assignment (6.6.4.3) must not occur within a single statement in a FORALL construct. For example:

```
FORALL (J=1:20)
   A1(INDEX(J)) = A2(J)
END FORALL
```

   is allowed only if INDEX(1:20) contains no duplicate values. It is possible to assign or pointer assign to the same object in different statements in a FORALL construct.

7. A FORALL body construct must not be a branch target.

The triplet notation has an interpretation similar to that for section triplets (6.6.4.2); that is,

scalar-integer-expression : scalar-integer-expression : scalar-integer-expression

corresponds to

first value : last value : stride

The stride may be positive or negative, but not zero; if omitted, it is assumed to be 1.

It is normally the case that each index name in the triplet list appears in the subscript or section subscript list of the variable being assigned.

The scalar logical expression defines a mask. A reference to an index name may appear in the expression. For example:

```
FORALL (I=1:10, J=1:10, A(I)/=0.0 .AND. B(J)>0.0)
   . . .
END FORALL
```

An assignment statement in a FORALL body may be an array assignment statement:

```
REAL A(100,100)
   . . .
```

```
FORALL (I=1:N)
   A(I,:) = 1.0 / REAL(I)   ! A scalar value is broadcast
      . . .                 !   to each row of A
```

or a pointer assignment statement:

```
TYPE SCREW
   CHARACTER (30), POINTER :: HEAD_TYPE
   REAL LENGTH, THREAD
END TYPE SCREW

TYPE (SCREW) INVENTORY (500)
REAL THREADS (100)
CHARACTER (30), TARGET :: HEAD_TYPES(5)
   . . .
FORALL (I=1:500, INVENTORY(I)%LENGTH > .05)
   INVENTORY(I)%HEAD_TYPE => HEAD_TYPES(MOD(I-1,5)+1)
   ! Subscripts for HEAD_TYPES are 1,2,3,4,5,1,2,3,4,5, ...
   INVENTORY(I)%THREAD = THREADS((I-1)/5+1)
   ! Subscripts for THREADS are 1,1,1,1,1,2,2,2,2,2,3,3,3,3,3, ...
END FORALL
```

or a defined assignment (12.5.4.3).

7.5.7.2 Execution of a FORALL Construct

There are three steps in the execution of a FORALL construct:

1. determination of the values for index name variables
2. evaluation of the mask expression, if there is one
3. execution of the body constructs

Determination of the values for index name variables. The scalar integer expressions in a triplet are evaluated; they may be evaluated in any order. If necessary, they are converted to the kind of the index name. They determine the set of values the index may assume. If the expressions are designated by m_1, m_2, and m_3 (where m_3 has the value 1 if not present), the number of values in the set is determined by the formula $(m_2 - m_1 + m_3) / m_3$. If this number, call it n, is less than or equal to zero, the execution of the construct is complete (like the DO construct, the body is not executed). Otherwise, the set of values for the index name is $m_1 + (k - 1) \times m_3$, where $k = 1, 2, ..., n$. The set of combinations of index values is determined by the Cartesian product of the sets of values defined by each triplet specification.

Evaluation of the mask expression. If there is no mask expression, it is as if it were present with the value true. Otherwise, the expression is evaluated for each combination of index values. The **active combination of index values** is then the subset of all possible combinations (determined in step 1) for which the mask expression has the value true.

Execution of the body constructs. FORALL body constructs are executed in the order in which they appear. Each of these constructs is executed for all active combinations of index values and may be an assignment statement, a pointer assignment statement, a WHERE construct or statement, or a nested FORALL construct or statement.

1. *Assignment statements*. An assignment statement has the form

   variable = expression

   Execution of such a statement within a FORALL construct causes evaluation of the expression on the right-hand side and all expressions within the variable for all active combinations of the index values. These evaluations may be done in any order. After all of these evaluations have been done, each expression value is assigned to the appropriate variable. The assignments may occur in any order. If the assignment is a defined assignment (12.5.4.3), the subroutine that defines the assignment must not contain a reference to any variable that becomes defined by the statement or any pointer that becomes associated by the statement.

2. *Pointer assignment statements*. A pointer assignment statement has the form

   pointer-object => target

   Execution of such a statement within a FORALL construct causes evaluation of all expressions within the target and the pointer object, the determination of any pointers within the pointer object, and the determination of the target for all active combinations of the index values. These evaluations may be done in any order. Afterward, each pointer object is associated with the corresponding target. These associations may be done in any order. The pointer-object may be either a data pointer or a procedure pointer.

3. *WHERE constructs and statements*. Each statement in a WHERE construct (7.5.6.1) within a FORALL construct is executed in sequence. When a WHERE statement, WHERE construct statement, or masked ELSEWHERE statement is executed, the statement's mask expression is evaluated for all active combinations of index values as determined by the outer FORALL construct (or constructs, if nested) and masked by any masks from outer WHERE constructs. The assignment statement within a WHERE statement and any assignment statements within a WHERE construct are then executed for all active combinations of index values masked by the new control mask in effect for that statement. For example,

   ```
   INTEGER A(5,4)
      . . .
   INT_WHERE: FORALL (I=1:5)
      WHERE (A(I,:) > I) A(I,:) = I
   END FORALL INT_WHERE
   ```

   If A has the initial value

$$\begin{bmatrix} 0 & 0 & 1 & 2 \\ 1 & 2 & 3 & 0 \\ 2 & 4 & 0 & 6 \\ 1 & 9 & 3 & 6 \\ 8 & 8 & 8 & 8 \end{bmatrix}$$

after execution, it will have the value

$$\begin{bmatrix} 0 & 0 & 1 & 1 \\ 1 & 2 & 2 & 0 \\ 2 & 3 & 0 & 3 \\ 1 & 4 & 3 & 4 \\ 5 & 5 & 5 & 5 \end{bmatrix}$$

4. *FORALL constructs and statements*. Execution of an inner FORALL construct or FORALL statement causes the evaluation of the expressions in the triplet list of the inner header for all active combinations of the index values of the outer FORALL construct. The set of combinations of index values for the inner FORALL is the union of the sets defined by these expressions for each active combination of the outer index values. The mask is then evaluated for all combinations of the index values of the inner construct or statement to produce a set of active combinations for the statement or statements it controls, which are executed for each active combination of the index values. For example,

```
INTEGER A(3,3)
   . . .
OUTER: FORALL (I=1:N-1)
   INNER: FORALL (J=I+1:N)
      A(I,J) = A (J,I)
   END FORALL INNER
END FORALL OUTER
```

If N is 3 and A has the initial value

$$\begin{bmatrix} 0 & 3 & 6 \\ 1 & 4 & 7 \\ 2 & 5 & 8 \end{bmatrix}$$

after execution, it will have the value

$$\begin{bmatrix} 0 & 1 & 2 \\ 1 & 4 & 5 \\ 2 & 5 & 8 \end{bmatrix}$$

The transpose of the lower triangle of array A (the section below the main diagonal) is assigned to the upper triangle of A.

7.5.7.3 FORALL Statement

The FORALL statement allows a single assignment or pointer assignment statement to be controlled by a set of index values and an optional mask expression.

The form of the FORALL statement (R759) is:

```
FORALL ( forall-triplet-specification-list [ , scalar-logical-expression ] ) &
   forall-assignment-statement
```

where a forall assignment statement (R757) is one of:

```
assignment-statement
pointer-assignment-statement
```

The FORALL statement is equivalent to the FORALL construct

```
FORALL ( forall-triplet-specification-list [ , scalar-logical-expression ] )
   forall-assignment-statement
END FORALL
```

The effect of the previous example of nested FORALL constructs can be achieved with a single FORALL statement:

```
FORALL (I=1:N-1, J=1:N, J>I) A(I,J) = A(J,I)
```

8 Block Constructs and Execution Control

- A **Block** is a bounded sequence of executable constructs and statements that is treated as a unit. It may be empty.
- A **Block Construct** has an initial statement and a terminal statement; it contains zero or more blocks and the statements that bound the blocks. It is used to control execution or simply to define a region of code.
- The **ASSOCIATE Construct** allows a named entity, the associate name, to be associated with a variable or expression during the execution of its single block.
- The **IF Construct** contains one or more blocks; at most one is chosen for execution. The choice is based on the value of a logical expression.
- The **CASE Construct** contains zero or more blocks; at most one is selected for execution. The selection is based on the value of an integer, character, or logical expression.
- The **SELECT TYPE Construct** contains zero or more blocks; at most one is selected for execution. Rather than a value, as in the CASE construct, the selection is based on the dynamic type of a variable or expression.
- The **DO Construct** contains a single block that is executed repeatedly. There are multiple forms for controlling the execution. A **CYCLE statement** is permitted at any point to start the next execution of the block. An **EXIT statement** terminates the repetition.
- The **IF Statement** permits the execution of a single statement if the contained logical expression evaluates to true.
- The **GO TO Statement** transfers control to a labeled statement.
- The **CONTINUE Statement** has no effect on execution.
- The **STOP Statement** causes termination of the execution of the program.
- The **Computed GO TO Statement**, the **Arithmetic IF Statement**, and the **nonblock DO** are obsolescent features that use labels.

This chapter describes five block constructs, four of which are execution control constructs. It also describes the executable statements that are used to alter the normal execution sequence. The block constructs that are control constructs are the IF construct, the CASE construct, the SELECT TYPE construct, and the DO construct. Individual statements that alter the normal execution sequence include the EXIT and CYCLE statements that are special statements for DO constructs, branch statements such as the

J.C. Adams et al., *The Fortran 2003 Handbook*,
DOI: 10.1007/978-1-84628-746-6_8, © Springer-Verlag London Limited 2009

GO TO statement, and a statement that causes execution to cease, the STOP statement. The fifth block construct described in this chapter is the ASSOCIATE construct. Its single block defines a region of the program in which an associate name may be used instead of a longer or more complicated variable or expression.

There are two other constructs that look like control constructs, but are really forms of assignment. These are the WHERE construct (7.5.6), which somewhat resembles an IF construct and the FORALL construct (7.5.7), which somewhat resembles a DO construct.

With any of the block constructs, construct names may be used to identify the constructs and also to identify which DO constructs, particularly in a nest of DO constructs, are being terminated or cycled when using the EXIT or CYCLE statements.

8.1 Blocks and Construct Names

A **block** (R801) has the form:

[execution-part-construct] ...

A block is treated as a whole. Not every statement or construct in a block need be executed; for example, a branch statement early in the block may prevent subsequent statements in the block from being executed. This is still a complete execution of the block.

A control construct consists of zero or more blocks and the control logic that explicitly or implicitly encloses these blocks. A construct has an initial statement and a terminal statement. In constructs that have more than one block, there are additional statements between blocks that determine which block is chosen for execution. The control for the DO construct determines how many times its block will be executed. An example of a named executable construct controlling a block of statements is:

```
INNER: IF (I<=1) THEN    ! Initial statement of the IF construct
   X = 1.2*I             ! First statement of the block
   Y = COS(X)            ! Final statement of the block
END IF INNER             ! Terminal statement of the IF construct
```

All of the block constructs (ASSOCIATE, CASE, DO, IF, and SELECT TYPE), as well as the FORALL and WHERE constructs, may have construct names. If a construct name is used, it must appear on the initial statement of the construct and a matching occurrence of the same name must appear on the terminal statement of the construct. If there is no construct name on the initial statement, the terminal statement must not have a construct name. If one of the internal control statements contains a construct name, it must be the same name as the one on the initial and terminal statements. The same construct name must not be used for different constructs in the same scoping unit.

Rules and restrictions:

1. The first statement or construct of a block is executed first. The statements of the block are executed in order unless there is a control construct or statement within the block that changes the sequential order.
2. A block, as an integral unit, must be completely contained within a construct.
3. If a block contains a construct, the construct must be completely contained within the block.
4. A block may be empty; that is, it may contain no statements or constructs at all.
5. A branching statement or control construct within a block that transfers to a statement or construct within the same block is permitted.
6. Exiting from a block may be done from anywhere within the block.
7. Branching to a statement or construct within a block from outside the block is prohibited. (Even branching to the first executable statement within a block from outside the block is prohibited.) An ENTRY statement must not appear in a block.
8. References to procedures are permitted within a block.

8.2 The ASSOCIATE Construct

The ASSOCIATE construct has one block in which associate names may be used instead of expressions or variables.

8.2.1 Form of the ASSOCIATE Construct

The form of the ASSOCIATE construct (R816) is:

```
[ associate-construct-name : ] ASSOCIATE ( association-list )
   block
END ASSOCIATE [ associate-construct-name ]
```

where an association has the form (R818):

```
associate-name => selector
```

and selector (R819) is one of:

```
expr
variable
```

Rules and restrictions:

1. An associate name must not be the same as another associate name in the same association list. If an associate name is the same as a name in the scoping unit of the construct, the name in the construct is interpreted as the associate name (16.1.3(6)).

2. If a selector is not permitted to appear in a variable definition context or is a variable with a vector subscript, the associated name must not appear in a variable definition context (16.3.1).

3. If a selector is a variable with the ALLOCATABLE attribute, it must be allocated. The associate name is associated with the data object and does not have the ALLOCATABLE attribute.

4. If a selector is a variable with the POINTER attribute, it must be pointer associated with a target. The associate name is associated with the target and does not have the POINTER attribute.

5. If the selector is an optional dummy argument, it must be present.

8.2.2 Execution of the ASSOCIATE Construct

The association between an associate name and a selector is established before the execution of the block. If the selector is not a variable, the expression is evaluated and the value of the expression is associated with the associate name. Because the association is established before the execution of the block, it is not affected by any subsequent changes to variables that were used in subscripts or substring ranges in the selector. This process is somewhat similar to what happens in a procedure call with the associate name taking the role of a local dummy argument.

During execution of the block, each associate name identifies the entity specified by its selector. This associating entity assumes the declared type and type parameters of its selector. If the selector is a variable of type character, a substring range may be appended to the associate name. If the selector is of derived type, a structure component may be appended. If the selector is an array, a subscript list or section subscript list may be appended to the name. If and only if the selector is polymorphic, the associating entity is polymorphic, in which case it assumes the dynamic type and type parameter values of the selector. The associating entity has the ASYNCHRONOUS or VOLATILE attribute if and only if the selector is a variable and has the attribute. The associating entity has the TARGET attribute if and only if the selector is a variable and has either the TARGET or POINTER attribute. Each associating entity has the same rank as its selector. The lower bound of each dimension is the result of the intrinsic function LBOUND (13.3.1.4) applied to the corresponding dimension of the selector. The upper bound is one less than the sum of the lower bound and the extent.

An example with an expression as the selector is:

```
ASSOCIATE  ( D => (X-H)**2 + (Y-K)**2 )
   PRINT *, SQRT(D)
END ASSOCIATE
```

These type definitions and declaration are needed in the following three examples that have variables as the selectors:

```
TYPE MORE
   INTEGER K, L
END TYPE MORE
```

```
TYPE COLLECTION
   REAL X
   INTEGER J
   CHARACTER (80) C
   TYPE (MORE) MMM
END TYPE COLLECTION

TYPE ARRAY_INSIDE
   TYPE (COLLECTION) :: MISC (10, 10)
END TYPE ARRAY_INSIDE

TYPE (ARRAY_INSIDE)   AI
```

1. An example with a variable that is a structure component as the selector is:

```
ASSOCIATE  ( COMP => AI % MISC(N, M) % MMM )
   COMP % K = COMP % L + INDEX
END ASSOCIATE
```

2. An example with a variable of type character as the selector is:

```
ASSOCIATE  ( ANS => AI % MISC(N, M) % C )
   ANS (1 : 75) = "FALSE" // ANS (1 : 70)
END ASSOCIATE
```

3. An example with a variable that is an array section as the selector is:

```
ASSOCIATE  ( ARRAY => AI % MISC (I, : ) % X)  ! ARRAY  has 10 elements
   ARRAY (1) = ARRAY (2) + ARRAY (3)
END ASSOCIATE
```

Without the ASSOCIATE construct, it would be necessary to write:

```
AI % MISC (I, 1) % X = AI % MISC (I, 2) % X + AI % MISC (I, 3) % X
```

The following example illustrates several selectors:

```
ASSOCIATE (IX=>AI%MISC(1,1)%J, VAR=>AI%MISC(:, 1), Q=>EXP (P)*100.0)
   R = VAR % X(IX)*Q
END ASSOCIATE
```

8.3 Controlling Execution

There is an established execution sequence for action statements in a Fortran program. It is called the **normal execution sequence**. Normally, a program or subprogram begins with the first executable statement in that program or subprogram and continues with the next executable statement in the order in which these statements appear. However, there are executable control constructs and executable branching statements that cause statements to be executed in an order that is different from the order in which they appear in the program.

There are two basic ways to affect the execution sequence. One is to use an executable construct that selects a block of statements and constructs for execution. The second is to execute a statement that branches to a specific statement in the program. In almost all cases, the use of constructs will result in programs that are more readable and maintainable.

8.4 The IF Construct and the IF Statement

An IF construct selects at most one block of statements and constructs within the construct for execution. The IF statement controls the execution of only one statement. The arithmetic IF statement is not the same as the IF statement; it is a branching statement that is designated as obsolescent and is discussed in 8.8.4.

8.4.1 The IF Construct

The IF construct contains one or more executable blocks; at most one block is chosen for execution, after which the IF construct is completed and it terminates.

8.4.1.1 Form of the IF Construct

The form of the IF construct (R802) is:

```
[ if-construct-name : ] IF ( scalar-logical-expression ) THEN
   block
[ ELSE IF ( scalar-logical-expression ) THEN [ if-construct-name ]
   block ] ...
[ ELSE  [ if-construct-name ]
   block ]
END IF [ if-construct-name ]
```

Rules and restrictions:

1. At most one of the blocks in the construct is executed. It is possible that no block is executed.
2. Any ELSE statement must follow any ELSE IF statements.
3. Branching to an ELSE IF or an ELSE statement is prohibited.
4. Branching to an END IF is allowed from any block within the IF construct.

8.4.1.2 Execution of the IF Construct

The logical expressions in the bounding statements are evaluated in order; when one is found to be true, the block following it is executed, and the execution of the IF construct terminates. There may be no logical expressions that are true. In this case, the block following the ELSE statement is executed if there is one; otherwise, no block in the construct is executed.

Figure 8-1 indicates the execution flow for IF constructs.

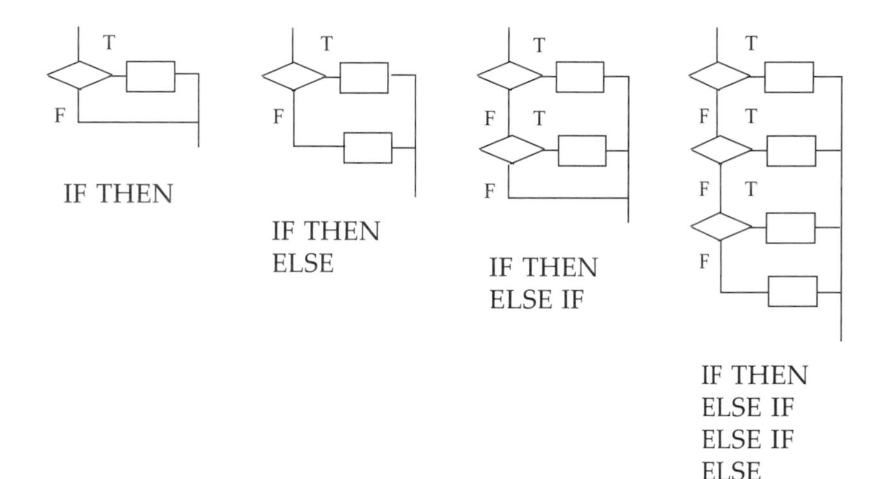

Figure 8-1 Execution flow for IF constructs.

An example of the IF construct is:

```
IF (I < J) THEN
   X = Y + 5.0
ELSE IF (I > 100) THEN
   X = 0.0
   Y = -1.0
ELSE
   X = -1.0
   Y = 0.0
END IF
```

If I is less than J, the statement X = Y + 5.0 is executed and execution proceeds following the END IF statement. If I is not less than J and if I is greater than 100, the two statements following the ELSE IF statement are executed and execution proceeds following the END IF statement. If neither of these conditions is true, the block after the ELSE statement is executed.

8.4.2 The IF Statement

The IF statement controls the execution of a single statement.

8.4.2.1 Form of the IF Statement

The form of the IF statement (R807) is:

```
IF ( scalar-logical-expression ) action-statement
```

Example:

```
IF (S < T) S = 0.0
```

8.4.2.2 Execution of the IF Statement

The scalar logical expression is evaluated. If true, the action statement is executed. If false, the action statement is not executed.

Rules and restrictions:

1. The action statement must not be an IF statement or an END statement.
2. If the logical expression contains a function reference, its evaluation may have side effects that modify the action statement. This is permitted.

A complete list of the action statements can be found in 2.5.3. Fundamentally, action statements change the definition state of variables or the condition of the input/output system, or are control statements. Examples of action statements are the assignment, WRITE, and GO TO statements. Specification statements such as type declaration statements, FORMAT statements, and ENTRY statements are not action statements.

8.5 The CASE Construct

The CASE construct, like the IF construct, consists of a number of blocks, of which at most one is selected for execution. The selection is based on the value of a scalar expression in the initial SELECT CASE statement; the value of this expression is called the **case index**. The block selected is the one for which the case index matches a case value in a preceding CASE statement. There is an optional DEFAULT CASE statement that contains no values, but, in effect, is considered to match all values not matched by any other case values in the construct. The types of the case index and case values are limited to the "discrete" types; namely integer, character, and logical. For other types, the IF construct is available, but for the discrete types, the CASE construct may be more expressive and more efficient in execution.

8.5.1 Form of the CASE Construct

The form of the CASE construct (R808) is:

```
[ case-construct-name : ] SELECT CASE ( case-expression )
[ CASE case-selector [ case-construct-name ]
    block ] ...
END SELECT [ case-construct-name ]
```

where case expression is a scalar expression. A case-selector (R813) is one of:

```
( case-value-range list )
DEFAULT
```

The forms of a case value range (R814) are:

```
case-value
case-value :
: case-value
case-value : case-value
```

where each case value is a scalar initialization expression of the same type as the case expression. Recall that an initialization expression is an expression that can be evaluated at compile time.

Rules and restrictions:

1. A CASE DEFAULT statement is optional. If it appears, the general form (R808) of the CASE construct does not require that such a CASE statement be the last CASE statement. (This is unlike the IF construct where the ELSE statement must be last.)

2. Within a particular CASE construct, the case expression and all case values must be of the same type. The kind type parameter values may be different unless the type is character. If the type is character, different character lengths are allowed, but the kind type parameter values must be the same.

3. The colon forms of the case values expressing a range may be used for any case value ranges; the case values must be of type integer and character (but not logical). For example, a CASE statement of the form

   ```
   CASE ('BOOK':'DOG')
   ```

   would select all character strings that collate between BOOK and DOG inclusive, using the processor-dependent collating sequence for the default character type.

4. Overlapping case values and case ranges are prohibited.

An example of the CASE construct is:

```
FIND_AREA: &  ! Compute the area with a formula
              ! appropriate for the shape of the object
              ! CIRCLE, SQUARE, and RECTANGLE  are named constants.
   SELECT CASE (OBJECT)
     CASE (CIRCLE)  FIND_AREA
         AREA = PI * RADIUS ** 2
     CASE (SQUARE)  FIND_AREA
         AREA = SIDE * SIDE
     CASE (RECTANGLE)  FIND_AREA
         AREA = LENGTH * WIDTH
```

```
    CASE DEFAULT  FIND_AREA
        PRINT *, "Object  not  recognized."
END SELECT  FIND_AREA
```

8.5.2 Execution of the CASE Construct

The case index (the scalar expression) in the SELECT CASE statement is evaluated and compared with the case values in the CASE statements preceding the blocks. The case index must match at most one of the selector values. The block following the case matched is executed, and the CASE construct terminates. If no match occurs and the CASE DEFAULT statement is present, the block after the CASE DEFAULT statement is selected. If there is no CASE DEFAULT statement, the CASE construct terminates with no block selected for execution. If the case value is a single value, a match occurs if the index is equal to the case value (determined by the rules used in evaluating the equality or equivalence operator (7.2.7.1.2). If the case value is a range of values, there are three possibilities to determine a match depending on the form of the range:

Case value range	Condition for a match
case-value_1 : case-value_2	$\text{case-value}_1 \le \text{case-index} \le \text{case-value}_2$
case-value :	$\text{case-value} \le \text{case-index}$
: case-value	$\text{case-value} \ge \text{case-index}$

Rules and restrictions:

1. There must not be case values that would select more than one block.
2. Branching to the END SELECT statement is allowed only from within the construct.
3. Branching to a CASE statement is prohibited; branching to the SELECT CASE statement is allowed, however.

Figure 8-2 illustrates the execution flow for a CASE construct.

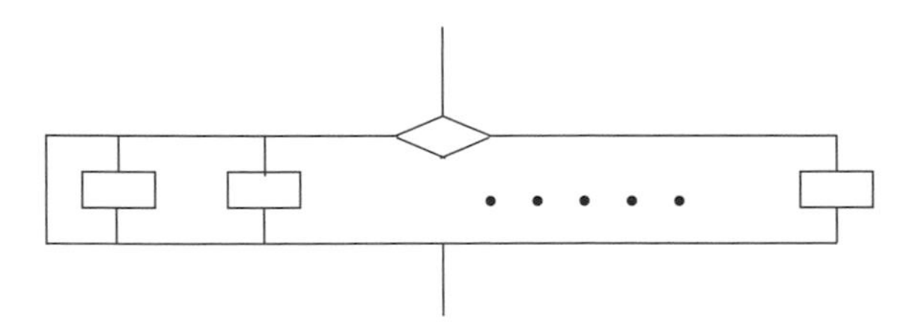

Figure 8-2 Execution flow for a CASE construct.

Example:

```
INDEX = 2
SELECT CASE (INDEX)
CASE (1)
   X = 1.0
CASE (2)
   X = 2.0
CASE DEFAULT
   X = 99.0
END SELECT
```

The case expression INDEX has the value 2. The block following the case value of 2 is executed; that is, the statement X = 2.0 is executed, and execution of the CASE construct terminates.

Example:

```
COLOR = 'GREEN'
SELECT CASE (COLOR)
CASE ('RED')
   STOP
CASE ('YELLOW')
   CALL STOP_IF_YOU_CAN_SAFELY
CASE ('GREEN')
   CALL GO_AHEAD
END SELECT
```

This example uses selectors of type character. The expression COLOR has the value GREEN, and therefore the procedure GO_AHEAD is executed. When it returns, the execution of the CASE statement terminates.

8.6 The SELECT TYPE Construct

The SELECT TYPE construct, like the CASE construct, consists of a number of blocks; at most one is selected for execution. The selection is based on the dynamic type of an expression. An optional name may be associated with the expression in the same way as for the ASSOCIATE construct (8.2). If this option is used, it is as if the SELECT TYPE construct without the option appeared inside an ASSOCIATE construct with that selector.

8.6.1 Form of the SELECT TYPE Construct

The form of the SELECT TYPE construct (R821) is:

```
[ select-construct-name : ] SELECT TYPE ( [ associate-name => ] selector )
   [ type-guard [ select-construct-name ]
      block ] . . .
END SELECT [ select-construct-name ]
```

where type-guard (R823) is one of:

```
TYPE IS ( type-spec )
CLASS IS ( derived-type-spec )
CLASS DEFAULT
```

Rules and restrictions:

1. The selector must be polymorphic.
2. If the selector is not a named variable, associate-name => must appear.
3. If the selector is not permitted to appear in a variable definition context (16.3.1) or is a variable that has a vector subscript, associate-name must not appear in a variable definition context.
4. Each length type parameter in a type-spec or derived-type-spec must be assumed.
5. A type-spec or derived-type-spec must not specify a sequence derived type or a type with the BIND attribute.
6. If the selector is not unlimited polymorphic, the type-spec or derived-type-spec must specify an extension of the declared type of selector.
7. For a given SELECT TYPE construct, the same type and kind type parameter values must not be specified in more than one TYPE IS type-guard and must not be specified in more than one CLASS IS type-guard.
8. For a given SELECT TYPE construct, there must be at most one CLASS DEFAULT type-guard.

If an associate name is specified, it is the associate name of the construct; otherwise, the associate name of the construct is the name of the variable that is the selector.

8.6.2 Execution of the SELECT TYPE Construct

The SELECT TYPE construct is used to select for execution the most appropriate block of code for the particular dynamic type and type parameter values of the selector, if such a block is provided.

If the selector is not a variable, the selector expression is evaluated. At most one block in the construct is selected for execution. It is the block following the type guard with a matching type-spec or the DEFAULT block if there is one. A TYPE IS type guard matches if the dynamic type and type parameter values of the selector are the same as those of the type-spec. A CLASS IS type guard matches if the dynamic type of the selector is an extension of the specified type and the kind type parameter values specified by the type-spec are the same as the corresponding type parameter values of the dynamic type of the selector.

The block to be executed is selected as follows:

1. If the type-spec in a TYPE IS type guard matches that of the selector, the following block is executed.

2. Otherwise, if exactly one type-spec in a CLASS IS type guard matches the type of the selector, the following block is executed.

3. Otherwise, if several type-specs in CLASS IS type guards match, one of these will specify a type that is an extension of all the types specified in the others; the block following that type guard is executed.

4. Otherwise, if there is a CLASS DEFAULT type guard, the block following it is executed.

During execution of the chosen block, the associate name identifies an entity that is associated with the selector. In the block following a TYPE IS type guard, the associating entity is not polymorphic (5.2), but has the type named in the type-spec and the type parameter values of the selector.

In the block following a CLASS IS type guard, the associating entity is polymorphic and has the same declared type as the selector. The type parameter values are those of the declared type of the selector.

In the block following a CLASS DEFAULT type guard, the associating entity is polymorphic and has the same declared type as the selector. The type parameter values are those of the declared type of the selector.

The other attributes of the associating entity: rank, ASYNCHRONOUS, OPTIONAL, TARGET, and VOLATILE are as described for the ASSOCIATE construct (8.2).

Figure 8-2 illustrates the execution flow of a SELECT TYPE construct as well as that of a CASE construct; however, the process of selecting a block is not the simple match of a CASE construct. Figure 8-3 illustrates the block selection process for a SELECT TYPE construct. A TYPE IS guard matches if the dynamic type and type parameters values of the selector match the guard. A CLASS IS guard matches if the dynamic type of the selector is an extension of the type of the guard and the kind type parameter values match.

An example of the SELECT TYPE construct:

```
TYPE  END_PTS
   REAL :: X1, Y1, X2, Y2
END TYPE  END_PTS

TYPE  LINE
   TYPE (END_PTS) PTS
END TYPE  LINE

TYPE, EXTENDS (LINE) :: LINE_W
   REAL :: WIDTH
END TYPE  LINE_W
```

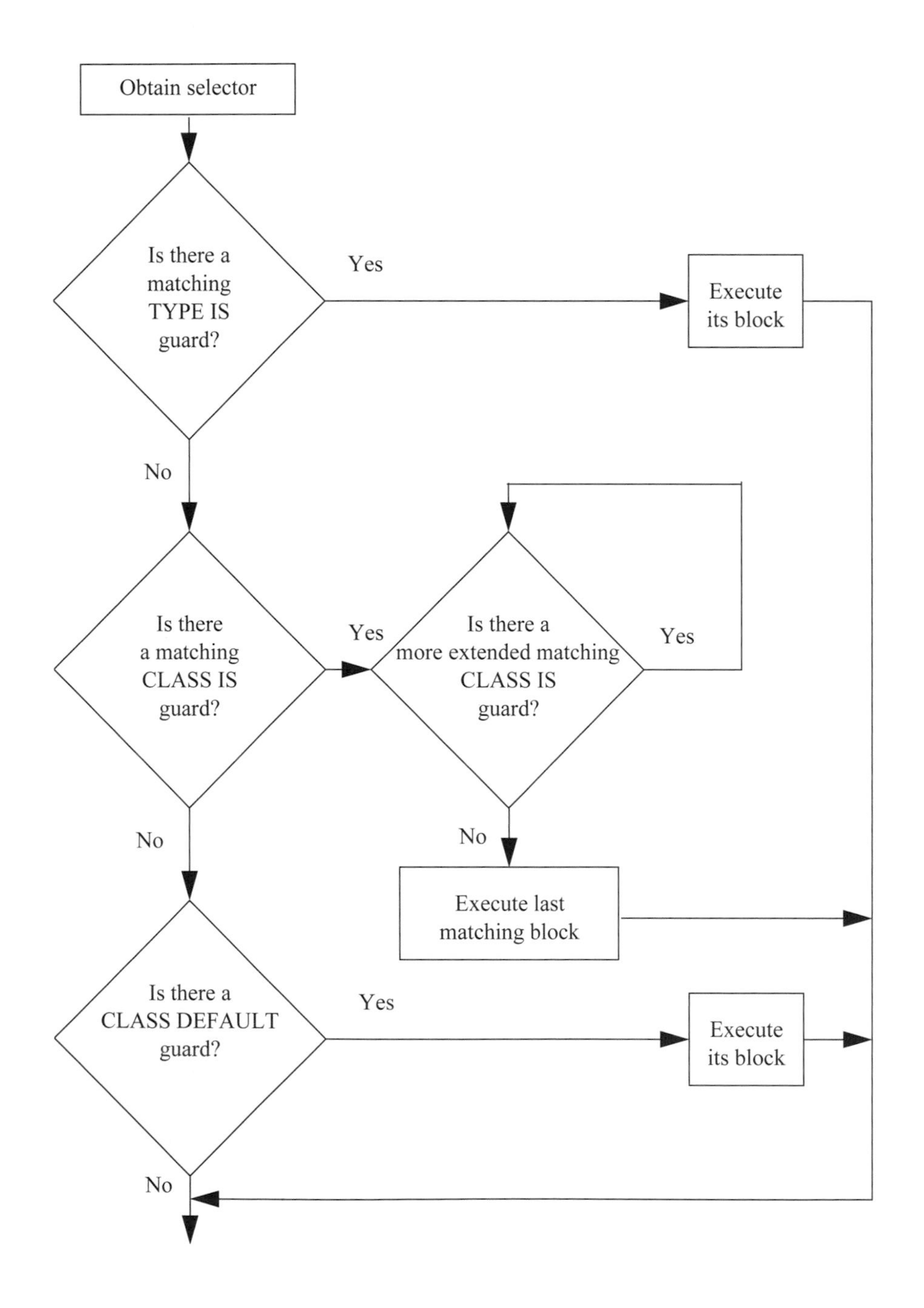

Figure 8-3 Block selection process for a SELECT TYPE construct. A TYPE IS guard matches if the dynamic type and type parameters values of the selector match the guard. A CLASS IS guard matches if the dynamic type of the selector is an extension of the type of the guard and the kind type parameter values match.

```
TYPE, EXTENDS(LINE_W) :: LINE_WC
   CHARACTER (10)  :: COLOR
END TYPE  LINE_WC

TYPE, EXTENDS (LINE_WC) :: LINE_WCS
   INTEGER STYLE
END TYPE LINE_WCS

TYPE (LINE_W), TARGET :: LW
TYPE (LINE_WCS), TARGET :: LWCS
CLASS (LINE), POINTER :: ANY

ANY => LW
SELECT TYPE  (ANY)
   CLASS IS (LINE)
      CALL DRAW (ANY % PTS)
   TYPE IS (LINE_W)
      CALL SET_W (ANY % WIDTH)                 ! Block selected
      CALL DRAW (ANY % PTS)
   TYPE IS (LINE_WC)
      CALL SET_W C (ANY % WIDTH, ANY % COLOR)
      CALL DRAW (ANY % PTS)
END SELECT

ANY => LWCS
SELECT TYPE (A => ANY)
   CLASS IS (LINE)
      CALL DRAW (A % PTS)
   CLASS IS (LINE_W)
      CALL SET_W (A % WIDTH)
      CALL DRAW (A % PTS)
   CLASS IS (LINE_WC)
      CALL SET_WC (A % WIDTH, A % COLOR)
      CALL DRAW (A % PTS)
   CLASS IS (LINE_WCS)
      CALL SET_S (A % STYLE)                     ! Block selected
      CALL SET_WC (A % WIDTH, A % COLOR)
      CALL DRAW (A % PTS)
END SELECT
```

8.7 The DO Construct

The DO construct controls the number of times its single block is executed. The number may be zero. There are three steps in the execution of a DO construct:

1. First, if execution of the DO construct is controlled by a DO variable, the expressions that determine the number of times the block is to be executed are evaluated.

2. Next, a decision is made as to whether the block is to be executed.

3. Finally, if appropriate, the block is executed; the DO variable, if present, is updated; and steps 2 and 3 are repeated.

There are three ways of controlling a loop: one involves a loop variable that is incremented with a prescribed value (which may be negative) a certain number of times as can be calculated from the initial DO statement; the second involves a WHILE condition; and the third is the simple DO, sometimes called a "DO forever". The execution of the simple DO construct must be terminated by executing a statement, such as an EXIT statement, that transfers control out of the DO block.

There is another DO—the nonblock DO (8.7.5). Except for one special form of the nonblock DO, it is obsolescent. The standard treats this special form of the nonblock DO as if it were a block construct. The block DO contains all of the functionality of the nonblock DO. Indeed, both forms permit DO WHILE and DO forever loop control. The primary difference between the two forms is that the block DO construct is always terminated by an END DO statement whereas the nonblock DO terminates with an action statement, and it may share a termination statement with another DO statement.

The first statement of a DO construct or nonblock DO is called a **DO statement**.

An example of a block DO construct is:

```
DO I = 1, N
   SUM = SUM + A(I)
END DO
```

An example of a nonblock DO (which is obsolescent) to perform the same computation is:

```
   DO 10 I = 1, N
10 SUM = SUM + A(I)
```

8.7.1 Form of the Block DO Construct

The form of the block DO construct (R826) is:

```
[ do-construct-name : ] DO [ label ] [ loop-control ]
           block
[ label ] END DO [ do-construct-name ]
```

where the two forms of loop control (R830) are:

```
[ , ] do-variable = scalar-integer-expression , scalar-integer-expression &
    [ , scalar-integer-expression ]

[ , ] WHILE ( scalar-logical-expression )
```

The first statement of a DO construct or nonblock DO is called a **DO statement**.

Rules and restrictions:

1. The DO variable must be a scalar variable of type integer.
2. The expressions must be scalar integer expressions.

3. If a label appears in the initial statement, the terminal statement must be identified with the same label.

Although a DO construct can have both a label and a construct name, use of both is not in the spirit of modern programming practice where the use of labels is minimized.

Examples:

```
SUM = 0.0
DO I = 1, N
  SUM = SUM + X (I) ** 2
END DO

FOUND = .FALSE.
I = 0
DO WHILE (.NOT. FOUND .AND. I < LIMIT )
  IF (KEY == X (I))  THEN
    FOUND = .TRUE.
  ELSE
    I = I + 1
  END IF
END DO

NO_ITERS = 0
DO
  ! F and F_PRIME are functions
  X1 = X0 - F (X0) / F_PRIME (X0)
  IF (ABS(X1-X0) < SPACING (X0) .OR. &
      NO_ITERS > MAX_ITERS )  EXIT
  X0 = X1
  NO_ITERS = NO_ITERS + 1
END DO

LOOP: DO I = 1, N
         Y(I) = A*X(I) + Y(I)
      END DO LOOP

   INNER_PROD = 0.0
   DO 10 I = 1, 10
     INNER_PROD = INNER_PROD + X (I) * Y (I)
10 CONTINUE
```

8.7.2 Execution of DO Constructs

There are three forms of the DO construct, each with its own loop control and rules for execution. These forms are: a DO construct with an iteration count, a DO construct with WHILE control, and a simple DO construct. Each form of the DO construct may contain executable statements that alter the sequential execution within the DO block; in addition, some statements terminate the DO construct as described in 8.7.3.

8.7.2.1 Execution of the DO Construct with an Iteration Count

In this case, an iteration count controls the number of times the DO block is executed.
The form of loop control using an iteration count is:

$expression_1$, $expression_2$ [, $expression_3$]

Examples of the DO statement are:

```
DO 10 I = 1, N
DO, J = -N, N
DO K = N, 1, -1
```

8.7.2.1.1 The Iteration Count

An iteration count is established for counting the number of times the program executes the DO block. This is done by evaluating the expressions $expression_1$, $expression_2$, and $expression_3$, and converting these values to the kind of the DO variable. Let m_1, m_2, and m_3 be the values obtained. The value of m_3 must not be zero. If $expression_3$ is not present, m_3 is given the value 1. Thus:

m_1 is the initial value of the DO variable
m_2 is the limiting value
m_3 is the DO variable increment

The iteration count is calculated from the formula:

$$\text{MAX}\left((m_2 - m_1 + m_3) / m_3, 0\right)$$

Note that the iteration count is 0 if:

$m_1 > m_2$ and $m_3 > 0$
or
$m_1 < m_2$ and $m_3 < 0$

8.7.2.1.2 The Execution Steps

The steps that control the execution of the DO construct with an iteration count are:

1. The DO variable is set to m_1, the initial value.

2. The iteration count is tested. If it is 0, the DO construct terminates.

3. a) If the iteration count is not 0, the DO block is executed.

   b) The iteration count is decremented by 1, and the DO variable is incremented by m_3. Steps 2 and 3 are repeated until the iteration count is 0.

After termination, the DO variable retains its last value, the one that it had when the iteration count was tested and found to be 0.

The DO variable must not be redefined or become undefined during the execution of the DO block. Note that changing the variables used in calculating the iteration count during the execution of the DO construct does not change the iteration count; it is fixed each time the DO construct is entered.

Example:

```
N = 10
SUM = 0.0
DO I = 1, N
   SUM = SUM + X (I)
   N = N + 1
END DO
```

The loop is executed 10 times; after execution I = 11 and N = 20.

Example:

```
DO I = 1, 9, 3
    K(I) = 0
END DO
```

The loop is executed 3 times; after execution I = 10.

Example:

```
X = 20.0
DO I = 1, 2
   DO J = 1, 5
      X = X + 1.0
   END DO
END DO
```

The inner loop is executed 10 times. After completion of the outer DO construct, J = 6, I = 3, X = 30.0.

If the second DO statement had been

```
DO J = 5, 1
```

the inner DO construct would not have executed at all; X would remain equal to 20; J would equal 5, its initial value; and I would be equal to 3.

8.7.2.2 Execution of the DO Construct with WHILE Control

The DO WHILE form of the DO construct specifies that the DO block will be repeated while a specified condition remains true.

The form of WHILE control (R830) is:

```
WHILE ( scalar-logical-expression )
```

Examples of the DO statement with WHILE control are:

```
DO WHILE( K >= 4 )
DO 20 WHILE( .NOT. FOUND )
DO, WHILE( A(I) /= 0 )
```

Prior to each execution of the DO block, the logical expression is evaluated. If it is true, the block is executed; if it is false, the DO construct terminates.

```
SUM = 0.0
I = 0
DO WHILE (I < 5)
   I = I + 1
   SUM = SUM + I
END DO
```

The loop would execute 5 times, after which SUM = 15.0 and I = 5.

8.7.2.3 Execution of the Simple DO Construct

A DO construct without any loop control allows statements in the DO block to be repeated until the DO construct is terminated explicitly by some statement within the block. When the end of the block is reached, the first executable statement of the block is executed next.

The form of the simple DO statement (R829) is:

```
DO [ label ]
```

Example:

```
DO
   READ *, DATA
   IF (DATA < 0) STOP
   CALL PROCESS (DATA)
END DO
```

The DO block executes repeatedly until a negative value of DATA is read, at which time the DO construct (and the program, in this case) terminates.

8.7.3 Altering the Execution Sequence within the DO Block

There are two special statements that may appear in the block of any DO construct that alter the execution sequence in a special way. One is the EXIT statement; the other is the CYCLE statement. Other statements, such as branch statements, RETURN statements, and STOP statements also alter the execution sequence but are not restricted to DO constructs as are the EXIT and CYCLE statements.

8.7.3.1 The EXIT Statement

The EXIT statement immediately causes termination of the DO construct. No further action statements within the block are executed. It may appear in either the block or nonblock form of the DO construct, except that it must not be the DO termination action statement of the nonblock form.

The form of the EXIT statement (R844) is:

```
EXIT [ do-construct-name ]
```

Rules and restrictions:

1. The EXIT statement must be within a DO construct.
2. If the EXIT statement has a construct name, it must be within the DO construct with the same name; when it is executed, the named DO construct is terminated as well as any DO constructs containing the EXIT statement and contained within the named DO construct.
3. If the EXIT statement does not have a construct name, the innermost DO construct in which the EXIT statement appears is terminated.

Example of the use of the EXIT statement:

```
LOOP_8 : DO
   . . .
   IF (TEMP == INDEX) EXIT LOOP_8
   . . .
END DO LOOP_8
```

The DO construct has a construct name, LOOP_8; the DO block is executed repeatedly until the condition in the IF statement is met, when the DO construct terminates.

8.7.3.2 The CYCLE Statement

In contrast to the EXIT statement, which terminates execution of the DO construct entirely, the CYCLE statement interrupts the execution of the DO block and begins a new cycle of the DO construct, with appropriate adjustments made to the iteration count and DO variable, if any. It may appear in either the block or nonblock form of the DO construct, except it must not be the DO termination action statement of the nonblock form. When the CYCLE statement is in the nonblock form, the DO termination action statement is not executed.

The form of the CYCLE statement (R843) is:

```
CYCLE [ do-construct-name ]
```

Rules and restrictions:

1. The CYCLE statement must be within a DO construct.
2. If the CYCLE statement has a construct name, it must be within the DO construct with the same name; when it is executed, the execution of the named DO construct is interrupted, and any DO construct containing the CYCLE statement and contained within the named DO construct is terminated.
3. If the CYCLE statement does not have a construct name, the innermost DO construct in which the CYCLE statement appears is interrupted.

4. The CYCLE statement may be used with any form of the DO construct and causes the next iteration of the DO block to begin, if permitted by the condition controlling the loop. Upon interruption of the DO construct, if there is a DO variable, it is updated and the iteration count is decremented by 1. Then, in all cases, the processing of the next iteration begins.

Example:

```
DO
   . . .
   INDEX = . . .
   . . .
   IF (INDEX < 0) EXIT
   IF (INDEX == 0) CYCLE
   . . .
END DO
```

In the above example, the loop is executed as long as INDEX is nonnegative. If INDEX is negative, the loop is terminated. If INDEX is 0, the latter part of the loop is skipped.

Figure 8-4 illustrates the execution flow for various DO constructs, some with EXIT and CYCLE statements.

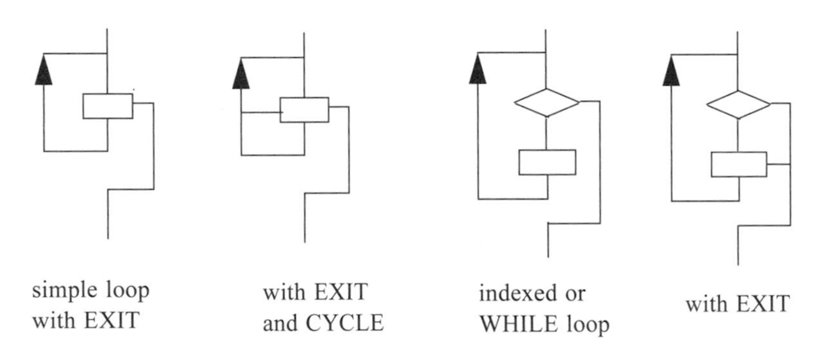

Figure 8-4 Execution flow for DO constructs

8.7.4 Terminating a DO Construct

A DO construct terminates when any of the following situations occur:

1. The iteration count is zero at the time it is tested.
2. The WHILE condition is false at the time it is tested.
3. An EXIT statement is executed that causes an exit from the DO construct or any DO construct containing the DO construct.

4. A CYCLE statement is executed that causes cycling of any DO construct containing the DO construct.

5. There is a transfer of control out of the DO construct.

6. A RETURN statement in the DO construct is executed.

7. The program terminates for any reason.

8.7.5 Form of the Nonblock DO

Except for one special form, the nonblock DO is obsolescent; it always uses a label to specify the terminal statement and construct names are not allowed. The form (R835) is:

```
[ DO label [ loop-control ]
    [ execution-part-construct ] ...] ...
    DO label [ loop-control ]
       [ execution-part-construct ] ...
label action-statement
```

Rules and restrictions:

1. Each occurrence of label must refer to the same label.

2. An action statement that terminates a nonblock DO must not be a CYCLE statement, an EXIT statement, a GO TO statement, a RETURN statement, a STOP statement, an arithmetic IF statement, or an END statement for a program or subprogram.

If more than one DO statement appears, the terminal statement is shared. In this case, it is permitted to branch to the terminal statement only from within the body of the innermost DO loop. Likewise, a CYCLE statement may appear only in the innermost loop. Recall that when a CYCLE statement is executed in a nonblock DO, the terminal action statement is not executed (8.7.3.2). EXIT statements are allowed in an outer loop body to exit from that outer loop.

Prior to Fortran 90, the nonblock DO was the only loop facility in the language. At that time, it was considered good programming practice for each loop to terminate with its own labeled CONTINUE statement, for example:

```
DO 10 I = 1, N
   . . .
   DO 20 J = 1, M
      . . .
   20 CONTINUE
10 CONTINUE
```

This was the safest way to create a nest of loops and avoid errors that a programmer might make when more than one loop terminates on the same action statement. Therefore, when the more error prone forms were made obsolescent, this special form was not. The standard treats this form as if it were a block construct, but it does not have a terminal statement particular to the form, construct names may not appear, and the body of the loop is not a block. However, the standard treats the body as a block and the rules that apply to the block form are extended to this form. The lack of construct names limits the functionality of the form in nested loops, as any EXIT or CYCLE statements can refer only to the innermost loop in which they appear—never to an outer loop. The WHILE form of loop control and the absence of loop control (the simple loop) are permitted in the nonblock DO.

Examples:

```
PROD = 1.0
DO 10 I = 1, N
10 PROD = PROD*P(I)

DO 10 I = 1, N
DO 10 J = 1, N
10 HILBERT(I,J) = 1.0 / REAL(I+J-1)

FOUND = .FALSE.
I = 0
DO 10 WHILE(.NOT. FOUND .AND. I < LIMIT)
   I = I + 1
10 FOUND = KEY == X(I)

DO 20 I = 1 ,L
   DO 20 J = I+M
      DO 20 K = 1, N
         IF (T(I, J, K) .GT. ALLOWED) GO TO 999
20 CONTINUE
```

8.7.6 Conversion from the Nonblock to the Block Form

There is a relatively straightforward translation from obsolescent nonblock loops to the block form. See Table 8-1.

8.8 Branching

Branching is a transfer of control from the current statement to another statement or construct in the program unit. A branch alters the execution sequence. This means that the statement or construct immediately following the branch is usually not executed. Instead, some other statement or construct is executed, and the execution sequence proceeds from that point. The terms **branch statement** and **branch target statement** are used to distinguish between the transfer statement and the statement to which the transfer is made.

The GO TO statement is used to transfer to a statement in the execution sequence that is usually not the next statement in the program, although this is not prohibited.

The statements that may be branch target statements are those classified as action statements plus the initial statements for the ASSOCIATE, CASE, DO, FORALL, IF, SELECT TYPE, and WHERE constructs, and a few additional statements in limited situa-

Table 8-1 Conversion of nonblock DO loops to block loops.

Nonblock loop	Block loop

<pre>DO 10 I = 1, N
   . . .
   IF (. . .) GO TO 20
   . . .
   DO 10 J = 1, M
      . . .
      IF (. . .) GO TO 10
      . . .
10 CONTINUE
20 . . .</pre>
	<pre>DO I = 1, N
	   . . .
	   IF (. . .) EXIT
	   . . .
	   DO J = 1, M
	      . . .
	      IF (. . .) CYCLE
	      . . .
	   END DO
	END DO</pre>

<pre>DO 10 I = 1, N
   . . .
   IF (. . .) GO TO 20
   . . .
   DO 10 J = 1, M
      . . .
      IF (. . .) GO TO 10
      . . .
10   A (I, J) = . . .
20 . . .</pre>
	<pre>DO I = 1, N
	   . . .
	   IF (. . .) EXIT
	   . . .
	   DO J = 1, M
	      . . .
	      IF (. . .) GO TO 10
	      . . .
	10 A(I, J) = . . .
	   END DO
	END DO</pre>

tions. However, it is not permitted to branch to a statement within a block from outside the block. The additional statements that may be branch targets in limited contexts are:

1. an END ASSOCIATE statement provided the branch is taken from within the ASSOCIATE construct
2. an END SELECT statement, provided the branch is taken from within the CASE or SELECT TYPE construct
3. an END DO statement provided the branch is taken from within the DO construct
4. an END IF statement provided the branch is taken from within the IF construct
5. a DO termination action statement, provided the branch is taken from within the innermost DO body, but this use is obsolescent

In addition to the statements described in this section, branching also may be caused by a CALL statement that has an alternate return or an input/output statement that has an END, ERR, or EOR specifier.

8.8.1 Use of Labels in Branching

A statement label is a means of identifying the branch target statement. Any statement in a Fortran program may have a label. However, if a branch statement refers to a statement label, some statement in the program unit must have that label, and the statement label must be on an allowed branch target statement (8.8). The labeled branch target statement must be in the same scoping unit as the branch statement (that is, in the same program unit, excluding labels on statements in internal procedures, derived-type definitions, and interface blocks).

As described in 3.2.5, a label is a string of from one to five decimal digits; leading zeros are not significant. Note that labels can be used in both free and fixed source forms.

8.8.2 The GO TO Statement

The GO TO statement is an unconditional branch statement.

8.8.2.1 Form of the GO TO Statement

The form of the GO TO statement (R845) is:

```
GO TO label
```

8.8.2.2 Execution of the GO TO Statement

When the GO TO statement is executed, the next statement that is executed is the branch target statement identified with the label specified. Execution proceeds from that point. For example:

```
GO TO 200      ! This is an unconditional branch
               !   and always goes to 200.

X = 1.0        ! Because this statement is not labeled and follows
               !   a GO TO statement, it is not reachable.

GO TO 10
GO TO 010      ! 10 and 010 are the same label.
```

8.8.3 The CONTINUE Statement

The form of the CONTINUE statement (R848) is:

```
CONTINUE
```

Normally, the statement has a label and is used for DO termination; however, it may serve as some other place holder in the program or as a branch target statement. It may appear without a label. The statement by itself does nothing and has no effect on the execution sequence or on program results. Examples are:

```
100 CONTINUE
CONTINUE
```

8.8.4 The Arithmetic IF Statement

The arithmetic IF statement is a three-way branch statement based on an arithmetic expression.

The form of the arithmetic IF statement (R847) is:

```
IF ( scalar-numeric-expression ) label , label , label
```

Rules and restrictions:

1. The same label may appear more than once in an arithmetic IF statement.
2. The numeric expression must not be of type complex.
3. Each statement label must be the label of a branch target statement in the same scoping unit as the arithmetic IF statement itself.

Execution begins with the evaluation of the expression. If the expression is less than zero, the branch is to the first label; if equal to zero, to the second label; and if greater than zero, to the third label. Note that both negative and positive zero are equal to zero. The arithmetic IF statement is obsolescent.

8.8.5 The Computed GO TO Statement

The computed GO TO statement transfers to one from a list of branch targets based on the value of an integer expression. The computed GO TO statement is obsolescent; the CASE construct provides a similar functionality in a more structured form.

The form of the computed GO TO statement (R846) is:

```
GO TO ( label-list ) [ , ] scalar-integer-expression
```

Examples:

```
GO TO ( 10, 20 ), SWITCH
GO TO ( 100, 200, 3, 33, 100 ),  2*I-J
```

Rules and restrictions:

1. If there are n labels in the list and the expression has the value i between 1 and n, the value identifies the ith statement label in the list. A branch to the statement with that label is executed.
2. If the value of the expression is less than 1 or greater than n, no branching occurs.
3. Each label in the list must be the label of a branch target statement in the same scoping unit as the computed GO TO statement.
4. A label may appear more than once in the list of branch targets.

Example:

```
      SWITCH = . . .
      GO TO (10, 11, 10) SWITCH
      Y = Z
   10 X = Y + 2.
       . . .
   11 X = Y
```

If SWITCH has the value 1 or 3, the assignment statement labeled 10 is executed; if it has the value 2, the assignment statement labeled 11 is executed. If it has a value less than 1 or greater than 3, the assignment statement Y = Z is executed, because it is the next statement after the computed GO TO statement, and the statement with label 10 is executed next.

8.9 The STOP Statement

This statement causes normal termination (2.3.1) of the program. At normal termination, all input/output units that are connected are closed (9.8). Finalization (4.4.11.3) does not occur.

The forms of the STOP statement (R849) are:

> STOP [scalar-character-constant]
> STOP digit [digit [digit [digit [digit]]]]

Rules and restrictions:

1. The character constant or list of digits identifying the STOP statement is optional and is called a **stop code**.

2. The character constant must be of default character type. It can have a kind value provided it is the value for the default character type.

3. When the STOP code is a string of digits, leading zeros are not significant; 10 and 010 are the same STOP code.

The stop code is accessible following program termination. This might mean that the processor prints this code or sets a program status return code for the operating system. Examples are:

```
STOP
STOP "Error #823"
STOP 20
```

In addition to normal termination, if exceptions (14.3.2) are supported in a scoping unit and any exception is signaling when a STOP statement is executed, the processor must issue a warning indicating which exceptions are signaling; this warning must be on the unit identified by the named constant ERROR_UNIT from the ISO_FORTRAN_ENV intrinsic module (13.6.1, 14.4).

Normal termination also occurs if the END statement of a main program is executed, but in this case, exceptions need not be reported.

9 Input and Output Processing

- **READ**, **WRITE**, and **PRINT statements** are used to transfer data to and from files.
- A **File** is either an external or internal file. An external file may be either a record file or a stream file. An internal file is a record file.
- A **Record** is a sequence of data values. The data may be formatted (converted to characters) or unformatted (not converted). There is also an end-of-file record.
- A **Stream** file is a sequence of file storage units that are not necessarily organized into records. It can be used compatibly with C.
- **File Connection** applies to external files only. The **OPEN** statement connects a file to a unit and determines connection properties. The **CLOSE** statement disconnects a file from a unit. The **INQUIRE** statement inquires about the connection properties.
- **Sequential Access** is a method of operating on records in sequence starting at the current file position.
- **Direct Access** is a method of operating on records by record number; records may be accessed in any order. All records are of the same length.
- **Stream Access** is a method of operating on files that reads or writes file storage units sequentially starting at the current file position.
- **Data formatting** can be explicit or implicit. List-directed and namelist input/output are the implicit forms.
- **Advancing Input/Output** leaves a file positioned between records. **Nonadvancing Input/Output** leaves the file positioned within a record.
- **Asynchronous Input/Output** allows computations to proceed while data is being transferred. A subsequent WAIT statement synchronizes processing by waiting until the input/output operation is complete.
- An **Internal File** is a character variable that is used in place of an external file in a data transfer statement. The transfer is memory to memory.
- **User-Defined Input/Output** subroutines allow specialized processing of derived-type objects.

Fortran input/output statements are designed to accommodate a wide variety of tasks—reading characters from a terminal, reading or writing disk files, efficiently transferring huge data files, even transferring data from attached devices such as Geiger counters. Also, the editing capabilities of the data transfer statements for internal

J.C. Adams et al., *The Fortran 2003 Handbook*,
DOI: 10.1007/978-1-84628-746-6_9,

files are so powerful that in conjunction with the character intrinsic functions, they effectively form a string processing language. Each of these tasks is accomplished using the Fortran input/output statements described in this chapter.

The input/output statements are:

READ
PRINT
WRITE
OPEN
CLOSE
INQUIRE
BACKSPACE
ENDFILE
REWIND
WAIT
FLUSH

The READ statement is a **data transfer input statement** that provides a means for transferring data from an external file to internal storage or from an internal file to internal storage through a process called **reading**. The WRITE and PRINT statements are both **data transfer output statement**s that provide a means for transferring data from internal storage to an external media or from internal storage to an internal file. This process is called **writing**. The OPEN and CLOSE statements are both **file connection statement**s. The INQUIRE statement is used to make inquiries about file and connection properties. The BACKSPACE, ENDFILE, and REWIND statements are **file positioning statement**s.

Most input/output happens synchronously with program execution when a data transfer statement is executed. Asynchronous input/output allows the processor to start the data transfer and then continue computations until a subsequent wait operation synchronizes the input/output operation. This is often called "buffered input/output"; it requires some care to overlap the transfers with the computations. The most common use of asynchronous input/output is for transferring parts of large data sets while computations proceed on other parts.

Programmers need to be careful about system-dependent input/output limitations, especially when doing input/output to devices that have unusual properties, such as terminals, pipes, or devices like Geiger counters. The processor is not required to perform any input/output operation that cannot be supported by the processor or the input/output device. This and other restrictions are described in 9.12.

9.1 Basic Input/Output Concepts

Collections of data are stored in either stream files or record files. In a record file, the data are organized into a series of records; most data transfer statements process entire records. In a stream file, the data consists of a sequence of file storage units (9.1.2, 9.5.1.3); most stream data transfer statements process only a few storage units at a time. The file storage unit is also the basic unit of record length for unformatted records and for the length of any external file. Because stream files are defined in terms of file storage units, most of the discussion of file storage units is embedded in the discussion of stream files. The standard recommends that the file storage unit be an 8-bit octet; most implementations follow that recommendation. The value is given by the named constant FILE_STORAGE_SIZE from the ISO_FORTRAN_ENV intrinsic module (13.6.1). A Fortran record could be a line on a computer terminal or printout, or a logical record on a magnetic tape or disk file. Most processors add a few hidden control bytes at the beginning and end of each record to help with the organization; these control bytes are not part of the Fortran record.

Historically, files were defined in a way that made magnetic tape easy to process. The tape unit had a physical concept of a "record" and a tape was processed one record at a time in sequence. BACKSPACE and REWIND statements allowed for basic repositioning within a file. Obviously, only one reel of tape (one file) could be on a tape unit at a time. An OPEN statement was analogous to mounting a tape on a unit and a CLOSE statement was analogous to dismounting the tape and freeing the unit for another use. As both Fortran and input/output systems have improved, more capabilities have been added to Fortran's file processing. But, basic record oriented file processing remains as the cornerstone for most input/output. Modern disk oriented file systems often use a somewhat different concept of a "record". Rather than do the physical input/output one record at a time, they usually block records into natural disk chunks (sectors or tracks) in order to do the input/output efficiently. One chunk might contain several small records, or a large record might be split into many chunks.

Because stream and record files might have different physical representations on a processor, it is not always possible to OPEN a file for stream input/output in one part of a program and for record input/output in different part of the program or with a different program. Similarly, direct-access and sequential-access files often have different representations and access might be limited to only one method. The properties of files and records do not depend on how they are stored on the hardware.

When phrases such as "file properties" or "file is a record file" are used, they are a shorthand to mean that not only does the file have that property, but also that it has been connected in a way that lets that property be used. Files can often be viewed in more than one way; an OPEN statement (9.3) lets a programmer choose or restrict the particular properties of interest. The set of allowed properties is sometimes restricted by the operating system and sometimes also by permissions from the owner of the file. Some programs might be allowed to read and write a particular file, while other programs might be allowed only to read the file.

9.1.1 Record Files

A **record file** is a sequence of records; it can be represented schematically with each box representing a record as shown in Figure 9-1. Although the boxes in the figure are the same size, records do not always have to have the same size—some or all of them can be empty and a file does not even have to have any records.

There are three kinds of records: formatted, unformatted, and end-of-file. Formatted and unformatted records are collectively called data records. A **data record** is a sequence of values; it can be represented schematically as a collection of small boxes,

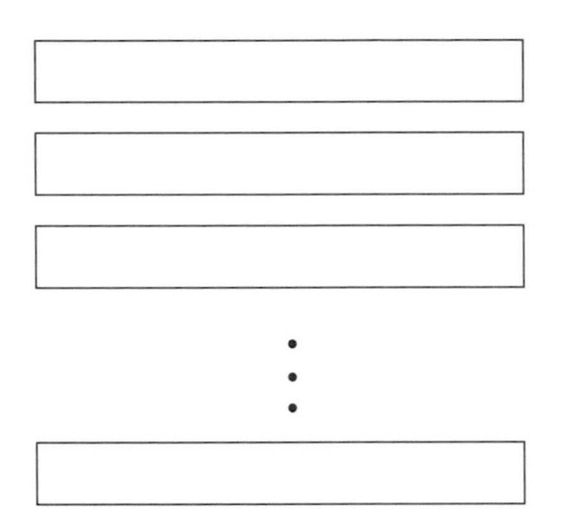

Figure 9-1 A record file

each containing a value, as shown in Figure 9-2. A record can be empty and contain no values; such records can still be read or written.

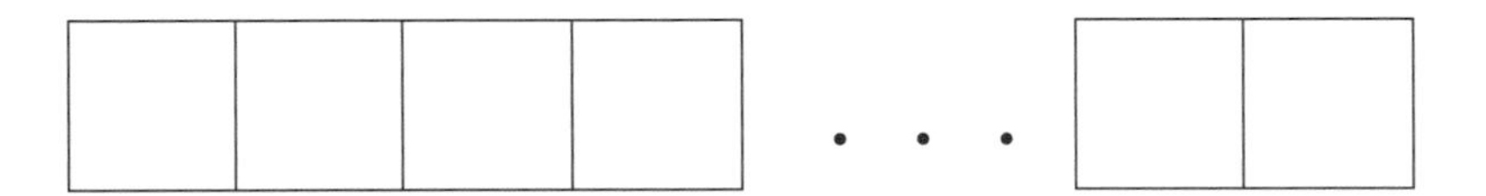

Figure 9-2 Data records

The records of a file must either be all formatted or be all unformatted, except that the file may have an end-of-file record as the last record. A formatted record is read or written by a formatted data transfer input/output statement, and an unformatted record is read or written by an unformatted data transfer input/output statement. A record file will either be a formatted file or an unformatted file and cannot be opened in the other mode.

A **formatted record** is one that contains a sequence of characters. It might be created by a person typing at a terminal or by a Fortran program that converts values into character strings that form human readable representations of those values. When formatted data is read into the computer, the characters are converted to the computer's internal representation of values, which is usually a binary representation. Character values may also be converted from one character representation in the record to another internal representation. The length of a formatted record is the number of characters in it; the length may be zero. A processor may prohibit use of some control characters (3.1) in formatted files to avoid conflicts with record markers or the file structure.

For example, a record containing the four character values "6", ",", "1", and "1" might represent the two numbers, 6 and 11. In this case, the record might be represented schematically as shown in Figure 9-3.

6	,	1	1

Figure 9-3 A formatted record

An **unformatted record** is one that contains only unformatted data, usually represented just as it is stored in computer memory. Unformatted records usually are created by running a Fortran program, although with the knowledge of how to form the bit patterns correctly, they could be created by other means. Unformatted data often require less space on an external device. Also, it is faster to read and write because no conversion is required. However, it is not suitable for reading by humans and frequently it is not suitable for easily transferring data from one computer to another because the internal representation of values is machine dependent. The length of an unformatted data record is measured in processor-dependent units called file storage units (9.1.5.3); it may be zero. The length of an unformatted record that is produced by a particular output list may be determined by the INQUIRE statement (9.9.3). For example, if integers are stored using a binary representation, an unformatted record, consisting of two integer values, 6 and 11, might look like Figure 9-4.

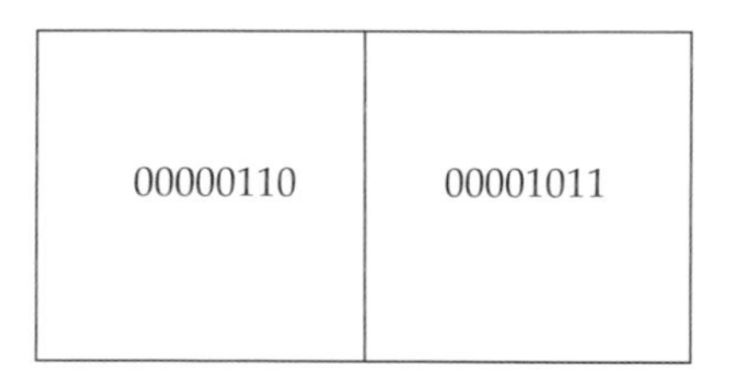

Figure 9-4 An unformatted record

The third kind of record is the **end-of-file record**. There is at most one end-of-file record in a file and is always the last record of the file. It marks the end of the file. It may be written explicitly by using the ENDFILE statement. It also may be written implicitly with a file positioning statement (REWIND or BACKSPACE statement) or by closing the file. End-of-file records can be read or written only when the file is connected for sequential access. An end-of-file record need not have a physical representation in the file, it might just be a bookkeeping entry in the file management system. Its form is processor dependent.

9.1.2 Stream Files

A **stream file** is a sequence of file storage units. Depending on the file, a stream file may be connected for either formatted or unformatted access.

Stream files, as the name implies, need not have the record orientation that record files have. This gives them several advantages over record files. They are designed to interoperate with C, but can work with almost any nonFortran file structure. They have a more intuitive structure for some applications, for example, standardized file interchange formats such as graphic image files. Because they have no required record structure, they have no maximum record length. Their random access is more flexible than Fortran's direct-access files. The cost of these advantages is that the programmer often must use more care and handle more of the small details when using stream files.

When a stream file is connected for unformatted access, there is no concept of records. Each file storage unit may be individually read or written. File storage units each have a unique position number. Some stream files can be positioned to specific places by using the position number; others, for example an input stream from a device such as a data encoder, cannot be positioned.

When creating a formatted stream file, the programmer can use the NEW_LINE intrinsic function to insert record markers into the file, which gives the file a record structure in addition to a stream structure. Each character in a formatted stream file has a unique position number. However, because of the possibility of record markers, not all position numbers necessarily correspond to characters in the file. The result of an INQUIRE statement with a POS specifier must be used if a formatted stream file is to be positioned other than at its initial point. (9.1.4, 9.4.2, 9.9.1).

9.1.3 External and Internal Files

There are two broad classes of files: those that are located on an external device such as a disk or magnetic tape, and those contained in character variables internal to the program.

The use of these files is illustrated schematically in Figure 9-5.

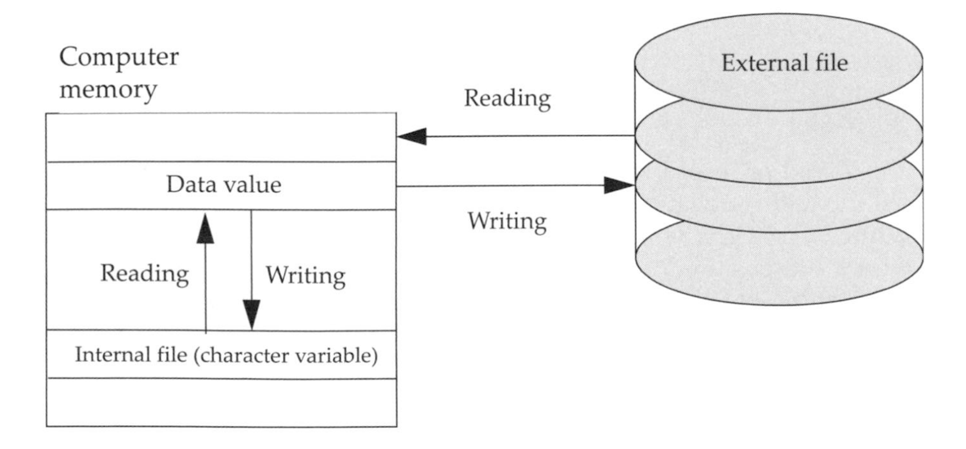

Figure 9-5 External and internal files

9.1.3.1 External Files

External files are located on devices such as tapes, disks, or computer terminals, as opposed to program memory. For each external file, there is a set of allowed access methods, a set of allowed forms, a set of allowed actions, and a set of allowed record lengths. How these characteristics are established is not described by the standard, but usually is determined by a combination of requests by the creator of the file, by actions of the operating system and by details of the device the file is on. Each of these characteristics is discussed later in this chapter. An external file connected to a unit has the position property. Unless the file is positioned at the beginning or end of the file, a record file is positioned at the current record (at the beginning or end), or in some cases, is positioned within the current record, and a stream file is positioned between file storage units. A file may have a name, but the allowed forms of a file name depend on the processor.

9.1.3.2 Internal Files

An internal file is a character variable of either default, ASCII, or ISO 10646 kind. These character values may be created using all the usual means of assigning character values, or they may be created with an output statement specifying the variable as an internal file. Data transfer to and from internal files is described in detail in 9.5.1.5. Data transfer to and from an internal file must use formatted sequential-access input/output statements; explicit, list-directed, and namelist formatting are allowed.

The initial value of any connection property (9.3.3) for an internal file is the default value an external file would have if it were opened with no corresponding specifier. An internal file acts as if it is connected just prior to execution of the input/output statement and closed at the end of execution.

Because of this, file connection, file positioning, and file inquiry are irrelevant and cannot be used with internal files. If the internal file variable is a scalar, the file has just one record; if the variable is an array, the file has one record for each element of the array. The order of the records is the order of the elements in the array. If the variable is an array section, it cannot have vector valued subscripts. The length of each record is the length of one array element.

Internal files provide a powerful way to convert numeric data to or from characters. Common usage includes constructing formats where the number of items depends on run-time values, creating file names from data values, and parsing input strings.

9.1.3.3 Existence of Files

Certain external files are made known to the processor for an executing program. These files are said to **exist** for the program. There are circumstances where a file does not exist for a program. A file might not exist because it is not anywhere on the disks accessible to a system. A file might not exist for a particular program because the user of the program is not authorized to access the file. For example, Fortran programs usually are not permitted to access files belonging to other users or system files, such as the operating system files. A file which is preconnected (9.1.6.2) does not exist if no data has been written to the file. For the most part preconnection of a file and its existence are obsolete concepts; the combination is a left over from early versions of Fortran before the OPEN statement was added.

In addition to files that are made available to programs by the processor, programs may create files needed during program execution. When the program creates a file with an OPEN statement, it exists, even if no data has been written into it. A file no longer exists after it has been deleted. Any of the input/output statements may refer to files that exist for the program at that point during execution. Some of the input/output statements (INQUIRE, OPEN, CLOSE, WRITE, PRINT, FLUSH, REWIND, and ENDFILE) may refer to files that do not exist. READ, BACKSPACE, and WAIT statements can refer only to files that exist. A file that does not yet exist but has been preconnected and is referenced by a WRITE, PRINT, or ENDFILE statement will be created and data put into that file, unless an error condition occurs.

File existence does not apply to internal files. Any character variable has the potential to be an internal file. The only operations available on internal files are READ and WRITE.

9.1.4 File Position

Each connected file has a **position**. During the course of program execution, as data records are read or written, the file position changes. Also, there are other Fortran statements that cause the file position to change; an example is the BACKSPACE statement. The action produced by the input/output statements is described in terms of the file position.

File position becomes indeterminate when an error condition occurs. The programmer cannot rely on the file being in any particular position and must do something, usually a REWIND or CLOSE operation, to put the file in a known state.

The **initial point** of a file is the point just before the first record. The **terminal point** of a file is the point just after the last record. If the file is empty, the initial point and the terminal point are the same. Initial and terminal points of a file are illustrated in Figure 9-6.

A file may be positioned between records. In the example pictured in Figure 9-7, the file is positioned between records 2 and 3. In this case, record 2 is the **preceding record** and record 3 is the **next record**. Of course, if a file is positioned at its initial point, there is no preceding record, and there is no next record if it is positioned at its terminal point.

If a record file is positioned within a record as shown in Figure 9-8, that record is the **current record**, the preceding record is the record immediately previous to the current record, and the next record is the record immediately following the current record. The first record has no preceding record and the last record has no next record. If the file is not positioned within a record, there is no current record.

When there is a current record, the file is positioned at the initial point of the record, between values in a record, or at the terminal point of the record as illustrated in Figure 9-9.

An internal file is always positioned at the beginning of the first record just prior to data transfer.

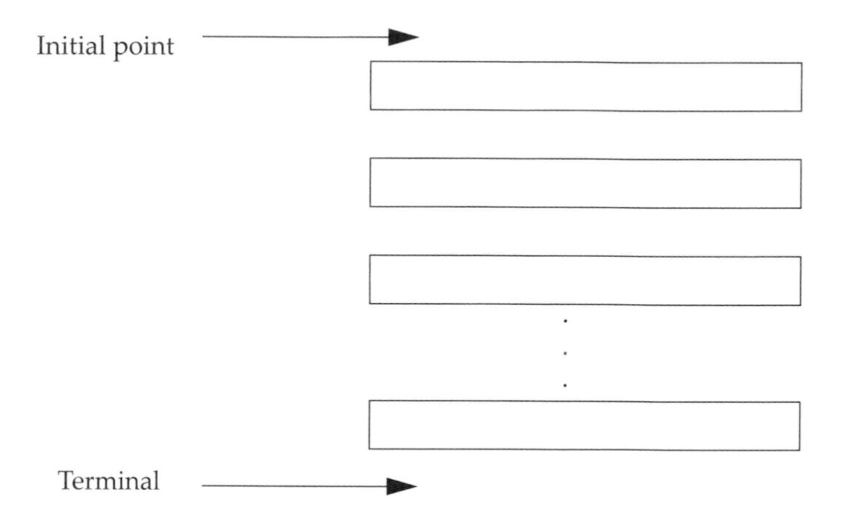

Figure 9-6 Initial and terminal points of a file

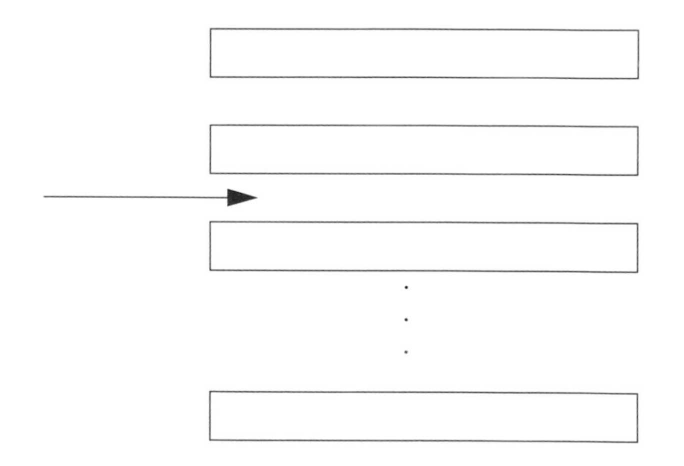

Figure 9-7 A file positioned between records

Advancing input/output is record oriented and leaves a file positioned between records even if an input operation does not completely consume a record. If an error condition occurs, the file position is indeterminate.

In contrast with advancing input/output, nonadvancing input/output usually leaves the file positioned within a record. The position of a nonadvancing file is never changed following a data transfer (although it will change during the data transfer), unless an error, end-of-file, or end-of-record condition occurs while reading the file. The file position is indeterminate following an error condition when reading a file.

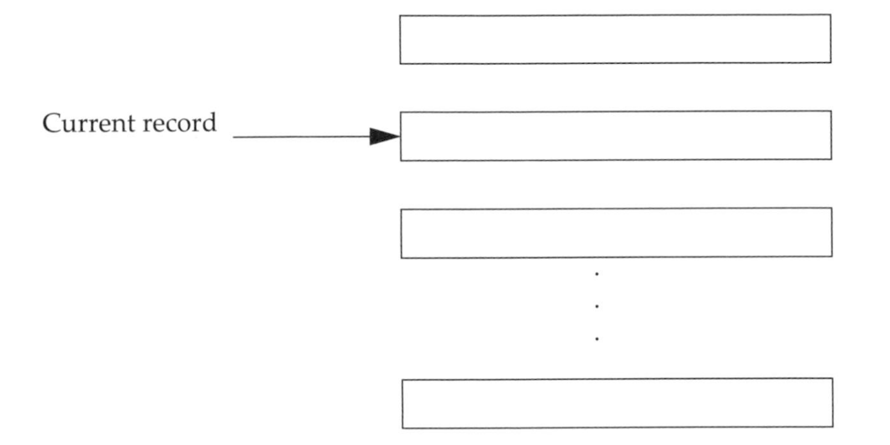

Figure 9-8 A file positioned with a current record

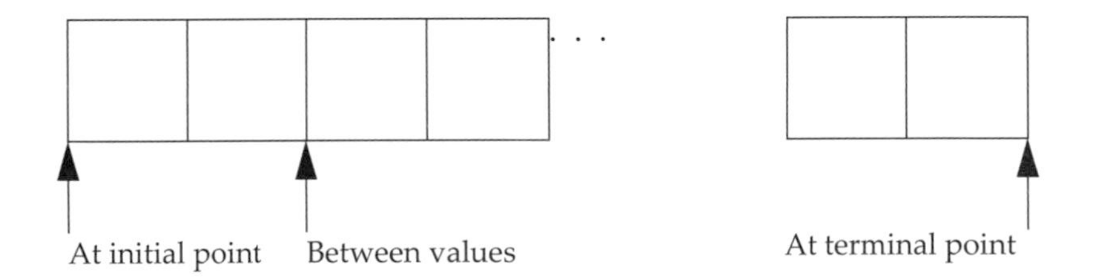

Figure 9-9 Positions within a record of a file

When a nonadvancing input operation is performed, the end-of-record condition occurs only if that input attempts to read past the last character in the record. The file can be positioned after the last character of the record without causing an end-of-record indication. A subsequent nonadvancing input operation causes an end-of-record condition to occur, and positions the file after the end of the record. If another read operation is executed after the end-of-record condition occurs and the record was the last record of the file, an end-of-file condition occurs. An advancing output operation often is used after a series of nonadvancing output operations to terminate processing of the current record.

9.1.5 File Access Methods

There are three access methods:

1. sequential access
2. direct access
3. stream access

Some files may be accessed by any of the methods; other files may be restricted to only some access methods. For example, a printer is restricted to sequential writing. While a file is connected, it has a particular access method, determined by the way it was connected (see OPEN). A file cannot be connected for more than one access method simultaneously; for example, if a file is connected for direct access, it must be disconnected with a CLOSE statement and reconnected specifying stream or sequential access before it can be referenced in a stream- or sequential-access data transfer statement. Note that stream- and sequential-access input/output statements can look the same but what they do depends on the file connection access method, whereas direct-access input/output statements require a REC specifier which is not allowed in either stream- or sequential-access statements.

In summary, the file access method used to read or write a file is not a property of the file itself, but is a property of the particular connection. However, the set of allowed access methods for a file is a property of the file, usually determined when the file is created, although that is operating system dependent.

9.1.5.1 Sequential Access

Sequential access is a method of accessing the records of a record file in order. When connected for sequential access, the records of the file can be read or written only with sequential input/output statements. Sequential access to the records in the file typically begins with the first record of the file and proceeds sequentially to the next records, record-by-record, as illustrated in Figure 9-10. The records are accessed serially as they appear in the file. It is not generally possible to begin at some particular record within the file without reading down to that record in sequential order; however, see 9.3.3 for the POSITION specifier in an OPEN statement.

When a file is being accessed sequentially, the records are read and written in order of their record number even if the records were written in any arbitrary order using direct access.

The records in a sequential file are either all formatted or all unformatted records, except for the end-of-file record, if there is one.

9.1.5.2 Direct Access

Direct access is a method of accessing the records of an external record file in arbitrary order. While a file is connected for direct access, records can be read or written only with direct-access input/output statements. The records are selected by a record number, which must be positive. The records, which must all be the same size, may be read or written in any order. Therefore, it is possible to write record number 47 first, then number 13. In a new file, this produces a file represented by Figure 9-11. Either record may be written without first writing the other. However, direct-access reads are restricted to records that have been written.

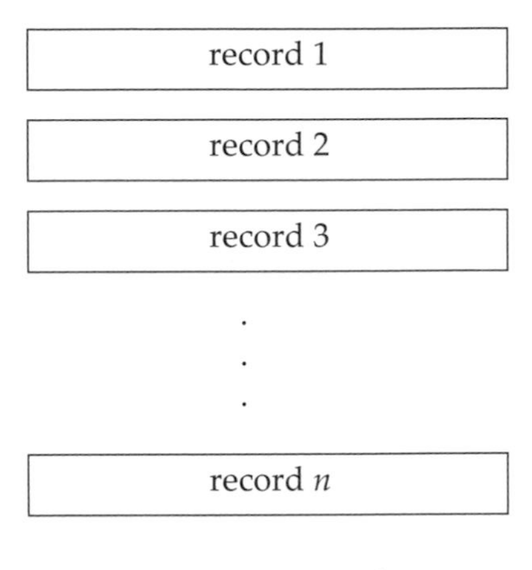

Figure 9-10 Sequential access

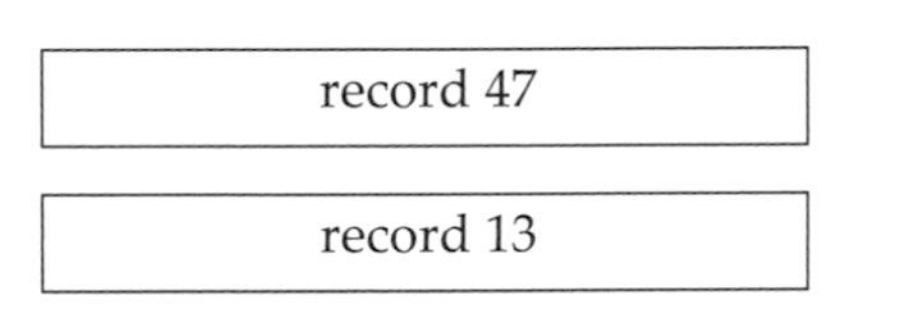

Figure 9-11 A file written using direct access

Some files can be accessed using either direct or sequential access. A direct-access file does not have an end-of-file record. If the file can also be accessed sequentially and has an end-of-file record, the end-of-file record is not considered part of the direct-access file.

Because all of the records of a direct-access file are the same size, many processors store the records in order and pad with empty records where needed. In the example above, a processor could write 46 empty records, record 47 and then rewrite record 13. This is neither required nor forbidden by the standard.

Records in a direct-access file cannot be deleted; however, unlike sequential-access records, they can be rewritten in place.

Records in a direct-access file cannot be read or written using list-directed, namelist, or nonadvancing input/output.

The records in a direct-access file are either all formatted or all unformatted.

9.1.5.3 Stream Access

Stream access is a method of accessing the file storage units of an external stream file. When a file is connected for stream access, its file storage units can be read or written only with stream-access input/output statements. Each file storage unit is uniquely identified by a positive integer position number. Some stream files can be positioned by using the position number of a file storage unit, others, such as a data sampler connected to a Geiger counter, cannot. Other properties of the file and the interpretation of the position number depend on whether the file is connected for formatted or unformatted access.

For a formatted file, some of the file storage units may contain record markers. On many systems, the record markers are multibyte codes, often a combination of carriage-return and line-feed. If they do, this imposes a record structure on top of the stream structure. If there are records, they may be any length. The size of the external device usually limits the record or file size. If there is no record marker at the end of the last record in a file, the record is, by definition, nonempty and also incomplete (9.5.1.2).

Because some of the file storage units of a formatted file might be record markers and not data, some positive integers might not correspond to data. Because of this, a formatted stream file can be repositioned only to a file storage unit that was previously written. Just as with output to a nonstream file, if data is written to this point, it overwrites the existing data and the highest position number written becomes the terminal point of the file. The POS specifier (9.9.2) in an INQUIRE statement can be used to find the current position number, which can be used to subsequently reposition the file. Values returned from an INQUIRE, or the value one, are the only values that can be used to reposition a formatted stream file.

A processor may prohibit use of some control characters (3.1) in formatted stream files to avoid conflicts with record markers or the file structure. If a POS specifier is present in a formatted stream output statement and no characters are written, the only effect is to position the file.

For an unformatted file, the position number is the sequence number, beginning with one, of the file storage units in the file; each position number is one more than that of the preceding file storage unit. If the file can be positioned, then the file storage units do not have to be read or written in order. The only requirement in the standard is that a file storage unit cannot be read before it has been written. In this respect, unformatted stream files are similar to direct-access files. However, unlike direct access, a processor might prohibit attempts to write a file storage unit if all previous file storage units have not been written.

9.1.6 Units

Input/output statements refer to a particular file by specifying an **input/output unit**. An input/output unit is either an external unit or an internal unit and a unit is specified by a scalar integer expression, an asterisk (*), or a character variable name.

The number of available external units and their numbering are processor dependent. A particular value identifies one and only one external unit in all program units in a Fortran program. In an input data transfer statement, an asterisk refers to the numbered unit INPUT_UNIT. In an output data transfer statement, an asterisk refers to the numbered unit OUTPUT_UNIT. The value INPUT_UNIT and * (in an input statement) both refer to the **default input unit**. The value OUTPUT_UNIT and * (in an output statement) both refer to the **default output unit**. Note that external unit numbers must be nonnegative unless they are equal to one of the special unit names. This allows, but does not require, the processor to use negative values to identify special units.

Traditionally, the asterisk referred to unit 5 for input and unit 6 for output on most processors. However, that was not a universal convention. To avoid existing differences between processors, Fortran 2003 introduced the parameter names, INPUT_UNIT, OUTPUT_UNIT, and ERROR_UNIT in the ISO_FORTRAN_ENV module (13.6.1). Using INPUT_UNIT and OUTPUT_UNIT, rather than 5 or 6, has a number of advantages beyond enhanced portability. Not all systems behaved the same way if either unit 5 or 6 were closed, possibly reopened connected to a different file, and then an input/output statement using an asterisk unit was executed. Because most processors try to maintain compatibility with past practice, doing an OPEN or CLOSE on the special files or units 5 or 6 should be avoided. The auxiliary input/output statements, such as OPEN, INQUIRE, and REWIND, do not allow an asterisk for the unit specifier.

The C standard defines three files, standard input, standard output and standard error. There is no requirement that these files correspond to the similarly named Fortran files. However, that is the most likely implementation and Fortran subprograms that read or write INPUT_UNIT or OUTPUT_UNIT should be able to interoperate with C functions that read or write standard input or standard output. It is probably best to avoid ADVANCE="NO" or tabbing if a C function will be processing the same record.

An internal unit is a character variable of either default, ASCII, or ISO 10646 kind; it cannot have a vector-valued subscript. An internal unit is also identified by a unit number if it is the active file in a child input/output statement (9.5.1.4). The phrases "internal unit" and "internal file" are often used interchangeably.

9.1.6.1 Unit Existence

The collection of unit numbers that can be used in a program for external files is determined by the processor, but most processors support units numbered between 1 and 99. An INQUIRE statement (9.9) can be used to find out if a particular unit number is available. The unit numbers that may be used are said to **exist**. Some unit numbers on some processors are always used for data input (for example, unit 5), others are always used for output (for example, unit 6); other values less than 10 are sometimes special cased by the processor. There may be certain unit numbers that are never allowed for user files because they are restricted by the operating system. For portable programming, an INQUIRE statement should be used to determine if a unit exists and unit numbers should be greater than 10 and less than 100. Input/output statements must refer to units that exist, except for the CLOSE or INQUIRE statement.

9.1.6.2 Establishing a Connection to a Unit

In order to transfer data to or from an external file, the file must be connected to a unit. An internal file is always connected to the unit that is the character variable. There are two ways to establish a connection between a unit and an external file:

1. execution of an OPEN statement in the executing program

2. preconnection by the operating system

Only one file can be connected to a unit at any given time and vice versa. If the unit is disconnected from the file with a CLOSE statement, it may be reconnected to another file or to the same file. A file that is not connected to a unit cannot be used in any statement, except the OPEN or INQUIRE statements.

Some units may be **preconnected** to files by the operating system without any action necessary by the program. For example, on most systems, units 5 and 6 are preconnected to the default input and default output files, respectively. Preconnection of units also may be done by the operating system if requested by the user in the operating system command language. In either of these cases, the user program does not use an OPEN statement to connect the file; it is preconnected.

Preconnection is primarily a historical method designed to deal with limitations in early operating systems and versions of Fortran. Modern practice uses an OPEN statement to establish a connection. If preconnection must be used, an INQUIRE statement should be used to make sure the connection was made and that the appropriate properties are set for the file.

A standard-conforming program must not attempt any data transfer to or from a unit unless it has a file connected to it, either by preconnection or by an OPEN statement. However, it is a fairly common extension for processors to relax this requirement and treat a data transfer to a unit with no connection as if there had been an OPEN of an empty file with the appropriate properties for the transfer. The file is typically given a processor-dependent name like FORT.*nn* or TAPE.*nn*, where *nn* is the unit number. An INQUIRE by unit statement executed before the data transfer would indicate that the unit is not connected even though it will subsequently behave as if it were connected. Because this is nonstandard, it should be avoided and OPEN statements should always be used.

Once a file has been disconnected, the only way to reference it is by its name using an OPEN or INQUIRE statement. There is no means of referencing an unnamed file once it is disconnected.

9.2 Input/Output Statement Specifier Lists

The general scheme for Fortran input/output is to connect a file to a unit with an OPEN statement, perform actions on the file with READ, WRITE, PRINT, position (BACKSPACE, ENDFILE, or REWIND), WAIT, or FLUSH statements, and then terminate the connection with a CLOSE statement. At any step in the process, an INQUIRE statement may be used to inquire about properties of the file, unit, or connection. These input/output statements are described in the following sections. However, many of the specifiers are common to several statements and are described here. Specifiers unique to a particular statement are described in the statement description. Many specifiers are interrelated; if one appears, others may be either required or prohibited.

9.2.1 General form of an Input/Output Statement

The general form of an input/output statement is

```
statement-name ( specifier-list ) [ variable-list ]
```

The PRINT statement and some forms of the READ and file position statements do not have specifier lists. These exceptions are noted in the statement descriptions, but not in the individual specifier descriptions. See 9.2.1.1 below.

Some specifiers provide values which control how the statement will execute; others return values which describe the results of execution. Two general rules for specifiers are that no variable which provides a value may also appear in a context where it would return a value and that no variable which returns a value may appear more than once in the statement. This also applies to variables used as subscripts or substrings.

Rules and restrictions:

1. Most specifiers have one of two forms, either

   [keyword =] variable

   or

   [keyword =] expression

   Specifiers are usually referred to by their keyword name. For example

   RECL = *nn*

   is called a RECL specifier. Depending on context, *nn* can be either a variable and receive a value after the statement is processed or be an expression and provide a value when the statement begins execution.

   For the UNIT, FMT, and NML specifiers, the keyword is optional in some cases; for the other specifiers, it is required. If the UNIT= is omitted, the unit specifier must be the first in the list.

2. No specifier may appear more than once in a given specifier list.

3. Most specifiers for the input/output statements provide values to be used and may use expressions; others return values and must use variables. Because an expression can be just a simple variable, the interpretation and rules for something that looks like keyword = variable depends on context.

4. The value of the character expression for many of the specifiers must be one of the permitted values for that specifier. Trailing blanks in any specifier value are ignored. The value is interpreted without regard to case.

5. File names are processor dependent and are an exception to the above rule. They can be in upper or lower case and can have embedded blanks, but not trailing blanks.

6. Character results will be in upper case when they are values from specific lists. Results, such as file names or IOMSG variables, that are not restricted to a small set of values, are processor dependent.

7. Most specifiers for the INQUIRE statement return status results and must use variables.

8. Most specifiers are optional; a UNIT specifier, but not necessarily the UNIT=, is required in all of the input/output statements, except for the inquire by output list form.

9. A variable which provides a value to the statement must not be the same as any other variable which returns a value, must not be used as a subscript or substring for any other entity in the statement, nor be associated with any variable which is in the input or output list of a data transfer statement.

10. The value of any variable or expression in a specifier, including its subscript values and substring bounds, must not affect nor be affected by the definition or evaluation of any other entity appearing in the statement.

The last rule allows the processor to evaluate terms in the statements in any order. If, for example, the IOMSG variable is subscripted, the processor need not evaluate the subscript expressions until it detects an error. This allows the normal error-free case to be processed somewhat more efficiently.

9.2.1.1 Abbreviated Form of Input/Output Statements

For mostly historical reasons Fortran has abbreviated forms of several of the input/output statements. Originally, input/output was simple. The card reader had no user selectable properties. The "Job Control Language" for mounting a tape was a handwritten note fastened to the card deck with a rubber band. The input/output statements had a correspondingly simple syntax. As operating systems and input/output devices matured, the syntax also grew to accommodate the new options. For compatibility, the original abbreviated forms are still part of Fortran and are often sufficient for quick and dirty programs. For mature applications, it is usually better practice to use the general form described above.

Each abbreviated form is exactly equivalent to a general form with a limited specifier list. The abbreviated forms will be used in some of the examples, but will generally not be described in detail since they are merely special cases of the general form. Descriptions and rules are usually given in terms of the general form. The corresponding rules for the special forms can be easily deduced. Table 9-1 shows the abbreviated form and the corresponding general form.

9.2.1.2 Recursive Input/Output Statements

A **recursive input/output statement** is an input/output statement executed while another input/output statement is executing. This can occur in a defined input/output subroutine or if a function referenced in an input/output statement executes any input/output statement.

Table 9-1 Abbreviated and general forms of some input/output statements

Abbreviated Form	Equivalent statement
BACKSPACE scalar-integer-expression	BACKSPACE (scalar-integer-expression)
ENDFILE scalar-integer-expression	ENDFILE (scalar-integer-expression)
FLUSH scalar-integer-expression	FLUSH (scalar-integer-expression)
READ format [, input-item-list]	READ (* , format) [input-item-list]
PRINT format [, output-item-list]	WRITE (* , format) [output-item-list]

Rules and restrictions:

1. The recursive input/output statement must not reference an external unit unless it is a child data transfer statement that references its parent external unit.
2. A recursive input/output statement must not modify any existing internal unit, except that a WRITE statement may modify its own internal unit.
3. The restrictions apply to all input/output statements, not just data transfer statements.

Note that an asynchronous input/output statement finishes execution before the next statement begins execution, even though the data transfer might not be complete. Thus, input/output statements executed while the transfer is taking place are not recursive.

These rules recognize the common implementation technique of doing internal transfers directly to or from the internal unit while also allowing the processor to use a single "system buffer" to hold intermediate results for transfers to or from external files. Allowing recursive output to different internal files makes it easy to use a function to format things like dates and times directly in a list.

Example

```
write(unit, *) pretty_date(day_number)
```

where pretty_date is an external function that converts a day number into month, date, and name of the day using internal write statements.

9.2.2 The UNIT Specifier

The UNIT specifier identifies the unit on which the action will be performed. The forms is one of

[UNIT =] scalar-integer-expression	external or child unit
[UNIT =] *	default external unit
[UNIT =] character-variable	internal unit

Rules and restrictions:

1. An external unit is required in OPEN, CLOSE, position, FLUSH, and WAIT statements; it is optional in an INQUIRE statement
2. A UNIT specifier is required in READ and WRITE statements.
3. If the optional UNIT= is absent, the unit specifier must be first in the specifier list.

An external unit is specified by a scalar integer expression or *. The scalar integer expression value must be zero, positive, or equal to one of the values of the named constants INPUT_UNIT, OUTPUT_UNIT, or ERROR_UNIT of the ISO_FORTRAN_ENV module (13.6.1).

Specifying a unit by * is shorthand for either INPUT_UNIT or OUTPUT_UNIT. These units are preconnected for formatted sequential access. The units associated with * are processor dependent. On many systems, units 5 and 6 are preconnected for input and output respectively. On other systems, INPUT_UNIT and OUTPUT_UNIT might be negative, preventing clashes with programmer selectable unit numbers.

There is another unit preconnected for formatted sequential output, ERROR_UNIT. The purpose is for error reporting, although the standard itself does not mandate that any processor detected errors be reported on that unit. The value of ERROR_UNIT might be the same as OUTPUT_UNIT.

If the unit specified is a character variable, the statement must be a READ or WRITE statement and the transfer is called an internal read or write (9.5.1.5). In effect, the character variable acts as the source or destination, respectively, for the data being transferred. The character variable must not have any vector-valued subscripts; its kind must be either default character, ASCII character, or ISO 10646 character.

If a parent data transfer statement (9.5.1.4) references an external unit, the value of the unit argument to the defined input/output subroutine will be value of the external unit. If the parent data transfer statement is an internal read or write, then the unit value will be a processor-dependent negative value. Thus, a child data transfer statement that accesses the defined input/output subroutine's unit argument might do internal input or output even though the unit is specified by an integer expression.

The relationship between a specific unit number and a file is established either by an OPEN statement (9.3) or by preconnection (9.1.6.2). Once established, all references throughout the program to the same unit number will refer to the same connected file. In effect, file unit numbers and the file they refer to are global to the entire program. The connection may be broken by a CLOSE statement (9.8) and then the unit may again be connected to a file by a subsequent OPEN statement.

All of the input/output statements may refer to units that exist; INQUIRE, CLOSE, and WAIT statements may also refer to units that do not exist. The allowed values for the unit specifier are processor dependent; numbers in the range 10 to 99 are generally safe to use portably. An INQUIRE by unit statement (9.9.1) can determine if a unit number is valid for a particular processor.

9.2.3 Error and Exception Handling Specifiers

These specifiers allow for recovery from some input/output errors or other exceptional conditions. Control will branch to a labeled statement and/or status variables will be given a value.

END

Allowed in WAIT and READ statements

EOR

Allowed in WAIT and READ statements

ERR

Allowed in all input/output statements

IOMSG

Allowed in all input/output statements

IOSTAT

Allowed in all input/output statements

Rules:

1. The label must be the label of a branch target statement in the current scoping unit. The normal rules for transferring into a block apply.
2. If an error is detected, control branches to the ERR label. The position of the file becomes indeterminate.
3. If an end-of-file is encountered during input, control branches to the END label.
4. If an end-of-record is encountered during nonadvancing input, control branches to the EOR label.
5. If an error, end-of-file, or end-of-record occurs, the IOSTAT variable is set to a processor-dependent nonzero value, as described in 9.6. Otherwise, it is set to zero, indicating a successful operation.
6. If an error, end-of-file, or end-of-record occurs, the processor will assign an explanatory message to the IOMSG variable. Otherwise, this variable is unchanged.
7. In a data transfer statement, the variable in an IOSTAT or IOMSG specifier must not be the same as or associated with any entity in the input/output item list or in the namelist group or with the variable specified in the SIZE specifier.

8. If a variable in an IOSTAT or IOMSG specifier is an array element, its subscript values must not be affected by the data transfer, by any implied-do processing, or with the definition or evaluation of any other specifier in the control specifier list.

9. If more than one error occurs during execution of an input/output statement, it is processor dependent which error is reported.

10. A statement might cause multiple operations to be implicitly performed on a unit. Examples include an OPEN statement performing a close operation, or a READ statement performing a wait operation. Errors and exceptions from any of these operations will be processed as if they were caused by the current statement. The processor does not go back to the initial statement and set its error variables or use its branch labels.

If an endfile, error, or end of record occurs and there is neither a corresponding branch label nor an IOSTAT specifier, program execution is terminated. Note that the presence of an IOMSG specifier does not, by itself, prevent program termination. The set of errors detected is processor dependent. The details of error and exception processing are described under the various input/output statements and also in 9.6.

9.2.4 Changeable Connection Mode Specifiers

These mode specifiers control processing of characters during formatted input/output. They are allowed only for files connected for formatted input/output.

They are all set on the initial open of a formatted file, either with explicit specifiers in the OPEN statement or with a default value. These modes remain in effect until a subsequent reopen for the same unit gives them new values. A READ or WRITE data transfer statement may change them temporarily for that specific statement execution. Most also have associated format specifiers that may change the mode as data is being transferred. With each, there is a corresponding INQUIRE specifier, which returns the mode currently in effect.

When a data transfer statement terminates, the changeable modes, as well as the scale factor (10.9.5) revert to the values in effect immediately before the statement began execution. Values set during one nonadvancing input/output statement do not propagate forward into subsequent input/output to the same record. Similarly, values set by a child data transfer statement do not propagate back to the parent data transfer statement. Values set by the parent do propagate down to the child data transfer statement (where they can be temporarily changed) because the parent does not terminate when it invokes the child statement. If a child statement changes the modes, they reset to the parent's values when the child input/output statement completes.

These specifiers may appear in OPEN, READ, and WRITE statements as

```
keyword = scalar-default-character-expression
```

and in INQUIRE statements as

```
keyword = scalar-default-character-variable
```

Some specifiers are not allowed in READ statements, others are not allowed in WRITE statements, as noted below. The changeable modes set by the OPEN statement are ignored in these cases.

BLANK (not allowed in WRITE statements)

Controls the interpretation of blanks in numeric fields during explicitly formatted input (10.2)

NULL	ignore all such blanks (default)
ZERO	interpret all such blanks except leading blanks as zeros

A field of all blanks evaluates to zero in both cases. This mode only applies to explicitly formatted input. List-directed and namelist input use a blank character as a delimiter, not as part of a numeric value.

Related edit descriptors: -BN, BZ

Related INQUIRE specifier: BLANK

DECIMAL

Sets the character used as the decimal point during numeric conversion (10.5)

COMMA	the decimal symbol is ","
POINT	the decimal symbol is "." (default)

Note that when the decimal mode is COMMA, the separator used between the real and imaginary part of a complex value and between values in list-directed and namelist input/output is a “;” rather than a “,”. This separator is referred to as a **CS symbol** in this book.

Related edit descriptors: DC, DP

Related INQUIRE specifier: DECIMAL

DELIM (Not allowed in READ statements)

Sets the character used to delimit character values in list directed and namelist output (10.10, 10.11)

APOSTROPHE	use the apostrophe (') as the delimiting character
QUOTE	use the quotation mark (") as the delimiting character
NONE	use no delimiter (default)

If the DELIM value is APOSTROPHE, each occurrence of an apostrophe within the character value is doubled; if the DELIM specifier is QUOTE, each occurrence of a quote within the character value is doubled. If a DELIM specifier appears, the statement must perform list-directed or namelist output.

Related edit descriptors: none

Related INQUIRE specifier: DELIM

PAD (Not allowed in WRITE statements)

Controls padding of short records with blanks during formatted input when the input item list and format specification require more characters than the record contains.

YES	use blank padding (default)
NO	requires that the input record contain the data indicated by the input list and format specification

Rules:

1. The blank padding character used for nondefault character types is processor dependent.
2. If this specifier has the value YES and an end-of-record condition occurs, the data transfer behaves as if the record were padded with sufficient blanks to satisfy the input item and the corresponding data edit descriptor.

Related edit descriptors: none

Related INQUIRE specifier: PAD

ROUND

Sets the rounding mode used for formatted input/output processing (but not for general computation) (10.9.7)

UP	the rounding mode is up
DOWN	the rounding mode is down
ZERO	the rounding mode is to zero
NEAREST	the rounding mode is to nearest
COMPATIBLE	the rounding mode is historically compatible
PROCESSOR_DEFINED	the rounding mode is processor defined

The default value is processor dependent if not specified in an OPEN statement. It must be one of the above values, but need not be PROCESSOR_DEFINED.

In the following rules, for input, “representable value” means internal hardware representation; for output it means the character value produced in the specified field.

Rules:

1. If the mode is UP, the resulting value will be the smallest representable value that is greater than or equal to the original value.
2. If the mode is DOWN, the resulting value will be the largest representable value that is less than or equal to the original value.
3. If the mode is ZERO, the resulting value will be the closest representable value which is no greater in magnitude than the original value.
4. If the mode is NEAREST, the resulting value will be the closest of the two nearest representable values if one is closer than the other; otherwise it is processor dependent which is chosen. On a processor supporting IEEE input and output, NEAREST will pick the even value.
5. If the mode is COMPATIBLE, the result is the same as for NEAREST unless the original value is halfway between the two representable values. In such a case, the value away from zero is selected.
6. If the mode is PROCESSOR_DEFINED, the rounding mode need not correspond to any of the other modes.
7. On processors that support the IEEE rounding conventions (14.2.8), UP, DOWN, ZERO, and NEAREST correspond to the respective IEEE rounding modes.

Related edit descriptors: RU, RD, RZ, RN, RC, RP

Related INQUIRE specifier: ROUND

SIGN (Not allowed in READ statements)

Controls whether or not a plus sign that is optional in numeric output (10.9.4) will appear

PLUS	the optional plus sign will appear
SUPPRESS	the optional plus sign will not appear
PROCESSOR_DEFINED	appearance of the optional plus sign is processor dependent (default)

Related edit descriptors: S, SP, SS

Related INQUIRE specifier: SIGN

9.3 The OPEN Statement

The OPEN statement establishes a connection between a unit and an external file and specifies the connection properties or it may modify the existing connection properties.

Once an OPEN statement connects a file to a unit, the connection of the unit to the file is valid in the main program or any subprogram, unless a CLOSE statement affecting the connection is executed.

If a file exists, the OPEN statement must specify properties that are consistent with the file properties. For example, ACCESS cannot specify stream unless the stream-access method is one of the allowed access methods for the file. If the file does not exist, the OPEN statement will create it with a set of properties that include those specified by the OPEN statement. It is processor dependent if other properties are also allowed. For example, a file opened for direct access might be created to also allow sequential access.

If a file is already connected to one unit, it must not be connected to a different unit.

9.3.1 Connecting a File to a Unit

In what is probably the most common situation, the unit is not connected to a file and the OPEN statement connects an external file to a unit. If the file does not exist, it is created. If the unit is connected to a file and if the FILE specifier is absent or names the same connected file, the OPEN statement may be used to change the changeable connection modes as described in 9.2.4. If new values for these specifiers are provided, they will be used in subsequent data transfer statements; otherwise, the old ones will be used. This is often called a reopen, although there is no special syntax to indicate this. The reopen may also contain ERR, IOMSG, and IOSTAT specifiers. They apply only to the OPEN statement being executed; after that, the values of these specifiers have no effect. If no ERR or IOSTAT specifiers appear in the new OPEN statement, an error condition will terminate the execution of the program. If the unit was preconnected to a file that does not exist and the FILE specifier is absent or names the connected file, the OPEN statement also creates the file. A reopen statement can only be used to alter the changeable modes; any other specifiers must be consistent with the existing connection. If there is a STATUS specifier, it must specify OLD.

If the FILE specifier specifies a different file, the effect is as if a CLOSE statement (9.8) without a STATUS specifier is executed on the unit and the OPEN statement is then executed.

If the unit is preconnected to a file that does not exist and there is no FILE specifier, the OPEN statement creates the file and establishes properties of the connection.

9.3.2 Form of the OPEN Statement

The form of the OPEN statement (R904) is:

```
OPEN ( connection-specifier-list )
```

where the forms of a connection specifier (R905) are:

```
[ UNIT = ] scalar-integer-expression
ACCESS = scalar-default-character-expression
ACTION = scalar-default-character-expression
ASYNCHRONOUS = scalar-default-character-expression
BLANK = scalar-default-character-expression
DECIMAL = scalar-default-character-expression
DELIM = scalar-default-character-expression
ENCODING = scalar-default-character-expression
ERR = label
FILE = file-name-expression
FORM = scalar-default-character-expression
IOMSG = scalar-default-character-variable
IOSTAT = scalar-integer-variable
PAD = scalar-default-character-expression
POSITION = scalar-default-character-expression
RECL = scalar-integer-expression
ROUND = scalar-default-character-expression
SIGN = scalar-default-character-expression
STATUS = scalar-default-character-expression
```

Rules and restrictions:

1. Note that the form UNIT = * for the unit specifier is not permitted in the OPEN statement. However, the values INPUT_UNIT, OUTPUT_UNIT, and ERROR_UNIT may be used to reopen these files and alter their changeable modes. It is also possible, in principle, to connect different files to these units; however, whether that is allowed or the way that it will interact with subsequent uses of the * unit is processor dependent.

2. If the last data transfer to a unit connected for sequential access to a particular file was an output data transfer statement, an OPEN statement for that unit connecting it to a different file writes an end-of-file record to the original file as part of the implicit close processing.

3. See 9.2 for the general rules and restrictions for input/output statement specifier lists.

Examples are:

```
OPEN (STATUS = "SCRATCH", UNIT = 9)
OPEN (8, FILE="PLOT_DATA", RECL=80, ACCESS="DIRECT")
```

9.3.3 The Connection Specifiers

The OPEN statement specifies the connection properties between the file and the unit, using keyword specifiers, which are described in this section. Table 9-2 gives the possible values for the specifiers in an OPEN statement that have a restricted set of options and their default values when the specifier is omitted.

Table 9-2 Values for some keyword specifiers in an OPEN statement

Specifier keyword	Possible values	Default value
ACCESS	DIRECT, SEQUENTIAL, STREAM	SEQUENTIAL
ACTION	READ, WRITE, READWRITE	Processor dependent
ASYNCHRONOUS	YES, NO	NO
BLANK	NULL, ZERO	NULL
DECIMAL	COMMA, POINT	POINT
DELIM	APOSTROPHE, QUOTE, NONE	NONE
ENCODING	UTF-8, DEFAULT	DEFAULT
FORM	FORMATTED, UNFORMATTED	Note 1
PAD	YES, NO	YES
POSITION	ASIS, REWIND, APPEND	ASIS
ROUND	UP, DOWN, ZERO, NEAREST, COMPATIBLE, PROCESSOR_DEFINED	Processor dependent
SIGN	PLUS, SUPPRESS, PROCESSOR_DEFINED	PROCESSOR_DEFINED
STATUS	OLD, NEW, UNKNOWN, REPLACE, SCRATCH	UNKNOWN

Note 1. FORMATTED for sequential access, UNFORMATTED for direct access

UNIT

Described in 9.2.2

A unit specifier is required.

ACCESS

Specifies the access method (9.1.5) for the connection

DIRECT	direct access
SEQUENTIAL	sequential access
STREAM	stream access

Rules:

1. If the ACCESS value is DIRECT, a RECL specifier also must appear.

Related INQUIRE specifier: ACCESS

ACTION

Specifies the actions allowed for the connection

READ	WRITE, PRINT, and ENDFILE statements are prohibited
WRITE	READ statements are prohibited
READWRITE	any input/output statement is permitted

Rules:

1. The default value is processor dependent.
2. If READWRITE is an allowed ACTION value for the file, READ and WRITE are allowed.

Related INQUIRE specifiers: ACTION, READ, WRITE, and READWRITE

ASYNCHRONOUS

Specifiers whether or not asynchronous input/output is allowed

YES	asynchronous input/output is allowed
NO	asynchronous input/output is not allowed (default)

Related INQUIRE specifier: ASYNCHRONOUS

BLANK

Described in 9.2.4

DELIM

Described in 9.2.4

ENCODING

Specifies the character encoding method for formatted files

UTF-8	ISO/IEC 10646-1:2000
DEFAULT	processor dependent (default)

Rules:

1. If UTF-8 is specified, the file is called a Unicode file and all the characters in the file must be ISO 10646 character kind.
2. UTF-8 must not be specified if the processor does not support ISO 10646 character kind.

3. The connection must be for formatted input/output.

Related INQUIRE specifier: ENCODING

ERR

Described in 9.2.3

FILE

Specifies the name of the file being connected

expression	the name of the file to be connected. It is called the file name expression

Rules:

1. Trailing blanks in the name are ignored.
2. The name must be a file name allowed by the processor.
3. The interpretation of case is processor dependent; for example, the processor may distinguish file names by case or it may interpret the name all in uppercase or lowercase letters.
4. The FILE specifier must appear if the STATUS value is NEW or REPLACE.
5. If the STATUS value is OLD, the FILE specifier must appear unless the unit is already connected to a unit that exists.
6. If the STATUS specifier has the value SCRATCH, the FILE specifier must not appear.

Related INQUIRE specifiers: NAME and NAMED

FORM

Specifies formatted or unformatted input/output

FORMATTED	the connection will be for formatted access
UNFORMATTED	the connection will be for unformatted access

Rules:

1. The default value is UNFORMATTED, if the file is connected for direct or stream access.
2. The default value is FORMATTED, if the file is connected for sequential access.

Related INQUIRE specifiers: FORM and FORMATTED

IOMSG

Described in 9.2.3

IOSTAT

Described in 9.2.3

PAD

Described in 9.2.4

POSITION

Specifies the initial position of the file

ASIS	file position is unchanged for a connected file and is unspecified for a file that is not connected (default)
REWIND	file is positioned at its initial point
APPEND	file is positioned either at the terminal point or just before an end-of-file record, if there is one

Rules:

1. The default value is ASIS, permitting an OPEN statement to change other connection properties of a file that is already connected without changing its position.
2. The file must be connected for sequential or stream access.
3. If the file is new, it is positioned at its initial point.

Related inquiry specifier: POSITION

RECL

Specifies record length for direct- or sequential-access files

The expression specifies the length of each direct-access record or the maximum record length if the access method is sequential.

Rules:

1. The expression value must be positive.
2. The RECL specifier must appear for a file connected for direct access.
3. The default maximum value is processor dependent for a file connected for sequential access.
4. The RECL specifier must not appear for a file connected for stream access.

5. If the file is connected for formatted input/output, the length is the number of default characters for a record that contains only default characters and the length is processor dependent for a record that contains nondefault characters.

6. If the file is connected for unformatted input/output, the length is measured in file storage units (9.1.5.3). On most systems, a file storage unit is a byte.

Related INQUIRE specifier: RECL

See also inquire by output list (9.9.3)

ROUND

Described in 9.2.4

SIGN

Described in 9.2.4

STATUS

Specifies the file existence before the connection is made

OLD	the file must exist
NEW	the file must not exist; it will be created
UNKNOWN	the file has a processor-dependent status (default)
REPLACE	if the file does not exist, it is file is created and given a status of OLD; if the file does exist, it is deleted, a new file is created with the same name, and the file is given a status of OLD
SCRATCH	file is created and connected to the specified unit; it is deleted when the program terminates normally or a CLOSE statement is executed on that unit

Rules:

1. Scratch files must be unnamed; that is, the STATUS value must not be SCRATCH when a FILE specifier appears. The term scratch file refers to this temporary file.

Note that, if the STATUS value is REPLACE, the specifier itself in this statement is not changed to OLD; the file status will be OLD when the OPEN statement terminates.

Related INQUIRE specifier: none

9.4 Data Transfer Statements

When a unit is connected, data may be transferred by reading and writing to the file associated with the unit. The transfer may occur to or from internal or external files.

The **data transfer statements** are the READ, WRITE, and PRINT statements.

9.4.1 General Form for Data Transfer Statements

The general forms for data transfer statements are (R910, R911, R912):

```
READ ( io-control-specifier-list ) [ input-item-list ]
READ format [ , input-item-list ]

WRITE ( io-control-specifier-list ) [ output-item-list ]

PRINT format [ , output-item-list ]
```

Input and output items are described in 9.4.4. The format is described as part of the FMT specifier in 9.4.3. io-control specifiers, described below, provide detailed specifications for the data transfer.

9.4.2 The Input/Output Control Specifiers

The forms of the input/output control specifier (R913) are:

```
[ UNIT = ] io-unit
[ FMT = ] format
[ NML = ] namelist-group-name
ADVANCE = scalar-default-character-expression
ASYNCHRONOUS = scalar-character-initialization-expression
BLANK = scalar-default-character-expression
DECIMAL = scalar-default-character-expression
DELIM = scalar-default-character-expression
END = label
EOR = label
ERR = label
ID = scalar-integer-variable
IOMSG = scalar-default-character-variable
IOSTAT = scalar-integer-variable
PAD = scalar-default-character-expression
POS = scalar-integer-expression
REC = scalar-integer-expression
ROUND = scalar-default-character-expression
SIGN = scalar-default-character-expression
SIZE = scalar-integer-variable
```

The ADVANCE, ASYNCHRONOUS, BLANK, DECIMAL, DELIM, PAD, ROUND, and SIGN specifiers have a limited list of allowed character values. The allowed values are case independent and any trailing blanks are ignored.

When used in a READ or WRITE statement, the changeable connection mode specifiers (9.2.4)—BLANK, DECIMAL, DELIM, PAD, ROUND, and SIGN—may temporarily change those connection modes for that specific statement execution. The BLANK, DECIMAL, ROUND, and SIGN modes may also be temporarily changed during format processing. The connection modes usually revert to the OPEN modes when execution of the statement completes; when a defined input/output subroutine terminates, the changeable connection modes revert to the modes in effect when the subroutine began execution.

The UNIT specifier, with or without the keyword UNIT, is called a **unit specifier**; the FMT specifier, with or without the keyword FMT, is called a **format specifier**; and the NML specifier, with or without the keyword NML, is called a **namelist specifier**.

Use of many of the control specifiers is prohibited or required depending on the file properties or other specifiers used. For example, if an ID specifier appears, then an ASYNCHRONOUS specifier with the value YES must also appear and the unit must have been opened for asynchronous input/output.

9.4.3 Specifiers for Data Transfer Statements

This section describes the form and effect of the control information specifiers that are used in the data transfer statements. Table 9-3 summarizes the allowed values for some specifiers.

UNIT

Described in 9.2.2

A unit specifier is required.

FMT

Specifies the format specification

default character expression	explicit format specification (10.2)
*	list-directed formatting (10.10)
label	the statement label of a FORMAT statement containing the explicit format specification (10.1.1)

Rules:

1. The keyword FMT may be omitted if the format specifier is the second specifier in the control information list and the unit, without the UNIT keyword, is the first; otherwise, it is required.
2. If a format specifier appears, a namelist specifier must not appear.
3. A format specifier without the FMT= is used in a PRINT statement and the abbreviated form of the READ statement.

4. The default character expression must produce a valid format specification, including the enclosing parenthesis (10.2). It may be a character literal constant. If it is an array, it is treated as if all elements of the array were concatenated in array element order.

5. The label must be the label of a FORMAT statement in the same scoping unit as the data transfer statement.

6. List-directed formatting must not be used with files connected for direct access.

NML

Specifies the namelist group name

namelist group name	the name of a namelist group declared in a NAMELIST statement

Rules:

1. The namelist group name identifies the list of data objects to be transferred.

2. If a namelist specifier appears, a format specifier must not appear.

3. The keyword NML may be omitted if the namelist specifier is the second specifier in the control information list and the unit, without the UNIT keyword, is the first; otherwise, it is required.

4. There must not be an input or output list.

5. The NML specifier may appear in a sequential- or stream-, but not direct-access data transfer statement.

ADVANCE

Specifies whether advancing or nonadvancing input/output is performed

NO	nonadvancing data transfer
YES	advancing data transfer (default)

Rules:

1. If an ADVANCE specifier appears in the control information list, the data transfer must be a formatted sequential or formatted stream data transfer statement.

2. List-directed or namelist input/output is not allowed and neither is data transfer to or from an internal file.

3. If an EOR or SIZE specifier appears in the control information list, an ADVANCE specifier must also appear with the value NO.

ASYNCHRONOUS

Specifies whether asynchronous or synchronous input/output is performed

NO	synchronous input/output data transfer (default)
YES	asynchronous input/output data transfer

Rules:

1. Both synchronous and asynchronous input/output are permitted on files opened for asynchronous input/output.
2. Asynchronous input/output can be specified only for external files opened for asynchronous input/output; that is, asynchronous input/output is not permitted for internal files nor for preconnected files accessed without an OPEN statement.
3. If an ID specifier appears, there must be an ASYNCHRONOUS specifier with the value YES.
4. Unlike most specifiers, the ASYNCHRONOUS value must be an initialization expression. This allows the processor to make compile time decisions about input/output transfers and determine if there is a set of variables involved in asynchronous input/output.
5. A variable is said to be **involved in asynchronous input/output** if it is an item in the input/output list or namelist group or is the SIZE variable. This set of variables is called the **pending input/output storage sequence** by the standard. They, and their base objects, are implicitly given the asynchronous attribute (5.7.5); this applies to the scoping unit containing the input/output statement and may be confirmed by an explicit asynchronous declaration. Any variable that is associated, totally or partially, with a pending input/output storage sequence is also involved. The standard refers to these variables as a **pending input/output storage sequence affector**.

BLANK

Described in 9.2.4

DECIMAL

Described in 9.2.4

DELIM

Described in 9.2.4

END

Described in 9.2.3

Rules:

1. The END specifier may appear in a sequential- or stream-, but not direct-access READ statement.

EOR

Described in 9.2.3

Rules:

1. The EOR specifier may appear only in a nonadvancing READ statement.

ERR

Described in 9.2.3

ID

Returns a value which identifies this asynchronous input/output operation

integer variable	identifier of the asynchronous data transfer

Rules:

1. An ID specifier may appear only in an asynchronous input/output statement.
2. The variable becomes defined with a processor-dependent value, called the identifier of the asynchronous data transfer. This value may be used in subsequent WAIT or INQUIRE statements.
3. If an error occurs during execution of the data transfer statement, the variable becomes undefined.
4. A child data transfer statement must not have an ID specifier.

IOMSG

Described in 9.2.3

IOSTAT

Described in 9.2.3

PAD

Described in 9.2.4

POS

Specifies the file position for an external file connected for stream input/output

integer	file position

Rules:

1. The UNIT specifier must identify an external unit connected for stream input/output.
2. The value is specified in file storage units (9.1.5.3).
3. For a formatted file, the value must either be 1 or be a value returned from a POS specifier in an INQUIRE statement.
4. A POS specifier must not appear in a child data transfer statement.
5. It is processor dependent whether or not a particular stream file supports positioning or positioning in a particular direction; for example, devices such as Geiger counters usually are not positionable.

REC

Specifies the record number for a direct-access data transfer

integer expression	the record number to be read or written

Rules:

1. The value of the scalar integer expression must be positive.
2. The REC specifier must appear in a nonchild data transfer statement with a unit that is connected for direct access; it must not appear in other data transfer statements.
3. The REC specifier must not appear in a child data transfer statement, even if the parent is a direct-access transfer statement. The child data transfer statement is, however, a direct-access data transfer statement.

ROUND

Described in 9.2.4

SIGN

Described in 9.2.4

SIZE

Returns the number of characters read during nonadvancing input

nonnegative integer	returns the number of characters read

Rules:

1. The SIZE specifier may appear only in a nonadvancing READ statement.

2. Blanks inserted as padding characters when the PAD mode is YES for the connection (9.2.4) are not counted.
3. The variable specified in the SIZE specifier must not be the same as or associated with any entity in the input/output item list or in the namelist group or with the variable specified in the IOSTAT specifier, if one appears.
4. For a synchronous nonadvancing input statement, the SIZE variable becomes defined when execution of that statement completes.
5. For asynchronous nonadvancing input, the SIZE variable becomes defined when the corresponding wait operation completes. The variable is in the set of variables involved in asynchronous input/output and must not be referenced or defined until the transfer has been completed.

Table 9-3 Allowed values for some keyword specifiers in a data transfer statement

Specifier keyword	Possible values	Default value when not specified in OPEN
ADVANCE	YES, NO	YES
ASYNCHRONOUS	YES, NO	NO
BLANK	NULL, ZERO	NULL
DECIMAL	COMMA, POINT	POINT
DELIM	APOSTROPHE, QUOTE, NONE	NONE
PAD	YES, NO	YES
ROUND	UP, DOWN, ZERO, NEAREST, COMPATIBLE, PROCESSOR_DEFINED	Processor dependent
SIGN	PLUS, SUPPRESS, PROCESSOR_DEFINED	PROCESSOR_DEFINED

Note. If a changeable edit mode specifier does not appear, the mode determined when the file was opened is used.

9.4.4 The Input/Output Item List

The input/output item list consists effectively of a list of variables in a READ statement or a list of expressions in a WRITE or PRINT statement. In addition, in any of these statements, the input/output item list may contain an input/output implied-do list. Note that in an output list an expression can be a variable or a constant, as well as a more general expression.

The forms of an input item (R915) are:

```
variable
io-implied-do
```

and the forms of an output item (R916) are:

```
expression
io-implied-do
```

where the form of an input/output implied-do (R917) is:

```
( io-implied-do-object-list , io-implied-do-control )
```

and the forms of an input/output implied-do object (R918) are:

```
input-item
output-item
```

and the form of an input/output DO control (R919) is:

```
do-variable = scalar-integer-expression , &
   scalar-integer-expression [ , scalar-integer-expression ]
```

Rules and restrictions:

1. The DO variable must be a named scalar integer variable.
2. The DO variable must not be one of the input items in the implied-do nor be associated with any of them. However, the DO variable may appear in an output list.
3. Two nested implied-dos must not have the same (or associated) DO variables.
4. An input item must be a variable or an implied-do whose objects are ultimately variables. Similarly, an output item must be an expression or an implied-do whose objects are ultimately expressions.
5. For an input/output implied-do, the loop is initialized, executed, and terminated in the same manner as for the DO construct (8.7.2.1). Its iteration count is established at the beginning of processing of the items that constitute the input/output implied-do.
6. An array appearing without subscripts in an input/output list is treated as if all elements of the array were listed in array-element order. For example, if UP is an array of shape [2 3],

   ```
   READ *, UP
   ```

   is the same as

   ```
   READ *, UP(1,1), UP(2,1), UP(1,2), UP(2,2), UP(1,3), UP(2,3)
   ```

   For an array section or array expression appearing in the list, the order is that of the array object that the section or expression represents (6.6.6).
7. When a subscripted array is an input item, it is not permitted to have the transfer of one value to the array affect another part of the input item. This normally applies to vector valued subscripts and is similar to the rule that a vector-valued assignment

statement cannot make multiple assignments to the same element. Consider the following READ statements:

```
INTEGER  A(100), V(10)

! Suppose V's elements are defined with values
!   in the range 1 to 100.

READ *, A(A)              ! always invalid
READ *, A(A(1):A(9))      ! valid if the range A(1) to A(9)
                          ! does not include either 1 or 9
READ *, A(V)              ! valid if V has no repeated elements
```

8. An assumed-size array must not appear in an input/output list, unless a subscript, a subscript triplet specifying an upper bound, or a vector subscript appears in the last dimension.

9. A derived-type list item will be processed by defined input/output (9.5.1.4) if an appropriate subroutine is accessible and, for explicitly formatted input/output, if a DT edit descriptor is encountered in the format.

10. In formatted input/output, if a derived-type list item is not processed by defined input/output, it is treated as if, in place of the structure, all components were listed in the order of the components in the derived-type definition. For example, if FIRECHIEF is a structure of type PERSON defined in 4.5,

    ```
    READ *, FIRECHIEF
    ```

    is the same as

    ```
    READ *, FIRECHIEF % AGE, FIRECHIEF % NAME
    ```

    Note that if a component is itself a derived-type list item, it will either be processed by defined input/output or recursively expanded.

11. In unformatted input/output, if a derived-type list item is not processed by defined input/output, and one or more of its components would be processed by defined input/output, it is treated as if its components were listed in declaration order. Otherwise, it is treated as a single object. It is not necessarily processed as if all components appeared in the order given in the derived-type definition, even if it is of a sequence type. Padding might be added or components might be word aligned. In the case where such a list item is treated as a single object, it must not have any allocatable or pointer components.

12. All components of a derived-type list item which is expanded must be accessible in that scoping unit and they must not have either the pointer or allocatable attribute.

13. An allocatable variable may be an input or output list item, but it must be allocated at the time the data transfer statement is executed.

14. A pointer may be an input or output list item, but it must be associated with a target at the time the data transfer statement is executed.

15. A variable in an input list must not be a procedure pointer and the value of an expression in an output list must not be a procedure pointer.

16. If the file is an internal file of kind:

default character	all input/output character list items must be of default character kind
ISO 10646	all input/output character list items must be of default, ISO 10646, or ASCII kind
ASCII	all input/output character list items must be of ASCII kind

17. For input from a Unicode file, characters read into a character variable of the ASCII kind must have a position in the ISO 10646 table of 127 or less and characters read into a default character variable must be representable in the default character set.

18. During input from a non-Unicode file, characters read into a character variable must have the same kind parameter as the character variable and characters that correspond to a logical or numeric variable must be of default character kind.

19. During output to a Unicode file, all characters transmitted from a character variable or character string edit descriptor must be of ISO_10646 kind or the result is processor dependent.

20. During output to a non-Unicode file, characters from a character string edit descriptor, a default character kind variable, or a logical or numeric variable will be of default type.

21. A constant or an expression other than a variable must not appear as an input list item, but may appear as an output list item.

22. An input list item, or an entity associated with it, must not contain any portion of an established format specification.

23. On output, every entity whose value is to be written must be defined. However, entities processed by a defined output subroutine need not be defined if the routine does not access their values.

24. If a list item is polymorphic, it must be processed by defined input or output.

25. Function references in an input or output list and defined input/output subroutines are subject to the rules for recursive input/output statements (9.2.1.2).

9.4.5 General Data Transfer Restrictions

Fortran supports a wide variety of file types, access methods, data formatting, and transfer properties. Not all of these features can interact with each other. The major restrictions and limitations are:

1. An internal file cannot be used for direct access or stream access, it cannot be used with an ADVANCE specifier (even if the specifier is YES), and it cannot be used with asynchronous input/output.

2. A file opened for direct, sequential, or stream access can only be read or written with direct, sequential or stream data transfer statements, respectively. Note that the forms of a sequential or stream data transfer statement can be identical. There is no explicit STREAM or SEQUENTIAL specifier.

3. A direct-access file cannot be used with list-directed or namelist data transfer, nor with an ADVANCE specifier (even if it specifies YES).

4. List-directed and namelist formatting cannot be specified if an ADVANCE specifier appears.

5. Synchronous data transfer can be performed on a file opened for asynchronous input/out, but the converse is not true.

9.4.6 Printing of Formatted Records

Previous standards included a concept of "carriage control", sometimes called "printing". If a file was "printed", the characters in column one were not actually printed; rather, they provided carriage control (double space, skip to top of the next page, etc.) as each line was printed. Because these actions were highly processor dependent and were primarily dependent on hardware characteristics of mechanical printers, the concept has been deleted from Fortran 2003. Removing support from the standard had no practical effect on compilers, libraries, or standard conforming programs. However, for backward compatibility, namelist and list-directed output generally produce output with a blank in the first character position. Many existing programs also put a blank in column one as a safety feature.

As a carry over from printing, many texts and programmers refer to a "line" when discussing output, or even input. A better term is "record" since it is independent of any physical medium.

9.5 Execution Model for Data Transfer Statements

When a data transfer statement is executed, it is as if these steps are followed in the order given:

1. Determine the direction of data transfer. A READ statement indicates that data is to be transferred from a file. A WRITE or PRINT statement indicates that data is to be transferred to a file.

2. Identify the unit. The unit identified by a data transfer input/output statement must be connected to a file when execution of the statement begins. Note that the unit may be preconnected and that internal files are always connected.

3. If the transfer is a WRITE or PRINT statement to a file that is preconnected and does not exist, the file is created unless an error occurs. It is a common extension to create a file even though it is not preconnected. Processors usually select a reasonable set of attributes that allow the transfer statement to execute. Some processors will also create a file used for input and supply an immediate end-of-file indication.

4. If the data transfer is synchronous, perform a wait operation for all pending asynchronous input/output operations for the unit. If an error, end-of-record, or end-of-file occurs, skip to step 10 below.

5. Establish the format, if one is specified. It specifies list-directed, namelist, or explicitly formatted data transfer.

6. If the statement is not a child data transfer statement, position the file prior to transferring the data and, if the transfer is also formatted, set the left tab limit. The position depends on the method of access (sequential, direct, or stream) and is described in 9.5.2.

7. If the transfer is synchronous, transfer data between the file and the entities specified by the input/output item list (if any).

   If the transfer is asynchronous, establish the set of storage units involved in the input/output list and initiate the transfer. See 9.4.4 for special cases of some READ statements. The transfer might be completed before execution of the statement is completed. If so, error, end-of-record, or end-of-file conditions might be detected by this statement; otherwise, they will be detected when a wait operation is performed. They will be detected only at one of these occurrences. Note that child data transfer statements cannot be asynchronous.

8. Determine if an error, end-of-record, or end-of-file condition exists. If one of these conditions occurs, the status of the file and the input/output items is specified in 9.6. Also, if one of these conditions would occur during a wait operation for a pending asynchronous transfer that did not have an ID specifier, it might occur here. In effect, the processor is allowed to maintain a queue of pending asynchronous data transfers for each unit and, at any subsequent asynchronous data transfer for that unit, examine some or all of the pending transfers that did not have an ID specifier. The pending transfers with an ID specifier are not examined until there is an synchronous transfer statement for the unit or there is a WAIT for that ID specifier.

9. Position the file after transferring the data (9.5.3) unless it is a child data transfer. The file position depends on whether one of the situations in step 4 above occurred

or if the data transfer was advancing or nonadvancing. Asynchronous files are positioned as if the transfer had finished.

10. Define the variables specified in the IOSTAT and SIZE specifiers, if they appear. See the description of these specifiers in the READ and WRITE data transfer statements in 9.4.3.

11. If ERR, END, or EOR specifiers appear in the statement, transfer to the branch target corresponding to the condition that occurs. If an IOSTAT specifier appears in the statement and the label specifier corresponding to the condition that occurs does not appear in the statement, the next statement in the execution sequence is executed. Otherwise, the execution of the program terminates. See the descriptions of these label specifiers in 9.2.3. Details of error and exception processing are given in 9.6.

9.5.1 Data Transfer

Data are transferred between records in the file and entities in the input/output list (9.4.4) or namelist. The list items are processed in the order of the input/output list for all data transfer input/output statements except namelist data transfer statements. The list items for a namelist formatted data transfer input statement are processed in the order of the entities specified within the input records. The list items for a namelist data transfer output statement are processed in the order in which the data objects are specified in the namelist group object list.

The next item to be processed in the input or output item list is called the **next effective item**, which is used to determine the interaction between the input/output item list and the format specification, if any (10.2).

Zero-sized arrays and implied-do lists with zero iteration counts are ignored in determining the next effective item, but zero-length character entities are effective list items.

The value of an item that appears early in an input/output list may affect the processing of an item that appears later in the list. In the example,

```
READ (10) N, X(N), (A(I), I = 1, N)
```

the new value of N is used for the subscript of X and the.implied-do bound.

Additional rules governing interaction of variables in data transfer specifier lists and the actual input/output list are given in 9.12. In general, a variable appearing as a specifier should not also appear in the input or output list.

Rules and restrictions:

1. Values for a list item are completely transferred before processing of a subsequent list item begins.

2. Before beginning the input/output processing of a particular list item, all values needed to determine which entities are specified by the list item are evaluated first. For example, the subscripts of a variable in an input/output list are evaluated before any data is transferred.

3. If a namelist entity is specified more than once in the input, the last value encountered is the value used.

4. If a list item is a pointer, data is transferred between the file and the target. For an input item, the data is transferred to the associated target. For an output item, the target associated with the pointer must be defined, and the value of the target is transferred.

5. On output, every entity whose value is to be written must be defined.

9.5.1.1 Unformatted Data Transfer

Unformatted data transfer moves data to or from an external file directly, without editing or formatting. Unformatted transfers are generally used for fast input/output, especially with large arrays, and to preserve exact values. The output is usually not in a human readable form and often cannot be interchanged between different computer systems.

An unformatted data transfer statement does not have a FMT or NML specifier.

The file may be connected for direct, sequential, or stream access, but must not be an internal file. If the file is connected for either sequential or direct access, exactly one record is written. Unformatted stream-access files do not have records; they are a sequence of file storage units.

Values are read or written as a contiguous series of file storage units, beginning immediately after the current file position. As each value is transmitted, the file position is moved to a point immediately after the last transmitted value.

Both intrinsic and derived-type entities may be transmitted and, if an appropriate defined input/output subroutine is available, derived-type entities will be processed by the defined input/output subroutine. If a derived-type entity is not processed by a defined input/output subroutine, the order of the components is processor dependent (9.4.4(11)).

For unformatted sequential output, a record will be written with a length which sufficient to hold the values. The processor is allowed to pad the record length to a convenient size. The necessary record length must not exceed the length specified by either the processor-dependent default length, or the value from the RECL specifier in the OPEN statement.

If the file is connected for unformatted direct access, the actual record length must not exceed the value specified by the RECL specifier in the OPEN statement. If the record would be shorter than this length, it will be padded with undefined values.

For unformatted sequential input, the input list must not specify more values than the record contains. If the file storage units do not contain values with the same type and kind as the input list item, then, except for complex and character data, the result is processor-dependent. Two real values, of the appropriate kind, may be used as input for a complex entity. A character list item of default kind and length N will accept N default characters from the input, even if they were not written as a single string of length N.

9.5.1.2 Formatted Data Transfer

Formatted data transfer moves data to or from an external or internal file with editing. The editing may be done with explicit format control (10.2) or implicitly with list-directed (10.10) or namelist formatting (10.11). Formatted data transfer reads or writes to the current record, and possibly to additional records. The output is generally human readable and portable among different kinds of computer systems. Accuracy is often lost when real or complex values are edited.

A formatted data transfer statement has either a FMT or NML specifier. The FMT= or NML= characters may be omitted if the corresponding specifier is the second item in the control list. If the format specifier is a FORMAT statement label or a character expression, the statement is an explicitly formatted data transfer statement. If the format specifier is an *, the statement is a list-directed formatted data transfer statement. If there is a NML specifier, the statement is a namelist formatted data transfer statement.

The file may be connected for direct, sequential, or stream access or may be an internal file. For direct-access files, the record number is increased by one as each record is processed. Both intrinsic and derived-type entities may be transmitted. For explicitly formatted transfers, if a DT edit descriptor is encountered and an appropriate defined input/output subroutine is available, derived-type entities will be processed by the defined input/output subroutine. If a derived-type entity is not processed by a defined input/output subroutine, the elements are processed in the order they were declared in the type definition (9.4.4).

During output, the record length must not be larger than the RECL specifier value from the OPEN statement or the record length of an internal file. If the file is a direct-access file or an internal file, blank characters will be added as necessary to fill the record.

During advancing input, if the pad mode is NO, the format and input list must not require more characters than are in the record. If the pad mode is YES, blank characters will be supplied by the processor until the input list is completed.

During nonadvancing input, if the pad mode is NO and the list and format require more characters than the record contains, an end-of-record condition will be signaled, unless the record is an incomplete stream record, in which case an end-of-file condition will be signaled. If the pad mode is YES, blank characters will be supplied by the processor until the input list is completed and then end-of-record will be signaled, unless the record is an incomplete stream record, in which case an end-of-file condition will be signaled. An incomplete record can only occur as the last record in a stream file.

List-directed and namelist data transfers are processed with an implicit format. The details are in 10.10 and 10.11.

9.5.1.3 Asynchronous Data Transfer

If an external file is opened for asynchronous data transfer and if a data transfer statement has an ASYNCHRONOUS specifier with the value YES, then the transfer statement is an **asynchronous transfer statement**. Synchronous data transfer statements perform all of the input or output operations during execution of the statement. An asynchronous data transfer statement is designed to allow the transfer to be initiated by the statement and to proceed asynchronously while other computations take place.

It is the programmer's responsibility to ensure that variables referenced in the transfer statement are not referenced or defined during the transfer. A wait operation must be performed before any of the variables can be used. Details of the usage rules are given in 9.4 under the ASYNCHRONOUS specifier.

Processors are not required to perform the operation asynchronously; they are allowed to effectively ignore the ASYNCHRONOUS attribute during the data transfer. This gives the processors the freedom to optimize some transfers while not imposing a burden on transfers that will not benefit from asynchronous transfers. As a general rule, the only transfers that are likely to benefit are unformatted transfers of large arrays with either no explicit subscript list or with very simple subscripts. Because of the high overhead of doing editing, formatted transfers are unlikely to benefit from asynchronous transfer. Also, because of the potential overhead and different implementation schemes, the ASYNCHRONOUS specifier must be an initialization expression, known at compile time, with an value of YES or NO.

Records and file storage units are processed by asynchronous input or output statements in the same order that they would be processed by synchronous transfer statements.

An ID specifier may also be given in the data transfer statement. If one appears, it is given a processor-dependent value upon completion of the transfer statement. This value can be used in subsequent INQUIRE or WAIT statements to determine the status of the transfer. If there is no ID specifier, there is no way to inquire about or wait for a this specific transfer. However, a wait or an inquire can be done on a file basis. If there is an error during the data transfer statement, the ID specifier becomes undefined.

9.5.1.4 Defined Input/Output

Defined input/output allows a programmer to override the default processing of derived-type objects. Like defined assignment and defined operations, defined input/output may be automatically invoked. This happens if a data transfer is executed for a derived-type object, an appropriate defined input/output subroutine is accessible, and, for explicit formatted transfers, if a DT edit descriptor is encountered in the format. Defined input/output is called "user-defined derived-type input/output" in the standard. If an appropriate defined input/output subroutine is not available, then the data object is expanded into its components, as described in 9.4.4. If a component is itself a structure, it will be processed by defined input/output or expanded into components. The expansion continues recursively until the ultimate components are reached. If an ultimate component has either the ALLOCATABLE or POINTER attribute, it must be processed by defined input/output. See 9.4.4 for more details about when structures are expanded. Defined input/output subroutines must match a dtio-generic-spec (R1208, 12.5.4.4) and are selected based on the form of the input/output statement and the type of the data element.

Defined output is similar to replacing the output list items with function references and, if necessary, modifying the format.

Defined Input/Output Data Transfer

Once a defined input/output subroutine has been selected, the processor calls the subroutine with an appropriate argument list. The defined input/output subroutine manages all of the data transfer for the data object. Normally the defined input/output subroutine will itself contain data transfer statements that read or write to the current record or file position. However, they need not do any actual data transfers. A defined output subroutine for the employee%department field might not output anything if the FIRED variable is true.

A data transfer statement which causes a defined input/output subroutine to be invoked is called the **parent data transfer statement**. A data transfer statement that is executed during execution of the invoked defined input/output subroutine (including any subprograms that it invokes) and that also specifies the unit passed into the defined input/output subroutine is called a **child data transfer statement**. A child data transfer statement may also be a parent data transfer if it causes a defined input/output subroutine to be invoked. The process can be directly or indirectly recursive, if so, then the defined input/output subroutine must also be declared RECURSIVE. Any restrictions on the defined input/output subroutine also apply to any procedures invoked while it is active.

A child data transfer statement transfers data just like any other data transfer statement except that there is no file positioning before the transfer nor does an unformatted transfer cause any file positioning after the transfer. The child data transfer statement starts transferring data at the current position of the file, as determined from the last parent transfer or record positioning format descriptor. This need not be at the beginning of a record.

The restrictions on recursive input/output allow a child data transfer statement to perform data transfers on its argument file. This is an exception to the general rule that a recursive input/output statement must not reference an external unit (9.2.1.2). The defined input/output subroutine may also use data transfer statements with internal files as it processes the derived-type data, perhaps to format a date and time code stamp into human readable months and days. But, a defined input/output subroutine cannot execute any input/output statement on a different external file.

A defined input/output subroutine can be directly invoked, just like any other subroutine, in which case the data transfer statements are not child data transfer statements (although they might become parent or child data transfer statements as the subroutine executes).

A PRINT statement may be a child data transfer statement; however, since the child data transfer statement must reference the external unit passed into the defined input/output subroutine, a PRINT statement can be a child only to another PRINT statement or to a WRITE statement specifying the default output unit.

Defined Input/Output Subroutines

Defined input/output subroutines can be specified for formatted or unformatted input or output, giving four possible subroutines for any defined type. The subroutines can be specified for either an extensible CLASS or for a particular TYPE. Only the subroutines used must be provided, although it is likely that a module that provides a definition for a type would provide all four. Generic resolution (12.5.4.4, 12.8) takes place based on both the data type and the input/output statement.

Defined input/output subroutines are either specified with a generic specifier in either an interface block (12.5.4) or bound to a type with a generic binding (4.4.11.2). As with defined assignment or defined operators, the interface block would normally be in a module. The generic specification must be one of:

```
READ (FORMATTED)

READ (UNFORMATTED)

WRITE (FORMATTED)

WRITE (UNFORMATTED)
```

The subroutine must have an argument list that matches the appropriate one of the following:

```
subroutine read_formatted(dtv, unit, iotype, v_list, iostat, iomsg)

subroutine read_unformatted(dtv, unit, iostat, iomsg)

subroutine write_formatted(dtv, unit, iotype, v_list, iostat, iomsg)

subroutine write_unformatted(dtv, unit, iostat, iomsg)
```

The arguments must be declared with the following characteristics:

```
TYPE(derived-type-spec), INTENT(INOUT or IN) :: dtv
```

or

```
CLASS(derived-type-spec), INTENT(INOUT or IN) :: dtv

INTEGER, INTENT(IN)               ::  unit
CHARACTER(LEN=*), INTENT(IN)      ::  iotype
INTEGER, INTENT(IN)               ::  v_list(:)
INTEGER, INTENT(INOUT)            ::  iostat
INTEGER, INTENT(INOUT)            ::  iomsg
```

The actual subroutine names (read_unformatted, etc.) and the dummy argument names (dtv, unit, etc.) are arbitrary. The names above are suggestive of the use of the arguments and will be used in discussions, both here and in 10. The argument meanings are:

dtv — Input or output item. For a READ statement, dtv is argument associated with the input/output list item. It is INTENT (INOUT) to allow the defined input subroutine to modify only part of the argument values. For a WRITE or PRINT statement, dtv must be INTENT (OUT) and the value of the list item is passed as the argument. If the type is extensible (4.4.12), the CLASS form must be used; otherwise, the type form is used.

unit — Unit number to be used in the child data transfer statements. If the parent data transfer statement has a unit number, this will be the same value. If the parent data transfer statement is a WRITE statement with * for the unit or is a PRINT statement, the value will be the same as the named constant OUTPUT_UNIT. If the parent data transfer statement is a READ statement with an * for the unit or is an abbreviated form without a control specifier list, the value will be the same as the named constant INPUT_UNIT. Otherwise, the parent data transfer statement must access an internal file and the unit value will be a processor-dependent negative value. Because the unit value will be negative for internal files, the subroutine must not execute an INQUIRE statement until it has checked that the unit value is either not negative or one of the special values, INPUT_UNIT, OUTPUT_UNIT, or ERROR_UNIT.

iotype — Type of transfer for a formatted data transfer. The value will be one of LISTDIRECTED, NAMELIST, or DT concatenated with the character literal constant portion of the DT edit descriptor that corresponds to dtv. The character literal might have zero length (10.8).

v_list — Values from the v-list portion of the DT edit descriptor (10.8) in the order given. If the v-list portion is absent, v_list will be a zero-sized array. The subroutine may interpret elements of v_list as field widths, but it is not required to do so. It can attach any arbitrary interpretation (or even no interpretation) to both v_list and iotype.

iostat — Error reporting variable. If an error condition occurs, the subroutine must return a positive value for this argument. Otherwise, if an end-of-file condition occurs, the subroutine will return IOSTAT_END. Otherwise, if an end-of-record condition occurs, the subroutine will return IOSTAT_EOR. Otherwise, it will return zero.

iomsg — Error description variable. If the subroutine returns a nonzero value, the subroutine must include an error message in iomsg. Otherwise, it must leave it unchanged.

The defined input/output subroutine is given great freedom in what it classifies as an error. It is not limited to errors that occur during a child data input/output transfer statement. Data values that are improper, for example an author's royalty greater than $1,000,000, can cause iostat to be set to an error value.

If a defined input/output subroutine executes an input/output statement and the statement returns an error indication, that error need not be passed up to the parent data transfer statement. It is possible for the defined input/output subroutine to intercept the error, do the right thing, and then set iostat to zero. The subroutine must preserve the initial value of iomsg in this case. Similarly, the subroutine may return an error, end-of-file, or end-of-record status in iostat even though no exception was signaled by any input/output statement.

If the subroutine returns a nonzero value for iostat and the parent data transfer statement does not have an appropriate IOSTAT, EOR, END, or EOR specifier, the processor will terminate execution of the program as described in 9.6.

A formatted child data transfer statement may use explicitly formatted, namelist, or list-directed input/output regardless of the formatting in the parent data transfer statement.

The subroutine cannot reopen the unit to change any of the changeable edit mode specifiers (9.2.4); however, it can override them on a statement-by-statement basis either in the control list or with an explicit format specifier.

Rules and restrictions:

1. When a parent input/output statement is active:
   a. Other input/output statements must either refer to the unit specified by the unit argument or to an internal file.
   b. If the parent is a READ statement, only READ or INQUIRE statements may refer to the unit.
   c. If the parent is a WRITE or PRINT statement, only WRITE, PRINT, or INQUIRE statements may refer to the unit.
   d. An OPEN, CLOSE, BACKSPACE, ENDFILE, REWIND, FLUSH, or WAIT statement must not be executed (but an INQUIRE is allowed if the unit is an external unit).
2. Neither the parent nor child data transfer statement can be recursive.
3. The defined input/output subroutine, and any procedures it invokes, must not define, nor cause to become undefined, any storage location referenced by any item in the input/output list, the format (if there is one), nor any specifier in an active parent data transfer statement, except through the dtv argument.
4. A child data transfer statement must not specify an ID, POS, or REC specifier. The parent data transfer statement may have a POS or REC specifier, but not an ID specifier.
5. A child data transfer statement must not use positioning edit descriptors to position the record to the left of the position when the defined input/output subroutine was invoked.
6. If the parent data statement is formatted or unformatted, the child data transfer statements also must be formatted or unformatted, respectively.

Defined Input/Output Subroutine Resolution

A subroutine is suitable as a defined input/output subroutine if it has a generic specification that matches the direction of the data transfer—read or write, matches the form of the data transfer—formatted or unformatted, and has a dtv argument that is compatible with the effective list item type, according to the rules for argument association (12.6).

Defined input/output occurs if a structure item is in the input/output list and both of the following conditions are true:

1. The data transfer statement is a list-directed, namelist, unformatted data transfer statement, or it is an explicitly formatted input/output statement and the corresponding edit descriptor is DT.
2. A suitable defined input/output subroutine is accessible and the subroutine is either:
   a. a generic type-bound procedure of the declared type of the effective list item, or
   b. an accessible generic interface.

If 2a is true, the normal rules for selecting type-bound procedures based on the dynamic type of the list item apply (12.8)

If 2b is true (and 2a is false), the reference is to the procedure with the appropriate specific interface.

Defined Input/Output Example

```
MODULE  rational_stuff

   TYPE rational
      INTEGER  n, d
   END TYPE
   INTERFACE WRITE( FORMATTED )
      MODULE PROCEDURE write_rational_value
   END INTERFACE WRITE ( FORMATTED )

CONTAINS

   SUBROUTINE write_rational_value &
         (dtv, unit, iotype, v_list, iostat, iomsg)
      TYPE(rational), INTENT(IN)     ::  dtv
      INTEGER, INTENT(IN)            ::  unit
      CHARACTER(LEN=*), INTENT(IN)   ::  iotype
      INTEGER, INTENT(IN)            ::  v_list(:)
      INTEGER, INTENT(INOUT)         ::  iostat
      INTEGER, INTENT(INOUT)         ::  iomsg

      IF (SIZE(v_list) == 0) then
         !  use a default format
         IF ( dtv%d == 0 ) then
            WRITE ( unit, '( I0, "/", I0, " = " Infinity)', &
                  IOSTAT = iostat, IOMSG = iomsg), dtv%n, dtv%d
```

```
            WRITE ( unit, '(I0, "/", I0, " = " G20.10)', &
                  IOSTAT = iostat, IOMSG = iomsg)  &
                  dtv%n, dtv%d, dtv%n/dtv%d
         ENDIF
      ELSE
         ! use v_list to format the results
            . . .
      ENDIF
   END SUBROUTINE write_rational_value

END MODULE rational_stuff

PROGRAM defined_example

   USE  rational_stuff
   TYPE ( rational ) x
      . . .
   x = rational ( 2 , 3 )

   WRITE (25, '(A, DT)')  " x is ", x

END PROGRAM defined_example
```

The previous program will produce output similar to

```
x is 2/3 = bbbbb.6666666667bbbb
```

9.5.1.5 Transfer on Internal Files

An internal file is a character variable. Transferring data between machine representation and this character variable is called internal input/output. A formatted sequential-access input or output statement is used. Explicit, namelist, or list-directed formatting may be used. As a practical matter, namelist and list-directed output are problematic with internal files because the processor has so much latitude in formatting and record structure. The format is used to interpret the characters. The internal file and the internal unit are the same character variable.

With this feature, it is possible to read in a string of characters without knowing its exact format, examine the string, and then interpret it according to its contents.

In addition to the following, the data transfer follows the general rules for formatted data transfer (9.5.1.2).

Rules and restrictions:

1. The unit must be a character variable of either default, ASCII, or ISO 10646 kind that is not an array section with a vector subscript.

2. If the character variable is a scalar, the file consists of one record. If the character variable is an array or an array section, each element of the array or section is a record. The order of the records is array element order. The length of the record is the length of the scalar or of one array element.

3. If the character variable is allocatable, it must be allocated. If the character variable is a pointer, it must be associated with a target. It must be defined if it is used as an internal file in a READ statement.

4. For output data transfer, the format specification must not be part of the internal file or associated with the internal file or part of it.

5. If the number of characters written is less than the length of the record, the remaining characters are set to blank.

6. A record in an internal file is defined when the record is written. An input/output list item must not be in the internal file or associated with the internal file.

7. An input/output list item must not be in the internal file or associated with the internal file.

8. Before a data transfer occurs, an internal file is positioned at the beginning of the first record (that is, before the first character, if a scalar, and before the first character of the first element, if an array) unless it is also a child data transfer statement. This record becomes the current record.

9. Only formatted sequential access is permitted on internal files.

10. On input, an end-of-file condition occurs when there is an attempt to read beyond the last record of the internal file.

11. Initially, all connection mode specifiers have the value they would have if an OPEN statement using only default values were executed. Changeable mode specifiers may be changed with data transfer specifiers or with format control edit descriptors.

12. File connection, positioning, and inquiry must not be used with internal files.

13. Execution of a data transfer statement on an internal file terminates when:

    a. format processing encounters a data or colon edit descriptor, and there are no remaining elements in the input item list or output item list

    b. if list-directed processing is specified, the input item list or the output item list is exhausted; or on input, a slash (/) is encountered as a value separator

    c. on input, an end-of-file condition is encountered

    d. an error condition is encountered.

9.5.2 File Position Prior to Data Transfer

The file position prior to data transfer depends on the method of access: sequential, direct, or stream.

For sequential access on input, if there is a current record, the file position is not changed; this will be the case if the previous data transfer was nonadvancing. Otherwise, the file is positioned at the beginning of the next record and this record becomes the current record. Input must not occur if there is no next record (there must be an end-of-file record at least) or if there is a current record and the last data transfer statement accessing the file performed output.

If the file contains an end-of-file record, the file must not be positioned after the end-of-file record prior to data transfer. However, a REWIND or BACKSPACE statement may be used to reposition the file.

For sequential access on output, if there is a current record, the file position is not changed; this will be the case if the previous data transfer was nonadvancing. Otherwise, a new record is created as the next record of the file; this new record becomes the last and current record of the file and the file is positioned at the beginning of this record.

For direct access, the file is positioned at the beginning of the record specified. This record becomes the current record.

For stream access, if there is no POS specifier, the file position is not changed. If there is, the file is positioned immediately before the specified file storage unit.

Child data transfer statements do not generally position the file prior to transfer. They are discussed in detail in 9.5.1.4.

9.5.3 File Position After Data Transfer

After a data transfer operation completes, the file will be positioned as follows.

If an error condition exists, the file position is indeterminate. If no error condition exists, but an end-of-file condition exists as a result of reading an end-of-file record, the file is positioned after the end-of-file record.

If an end-of-file occurs during formatted stream input, the file position is not changed. If no error occurs during formatted stream output, the terminal point of the file is set to the highest-numbered position to which data was transferred. This might not be the current position if T or TL edit descriptors were used during the editing.

If no error occurs during unformatted stream output, the file position is not changed. If the current position exceeds the previous terminal point, the terminal point is set to the current value.

For nonadvancing input, if no error condition or end-of-file condition exists, but an end-of-record condition exists, the file is positioned after the record just read. If no error condition, end-of-file condition, or end-of-record condition exists, the file position is not changed. For nonadvancing output, if no error occurs, the file position is not changed.

In all other cases, the file is positioned after the record just read or written, and that record becomes the preceding record.

9.5.4 Termination of Data Transfer

Data transfer statements normally terminate execution when all of the items in the input/output list have been processed. The specific conditions are:

1. Unformatted or list-directed input/output has processed every list item

2. Formatted input/output has processed every list item and encounters a data edit descriptor or a colon edit descriptor in an explicit format

3. Namelist output exhausts the output item list
4. A slash value separator is encountered during namelist or list-directed input
5. An error or end-of-file condition occurs
6. An end-of-record occurs during nonadvancing input

9.6 Error and Other Conditions in Input/Output Statements

All of the input/output statements have optional specifiers which allow the programmer to detect, and possibly recover from, error conditions encountered during execution. Input data transfer statements can also detect end-of-file and end-of-record conditions. If an error condition and an end-of-file or end-of-record condition both occur, the error condition takes precedence and will be signaled; information on the other conditions will be lost.

9.6.1 Error Conditions

The set of error conditions which are detected is processor dependent. The standard does not specify any input/output errors. A processor is not required to diagnose any errors, and is free to treat errors as if they were features. Most compilers have a large list of errors they will detect. But the lists are not consistent from compiler to compiler. For example, some compilers will accept zero or one as logical input values, others will not.

If an error condition occurs, the following steps occur:

1. The file position becomes indeterminate.
2. If no ERR or IOSTAT specifier appears, program execution is terminated.
3. Processing of the input/output list, if there is one, terminates.
4. If the statement is a READ statement, or the error occurs during a wait operation for an asynchronous read operation, all of the items in the input list (or namelist group) become undefined.
5. If the statement is a data transfer statement, or the error occurs during a wait operation, all implied do variables in the data transfer statement become undefined.
6. If an IOSTAT specifier appears, the variable is assigned a processor-dependent positive integer value.
7. If an IOMSG specifier appears, the variable is assigned a processor-dependent explanatory message.
8. If the statement is a READ statement with a SIZE specifier, the variable is assigned a count of the number of characters transferred, as described in 9.4.3.
9. If an ERR specifier appears, control is transferred to the label specified.

Note that rules 4 and 5 are very severe. They are in part due to backward compatibility with early operating systems which could not provide any details if, for example, a magnetic tape broke, and also to allow modern operating systems to provide high speed input/output. A consequence of these rules is that it is impossible to read a record of unknown length and somehow determine how many values were correctly read in.

9.6.2 End-of-File Condition

An end-of-file condition occurs during an input data transfer, or during a wait operation associated with an input data transfer, when one of the following occurs:

1. The endfile record is encountered during sequential input,
2. An attempt is made to read beyond the end of an internal file, or
3. An attempt is made to read beyond the end of a stream file.

The end-of-file condition might occur when the data transfer statement begins to execute, or, for formatted input, it might occur when a new record is required. For asynchronous input, it is processor dependent whether the condition is detected during execution of the transfer statement or during the corresponding wait operation.

If an end-of-file condition occurs and no error condition occurred, the following steps occur:

1. If the file is an external record file, it is positioned after the end-of-file record.
2. If there is no END or IOSTAT specifier, program execution is terminated.
3. Processing of the input list, if there is one, terminates.
4. If the statement is a READ statement, or the end-of-file occurs during a wait operation for an asynchronous read operation, all of the items in the input list (or namelist group) become undefined.
5. If the statement is a READ statement, or the end-of-file occurs during a wait operation, all implied do variables in the READ statement become undefined.
6. If an IOSTAT specifier appears, the variable is assigned the processor-dependent negative integer value IOSTAT_END.
7. If an IOMSG specifier appears, the variable is assigned a processor-dependent explanatory message.
8. If an END specifier appears, control is transferred to the label specified.

9.6.3 End-of-Record Condition

An end-of-record condition occurs during nonadvancing input when there is an attempt to transfer from beyond the end of the current record. However, if the file is a stream file and the current record is also the last record, end-of file is signaled instead.

If an end-of-record condition occurs and no error conditioned occurred, the following steps occur:

1. If no EOR or IOSTAT specifier appears, program execution is terminated.
2. If the pad mode is YES, enough blanks are provided to satisfy the edit descriptors and the list items and the input list items become defined. Otherwise they become undefined.
3. Processing of the input/output list, if there is one, terminates.
4. The file is positioned after the current record.
5. If the statement is a READ statement, or the end-of-record occurs during a wait operation, all implied do variables in the READ statement become undefined.
6. If an IOSTAT specifier appears, the variable is assigned the processor-dependent negative integer value IOSTAT_EOR.
7. If an IOMSG specifier appears, the variable is assigned a processor-dependent explanatory message.
8. If a SIZE specifier appears, the variable is assigned a count of the number of characters transferred, as described in 9.4.3.
9. If an EOR specifier appears, control is transferred to the label specified.

The following program segment illustrates how to handle end-of-file and error conditions.

```
READ (FMT = "(E8.3)", UNIT=3, IOSTAT = IOSS) X

IF (IOSS < 0) THEN
   ! PERFORM END-OF-FILE PROCESSING ON THE
   ! FILE CONNECTED TO UNIT 3.
   CALL END_PROCESSING

ELSE IF (IOSS > 0) THEN

   ! PERFORM ERROR PROCESSING
   CALL ERROR_PROCESSING

END IF
```

The procedure END_PROCESSING is used to handle the case where an end-of-file condition occurs and the procedure ERROR_PROCESSING is used to handle all other error conditions, because an end-of-record condition cannot occur.

9.7 The WAIT Statement

The form of the WAIT statement (R921) is:

```
WAIT ( wait-specifier-list )
```

where the forms of a wait specifier (R922) are:

```
[ UNIT = ] scalar-integer-expression
END = label
EOR = label
ERR = label
ID = scalar-integer-expression
IOMSG = scalar-default-character-variable
IOSTAT = scalar-integer-variable
```

UNIT

Described in 9.2.2

END

Described in 9.2.3

EOR

Described in 9.2.3

ERR

Described in 9.2.3

ID

Specifies the asynchronous transfer to wait for

IOMSG

Described in 9.2.3

IOSTAT

Described in 9.2.3

If an ID specifier appears, the wait operation is for that specific pending transfer. The value must be the identifier of a pending transfer operation. If no ID specifier appears, wait operations will be performed for all pending transfers on the unit in a processor-dependent order.

A WAIT statement may specify a unit that does not exist, has no file connected to it, or was not opened for asynchronous input/output only if there is no ID specifier. This does not cause an error condition and has no effect on the unit.

If the pending data transfer is not a read, the END specifier is ignored. If the pending transfer is a write or advancing read, the EOR specifier is ignored.

9.7.1 The WAIT Operation

As the name implies, a wait operation waits for a previously initiated asynchronous input or output transfer to complete. There may be many pending transfers on many different external units. However, a wait operation is always specific to a particular transfer on a specific external unit. The wait may be explicitly performed with a WAIT or INQUIRE statement that has an ID specifier, or implicitly performed with a synchronous data transfer, CLOSE, or file positioning statement or WAIT or INQUIRE statement without an ID specifier.

If a wait operation is performed implicitly, the initiating statement ultimately will perform wait operations for all pending asynchronous transfers for the unit, unless an error or exception occurs during a particular wait operation. The waits will be performed in a processor-dependent order.

After a wait operation for an input data transfer, the variables involved in the transfer become defined. Until the wait completes, the variables can be neither referenced nor defined. After a wait operation for an output operation, the variables are now redefinable. During the output operation, the variables may be referenced, but not redefined.

Once the data transfer is complete, the wait operation checks for an error, end-of-file, or end-of-record condition. If any are present, they are processed according to the IOSTAT, IOMSG, ERR, END, and EOR specifiers, as appropriate. It is important to note that the error handling is performed according to the specifiers in the statement that initiated the wait operation, not the ones that initiated the data transfer. Thus, a synchronous read will branch to its ERR label if a pending asynchronous operation on the same unit had an error or if there is an error in the current read operation. Frequent use of implicit waits makes it difficult to do error and exception processing.

9.8 The CLOSE Statement

Execution of a CLOSE statement terminates the connection of a file to a unit. Any connections not closed explicitly by a CLOSE statement are closed by the processor have an implicit close operation performed when the program terminates normally. Under certain circumstances, an OPEN statement (9.3) or normal program termination (2.3.1, 8.9) also perform implicit close operations. The implicit close is equivalent to a CLOSE statement with no STATUS specifier. The form of the CLOSE statement (R908) is:

```
CLOSE ( close-specifier-list )
```

where the forms of a close specifier (R909) are:

```
[ UNIT = ] scalar-integer-expression
IOMSG = scalar-default-character-variable
IOSTAT = scalar-integer-variable
```

ERR = label
STATUS = scalar-default-character-expression

Rules and restrictions:

1. A CLOSE statement may refer to a unit that is not connected or does not exist; such a CLOSE statement has no effect on any file.
2. A CLOSE statement performs a wait operation if there is a pending asynchronous data transfers for the unit.
3. If the last data transfer to a file connected for sequential access is an output data transfer statement, a CLOSE statement for a unit connected to this file writes an end-of-file record to the file.
4. After a unit has been disconnected by a CLOSE statement, it may be connected again to the same or a different file. Similarly, after a file has been disconnected by a CLOSE statement, it may also be connected to the same or a different unit, provided the file still exists.
5. See 9.2 for the general rules and restrictions for input/output statement specifier lists.

Examples are:

```
CLOSE (ERR = 99, UNIT = 9)
CLOSE (8, IOSTAT = IR, STATUS = "KEEP")
```

9.8.1 The CLOSE Specifiers

This section describes the form and effect of the specifiers that may appear in a CLOSE statement.

UNIT

Described in 9.2.2

ERR

Described in 9.2.3

IOMSG

Described in 9.2.3

IOSTAT

Described in 9.2.3

STATUS

Determines whether or not the file will exist after being closed.

KEEP	the file will exist after it is closed
DELETE	the file will not exist after it is closed

Rules:

1. The default value is DELETE, if the unit was opened with a STATUS value of SCRATCH.
2. The default value is KEEP, if the unit was opened with any other value of the STATUS specifier.
3. KEEP must not be specified if the unit was opened with a STATUS specifier of SCRATCH.
4. If KEEP is specified for a file that does not exist, the file does not exist after the CLOSE statement is executed.

9.9 The INQUIRE Statement

There are three kinds of INQUIRE statements (R929): inquiry by unit, inquiry by file, and inquiry by output list.

The form of an inquiry by unit or file is:

```
INQUIRE ( inquiry-specifier-list )
```

The form of an inquiry by an output list is:

```
INQUIRE ( IOLENGTH = scalar-integer-variable ) output-item-list
```

The inquire by unit inquires about the connection properties of the unit to its file.

The inquire by file inquires about what connection properties are allowed for the specific file.

The inquire by output list determines a value that could be used as a RECL specifier for a direct-access read or write for a specific output list

9.9.1 Inquire by File or Unit

An inquiry may be made about the existence, connection, access methods, or other properties of a file or unit. For each property inquired about, a scalar variable is supplied; that variable is given a value that answers the inquiry. The inquiry specifiers have the form of

```
keyword = variable
```

pairs in the INQUIRE statement.

An inquire by file can be made if the file is connected or unconnected. An inquire by unit can be made if the unit has a file connected to it or not. If a file is connected to a unit, inquiry by unit or file returns information about the file and the connection. However, some specifiers return information about the file only, not the specific connection. In an inquire by unit to a disconnected unit or by file to a file that is not connected, the OPENED specifier will return false and many other specifiers are not useful and return UNKNOWN or UNDEFINED. In some cases, inquiry by file for a disconnected file will give properties from the file catalog. Details are given in the specific specifier sections below.

An inquire by unit can be made to a unit that does not exist. Because the set of allowed unit numbers is processor dependent, the EXIST specifier should be used in portable code.

The forms of an inquiry specifier (R930) are:

[UNIT =] scalar-integer-expression

FILE = scalar-default-character-expression

ACCESS = scalar-default-character-variable
ACTION = scalar-default-character-variable
ASYNCHRONOUS = scalar-default-character-variable
BLANK = scalar-default-character-variable
DECIMAL = scalar-default-character-variable
DELIM = scalar-default-character-variable
DIRECT = scalar-default-character-variable
ENCODING = scalar-default-character-variable
ERR = label
EXIST = scalar-default-logical-variable
FORM = scalar-default-character-variable
FORMATTED = scalar-default-character-variable
ID = scalar-integer-expression
IOMSG = scalar-default-character-variable
IOSTAT = scalar-integer-variable
NAME = scalar-default-character-variable
NAMED = scalar-default-logical-variable
NEXTREC = scalar-integer-variable
NUMBER = scalar-integer-variable
OPENED = scalar-default-logical-variable
PAD = scalar-default-character-variable
PENDING = scalar-default-logical-variable
POS = scalar-integer-variable
POSITION = scalar-default-character-variable
READ = scalar-default-character-variable
READWRITE = scalar-default-character-variable
RECL = scalar-default-integer-variable
ROUND = scalar-default-character-variable
SEQUENTIAL = scalar-default-character-variable
SIGN = scalar-default-character-variable
SIZE = scalar-integer-variable
STREAM = scalar-default-character-variable
UNFORMATTED = scalar-default-character-variable
WRITE = scalar-default-character-variable

Rules and restrictions:

1. An INQUIRE statement with an inquiry specifier list must have a UNIT specifier or a FILE specifier, but not both.
2. In an inquire by unit, the unit must be an external file unit.
3. The value given to a variable in an inquiry specifier is the value that would be obtained if the specified value were assigned to the variable using an intrinsic assignment statement.
4. An INQUIRE statement about a file unit may be executed even if the file or unit is not connected.
5. Except for the NAME specifier, the processor will return character values in uppercase. For the NAME specifier, the allowed characters used in the value returned are processor determined.
6. If an error condition occurs during the execution of an INQUIRE statement, all the inquiry specifier variables become undefined except the IOSTAT and IOMSG variables.
7. An INQUIRE statement performs a wait operation if there is a pending asynchronous data transfer for the unit or file. If there are any errors detected in the wait operation, the CLOSE statement will reflect these errors.
8. See 9.2 for the general rules and restrictions for input/output statement specifier lists.

Examples of the INQUIRE statement are:

```
INQUIRE (9, EXIST = EX)
INQUIRE (OPENED = OP, ACCESS = AC, FILE = "T123")
INQUIRE (IOLENGTH = IOLEN)  X, Y, CAT
```

9.9.2 Specifiers for Inquiry by Unit or File Name

This section describes the form and effect of the inquiry specifiers that may appear in the inquiry by unit and file forms of the INQUIRE statement. The UNIT, FILE, ERR, and ID specifiers provide input to the INQUIRE statement, the other specifiers return values based on the connection properties. Most specifiers are paired with connection specifiers as described in 9.3.3.

UNIT

Described in 9.2.2

When this specifier appears, the inquiry is referred to as "inquire by unit". Note, however, that if a file is connected to the unit, the inquire can also inquire about file properties.

FILE

Gives the name of the file being inquired about. When this specifier appears, the inquiry is referred to as "inquire by file".

expression	the name of the file

Rules:

1. The value of the scalar default character expression must be a file name acceptable to the processor. Trailing blanks are ignored. The interpretation of case is processor dependent.

2. The file name may refer to a file that is not connected or to one that does not exist.

ACCESS

Find the access method of the connection.

SEQUENTIAL	the connection is for sequential access
DIRECT	the connection is for direct access
STREAM	the connection is for stream access
UNDEFINED	the unit or file is not connected

ACTION

Find what action is allowed for this connection.

READ	access is limited to input only
WRITE	access is limited to output only
READWRITE	both input and output are allowed
UNDEFINED	the unit or file is not connected

ASYNCHRONOUS

Determine if the unit or file is connected for asynchronous input/output.

YES	asynchronous input/output is allowed
NO	asynchronous input/output is not allowed
UNDEFINED	the unit or file is not connected

BLANK

Determine the blank interpretation mode.

NULL	null blank interpretation is in effect
ZERO	zero blank interpretation is in effect
UNDEFINED	the unit or file is not connected for formatted input/output

DECIMAL

Find which character is used for the decimal symbol.

COMMA	the decimal mode is COMMA
POINT	the decimal mode is POINT
UNDEFINED	the unit or file is not connected for formatted input/output

DELIM

Find which character is used to delimit character values in list-directed or namelist output.

APOSTROPHE	an apostrophe is used as the delimiter
QUOTE	the quotation mark is used as the delimiter
NONE	there is no delimiting character
UNDEFINED	the unit or file is not connected for formatted input/output

DIRECT

Determine if direct access is allowed.

YES	direct access is an allowed access method for the file
NO	direct access is not an allowed access method for the file
UNKNOWN	the processor is unable to determine if direct access is allowed for the file

ENCODING

Find the character encoding used for the file.

UTF-8	the file is either connected for formatted input/output with UTF-8 encoding, or not connected but the processor can determine that the encoding form is UTF-8

UNDEFINED	the file is connected for unformatted input/output
UNKNOWN	the file is not connected and the processor cannot determine the encoding form
other	if the processor supports additional encoding forms, it is allowed to return other values

ERR

Described in 9.2.3

EXIST

Find whether a file or unit exists.

true	the file or unit exists
false	the file or unit does not exist

FORM

Determine if a connection is for formatted or unformatted input/output.

FORMATTED	the connection is for formatted input/output
UNFORMATTED	the connection is for unformatted input/output
UNDEFINED	the unit or file is not connected

FORMATTED

Determine if formatted input/output is allowed.

YES	formatted input/output is allowed for the file
NO	formatted input/output is not allowed for the file
UNKNOWN	the processor is unable to determine if formatted input/output is allowed for the file

ID

expression	an input value to the inquiry. It must be the identifier of a pending data transfer. The PENDING specifier value is affected by this specifier.

IOMSG

Described in 9.2.3

IOSTAT

Described in 9.2.3

NAME

Find the name of a file connected to a unit.

file name	the name of the file, if the file has a name

Rules:

1. The processor may return a name different from the one specified in the FILE specifier or in the OPEN statement, perhaps because a user identifier or directory path was added.
2. Whatever the name returned, it will be acceptable for use as a FILE specifier in an OPEN statement.
3. The interpretation of the case of letters used and characters allowed in a file name is determined by the processor.
4. If the unit is connected and the file does not have a name, or if the unit is not connected, the character variable value becomes undefined. See the NAMED specifier.

NAMED

Find out if a file has a name.

true	the file has a name
false	the file does not have a name or there is no connection

NEXTREC

Find the next record number of a direct-access file.

last record number + 1	the next record number for a file connected for direct access. The value is one more than the record number of the record most recently read or written
1	no records have been read or written since the connection was made
undefined value	the file is not connected for direct access or the file position is indeterminate because of a previous error condition

If there are any pending asynchronous data transfers, the value is computed as if all of the transfers had completed.

NUMBER

Find which unit is connected to a file.

unit number	the number of the unit connected to the file
–1	there is no unit connected to the file

OPENED

Determine if a file or unit is open (connected).

true	the file or unit is connected (that is, opened)
false	the file or unit is not connected (that is, not opened)

An inquire by file determines if the file is connected to some unit, an inquire by unit determines if the unit is connected to some file.

PAD

Find the pad mode for the connection.

NO	the pad mode is NO
YES	the pad mode is YES
UNDEFINED	the connection is not for formatted input/output or the file or unit is not connected

PENDING

Determine if an asynchronous input/output operation is pending.

true	asynchronous operation on the unit is pending
false	asynchronous operation on the unit is complete

If an ID specifier also appears, the test is on the specific asynchronous operation. Otherwise, all operations on the unit are tested. If the return value is false (the tested asynchronous input/output operation is not pending), the inquire will also perform the wait operation (9.7.1) for either the specific transfer or all previously issued transfers.

Because asynchronous input/output is highly processor dependent, processors can choose among a variety of ways of determining the result value.

1. It can assume an asynchronous operation to be pending until a wait operation has been executed; it would always return true until a wait operation removes the operation from the pending list.

2. The inquire could always wait for all asynchronous operations to complete; it would always return false.

3. The inquire could dynamically test the transfer and return true or false depending on the physical transfer status; however, the status could change before the inquire processing completes.

POS

Find the file position of a file connected for stream access.

position value	the number of the file storage unit immediately following the current position. If the file is at its terminal point, the value is one more than that of the highest-numbered file storage unit in the file.
undefined	the file is not connected for stream access or the position is indeterminate because of a previous error condition.

POSITION

Find the file position for a connection.

REWIND	the file was connected with its position at the initial point
APPEND	the file was connected with its position just before the endfile or at the terminal point
ASIS	the file was connected without changing its position
UNDEFINED	the file is not connected or is connected for direct access
processor dependent	the file has been repositioned since it was connected

This specifier is most useful before any action has been taken on the file. If any repositioning has occurred since the file was connected, the value returned is processor dependent, but it is not equal to REWIND unless the file is positioned at the initial point, and it is not equal to APPEND unless the file is positioned just before the endfile record or at the terminal point. Processors could provide information in the return value, such as the value of the record number, although depending on this in not portable.

READ

Determine if read is an allowed action for the file.

YES	READ is one of the allowed actions for the file

NO	READ is not one of the allowed actions for the file
UNKNOWN	the processor is unable to determine if READ is one of the allowed actions for the file

READWRITE

Determine if readwrite is an allowed action for the file.

YES	READWRITE is an allowed action for the file
NO	READWRITE is not an allowed action for the file
UNKNOWN	the processor is unable to determine if READWRITE is an allowed action for the file

RECL

Find the record length of the file.

record length	the length of each record of a file connected for direct access or the maximum record length of a file connected for sequential access
undefined value	the file is not connected or the connection is for stream access

Rules:

1. For a formatted file, the length is the number of default characters in a record. If a record contains nondefault characters, the number that can be in a record is processor dependent.
2. For an unformatted file, the length is in file storage units.

ROUND

Determine the input/output rounding mode.

UP	the mode is up
DOWN	the mode is down
ZERO	the mode is zero
NEAREST	the mode is nearest
COMPATIBLE	the mode is compatible
PROCESSOR_DEFINED	the mode is processor defined
UNDEFINED	the unit is not connected for formatted input/output

The value PROCESSOR_DEFINED will be returned only if the rounding mode is not one of the other above modes.

SEQUENTIAL

Find if sequential access is allowed for a file.

YES	sequential access is an allowed access method
NO	sequential access is not an allowed access method
UNKNOWN	the processor is unable to determine if sequential access is allowed

SIGN

Find the sign mode in effect.

PLUS	the sign mode is plus
SUPPRESS	the sign mode is suppress
PROCESSOR_DEFINED	the sign mode is processor defined
UNDEFINED	the unit is not connected for formatted input/output

SIZE

Find file size.

file size	size of the file in file storage units
–1	the file size cannot be determined

Rules:

1. If STREAM is an allowed access mode, the size is the highest-numbered file storage unit in the file.
2. If SEQUENTIAL or DIRECT is an allowed access method, the file size is processor dependent and may be different from expected due to record padding, etc.

STREAM

Determine if stream is an allowed access method for a file.

YES	stream access is an allowed access method
NO	stream access is not an allowed access method
UNKNOWN	the processor is unable to determine if stream access is allowed

UNFORMATTED

Determine if unformatted input/output is allowed.

YES	unformatted input/output is allowed for the file
NO	unformatted input/output is not allowed for the file
UNKNOWN	the processor is unable to determine if unformatted input/output is an allowed form for the file

WRITE

Determine if WRITE is an allowed action on a file.

YES	WRITE is an allowed action
NO	WRITE is not an allowed action
UNKNOWN	the processor is unable to determine if WRITE is an allowed action for the file

9.9.3 Inquire by Output List

For an inquire by output list, the output item list must be a valid output item list for an unformatted output statement. The length value returned in the scalar integer variable is the minimum number of file storage units required when used as the value of the RECL specifier in an OPEN statement. This value may be used in a RECL specifier to connect a file whose unformatted records will hold the data indicated by the output list of the INQUIRE statement. The output list must not have any derived-type entities that require a defined input/output subroutine (9.5.1.4, 9.4.4).

9.9.4 Table of Values Assigned by the INQUIRE Statement

Table 9-4 summarizes the values assigned to the various variables by the execution of an INQUIRE statement.

9.10 File Positioning Statements

Execution of a data transfer statement usually changes the file position. In addition, there are three statements whose main purpose is to change the file position. Changing the position backwards by one record is called **backspacing** and is performed by the BACKSPACE statement. Changing the position to the beginning of the file is called **rewinding** and is performed by the REWIND statement. The ENDFILE statement writes an end-of-file record and positions the file after the end-of-file record.

The forms of the BACKSPACE statement (R923) are:

```
BACKSPACE scalar-integer-expression
BACKSPACE ( position-specifier-list )
```

Table 9-4 Allowed values returned for keyword specifier variables in an INQUIRE statement

	INQUIRE by file		INQUIRE by unit	
Specifier keyword	**Unconnected**	**Connected**	**Connected**	**Unconnected**
ACCESS	UNDEFINED	SEQUENTIAL or DIRECT		UNDEFINED
ACTION	UNDEFINED	READ, WRITE, or READWRITE		UNDEFINED
ASYNCHRONOUS	UNDEFINED	YES or NO		UNDEFINED
BLANK	UNDEFINED	NULL, ZERO, or UNDEFINED		UNDEFINED
DECIMAL	UNDEFINED	COMMA, POINT, or UNDEFINED		UNDEFINED
DELIM	UNDEFINED	APOSTROPHE, QUOTE, NONE, or UNDEFINED		UNDEFINED
DIRECT	YES, NO, or UNKNOWN			UNKNOWN
ENCODING	UTF-8, UNKNOWN, or other	UTF-8, UNDEFINED, or UNKNOWN		UNKNOWN
EXIST	true if file exists, false otherwise		true if unit exists, false otherwise	
FORM	UNDEFINED	FORMATTED or UNFORMATTED		UNDEFINED
FORMATTED	YES, NO, or UNKNOWN			UNKNOWN
IOLENGTH	RECL value for output item list			
IOMSG	error message or unchanged			unchanged
IOSTAT	0 for no error, a positive integer for an error			
NAME	file name (might not be the same as FILE value)		file name if named, else undefined	undefined value
NAMED	true		true if file named, false otherwise	false
NEXTREC	Undefined value	If direct access, next record number; else undefined		Undefined
NUMBER	–1	unit number		
OPENED	false	true		false
PAD	UNDEFINED	YES or NO		YES
PENDING	false	true or false		false

Table 9-4 Allowed values returned for keyword specifier variables in an INQUIRE statement

	INQUIRE by file		INQUIRE by unit	
Specifier keyword	**Unconnected**	**Connected**	**Connected**	**Unconnected**
POS	undefined value	file storage unit position number or undefined value		undefined value
POSITION	UNDEFINED	REWIND, APPEND, ASIS, or UNDEFINED		UNDEFINED
READ	YES, NO, or UNKNOWN			UNKNOWN
READWRITE	YES, NO, or UNKNOWN			UNKNOWN
RECL	undefined	if direct access, record length; else maximum record length		undefined
ROUND	UNDEFINED	UP, DOWN, ZERO, NEAREST, COMPATIBLE, PROCESSOR_DEFINED, or UNDEFINED		UNDEFINED
SEQUENTIAL	YES, NO, or UNKNOWN			UNKNOWN
SIGN	UNDEFINED	PLUS, SUPPRESS, PROCESSOR_DEFINED, or UNDEFINED		UNDEFINED
SIZE	the size or –1			–1
STREAM	YES, NO, or UNKNOWN			UNKNOWN
UNFORMATTED	YES, NO, or UNKNOWN			UNKNOWN
WRITE	YES, NO, or UNKNOWN			UNKNOWN

The forms of the REWIND statement (R925) are:

```
REWIND scalar-integer-expression
REWIND ( position-specifier-list )
```

The forms of the ENDFILE statement (R924) are:

```
ENDFILE scalar-integer-expression
ENDFILE ( position-specifier-list )
```

The forms of a position specifier (R926) are:

```
[ UNIT = ] scalar-integer-expression
ERR = label
IOMSG = scalar-default-character-variable
IOSTAT = scalar-integer-variable
```

Rules and restrictions:

1. The BACKSPACE, REWIND, and ENDFILE statements are used to position external files; the integer expression must be a unit number.

2. The files must be connected for sequential or stream access—not direct access.

3. If the last data transfer to a file connected for sequential access is an output data transfer statement, a BACKSPACE or REWIND statement for this file implicitly writes an end-of-file record.

4. A file positioning statement performs a wait operation if there are any pending asynchronous data transfers for the unit.

Example file positioning statements are:

```
BACKSPACE 9
BACKSPACE (UNIT = 10)
BACKSPACE (ERR = 99, UNIT = 8, IOSTAT = STATUS)
REWIND (ERR = 102, UNIT = 10)
ENDFILE (10, IOSTAT = IERR)
ENDFILE (11)
```

9.10.1 Specifiers for File Position Statements

This section describes the form and effect of the position specifiers that may appear in the file positioning statements.

UNIT

Described in 9.2.2

Rules:

1. There must be a file connected for sequential access to the unit.

ERR

Described in 9.2.3

IOMSG

Described in 9.2.3

IOSTAT

Described in 9.2.3

9.10.2 The BACKSPACE Statement

Execution of a BACKSPACE statement causes the file to be positioned before the current record if there is a current record, or before the preceding record if there is no current record. If there is no current record and no preceding record, the file position is not changed. If the preceding record is an end-of-file record, the file becomes positioned before the end-of-file record. If a BACKSPACE statement causes the implicit writing of an end-of-file record and if there is a preceding record, the file becomes positioned before the record that precedes the end-of-file record.

If the file is already at its initial point, a BACKSPACE statement has no effect. If the file is connected, but does not exist, backspacing is prohibited. Backspacing over records written using list-directed or namelist formatting is prohibited. A file connected for unformatted stream access cannot be backspaced.

Examples of BACKSPACE statements are:

```
BACKSPACE  ERROR_UNIT     ! ERROR_UNIT is an integer variable
                          !    or named constant.
BACKSPACE (10, IOSTAT = STAT)
```

9.10.3 The REWIND Statement

A REWIND statement positions the file at its initial point. Rewinding has no effect on the file position when the file is already positioned at its initial point. If a file does not exist, but it is connected, rewinding the file is permitted, but has no effect. Examples of REWIND statements are:

```
REWIND INPUT_UNIT         ! INPUT_UNIT is an integer variable
REWIND (10, ERR = 200)
```

9.10.4 The ENDFILE Statement

For a file connected for sequential access, an ENDFILE statement writes an end-of-file record as the next record and positions the file after the end-of-file record. The end-of-file record becomes the last record in the file. Any records that existed after the current position no longer exist. Reading or writing records after the end-of-file record is prohibited; it is necessary to execute a BACKSPACE or REWIND statement to position the file ahead of the end-of-file record before reading or writing the file. If the file is subsequently connected for direct access, only those records before the end-of-file record may be read. Records may be written after the last direct-access record as if the end-of-file record did not exist.

For a file connected for stream access, an ENDFILE statement sets the terminal position of the file to the current position. Only file storage units prior to the current position may be read. Stream output statements may be used to write additional data to the file. Note that stream files have no explicit end-of-file record.

An ENDFILE statement for a file that is connected but does not exist will create the file. If the file is connected for sequential access, it is created before the endfile record is written.

An ENDFILE statement must not be used for a unit connected for read only access.

Examples of ENDFILE statements are:

```
ENDFILE  OUTPUT_UNIT    ! OUTPUT_UNIT is an integer variable
ENDFILE (10, ERR = 200, IOSTAT = ST)
```

9.11 The FLUSH Statement

On most computer systems, data from an output statement is kept in memory in a buffer local to the output library and physically transferred to the output device when some trigger point is reached. This typically happens when the buffer is almost full, or when the amount of data is large enough to be efficiently transferred, or, obviously, at program termination. Sometimes the data needs to be used or monitored as it is generated. Perhaps as a progress bar on a screen, or to check for unusual events in the program operation. Similarly, during input, the input routines typically read a large amount of data, often a disk sector or track, into internal buffers and do memory to memory transfers during execution of the input statements. When reading data from a file that is being produced in real time the buffers need to be refreshed to get the latest data. From a language definition point of view, execution of an input/output statement causes the data to be transferred to or from a file. The standard does not specify details of the timing of the transfer; buffering is a normal optimization on most systems. The FLUSH statement is a prompt to the compiler that normal buffering is not desired at that point and there should be an immediate physical transfer of data.

9.11.1 Form of the FLUSH Statement

The forms of the FLUSH statement are

FLUSH scalar-integer-expression
FLUSH (flush-specifier-list)

The form of a flush specifier is

[UNIT =] scalar-integer-expression
ERR = label
IOMSG = scalar-default-character-variable
IOSTAT = scalar-integer-variable

Rules and restrictions:

1. The unit must be connected to a file. If the file does not exist, the flush operation has no effect—it does not create the file.

Examples:

```
FLUSH 10
FLUSH (UNIT = 10, IOSTAT = IERR)
```

9.11.2 Specifiers for the FLUSH Statement

UNIT

Described in 9.2.2

ERR

Described in 9.2.3

IOMSG

Described in 9.2.3

IOSTAT

Described in 9.2.3

The variable will be set to zero if the flush operation succeeds, to a processor-dependent positive value if there was an error, and to a processor-dependent negative value if a flush operation is not supported for the unit.

9.11.3 Execution of the FLUSH Statement

Because the standard cannot specify details of the file handling system, execution of a FLUSH statement is ultimately processor dependent. The intention is that data written to a file should be made available to other processes and that data written to a file by other processes be made available to the program in a timely fashion. The standard is unclear about whether or not a FLUSH statement performs a wait operation if asynchronous input/output is pending. It seems likely that it would perform an implicit wait. However, since the purpose of the FLUSH statement is to prohibit the processor from overlapping computation and input/output, it seems unlikely that a real program would execute a FLUSH statement for a unit that had any pending asynchronous input/output.

9.12 Restrictions on Input/Output Specifiers, List Items, and Statements

There are three major restrictions on the execution of input/output statements. These apply in addition to specific restrictions for specific statements. Usually the restrictions have the most impact on input/output data transfer statements, because they have the most generality; but the restrictions apply to all input/output statements. They are:

1. A processor is allowed to limit the operations that can be performed on a file or unit.

2. Input/output operations cannot be recursive, except for some special cases of child data transfers or internal files.

3. An action of an input/output statement cannot cause another entity in the input/output statement, or an active parent statement, to be changed.

There is also the minor restriction that a STOP statement cannot be executed while an input/output statement is active.

The program is not allowed to attempt any operation the processor does not support. It is not guaranteed that attempting to do an unsupported action will cause an error indication, although most processors catch most errors. Some limits are related to the physical nature of the file itself—the terminal cannot be rewound. Some are imposed by the underlying operating system—the file system might or might not let direct-access files be opened for sequential or stream access. These cases can usually be checked for by reading the documentation, or using a suitable INQUIRE statement. Other limits are harder to cope with—once the disk becomes full, you cannot do any more output.

A **recursive input/output statement** is one that is executed while another input/output statement is executing. A recursive input/output statement is one that appears in a function or subroutine invoked either during evaluation of an expression in the initial input/output statement or by defined input/output. Except for a child data transfer statement writing to its parent data transfer statement's unit, a recursive input/output statement must not refer to an external unit. This limits non-child recursive input/output statements to data transfer statements that operate on internal files. A defined input/output subroutine may execute an INQUIRE statement on its parent's unit as well as perform data transfers to the parent's unit or to internal files.

Most actions which could change the values of entities appearing in an input/output statement in more than one way are prohibited. These prohibitions basically allow a processor to do input/output efficiently without imposing an order on the way things are evaluated or defined. The rules apply to input/output statements directly and to any statement executed in a defined input/output subroutine or a function invoked during expression evaluation. The rules apply to variables explicitly named in an input out output list, as well as to variables in a namelist group. The rules are:

1. A function invoked during execution of an input/output statement must not change the value of any variable used anywhere in the input/output statement. This is simply an application of the general rule on side effects (7.3.2).

2. An input/output statement must not change the value of any format specification that is in use, including parts that have already been used or that haven't yet been reached during format processing.

3. The value of a specifier, including the values of any subscripts or substring bounds, in a control list must not depend on the value of any item in an input list nor on the value of an input/out implied-do variable.

4. The value of a specifier, including the values of any subscripts or substring bounds, in a control list or inquire list must not depend on any other specifier in the list.

5. In a data transfer statement, the variables specified by an IOSTAT, IOMSG, or SIZE specifier, including any subscripts, must not be associated with any entity in the input or output list nor with the do variable of an implied do.

6. In a data transfer statement, the values of the variables specified by an IOSTAT, IOMSG, or SIZE specifier, including any subscripts, must not be affected by the transfer itself, including implied do processing, and the evaluation of any other specifier in the control list.

7. Any variable that might become defined or undefined by an INQUIRE statement cannot appear more than once in the inquire list, nor can it be associated with any other variable in the list.

The net effect of these rules is that the processor is free to evaluate, or assign values to, variables in the control lists in any order.

10 Input and Output Editing

- **Explicit Formatting** uses a format specification either in a character expression or in a FORMAT statement. A label is required to reference a FORMAT statement.
- **Implicit Formatting**, used for list-directed or namelist input/output, does not use explicit format specifications; rather the processor supplies edit descriptors for each data item.
- A **Format Specification** is basically a list of edit descriptors enclosed in parentheses. There are data, control, and character string edit descriptors.
- **Data Edit Descriptors** determine how data is converted to or from the internal representation. There must be an appropriate data edit descriptor for each item in the input/output list. There are specific data edit descriptors for each intrinsic type, additionally, the G edit descriptor can be used for any intrinsic type. Alternatively, the DT edit descriptor may be used for derived-type items.
- **Control Edit Descriptors** create new records, specify spacing, position, blank interpretation, decimal point and algebraic sign display, and also may affect numeric conversions by specifying a scale factor or rounding mode.
- **Character String Edit Descriptors** specify the output of literal character strings delimited by quotes or apostrophes; they are not used on input.
- **List-Directed Formatting** is specified by an asterisk used as a format specifier. The conversion that occurs for an item in the input/output list is based on the type and value of the item.
- **Namelist Formatting** is indicated by a namelist specifier appearing in the input/output statement; there is no explicit input/output list. The names of the items in the namelist group appear as pairs, name=value, in both the input and the output. Comments may appear in the input. Conversion is based on the type and value of the item.

Data usually are stored in memory in binary form. For example, the integer 6 might be stored as 0000000000000110, where the 0s and 1s represent binary digits. On the other hand, formatted data records in a file consist of characters. Thus, when data is read from a formatted record, it must be converted from characters to the internal representation. Conversely, when data is written to a formatted record, it must be converted from the internal representation into a string of characters.

Although the form of the data read during formatted input is similar to that of Fortran's constants, the values are not actually Fortran constants. They do not have to follow all the rules for Fortran literal constants and they are not allowed to have kind

J.C. Adams et al., *The Fortran 2003 Handbook*,
DOI: 10.1007/978-1-84628-746-6_10,

parameters. The specific forms allowed are described in the appropriate section for each type of input. This chapter refers to the input items as "values", although "character string representation of the values before conversion" would be a more descriptive term.

There are three forms of formatted input/output: namelist (10.11), list-directed (10.10), and explicitly formatted (10.1). Namelist input/output reads or writes values in a form name=value. The variable names are specified in a namelist declaration. It is often used to provide simple human readable output for a large number of variables. On input, not all of the variables need to be specified; namelist input is often used to modify a few variables in a group while the others retain a default value. List-directed input/output reads or writes a comma separated list of values. It is often used for reading human prepared values, say from a terminal, or writing debugging output where the appearance of the output is not important. As the name implies, explicit formatting specifies where each value will appear and how it will look. Namelist and list-directed input/output use implicit formatting; the processor can read almost any appropriate form of data and is allowed to use "reasonable" forms for the output.

An explicit **format specification** provides the information necessary to determine how conversions are to be performed. The format specification is basically a list of **edit descriptors**. There must be a data edit descriptor for each data value in the input/output list of the data transfer statement. Control edit descriptors may be added to control the spacing and position within a record, create new records, specify the interpretation of blanks, set the rounding mode and scale factor, and change the display of the decimal point and optional plus sign. String edit descriptors transfer strings of characters from the format specifications to output records. Tables 10-1, 10-2, and 10-3 list all the edit descriptors and provide a brief description of each.

Table 10-1 Summary of control edit descriptors

Descriptor	Description	Default	Section
BN, BZ	Set blank mode	BN	10.9.6
DC, DP	Set decimal mode	DC	10.9.8
RC, RD, RN, RP, RU, RZ	Set rounding mode	RP	10.9.7
S, SP, SS	Set sign mode	S	10.9.4
T, TL, TR, X	Tab	N/A	10.9.1
/	End current record	N/A	10.9.2
:	Conditionally stop format processing	N/A	10.9.3
P	Set scale factor	0P	10.9.5

The format specifier may be a statement label that identifies a FORMAT statement, or it may be a character expression giving the format directly. Either method is called **explicit formatting**.

Table 10-2 Summary of data edit descriptors

Descriptor	Description: convert data of	Section
A	type character	10.7
B	type integer to/from a binary base	10.5.1
D	type real—similar to E edit descriptor	10.5.2.2
DT	derived type	10.8
E	type real with an exponent	10.5.2.2
EN	type real to engineering notation	10.5.2.2
ES	type real to scientific notation	10.5.2.2
F	type real with no exponent on output	10.5.2.1
G	any intrinsic types	10.5.1, 10.5.2.3, 10.6, 10.7
I	type integer	10.5.1
L	type logical	10.6
O	type integer to/from an octal base	10.5.1
Z	type integer to/from a hexadecimal base	10.5.1

Table 10-3 Summary of string edit descriptors

Descriptor	Description	Section
'text'	Transfer of a character literal constant to output record	10.9.9
"text"	Transfer of a character literal constant to output record	10.9.9

10.1 Explicit Formatting

As indicated above, explicit formatting information may be:

1. contained in a FORMAT statement

```
    PRINT 100, LIGHT, AND, HEAVY
100 FORMAT (F10.2, I5, E16.8)
```

2. given as the value of a character expression, which might be a literal constant or a simple variable or array

```
character (len=50) :: my_format
my_format = "(F10.2, I5, E16.8)"
print my_format, LIGHT, AND, HEAVY

PRINT '(F10.2, I5, E16.8)', LIGHT, AND, HEAVY
```

10.1.1 The FORMAT Statement

The form of the FORMAT statement (R1001) is:

```
FORMAT ( [ format-item-list ] )
```

A **format specification** (R1002) consists of the parentheses and the format item list (10.2).

The FORMAT statement must be labeled. The label is used in input/output statements to reference a particular FORMAT statement.

10.1.2 Character Expression Format Specifications

A character expression may be used in an input/output statement as a format specification. The leading part of the character expression must be a valid format specification including the parentheses; that is, the value of the expression must be such that the first nonblank character is a left parenthesis, followed by a list of valid format items, followed by a closing right parenthesis.

Rules and restrictions:

1. All character positions up to the closing right parenthesis in the character expression must be defined when the input/output statement is executed.

2. In a character expression, the closing right parenthesis may be followed by any characters, including parentheses. None of these trailing characters are relevant to the rules and restrictions in this section. For example, the following character string may be used as a format specification in a character string; the part after the first right parenthesis is ignored.

```
'(I5,E16.8,A5)  This (part (is (ignored)))))'
```

3. If the expression is a character array, the format is scanned in array element order. For example, the following format specification is valid (where A is a character array of length at least 6 and size at least 2):

```
A (1) = '(1X,I3,'
A (2) = ' I7, I9)'
PRINT  A, MUTT, AND, JEFF
```

4. If the expression is an array element, the format must be entirely contained within that element.

5. If the expression is a character variable, no part of it up to the closing right parenthesis may be redefined or become undefined during the execution of the input/output statement.

6. If the expression is a character literal constant, the normal rules for such constants apply. In particular, if it is delimited by apostrophes, two apostrophes must be written to represent each apostrophe in the format specification. If a format speci-

fication contains, in turn, a character constant delimited by apostrophes, there must be two apostrophes for each of the apostrophe delimiters, and each apostrophe within the character constant must be represented by four apostrophes (see the example below). If quotes are used for the string delimiters and quotes are used within the string, a similar doubling of the quote marks is required. One way to avoid problems is to use both character delimiters in the format specification, if possible. The best way to avoid the problem is to put the character expression in the input/output list instead of the format specification as shown in the last line of the following example.

```
PRINT '(''I can''''t hear you'')'
PRINT "('I can''t hear you')"
PRINT "(A)", "I can't hear you"
```

The processor must scan FORMAT statements for errors; however, when a character expression, including a simple constant, is used as a format specification, the processor is not required to detect at compile-time any syntax or constraint violations in the format specification. The reason for relaxing the requirements for detection of errors is that the format specification might not be complete or known until the data transfer statement is executed and therefore cannot be checked for validity until execution time. The same relaxation on the requirements for run-time error detection also applies to the use of deleted, obsolescent, and extended features used in format specifications.

10.2 Format Specifications

A format specification is built using data, control, or character string edit descriptors. Each data list item must have a corresponding data edit descriptor; other descriptors, which are optional, specify such things as spacing, tabulation, scale factors for real data, and printing of optional signs. Most edit descriptors must be separated by commas within a format specification; however, rule 2 below gives some exceptions.

Blanks may be used freely in format specifications without affecting the interpretation of the edit descriptors (except within the character string descriptor), both in free and fixed source forms. Named constants are not allowed in format specifications because they would create ambiguities in the interpretation of the format specifications. For example, if N12 were a named integer constant with value 15, the engineering format edit descriptor E N12.4 could be interprete as the edit descriptor EN12.4 or E15.4.

The forms of a format item (R1003) are:

```
[ r ] data-edit-descriptor
control-edit-descriptor
character-string-edit-descriptor
[ r ] ( format-item-list )
```

where r is a default integer literal constant and is called a **repeat factor**. If a repeat factor does not appear, it is as if it were there with a value of 1. The effect is as if the format item were repeated r times. Note that a parenthesized format item can be repeated and can contain repeated format items.

Rules and restrictions:

1. r must not have a kind value specified for it and it must be positive.
2. The comma between edit descriptors may be omitted in the following cases:
   a. between the scale factor (P) and the numeric edit descriptors F, E, EN, ES, D, or G, including a repeat factor if one appears
   b. before a new record indicated by a slash when there is no repeat factor
   c. after the slash for a new record
   d. before or after the colon edit descriptor
3. Blanks may be used as follows:
   a. before the first left parenthesis
   b. anywhere in the format specification; the blanks have no effect on the formatting, except within a character string descriptor

The following examples illustrate many of the edit descriptors that are described in detail in the next sections.

```
100 FORMAT (2(5E10.1, I10) / (1X, SP, I7, ES15.2))
110 FORMAT (I10, F14.1, EN10.2)
120 FORMAT (TR4, L4, 15X, A20)
130 FORMAT ("MORE SNOW")
140 FORMAT (9X, 3A5, 7/ 10X, 3L4)
```

10.2.1 Data Edit Descriptor Form

Data edit descriptors specify the conversion of values to and from the internal representation to the character representation in the formatted record of a file. The forms of the data edit descriptors (R1005) are:

```
I w[ . m]
B w[ . m]
O w[ . m]
Z w[ . m]
F w . d
E w . d[ E e]
EN w . d[ E e]
ES w . d[ E e]
G w . d[ E e]
```

```
L w
A [ w ]
D w . d
DT [ character-literal-constant ] [ ( v-list ) ]
```

where w, m, d, and e are default integer literal constants, and

w	determines the width of the field
m	is the least number of digits in the field
d	is the number of digits after the decimal symbol
e	is the number of digits in the exponent field
v	is a signed integer literal constant and its interpretation depends on the user supplied DTIO subroutine

Rules and restrictions:

1. w, m, d, e, v and character-literal-constant must not have a kind value.
2. e must be positive.
3. w must be zero or positive for the I, B, O, Z, and F edit descriptors; w must be positive for all other edit descriptors.
4. The I, B, O, Z, F, E, EN, ES, G, L, A, D, and DT edit descriptors indicate the manner of editing.

The detailed meanings of the data edit descriptors are provided in 10.5 through 10.8.

10.2.2 Control Edit Descriptor Form

Control edit descriptors determine the position, form, layout, and interpretation of characters transferred to and from formatted records in a file. The forms of a control edit descriptor (R1011) are:

```
T n       TL n    TR n     n X
[ r ] /
:
S         SP      SS
k P
BN        BZ
DC        DP
RC        RD      RN       RP      RU      RZ
```

where n and r are default integer literal constants, k is a signed default integer literal constant, and

k	is a scale factor
n	is a position in the record, relative to the left tab limit for T
n	is the number of spaces for X, TR, and TL
r	is a repeat factor

The control edit descriptors T, TL, TR, and X are called **position edit descriptors** (R1013). The control edit descriptors S, SP, and SS are called **sign edit descriptors** (R1015). The control edit descriptors BN and BZ are called **blank interpretation edit descriptors** (R1016). The control edit descriptors RC, RD, RN, RP, RU, and RZ are called **round edit descriptors** (R1017). The control edit descriptors DC and DP are called **decimal edit descriptors** (R1018).

Rules and restrictions:

1. n and r must be positive.
2. k, n, and r must not have a kind value specified for them.

The control edit descriptors are described in detail in 10.9.

10.2.3 Character String Edit Descriptor Form

Character string edit descriptors specify character strings to be transmitted to the formatted output record of a file. The form of the character string edit descriptor (R1019) is:

```
character-literal-constant
```

The character string edit descriptor is described in detail in 10.9.9.

10.3 Formatted Data Transfer

The format specification indicates how data are transferred by READ, WRITE, and PRINT statements. The data transfer typically involves a conversion of data values. The particular conversion depends on the data input or output item and the matching edit descriptor in the format specification.

An empty format specification ( ) is restricted to input/output statements with no effective items (9.5.1) in the input/output data item list.

The effect on input and output of an empty format specification depends on whether the data transfer is advancing or nonadvancing (9.1.4), and whether there is a current record (9.1.4). The effect is described by the following algorithm:

1. If the data transfer is advancing:
   a. if there is no current record, then:
      i. on input, skip the next record
      ii. on output, write an empty record
   b. if there is a current record, then:
      i. on input, skip to the end of the current record
      ii. on output, terminate the current record
2. If the data transfer is nonadvancing:
   a. if there is no current record, then:
      i. on input, move to the initial point of the next record
      ii. on output, create an empty record and move to its initial point
   b. if there is a current record, then:
      i. on input, there is no effect
      ii. on output, there is no effect

Example of nonadvancing input:

```
DO I = 1, 5
   READ (input_unit, '(80A1)', ADVANCE='NO') CHARS(I)(1:10)
ENDDO
READ (input_unit, '()', ADVANCE = 'YES')
```

The above program segment reads five character strings, each of length 10, from a single record and then advances to the end of the current record (step 1(b)(i) above).

The data and the edit descriptors are processed in a left-to-right fashion, except for repeated edit descriptors, which are processed until either the data items are exhausted or the repeat number is reached. A complex data item requires two data edit descriptors suitable for data items of type real; that is, two of the edit descriptors E, F, D, ES, EN, or G (they may be different).

Control edit descriptors and character edit descriptors do not have a corresponding data item in the list. The effect is directly on the record transferred.

10.3.1 Parentheses Usage

The effect of parentheses in a format specification depends on the nesting level of the parentheses. They are normally used to allow a series of edit descriptors to be used repeatedly or to control reversion, which is what happens when the end of a format is reached while there are still remaining data items. The algorithm for format reversion is:

1. When the rightmost right parenthesis of a complete format specification is encountered and there are no more data items, the input/output data transfer terminates.

2. When the rightmost right parenthesis is encountered and there are more data items, format control continues beginning at the left parenthesis corresponding to the last preceding right parenthesis in the specification, if there is one, with an implied slash (/) to cause a new record to begin. If there is no preceding right parenthesis, the reversion is to the beginning of the format. There must be at least one data edit descriptor to the right of the reversion point.

3. If there is a repeat factor preceding the left parenthesis, it is reused.

4. Reversion does not affect or "reset" the scale factor, the sign control, rounding, decimal display, or blank interpretation.

Example of format reversion:

```
CHR_FMT = "(I5, /, 4(3F10.2, 10X), E20.4)"
```

If the above character string were used in a formatted output data transfer statement, the first output data item must be an integer and will be the only item in the first record. The remaining items must be of type real or complex. Up to 13 values are printed in the second record, and then the remaining values are transferred to each new record, 13 at a time, until the data items are exhausted. All but the first and possibly the last record will have 13 values, 4 sequences of 3 values using an F10.2 edit descriptor and 10 blanks, followed by a value using the E20.4 edit descriptor.

A typical use of this format would be

```
PRINT CHR_FMT, N, (A(I), I = 1,N)
```

This behavior is described in more detail in the next section.

10.3.2 Correspondence between a Data-Edit Descriptor and a List Item

The best way to describe how this correspondence is determined is to think of two markers, one beginning at the first item of the input/output data item list and the other beginning at the first left parenthesis of the format specification. The input/output data item list is conceptually expanded by listing each element of an array, each component of a structure (unless the matching edit descriptor is a DT descriptor), each part (real and imaginary) of each item of type complex, and each iteration of each implied-do list. The expanded item list is called the **effective data item list**, and each item in the list is called an **effective item**. Zero-sized arrays yield no effective items. A zero-length character object is an effective item. The format specification is conceptually expanded for repeat factors and reversion. Because of interactions between data values, data types, and edit descriptors, the format can only be conceptually expanded before data items are processed. For example, an input value might be used in an implied do expression and, obviously, will not be known until some list items and edit descriptors have been processed. If the data item list is not empty, there must be at least one data edit descriptor in the format specification.

If the effective data list is empty, the format marker proceeds through the format specification. On output, string edit descriptors are processed (10.9.9). On either input or output, slash control edit descriptors are processed (10.9.2). When a data edit descriptor, a colon edit descriptor, or the outer right parenthesis is encountered, the input/output operation terminates.

If the effective data item list is not empty:

1. The format marker proceeds through the format specification until a data edit descriptor or right parenthesis is encountered. Any control edit descriptor or string edit descriptor encountered before the first data edit descriptor is encountered is interpreted according to its definition, each possibly changing the position within the record or the position within the file, or changing the interpretation of data in the record or conversion of data to the record.

2. The effective data item pointed to by the marker in the data item list is transferred and converted according to the data edit descriptor, and the markers in the data item list and format item list proceed to the next effective data item or format specification, respectively.

3. If there is a remaining effective list item, processing continues at step one, otherwise.

4. Format processing continues as in step 1 until a data edit descriptor, a colon edit descriptor, or the outer right parenthesis is encountered; at this point the input/output operation terminates.

To illustrate how this works, consider the following example:

```
INTEGER  A(3)
COMPLEX  C
TYPE RATIONAL
   INTEGER  N, D
END TYPE
TYPE (RATIONAL)  R
   . . .
PRINT &
   "('A and C appear on line 2, R appears on line 3.' &
      & / (1X, 3I5, 2F5.2) )",  A, C, R
```

The data item list is expanded as described above. The expanded data item list becomes:

```
A(1), A(2), A(3), REAL(C), AIMAG(C), R % N, R % D
```

The format specification is also expanded and becomes:

```
('A and C appear on line 2, R appears on line 3.'  &
   & / (1X, I5, I5, I5, F5.2, F5.2) )
```

A marker is established in the data item list, which initially points at the item A(1). A marker is also established in the format specification and initially points to the first left parenthesis. The marker in the format specification proceeds left to right to the first edit descriptor, which is the first I5. In so doing, it sees the string edit descriptor which is transferred to the output record, the slash edit descriptor which causes the previous record to terminate and to begin a new record, and the position edit descriptor which positions the record at the second character, blank filling the record. The item A(1) is then converted according to the I5 specification and the converted value is transferred to the output record. The format specification marker is moved right to the second I5 edit descriptor. The marker in the data item list is moved to A(2) and A(2) is converted and transferred to the output record. Similarly, A(3), the real part of C, and the imaginary part of C are converted and transferred to the output record. At this point, the format specification marker begins scanning after the second F5.2 edit descriptor looking for the next edit descriptor, the data item list marker is pointing at R % N. The first right parenthesis is encountered and the scan reverts back to the corresponding left parenthesis. The repeat factor in front of this parenthesis is 1 by default and is reduced by 1 to 0. The marker in the format specification proceeds right from the first right parenthesis, encountering the outermost right parenthesis and then reverts to the left parenthesis before the edit descriptor 1X. As a result, an implicit slash edit descriptor is interpreted, causing the previous output record to be completed and a new record to be started. The format specification marker scans right looking for a data edit descriptor, which is the first I5. In the process of the scan right, the position edit descriptor is interpreted, which positions the file at the second character of the next record (and blank fills the skipped characters). Finally, the N and D components of R are converted and transferred to the output record, using the first two I5 edit descriptors. The format specification marker moves to the third I5 edit descriptor. The data item list marker finds no further items, and the output operation terminates.

If A has the value [2 4 6], C has the complex value (1.0, 2.0), and R has the value RATIONAL(10, 20), then the completed output will be:

```
A and C appear on line 2; R appears on line 3.
bbbbb2bbbb4bbbb6b1.00b2.00
bbbb10bbb20
```

where *b* is a blank character.

Examples of writing a zero-sized array and zero-length character string using formatted output data transfer are:

```
REAL  A(10)
CHARACTER(LEN=10)  CHR, X(3)
WRITE( *, '()' )  A(1:0)
WRITE( *, '(A4)' )  CHR(4:3)
WRITE( *, '(A4)' )  X(1:3)(4:3)
```

An empty format specification is allowed for the first WRITE statement, because the array to be printed is a zero-sized array section. The format specification in the second WRITE statement is required to have at least one A edit descriptor, because the effective data item is a zero-length character string, not a zero-sized array. In the first case, an empty record is written, and, in the second case, a record consisting of four blank characters is written. The third WRITE statement will produce three records, each consisting of four blank characters.

10.4 File Positioning by Format Control

During a formatted input or output operation, there is a current record being processed. After each data edit descriptor is used, the file position within that record is following the last character read or written by the particular edit descriptor. On output, after a string edit descriptor is used, the file is positioned within that record following the last character written. (See the description of the control edit descriptors T, TL, TR, and X for any special positioning within the current record; see the description of the slash edit descriptor for special positioning within the file.) The remaining control edit descriptors do not affect the position within a record or within the file; they affect only the interpretation of the input characters or the form of the output character string or how subsequent edit descriptors are interpreted. The interpretation of the edit descriptors is not affected by whether the operation is an advancing or a nonadvancing input/output operation.

10.5 Numeric Editing

The edit descriptors that apply to numeric editing are I, B, O, Z, F, E, EN, ES, D, and G. The following rules apply to all of them.

Rules and restrictions:

On input:

1. Leading blanks are never significant.
2. Plus signs may be omitted in the input data.
3. A completely blank field is considered to be zero, regardless of the BLANK interpretation mode in effect.
4. Within a field, blanks are interpreted in a manner that depends on the BLANK interpretation mode in effect, unless the value is an IEEE infinity or NaN (10.5.2.4).
5. In numeric fields that have a decimal symbol and correspond to F, E, EN, ES, D, or G edit descriptors, the decimal symbol in the input field overrides the placement of the decimal symbol specified by the edit descriptor specification.
6. Data input can have more digits of significance than the processor can use to represent a number.
7. The lowercase exponent letters e and d are equivalent to the corresponding uppercase exponent letters and lower case letters are equivalent to upper case letters in IEEE infinity or NaN values (10.5.2.4).

On output:

1. Except for B, O, and Z editing, it is processor dependent whether a positive value, including a positive zero, has a plus sign, unless a sign edit descriptor is used to force the presence or absence of the plus sign (10.9.4).
2. Except for B, O, and Z editing, negative values will have a minus sign. For real data, a zero with a negative sign may be produced on output. This allows support for IEEE arithmetic processors that can represent negative zero.
3. The number is right justified in the field. Leading blanks will be inserted as needed.
4. Except when w is zero, if the number or the exponent is too large for the field width specified in the edit descriptor, the entire output field is filled with asterisks.
5. The processor will not produce asterisks if the output value fits in the output field when the optional characters are omitted.
6. w may be zero only for I, B, O, Z, or F editing. In this case, the processor will select the minimum field width which will contain the sign (if one is necessary) and value with no leading blanks.

10.5.1 Integer Editing

Integer editing converts integer values to or from strings of characters. The integer edit descriptors are:

```
I w[ . m]
B w[ . m]
O w[ . m]
Z w[ . m]
G w . d[ E e]
```

where:

w	determines the field width
m	is the minimum number of digits in the output field
d, e	have no effect on integer editing

They are the only edit descriptors that may be used with integer data (note that the G edit descriptor also may be used with other data types).

Rules and restrictions:

On both input and output:

1. For an integer input/output list item, the edit descriptor Gw.d[Ee] is the same as an Iw descriptor with the same value of w; w must be greater than zero.

On input:

1. w must not be zero.
2. m has no effect on an input field.
3. For the I edit descriptor, the character string in the file must be an optionally signed string of blanks and digits.
4. For the B, O, or Z edit descriptors, the character string must be an unsigned string of blanks and digits of binary, octal, or hexadecimal base, respectively (4.3.1.4).

   Example of formatted input:

   ```
         READ 100,  K, J, L
     100 FORMAT (I5, G8.0, O4)
   ```

   If the input field is (where *b* is a blank)

   bb-24*bbbbb*117*bb*77

   K is read using the integer I5 edit descriptor, J is read with a G8 edit descriptor, and L is read with an O4 edit descriptor. The resulting values of K, J, and L are –24, 117, and 63, respectively.

On output:

1. The output field generally consists of blank characters, possibly a sign, and a digits string. The sign and digits are right adjusted within the field. Some or all of these parts may be absent.
2. The value of m, if it appears, must not exceed the value of w unless w is zero.
3. If w>0, the Iw edit descriptor produces a field with a width of w characters consisting of leading blanks as needed, a minus sign if the value is negative or an optional plus sign, and a digit string in the rightmost characters of the field. If the value will not fit in w characters (with no optional plus sign), the field is filled with asterisks. If w=0, the processor will select the minimum field width to contain a minus sign if needed or an optional plus sign and the value with no leading spaces.
4. For Iw.m, with w>0, at least m digits, with leading zeros as necessary, are produced.
5. The I0.m edit descriptor produces a result with at least m digits, and as many more digits as are necessary to represent the integer value. The output field contains a minus sign if necessary and optionally a plus sign.

6. The Bw[.m], Ow[.m], and Zw[.m] edit descriptors follow the same rules and produce the same output form as Iw[.m] except that the digits are in the binary, octal, or hexadecimal system and no sign is produced.

7. For the special case of m=0 and a data value of zero, the output field consists of w blanks, unless w is also zero, in which case the field is one blank. No sign is produced, regardless of the sign control in effect.

Because the B, O, and Z edit descriptors do not allow a sign in the input field or produce one on output, depending on the processor, a negative value might be encoded in the digits. For example, –1 might be given as 80000000 using a hexadecimal edit descriptor. If the highest order bit is set, the interpretation of the digits with B, O, and Z edit descriptors is not specified by the standard.

Examples:

The statement

```
PRINT "(4I4)", 22, -444, 0, 55555
```

produces

```
bb22-444bbb0****
```

because 55555 will not fit in a field that is only 4 spaces wide, whereas the statement

```
PRINT "(4I0)", 22, -444, 0, 55555
```

produces

```
22-444055555
```

The processor selects the minimum field width necessary, although the compressing of the resulting values is difficult for humans to read.

The statement

```
PRINT "(4I4.3)", 22, -444, 0, 55555
```

produces

```
b022-444b000****
```

with at least 3 digits, including forced leading zeros for each value, whereas

```
PRINT "(4I0.3)", 22, -444, 0, 55555
```

produces

```
022-44400055555
```

and the fields are adjusted to allow all of the values to print with at least 3 digits displayed.

The following table shows the output displayed when internal values are printed with different integer edit descriptors.

Internal value	I6	I6.5	I6.3	I6.0	I0.6	I0.3	I0.0
1874	*bb*1874	*b*01874	*bb*1874	*bb*1874	001874	1874	1874
–1874	*b*-1874	-01874	*b*-1874	*b*-1874	-001874	-1874	-1874
0	*bbbbb*0	*b*00000	*bbb*000	*bbbbbb*	000000	000	*b*

Consider the particular case where 1874 is printed with the I6.5 edit descriptor. Then the minimal number of digits needed to represent the value is 4. Because the value of m is 5, the output field consists of one blank followed by a five-digit representation of 1874, in which one leading zero is needed to produce five digits. Also, note carefully the special case of zero and an I0.0 edit descriptor: it produces a single blank space, not a zero!

The I0 and I0.m are often used with internal output to produce a sequence of file names that follow operating system conventions. If C1 and C2 are character variables with a length of 7, the statements

```
WRITE (C1, "(A, I0.3)")   "file", 012
WRITE (C2, "(A, I3.3)")   "file", 012
```

will set the variables C1 and C2 to file12*b* and file012, respectively, producing file names with no internal blanks.

10.5.2 Real Editing

Real editing converts real (or complex) values to or from strings of characters. The F, E, EN, ES, D, and G edit descriptors specify editing for real and complex input/output list items. Two such edit descriptors are required for each complex data item (10.5.3).

The forms of the edit descriptors for real values are:

```
F w . d
E w . d[ E e]
EN w . d[ E e]
ES w . d[ E e]
D w . d
G w . d[ E e]
```

For output, the E, EN, and ES descriptors produce a field with a decimal symbol and an exponent field; the position of the decimal symbol depends on the descriptor used. The E descriptor writes no significant digits to the left of the decimal, the ES, (scientific), descriptor writes one digit to the left of the decimal, and the EN (engineering) descriptor writes one to three digits to the left of the decimal such that the exponent value is divisible by 3. The F descriptor writes values with a decimal symbol in a fixed place and no exponent field. The D descriptor was historically used for double precision values, but is now essentially the same as an E descriptor. The G descriptor will produce either F or E output, depending on which form fits "best" in the output field.

For input, all six descriptors may be used interchangeably.

The general rules for the format descriptors and the processing of numeric values are described below. The special case IEEE values, infinity and NaN, are described in 10.5.2.4.

10.5.2.1 F Editing

Fw.d editing converts to or from a string occupying w positions, except when w is zero.

Rules and restrictions:

On both input and output:

1. The value in the input field or the value transferred to the output field may be signed.
2. Rounding during the conversion is controlled by the ROUND mode in effect (10.9.7, 9.2.4).

On input:

1. w must not be zero.
2. d specifies the number of decimal places in the input value if a decimal symbol does not appear in the input field.
3. The input field may be
    a. an IEEE exceptional value (10.5.2.4)
    b. a string which has the form of an optionally signed integer or real literal constant, possibly including an exponent
    c. an optionally signed digit string followed by a sign followed by a digit string treated as an exponent
    d. an optionally signed digit string containing a decimal symbol followed by a sign followed by a digit string treated as an exponent
4. None of the constants or digit fields may contain an underscore or kind parameter.
5. Blanks usually may be inserted freely anywhere in the input field, although they may interact with the blank interpretation mode.
6. If the input field contains a decimal symbol, the value of d has no effect.
7. If there is no decimal symbol, it is as if a decimal symbol appeared in front of the rightmost d digits of the nonexponent part, treating blanks as 0 digits or as if they

were not there, according to the blank interpretation mode. If w > d, leading zeros are supplied as necessary.

For example, with the format specification F5.1, the input data item

```
1bb99
```

is treated as the real number 19.9 if the BLANK interpretation mode is NULL; it is treated as the real number 1009.9 if the BLANK interpretation mode is ZERO.

8. There may be more digits in the number than the processor can use.
9. The number may contain an E or D indicating an exponent value; a field with a D exponent letter is processed identically to the same field with an E exponent letter. If there is no exponent field on input, the number is processed as if the character string were followed by an exponent with the value –k where k is the scale factor established by a previous kP edit descriptor. If there is an exponent field, the scale factor kP is ignored.

On output:

1. d specifies the number of digits after the decimal symbol, d must be less than w unless w is zero.
2. If the value is an IEEE NaN or infinity it is processed as described in 10.5.2.4.
3. Except when w is zero, the form of the output field consists of w positions comprised of leading blanks, if necessary, and an optionally signed real value modified by the scale factor in effect (10.9.5) with a decimal symbol, rounded to d digits after the decimal symbol but with no exponent, underscore, or kind parameter. If the number is positive, the sign is optional and is controlled by the sign mode in effect.
4. Leading 0s do not appear unless the number is less than 1 in absolute value, in which case the processor optionally may place a 0 in front of the decimal symbol.
5. One zero will be output if no other digits would appear.
6. The form of the output field from an F0.d edit descriptor is the same as from an Fw.d edit descriptor, where w is the smallest value such that all of the nonblank characters that would be produced are produced and no optional blank or plus sign appears. However, if the sign mode is PLUS, the leading sign is not optional and will be produced.

Examples:

The statements

```
      READ (*, 100)  X, Y
  100 FORMAT (F10.2, F10.3)
```

and input field of

```
bbbb6.42181234567890
```

assigns to X and Y the values 6.4218 and 1234567.89, respectively. The value of d is ignored for X because the input field contains a decimal symbol.

The statement

```
PRINT "(A, F0.4, A)", "X", -12.34567, "X"
```

produces the output record

```
X-12.3457X
```

10.5.2.2 E, EN, ES, and D Editing

The Ew.d[Ee], Dw.d, ENw.d[Ee], and ESw.d[Ee] edit descriptors convert to and from a string occupying w positions. For these edit descriptors, the field representing the floating-point number contains w characters, including an optional exponent field.

Rules and restrictions:

On both input and output:

1. w is the field width, d is the number of places after the decimal, and e is the exponent width.

On input:

1. Each of the forms Ew.d[Ee], Dw.d, ENw.d[Ee], and ESw.d[Ee] is equivalent to an Fw.d edit descriptor. e has no effect on input.

On output:

The forms of the output field for a scale factor of zero are:

$[\pm]\ [0]\ .\ x_1x_2 \ldots x_d\ \ exp$ E or D

$[\pm]\ yyy\ .\ x_1x_2 \ldots x_d\ \ exp$ EN

$[\pm]\ z\ .\ x_1x_2 \ldots x_d\ \ exp$ ES

where:

- $\pm$ signifies a plus or a minus; the plus is optional for positive numbers and is controlled by the sign mode (9.2.4).
- $x_1x_2 \ldots x_d$ are the d digits after the decimal symbol of the rounded value.
- *yyy* (for EN) are the one, two, or three decimal digits representing the most significant digits of the value of the datum after rounding; that is, *yyy* is an integer such that $1 \le yyy < 1000$ or, if the output value is zero, $yyy = 0$.

- z (for ES) is a decimal digit representing the most significant digit of the value of the datum after rounding; that is, z is an integer such that $0 < z < 10$ or, if the output value is zero, $y = 0$.

- *exp* is a decimal exponent having one of the forms specified in Table 10-4, where each z_i is a decimal digit. For the EN descriptor, the exponent will be divisible by three and the number of digits in *yyy* adjusted accordingly.

Table 10-4 Forms for the exponent *exp* in E and D editing

Edit descriptor	Value of exponent *ae*	Form of *exp*
Ew.d ENw.d ESw.d	$\lvert ae \rvert \le 99$ $99 < \lvert ae \rvert \le 999$	$E \pm z_1 z_2$ or $\pm 0 z_1 z_2$ $\pm z_1 z_2 z_3$
Ew.dEe ENw.dEe ESw.dEe	$\lvert ae \rvert \le 10^e - 1$	$E \pm z_1 z_2 \ldots z_e$
Dw.d	$\lvert ae \rvert \le 99$ $99 < \lvert ae \rvert \le 999$	$D \pm z_1 z_2$ or $E \pm z_1 z_2$ or $\pm 0 z_1 z_2$ $\pm z_1 z_2 z_3$

2. The sign in the exponent will be produced.

3. A plus sign is used for zero exponents.

4. If the exponent exceeds 999 in magnitude, the forms with Ee must be used with a sufficiently large value of e to accommodate the exponent *exp*.

5. There is no way to force the processor to use a D in the exponent field.

6. For the Ew.d and Dw.d forms, a scale factor kP may be used to specify the number of digits to the left of the decimal symbol, with the exponent adjusted accordingly; that is, the scale factor k controls the decimal normalization. If $-d < k \le 0$, the output field contains the decimal symbol, exactly $\lvert k \rvert$ leading zeros, and $d - \lvert k \rvert$ significant digits. If $0 < k < d + 2$, the output field contains exactly k significant digits to the left of the decimal symbol and $d - k + 1$ significant digits to the right of the decimal symbol. Other values of k are not permitted; that is, those values of k that will produce no digits to the left of the decimal symbol or specify fewer than zero digits to the right of the decimal symbol. The scale factor has no effect on ES or EN editing.

7. The precise form of zero on output is not specified, except that it must contain a decimal symbol, d zero digits, and an exponent of at least 4 characters whose digits are not specified. However, the likely value of the exponent is zero. The EN and ES descriptors will produce exactly one zero to the left of the decimal.

Examples of real output:

If the value of Y is –21.2, Table 10-5 indicates the output for various edit descriptors when the following statements are executed.

```
    PRINT 105, Y
105 FORMAT (edit descriptor)
```

Table 10-5 Values produced by various E, EN, ES, and D edit descriptors

Edit descriptor	Values produced
E15.3	*bbbbb*-0.212E+02 or *bbbbbb*-.212E+02
3PD15.3	*bbbbb*-212.0E-01 or *bbbbb*-212.0D-01
EN15.3	*bbbb*-21.200E+00
ES15.3	*bbbbb*-2.120E+01

Table 10-6 shows the effects of the different edit descriptors for values with different magnitudes. The SS edit descriptor is used to suppress the optional plus sign in the displayed values.

Table 10-6 Comparison between E, EN, and ES edit descriptors

Internal value	Possible output from SS, E12.3	Output from SS, EN12.3	Output from SS, ES12.3
6.421	*bbbb*.642E+01	*bbb*6.421E+00	*bbb*6.421E+00
–.5	*bbb*-.500E+00	-500.000E-03	*bb*-5.000E-01
.00217	*bbbb*.217E-02	*bbb*2.170E-03	*bbb*2.170E-03
4721.3	*bbbb*.472E+04	*bbb*4.721E+03	*bbb*4.721E+03

Note that other forms are allowed for the E12.3 edit descriptor. A processor is allowed to produce a leading zero before the decimal point and may independently omit the E from the exponent and produce a 3 digit exponent field.

10.5.2.3 Generalized Editing of Real Data

Gw.d[Ee] converts to or from a string using generalized editing. Except for IEEE infinity and NaN values, the form for generalized editing is determined by the magnitude of the value of the number.

Rules and restrictions:

On input:

1. The Gw.d[Ee] edit descriptor is the same as the Fw.d edit descriptor (10.5.2.1).

On output:

1. Let N be the magnitude of a number to be printed using a G edit descriptor. If N = 0 or is approximately between 0.1 and 10^d, Table 10-7 specifies the form of the output. If N is outside this range, output editing with the edit descriptor kPGw.d[Ee] is the same as that with kPEw.d[Ee]. A kP scale factor has no effect if F editing is selected.

Table 10-7 The form of the output using a G edit descriptor for a number of magnitude N

Magnitude of datum	Equivalent conversion
$N = 0$	F(w – n).(d – 1), n("b")
$0.1 - R \times 10^{-d-1} \leq N < 1 - R \times 10^{-d}$	F(w – n).(d), n("b")
$1 - R \times 10^{-d} \leq N < 10 - R \times 10^{-d+1}$	F(w – n).(d – 1), n("b")
$10 - R \times 10^{-d+1} \leq N < 100 - R \times 10^{-d+2}$	F(w – n).(d – 2), n("b")
.	
.	
.	
$10^{d-2} - R \times 10^{-2} \leq N < 10^{d-1} - R \times 10^{-1}$	F(w – n).1, n("b")
$10^{d-1} - R \times 10^{-1} \leq N < 10^{d} - R$	F(w – n).0, n("b")

1. Note that zero is a special case.
2. n blanks are produced to the right of the digits. n is 4 for Gw.d output and e+2 for Gw.dEe.
3. n must be chosen such that w – n will be positive.
4. R is a rounding term whose value depends on the rounding mode in effect.

Rounding mode: *R*

COMPATIBLE: 0.5
NEAREST: 0.5 if the higher value is even; –.5 if the lower value is even
UP: 1
DOWN: 0
ZERO: 1 if the internal value is negative; 0 if the internal value is positive

Although the rules and forms appear complicated, the simple result is that a G format descriptor will produce a value in a simple F form if the decimal symbol fits in the field and use an exponential form for other values.

Table 10-8 shows the result if 123.456 is printed using different G formats.

Table 10-8 The result of printing 123.456 with G edit descriptors

Format	Selected format	Possible output field
`G 12.1`	`E 12.1`	*bbbbb*`0.1E+03`
`G 12.2`	`E 12.2`	*bbbb*`0.12E+03`
`G 12.3`	`F 8.3, "`*bbbb*`"`	*bbbb*`123.`*bbbb*
`G 12.4`	`F 8.4, "`*bbbb*`"`	*bbb*`123.5`*bbbb*
`G 12.5`	`F 8.5, "`*bbbb*`"`	*bb*`123.46`*bbbb*
`G 12.6`	`F 8.6, "`*bbbb*`"`	*b*`123.456`*bbbb*
`G 12.7`	`F 8.7, "`*bbbb*`"`	`123.4560`*bbbb*
`G 12.8`	`E 12.8`	`************`

Examples of G output:

The statement

```
PRINT "(G10.3E1)", 8.76E1
```

produces the output:

bbb`87.6`*bbb*

because *n* is 3 (=1+2) and the format reduces to F7.1,"*bbb*".

The statement

```
PRINT "(G10.3)", 8.76E10
```

produces the output

b`0.876E+11`

because the value is too large to be produced with an F field, so E10.3 is used.

Note that the leading zeros, an initial plus sign, and the form of the exponent field in the examples above are optional with the processor.

10.5.2.4 IEEE Exceptional Values

On processors that support the IEEE floating-point standard, the exceptional values infinity and NaN are treated separately from ordinary numeric values. Denormalized values are also IEEE exceptional values; however, they are treated as ordinary values for input/output. This applies to all real and complex editing.

Rules and restrictions:

On input:

The input field consists of an optional series of blanks followed by either:

1. an optional sign, the characters INF or INFINITY, and optional trailing blanks, which denotes a positive or negative IEEE infinity value, or
2. an optional sign, the characters NAN optionally followed by a string of zero or more alphanumeric characters enclosed in parenthesis, optionally followed by trailing blanks which denotes an IEEE NaN value. The effect of the optional sign is processor dependent. If the input field is either NAN or NAN() the input value is a quiet NaN; otherwise it is a processor-dependent NaN value.

The interpretation of NAN, INF, and INFINITY does not depend on the case of the characters. The values may be used only on processors that support IEEE infinity or NaN for the kind of the associated input variable.

On output:

1. For IEEE infinity
   a. if $w \le 2$ (3 if a sign is required), the output field consists of asterisks
   b. if $3 \le w \le 7$ ($4 \le w \le 8$ if a sign is required), the output field consists of blanks as necessary, a minus sign if the value is a negative infinity or an optional plus sign and the characters Inf right justified in the field
   c. if $w \ge 8$ (9 if a sign is required), the output field consists of blanks as necessary, a minus sign if the value is a negative infinity; or an optional plus sign and the characters Infinity right justified in the field
2. For IEEE NaN

   If $w < 3$ the output field consists of asterisks. Otherwise, the output field consists of blanks as necessary followed by the characters NaN optionally followed by up to w–5 processor dependent characters enclosed in parentheses. Some processors can give additional information about the NaN in the parenthetical string.

10.5.3 Complex Editing

Complex editing converts complex values to or from strings of characters and follows the rules for real editing (10.5.2). Editing of complex numbers requires two real edit descriptors, the first one for the real part and the second one for the imaginary part. Different edit descriptors may be used for the two parts. Control and character string edit descriptors may be inserted between the edit descriptors for the real and imaginary parts.

Example:

For the statements

```
COMPLEX CM(2)
READ "(4E7.2)", CM(1:2)
```

if the input record is:

*bb*55511*bbb*2146*bbbb*100*bbbb*621

the values assigned to CM(1) and CM(2) are 555.11 + 21.46*i* and 1 + 6.21*i*, respectively.

10.6 Logical Editing

Logical editing converts logical values to or from strings of characters. The edit descriptors used for logical editing are:

```
L w
G w . d[ E e]
```

Rules and restrictions:

On both input and output:

1. w is the field width and must be greater than zero.
2. Generalized logical editing Gw.d[Ee] follows the rules for Lw editing.

On input:

1. The input field for a logical value consists of any number of blanks, followed by an optional period, followed by T or F (which may be in upper or lower case), for a true and false value respectively, followed by any other characters, which are ignored.

Example: Using the READ statement:

```
READ "(2L8)", L1, L2
```

to read the input record:

.TRUE.*bb*.FALSE.*b*

will cause L1 and L2 to have the values true and false, respectively. The result would be the same if the input record were:

TUESDAY*bbb*FRIDAY

On output:

1. The output field consists of w – 1 leading blanks, followed by T or F, for a true and false value, respectively, of the output item.

Example of logical output:

```
PRINT "(2L7)"), L1, L2
```

If L1 and L2 are true and false, respectively, the output record will be:

*bbbbbb*T *bbbbbb*F

10.7 Character Editing

Character editing converts character values to or from strings of characters. The edit descriptors for character editing are:

```
A [w]
G w . d[ E e]
```

Rules and restrictions:

On both input and output:

1. w is the field width measured in characters and must be greater than zero.
2. A Gw.d[Ee] general edit descriptor is the same as an Aw edit descriptor for character data.
3. If w is omitted, the length of the character data object being transferred is used as the field width.
4. If the character datum is of nondefault kind, the character used for "blank padding" is processor dependent.

On input:

1. If w is greater than or equal to the length *len* of the character datum read, *len* rightmost characters of the input field are read.
2. If w is less than the length *len* of the character data read, the w characters of the character datum will be read from the input field and placed left justified in the character variable followed by *len* – w trailing blanks.
3. All characters in the input field must be the same kind as the input list variable.

On output:

1. If w exceeds the length *len* of the character datum written, w – *len* blank padding characters are written followed by *len* characters of the character datum.
2. If w is less than or equal to the length *len* of the character data written, the w leftmost characters of the character datum will appear in the output field.
3. All characters in the output field will be the same kind as the output list item.

Note that an A edit descriptor is very similar to an I0 edit descriptor.

For files connected for stream access, character output may cause the output to be split into more than one record if the data contains a new line character. A new line character is the character returned by the intrinsic function NEW_LINE, unless that character is a blank, in which case the processor does not support new line record termination for that character kind. Each character in the output item is written to the file in sequence until a new line character is encountered. At that point, the current record is terminated, as if by a slash edit descriptor (10.9.2), and a new record begins. There may be several new line characters in an output item.

Example: Given the declaration:

```
Character (Len=11)  ::  Name1, Name2, Name3
```

and an input file that contains the lines

Ralph*b*Smith*bbbb*
Ralph*b*Smith*bbbb*
Ralph*b*Smith*bbbb*

the following READ statements

```
Read  "(A)",  Name1
Read  "(A5)"),  Name2
Read  "(A15)",  Name3
```

would assign the values

Ralph*b*Smith
Ralph*bbbbbb*
h*b*Smith*bbbb*

to Name1, Name2, and Name3, respectively.

Example:

The statements:

```
CHARACTER (LEN = 14), PARAMETER :: SLOGAN = "SAVE THE RIVER"
PRINT  "(A)",  SLOGAN
PRINT "(A4)",  SLOGAN
PRINT "(A20)",  SLOGAN
```

produce the output records, respectively:

SAVE*b*THE*b*RIVER
SAVE
*bbbbbb*SAVE*b*THE*b*RIVER

10.8 Defined Editing

Defined editing, called user-defined derived-type editing in the standard, allows user written subroutines to control the input/output formatting for derived types, rather than the intrinsic derived-type input/output formatting. The edit descriptor for defined editing is:

```
DT [ char-literal-constant ] [ ( v-list ) ]
```

Neither the character literal constant nor any of the integer literal constants in v-list may have a kind parameter.

User defined input/output subroutines are described in 9.5.1.4. The string

```
"DT"//char-literal-constant
```

is passed to the iotype argument and the v-list constants are passed as an array to the v_list array argument. The user defined input/output subroutine uses these values to control the formatting and processing of the values. A DT edit descriptor can be used only if the corresponding list item is of derived type and there is an appropriate derived-type input/output subroutine accessible.

Examples of defined editing:

The edit descriptor `DT'my_type'(3,5,7,9)` will invoke the appropriate subroutine with the value of iotype set to DTmy_type; v-list will be [3 5 7 9].

For the plain edit descriptor `DT`, iotype will be set to DT, and v-list will be the empty array [ ].

Note also that derived-type values, for some types, can be read or written using formatted input/output without using defined input/output subroutines, in which case it is as if the derived-type components were listed in the input/output list and appropriate edit descriptors were provided for each value.

10.9 Control Edit Descriptors

No data is transferred or converted with the control edit descriptors. Control edit descriptors affect skipping, tabbing, scale factors, rounding, decimal symbol appearance, and optional signs. These edit descriptors may affect how the data is input or output using the subsequently processed data edit descriptors in the format specification.

10.9.1 Position Editing

Position edit descriptors control relative tabbing left or right in the record before the next list item is processed. The edit descriptors for tabbing are:

`T n`	tab to position n
`TL n`	tab left n positions
`TR n`	tab right n positions
`n X`	tab right n positions

The tabbing operations to the left are limited by a position called the **left tabbing limit**. This position is normally the first position of the current record or the current position in a stream file, but, if the previous operation on the file was a nonadvancing formatted data transfer, the left tabbing limit is the current position within the record before the current data transfer begins. If the file is positioned to another record during the data transfer, the left tabbing limit changes to the first position of the new record.

The Tn edit descriptor positions the record just before the character in absolute position n relative to the left tabbing limit. TRn and nX are equivalent and move right n characters from the current position. TLn moves left n characters from the current position, but is limited by the left tabbing limit. However, in a child data transfer statement, it is the programmer's responsibility to make sure that a TLn does not cause the record position to move to the left of the position at the time the child transfer started. A child data transfer cannot reprocess characters prior to where it started, whereas a parent data transfer can use TLn to reprocess characters.

Rules and restrictions:

On both input and output:

1. n must be a positive integer constant with no kind parameter.
2. Left tabbing is always limited so that even if left tabbing specifies a position to the left of the left tabbing limit, the record position is set to the left tabbing limit in the record.
3. The left tabbing limit in the record is determined by the position in the record before any data transfer begins for a particular data transfer statement.
4. If a file is positioned to another record during a particular data transfer statement, the left tabbing limit is the first position of the new record.
5. For internal files or external files that are not Unicode files, if any character that is skipped over by position editing is a nondefault character type, the position is processor dependent.

On input:

1. Moving to a position left of the current position allows input to be processed twice.
2. The X and Tr descriptors always move the position to the right and skip characters.
3. The file may be positioned arbitrarily far past the last character in the record; however, no characters may be transferred from this position, so it is not very useful.

On output:

1. The positioning does not transmit characters, and does not by itself cause the record to be shorter or longer.

2. Positions that are skipped and have not been filled previously behave as if they are blank filled if data are written subsequently.

3. Positions previously filled may be replaced with new characters, but are not blank filled if they are skipped subsequently using any of the position edit descriptors.

Example: If DISTANCE and VELOCITY have the values 12.66 and –8654.123,

```
      PRINT 100, DISTANCE, VELOCITY
  100 FORMAT (F9.2, 6X, F9.3)
```

produces the record:

*bbbb*12.66*bbbbbb*-8654.123

and

```
      PRINT 100, DISTANCE, VELOCITY
  100 FORMAT (F9.2, T7, F9.3)
```

produces the record:

*bbbb*12-8654.123

because T7 specifies the first position for VELOCITY as the seventh character in the record.

10.9.2 Slash Editing

The slash edit descriptor consists of the single slash character (/), optionally preceded by a repeat count. Data transfer to or from the current record is ended when a slash is encountered in a format specification. Multiple slashes or a slash with a repeat count skip multiple input records or create multiple output records.

Rules and restrictions:

On input:

1. If the file is connected for sequential or stream access, the file is positioned at the beginning of the next record. The effect is to skip the remainder of the current record.

2. For direct access, the record number is increased by one, and the file is positioned at the beginning of the record with this increased record number, if it exists; it becomes the current record.

3. A record may be skipped entirely on input.

On output:

1. If the file is connected for sequential or stream access, a new empty record is created after the current record, and the file is positioned at the beginning of the new record. In completing the previous record, no additional characters will be written to the file.

2. For direct access, the current record is blank filled, the record number is increased by one, and this record becomes the current record.

3. For an internal file, the current record is blank filled, and the file is positioned at the beginning of the next array element.

Example: If ALTER, POSITION, and CHANGE have the values 1.1, 2.2, and 3.3, respectively,

```
PRINT "(F5.1, /, 2F6.1)", ALTER, POSITION, CHANGE
```

produces two records:

```
bb1.1
bbb2.2bbb3.3
```

10.9.3 Colon Editing

The colon edit descriptor consists of the character colon (:). If the list of effective items in a formatted READ or WRITE statement is exhausted, a colon stops format processing at that point. If the list is not exhausted, the colon edit descriptor has no effect.

Example: If ALTER, POSITION, CHANGE, and DELTA have the values 1.1, 2.2, 3.3, and 4.4, respectively,

```
PRINT 100, ALTER, POSITION, CHANGE
100 FORMAT (10(F5.1, :, ","))
```

produces:

```
bb1.1,bb2.2,bb3.3
```

and

```
PRINT 100, ALTER, POSITION, CHANGE, DELTA
```

produces:

```
bb1.1,bb2.2,bb3.3,bb4.4
```

10.9.4 Sign Editing

Sign editing applies to the output data transfer of positive numeric values only. It controls the writing of the optional plus sign when an I, F, E, EN, ES, D, or G edit descriptor is used. A sign edit descriptor overrides the sign mode (9.2.4) set by the data transfer statement or the OPEN statement. The sign edit descriptors are listed in Table 10-9.

Table 10-9 Sign edit descriptors

Edit descriptor	Sign mode	Effect
S	Processor defined	Appearance of the optional plus sign is processor dependent.
SP	Plus	The optional plus sign will not appear.
SS	Suppress	The optional plus sign will appear.

Rules and restrictions:

1. The descriptors have effect until another sign edit descriptor is encountered in the format specification or execution of the input/output statement terminates.
2. The descriptors have no effect during formatted input data transfers.

Example: If SPEED(1) and SPEED(2) are 1.46 and 2.3412 respectively,

```
      PRINT  110,  SPEED (1:2)
  110 FORMAT (SP, 2F10.2)
```

produces the record:

```
bbbbb+1.46bbb+234.12
```

10.9.5 Scale Factors

The kP edit descriptor indicates scaling, where the scale factor k is a signed integer literal constant with no kind parameter.

The scale factor is zero at the beginning of a formatted input/output statement. When a kP descriptor occurs, the scale factor becomes k, and all succeeding numeric fields processed with an F, E, EN, ES, D, or G edit descriptor will use this scale factor until another kP edit descriptor occurs. Note that a comma between a kP descriptor and a following real edit descriptor is optional. Both 1P4E10.0 and 1P,4E10.0 are acceptable and have the same meaning.

The scale factor behaves the same way as the changeable mode values (9.2.4) with respect to nonadvancing and defined input/output.

Rules and restrictions:

On input:

1. The scale factor has no effect if the input field has an exponent.
2. If the input field has no exponent, the internal value is the external value divided by the scale factor 10^k. This is often the inverse of what is expected.

3. If UP, DOWN, ZERO, or NEAREST rounding mode is in effect, the scale factor is applied to the external value and then this value is rounded.

On output:

1. For the F edit descriptor, the external number equals the internal number multiplied by the scale factor 10^k.
2. For the E and D edit descriptors, the nonexponent part (significand) of the number appearing in the output is multiplied by 10^k and the exponent is reduced by k.
3. For the G edit descriptor, the output value is not affected by the scale factor if the number will print correctly with the appropriate F edit descriptor as described in Table 10-7. Otherwise, the scale factor for the G edit descriptor has the same effect as for the E edit descriptor.
4. EN and ES edit descriptors are not affected by a scale factor.
5. If UP, DOWN, ZERO, or NEAREST rounding mode is in effect, the internal value is converted using the rounding mode and the scale factor is applied to the rounded decimal result.

Example: If TREE has the value 12.96:

```
PRINT "(2PG10.1, TR3, F7.2)", TREE, TREE
```

produces:

```
bbb13.E+00bbb1296.00
```

Scale factors were traditionally used on input from punched cards to allow values with large exponents to be represented in as few columns as possible and on output to simulate the effects of the ES edit descriptor. Because the scale factor actually changes the input or output value with F editing, they should be used with care. This is often a problem with formats that mix E and F edit descriptors, such as: (1PE20.10, F10.0) or (F12.0, 1P, E20.10). In the first case, the 1P carries over to the F10.0. In the second, the 1P will carry back to the F12.0 if format reversion occurs and the format will be interpreted as (F12.0, 1PE20.10,/,1PF12.0…).

Examples of scaling:

If the external value 1.2345 is read with an F6.3 descriptor with various scale factors, the internal values will be:

-2P	123.45
-1P	12.345
0P	1.2345
1P	0.12345
2P	.012345

If a variable with the value 1.2345 is printed using F10.3 and E9.3 descriptors with various scale factors, the external values will be:

	F10.3	E9.3
-2P	0.012	0.001E+03
-1P	0.123	0.012E+02
0P	1.235	0.123E+01
1P	12.345	1.235E+00
2P	123.450	12.35E-01
3P	1234.500	123.5E-02
4P	12345.000	1235.E-03

10.9.6 Blanks in Numeric Fields

Blanks other than leading blanks may be ignored or interpreted as zero characters in numeric input fields as determined by the blank edit descriptors:

BN change the blank interpretation mode to NULL and treat nonleading blanks in numeric input fields as nonexistent

BZ change the blank interpretation mode to ZERO and treat nonleading blanks in numeric input fields as zeros

The interpretation is for input fields only when the field is processed using an I, B, O, Z, F, E, EN, ES, D, or G edit descriptor; output fields are not affected. The blank interpretation mode (9.2.4) from the data transfer statement or the OPEN statement is used if no BN or BZ descriptor is used.

Rules and restrictions:

On input:

1. If the blank interpretation mode is NULL, nonleading blanks are ignored and treated as if they were not in the input field, although they are counted in determining input field widths.
2. A field consisting of only blanks is given the value zero, regardless of the blank interpretation mode.
3. If the blank interpretation mode is ZERO, nonleading blanks are interpreted as zeros in succeeding numeric fields.
4. The BN and BZ edit descriptors remain in effect until a subsequent BN or BZ edit descriptor is encountered within the format or execution of the statement terminates.

Examples of blank interpretation:

```
    READ 100, N1, N2
100 FORMAT (BN, I5, BZ, I5)
```

If the input record is:

*b*9*b*9*b*8*b*8*b*8

the values assigned to N1 and N2 are 99 and 80808, respectively.

However, if the input record is

*bbbbbbb*5

the statement

```
READ ( 47, "(F8.4)" ) X
```

will assign the value 0.0005 to X regardless of the blank interpretation mode. When the blank interpretation mode is NULL, it is as if all of the non-blank characters were shifted to the right in the field before editing takes place. Once the conceptual shift has taken place, the normal editing rules apply.

10.9.7 Round Edit Descriptors

The round edit descriptors control the rounding mode (9.2.4) used for real and complex values with the D, E, EN, ES, F, and G edit descriptors. The rounding mode from the data transfer statement or OPEN statement is used if no round edit descriptors appear. The descriptors are shown in Table 10-10.

Table 10-10 Round edit descriptors

Edit descriptor	ROUND mode
RC	Compatible
RD	Down
RN	Nearest
RP	Processor defined
RU	Up
RZ	Zero

The round edit mode remains in effect until a subsequent round edit descriptor is encountered in the format or execution of the statement terminates. The effect of rounding is described in 9.2.4.

10.9.8 Decimal Edit Descriptors

The decimal edit descriptors control the decimal edit mode (9.2.4) used during the conversion of real and complex values with the D, E, EN, ES, F, or G edit descriptors. If there are no decimal edit descriptors, the decimal edit mode from the data transfer statement or the OPEN statement is used. The descriptors are shown in Table 10-11.

Table 10-11 Decimal edit descriptors

Edit descriptor	DECIMAL mode	
DC	COMMA	Use comma for the decimal symbol
DP	POINT	Use period for the decimal symbol

The decimal edit mode remains in effect until a subsequent decimal edit descriptor is encountered or execution of the statement terminates.

If the decimal edit mode is COMMA, complex values will use a semicolon (;) as the separator between the real and imaginary parts and list-directed and namelist input/output will also use a semi-colon as a separator rather than a comma. This is referred to as a **CS symbol**.

10.9.9 Character String Edit Descriptors

Character string edit descriptors are used to transfer characters to an output record. The character string edit descriptors must not be used on input. The character string edit descriptors use apostrophe and quote as delimiters, and are respectively:

```
' characters '
" characters "
```

Rules and restrictions:

On output:

1. The apostrophe and quote edit descriptors have the form of literal character constants with no kind parameter. The constants are placed in the output.

2. To print a quote in the output field when a quote is the delimiting character, use two consecutive quotes; to print an apostrophe in the output field when an apostrophe is the delimiting character, use two consecutive apostrophes.

3. The field width is the length of the character constant, but does not include the extra character for each pair of doubled apostrophes or quotes.

Example: If TEMP has the value 32.120001,

```
    PRINT 120, TEMP
120 FORMAT (' TEMPERATURE = ', F13.6)
```

produces the record:

```
bTEMPERATUREb=bbbbb32.120001
```

10.10 List-Directed Formatting

List-directed formatting is one of the implicit formatting methods in Fortran; list-directed formatting may occur with files connected for sequential or stream access and with internal files, but not for direct-access files. The input/output data transfer must be advancing and there cannot be an ADVANCE specifier, even with the value YES. Almost any form of a value that would be assignment compatible with the variable is suitable for list-directed input. The processor has great freedom to choose an output format and there is no programmer control over these choices.

The input/output statement uses an asterisk (*) instead of an explicit format specification. The editing is based on the type of the list item. The input and output data are free field, without the rigid placement of explicit formatting.

The rules in 9.5.1 for formatted input/output also apply to list-directed input/output.

Examples of list-directed input/output:

```
READ (input_unit, *) HOT, COLD, WARM

Print *, HOT, COLD, WARM
```

Rules and restrictions:

On both input and output:

1. Execution of a list-directed data transfer statement terminates when the input or output item list is exhausted or an error condition is encountered.

2. List-directed data consist of values and value separators.

3. As with explicit formatting, the CS symbol and decimal symbol depend on the decimal mode (9.2.4) in effect.

4. A derived-type variable is processed by defined input/output subroutines if an appropriate one is accessible; otherwise, it is expanded into a sequence of components.

5. List-directed formatting must not be specified for direct access or nonadvancing sequential-access data transfer.

6. If there are no effective list items and there is no current record, an input record is skipped or an output record with one blank is written. If there are no effective list items and there is a current record, the current record is skipped (the file is positioned at the end of the current record) or the current record is terminated at the current position. (Recall that a current record exists only if the previous input/output data transfer to the unit was nonadvancing.)

10.10.1 List-Directed Input

Input values are generally accepted as list-directed input if they are the same as those allowed for explicit formatting with an edit descriptor for a variable of the type of the list item. A series of values is separated by value separators. A series of identical values may be represented by a repeat count and a single value.

10.10.1.1 Value Separators

A value separator allowed in list-directed input is one of:

- the CS symbol, optionally preceded or followed by contiguous blanks
- one or more contiguous blanks between two values or following the last value
- a slash, optionally preceded or followed by contiguous blanks, which terminates the input

This allows a great deal of flexibility in data layout. Spaces can be used to improve human readability, or commas (or semicolons) can be used if that seems more natural. For example, the following two lines are equivalent for list-directed input.

```
2 4 6 8, who do we appreciate?     tommy, that's who
  2,  4,  6,  8,  who,  do,  we,  appreciate?,  tommy,  that's,  who
```

A list-directed input statement would specify four integer or real variables, four character variables, two variables that could be either character or logical, and a character variable. Note that the apostrophe in "that's" is not the start of a character value because it is not preceded by a value separator.

10.10.1.2 Values

The values allowed in list-directed input data are:

c	a data value, including a undelimited character string
null	a null value, specified by two consecutive CS separators (e.g., ,, or ;;)
*r***c*	*r* repetitions of the value *c*
*r**	*r* repetitions of the null value

where *r*, the repeat factor, is a nonzero digit string. Neither *r* nor *c* may have a kind parameter. *c*, if numeric, may be signed.

The type of the input variable determines how the value is interpreted. For example, if an input line is

```
17
```

then the input variable could be of type integer, real, or character. If the line contains

```
TRUE
```

then the input variable could be of type logical or character.

Rules and restrictions:

1. Execution of a list-directed input statement terminates when either an end-of-file is encountered or a slash (/) is encountered as a value separator.
2. Binary, octal, and hexadecimal values must not appear in list-directed input data.
3. Blanks are never interpreted as zeros.
4. If the input variable is of type integer, the value must be of a form suitable for the I edit descriptor.
5. If the input variable is of type real, the value must be of a form suitable for the F edit descriptor. If no decimal symbol appears in the value, the value has no fractional digits specified for it.
6. If the input variable is of type complex, the input must be a left parenthesis followed by the real and imaginary values, separated by a CS symbol, followed by a right parenthesis. Any number of blanks and end of records may be intermixed between the parts of the complex representation. Values of type complex include the real-imaginary separator and the parentheses.
7. If the variable is of type logical, the value must be of a form suitable for the L edit descriptor. Logical values must not have value separators in the optional characters following the T or F. That is, TTOO is allowed for the value true; T,TOO is not.
8. Embedded blanks are not allowed within a value, except within a delimited character value or within a complex value. Blanks may occur before or after the CS symbol, and before or after the parentheses.

   Examples are:

   ```
   "NICE DAY"
   (1.2, 5.666 )
   ```

9. An end of record must not occur within a value, except a complex value or a delimited character string: for a complex value, the end of record may occur between the real part and the CS symbol, or between the CS symbol and the imaginary part; for a character string, the end of record may occur anywhere in the string except between any consecutive (doubled) quotes or apostrophes in the string. The end of record does not cause a blank or any other character to become part of the character value. A complex or character value may be continued on as many records as needed.
10. The end of a record has the same effect as a blank, unless it occurs within a delimited character literal value.

11. Value separators may appear in any delimited default character constant. They are, however, not interpreted as value separators, but are characters in the delimited character constant.

12. If *len* is the length of the corresponding input list variable, and *w* is the number of effective characters in the character value, then if:

$len \leq w$	the leftmost *len* characters of the value are used
$len > w$	the *w* characters of the value are left justified in the input variable and the variable is blank filled on the right

For example, consider the code:

```
CHARACTER (2) NAME (2)
  . . .
READ *, NAME
```

where the input record is:

```
JONES, SMITH
```

After the READ statement, the values in NAME are JO and SM, because *len* (=2) is less than *w* (=5). Note that JONES is the first undelimited character constant in the list and is completely read before the second constant is read.

10.10.1.3 Undelimited Character Strings

In certain cases, delimiters are not required for character values on input. However, undelimited character strings impose certain requirements in order to distinguish them from other values. These requirements are:

1. The corresponding input variable must be of type default, ASCII, or ISO_10646 character.
2. The character string must contain at least one non-blank character and must not contain a blank, CS symbol, or slash (because they are value separators).
3. The character string must not be continued across a record boundary.
4. The first nonblank character must be neither an apostrophe nor a quote.
5. The leading characters must not be a string of digits followed immediately by an asterisk.

If all of these conditions are met, the character value represented by the character string is terminated by the first value separator or end of record. Apostrophes and quotes are not doubled in such a value. If any of these conditions are not met, the value must be delimited with either apostrophes or quotes and value separators may appear within the string.

10.10.1.4 Null Values

A null value is used to specify no change to the variable in the input list. A null value has one of the forms:

1. no value between separators, such as 10,,20
2. a nonblank value separator as the first entity in the record; for example, a record beginning with a CS symbol as the first nonblank character represents a null value, as in ,4.56. Note that leading blanks on the first line do not represent a null value unless they are followed by a CS symbol.
3. *r** followed by a value separator as in:

   ```
   7*,'TODAY'
   ```

   which represents 7 null values, followed by a single value TODAY.

Rules and restrictions:

1. An end of record does not represent a null value.
2. The null value does not affect the definition status or value of the corresponding list item.
3. For a complex variable, the entire value may be null, but not just one of the parts.
4. If a slash terminates input, the remaining characters in the record are ignored, and the remaining variables are treated as though null values had been read. This applies to any remaining items in an implied-do or to any remaining elements of an array, as well as to scalars.

Example of null input:

```
REAL AVERAGE (2)
READ (5, *)  NUMBER, AVERAGE
```

If the input record is:

```
b6,,2.418
```

the result is that NUMBER = 6, AVERAGE (1) is unchanged, and AVERAGE (2) = 2.418.

10.10.1.5 Repeated Values

If the *r*c* or *r** form appears in the input, it is as if the value *c* or the null value were repeated *r* times. The associated variables do not all have to be of the same type. For example, if the input record contains

```
2*17
```

the corresponding variables could be integer and real, because 17 is an allowable input form for both I and F editing. However, 17 is also an acceptable undelimited character string, so the corresponding input variables could also be of type character.

To make implementation easier and to avoid potential ambiguities when reading repeated undelimited character strings, the standard disallows mixing of character and noncharacter variables when reading a repeated value. That is, the variables corresponding to an r*c form must either all be of type character or must all be noncharacter. Thus, in the example above the data could not be read into a character and integer variable, nor into an integer and character variable.

10.10.2 List-Directed Output

List-directed output uses conventions similar to those used for list-directed input. The processor selects suitable formatting for items in the output list based on their type and value. The processor has great freedom in producing the output and there is no way to force the processor to use a specific form or style. List-directed output is often suitable for use as list-directed input; exceptions involve undelimited character strings and defined editing.

Rules and restrictions:

1. One or more blanks or a CS symbol, optionally preceded or followed by one or more blanks is used as a value separator.

2. The processor begins new records, as needed, at any point in the list of output items. A new record will not begin in the middle of a value, except as noted below for complex and character values. Each new record begins with a blank, except for delimited character values.

3. Slashes are not used as terminators.

4. The processor has the option of using the repeat factor *r*c* for two or more consecutive values that are identical.

5. The processor chooses reasonable widths and precisions for the output fields. There is no way for the programmer to control the output forms chosen.

Specific types are processed as follows:

Integer. The effect is as though an Iw edit descriptor were used.

Real. The effect is as though an 0PFw.d or an 1PEw.dEe edit descriptor were used. Which edit descriptor is chosen by the compiler depends on the magnitude of the number written and is processor dependent except that a value of zero will be formatted with an F edit descriptor.

Complex. The real and imaginary parts are enclosed in parentheses and separated by a CS symbol. If the length of the complex number is longer than an entire record, the processor may divide the real and imaginary parts between two consecutive records with the real part in the first record. The only blanks in a complex value will be an optional blank between the separator and the end-of-record and a required blank as the first character of the next record.

Logical. List-directed output produces T or F.

Character. The form of the output for character values depends on the value of the DELIM mode in effect (9.2.4).

1. If the value of the DELIM mode is NONE:
   a. Character values are not delimited.
   b. Character values are not surrounded by value separators.
   c. Only one quote or apostrophe is produced for each quote or apostrophe in the string transferred.
   d. A blank is inserted at the beginning of new records for a continued character value.
2. If the DELIM mode is QUOTE or APOSTROPHE:
   a. Character values are delimited with the specified delimiter quote or apostrophe.
   b. All values are surrounded by value separators.
   c. A character that is the same as the specified delimiter is doubled when written to the output record.
   d. No blank is inserted at the beginning of a continued record for carriage control in the case of a character value continued between records.

If the DELIM mode is NONE and if a character value contains blanks, CS symbols or slashes, the resulting output cannot be read using list-directed input because those characters will act as value separators.

Derived type. If an appropriate defined input/output subroutine is accessible, it is used; otherwise, the variable is expanded into a sequence of intrinsic values. Depending on the action of the defined output subroutine, the results may not be usable for list-directed input.

Example of list-directed output:

```
REAL                     :: TEMPERATURE = -7.6
INTEGER                  :: COUNT = 3
CHARACTER(*), PARAMETER :: PHRASE = "This isn't so"

OPEN( 10, DELIM = 'NONE' )
WRITE( 10, * )  TEMPERATURE, COUNT, PHRASE
```

The output record on unit 10 could be:

```
bb-7.6bbbb3bbbThis isn't so
```

where the actual spacing and representation of the -7.6 is processor dependent and might be -7.59999997 or -7.6000001.

10.11 Namelist Formatting

Namelist input/output uses a group name for a list of variables that are transferred. Before the group name can be used in the transfer, the list of variables must be declared in a NAMELIST statement (5.13). Using the namelist group name eliminates the need to specify the list of input/output items in the namelist data transfer statement. Name-list data transfer statements process the variables in the group as a set of name=value pairs.

Namelist input/output is convenient for initializing the same variables with different values in successive runs or for changing the values of a few variables among a large list of variables that are given default initial values. The formatting of the input or output record is not specified in the program; it is determined by the contents of the record itself or the items in the namelist group. Conversion to and from characters is implicit for each variable in the list. As with list-directed input, essentially any input form that would be assignment compatible with the variable is allowed. For output, the processor chooses the output forms and there is no programmer control.

During namelist data transfer, data are transferred with editing between the file and the entities specified by the namelist group name. The current record and possibly additional records are read or written.

When a namelist input/output statement is executed, every allocatable object in the group must be allocated and every pointer in the group must be associated with a target. But, subobjects of derived types that are processed by defined input/output do not need to be allocated or associated. If a namelist object is polymorphic or has an ultimate component that is either allocatable or a pointer, the object must be transferred via a defined input/output subroutine.

Except as noted below, the rules in 9.5.1 for formatted input/output also apply to namelist input/output.

Examples of namelist data transfer statements are:

```
READ(NML=NAME_LIST_23,IOSTAT=KN,UNIT=5)
WRITE(6,NAME_LIST_23,ERR=99)
```

Examples below use the namelist group MEETING defined as

```
NAMELIST /MEETING/  Joe,  Jake,  Jane
```

Namelist input and output consists primarily of an ampersand (&) immediately followed by a namelist group name followed by a sequence of name=value pairs followed by a slash (/).

Rules and restrictions:

1. Namelist input/output may span multiple records.

2. Execution of a namelist data transfer statement terminates when an error condition is encountered.

3. Namelist input/output may can be used with nonadvancing, asynchronous, and stream input/output.

10.11.1 Namelist Input

Namelist input may start with a series of namelist comments and blanks before the &group-name. Comments may also appear on the end of each line.

Rules and restrictions:

1. Execution of a namelist input statement terminates when an end-of-file is encountered, or a slash (/) is encountered as a value separator.

2. If any entities in the namelist group name have not been assigned a value when execution of the statement terminates, it is as if they had been assigned null values.

3. Blanks may precede the ampersand or the slash.

4. Each name, or parent name, if it a subobject designator, appearing in a name=value pair must be in the namelist group.

5. If an entity appears more than once within the input record for a namelist input data transfer, the last value is the one that is used.

6. A lowercase letter is the same as an uppercase letter and vice versa when used in the group name or object name.

   An example of namelist input is:

   ```
   READ (*, NML = MEETING)
   ```

The input might be:

```
&MEETING JAKE = 3500, JOE = 100, ! Jane on the next line
JANE = 0/
```

10.11.1.1 Names in Name=Value Pairs

There are rules and restrictions for the names used in the name=value pairs in namelist input.

Rules and restrictions:

1. The name=value pairs may appear in any order in the input.

2. The name=value pairs are evaluated serially, in left-to-right order.

3. During input, a name in the namelist group may be omitted in which case neither the value nor definition status of the object is changed.

4. A name in the namelist group may be repeated, in which case the value of the last name=value pair is used.

5. Each name must correspond with a name in the designated namelist group; a component name, if any, must also be the name of a component of the structure named in the namelist group.

6. When the name in the name=value pair is a subobject designator, it must not be a zero-sized array, zero-sized array section, or a zero-length character string, nor can it have a vector-valued subscript.

7. Optionally-signed integer literal constants with no kind parameters must be used in all expressions that appear in subscripts, section designators, or substring designators.

8. The name of a structure or a subobject designator may be the name in a name=value pair.

9. A name must not contain embedded blanks. A name in the name=value pair may be preceded or followed by one or more blanks.

10.11.1.2 Values in Name=Value Pairs

The value in a name=value pair must be in a form acceptable for a format specification for the type of the name, except for restrictions noted below.

Each value is a null value or one of these forms:

c	a data value (optionally signed if it is integer or real)
*r***c*	*r* repetitions of the value *c*
*r**	*r* repetitions of the null value

where *r* is a repeat factor and is a nonzero digit string. Neither *r* nor *c* may have a kind value. The value of *c* is interpreted as if it had the same kind value as the associated variable has.

Null values have the forms:

1. no value between value separators

2. no value between the equal sign and the first value separator

3. the *r** form, followed by one or more blanks.

Null values do not change the value of the named variable or its definition status. An entire complex value may be null; neither of the parts can be. The end of a record following a value separator does not specify a null value.

Rules and restrictions:

1. The form of a value must be acceptable to a format specification for an entity of the type of the corresponding variable, except as noted below; for example, the value *c* corresponding to a variable of type real can be of the forms:

```
1
1.0
1.0E0
```

   but cannot be of the forms:

```
(1.0,0.0)
A0
1.0EN-3
2.2_QUAD
1.0_2
```

2. Blanks are never zero, and embedded blanks must not appear in numeric or logical values. The exception is that a blank may appear as a character in a character value, or preceding or following the real or imaginary parts of a complex value. Blanks may appear before or after the equal sign.

3. The number of values following the equals sign must not be larger than the number of elements of the array when the name in the name=value pair is an array, or must not be larger than the ultimate number of components when the name in the name=value pair is that of a structure. Any array or component that is an array is filled in array element order.

   For the example:

```
TYPE PERSON
   INTEGER LEN
   CHARACTER (10) NAME
END TYPE PERSON
TYPE (PERSON) PRESIDENT, VICE_PRES
NAMELIST/PERSON_LIST/PRESIDENT, VICE_PRES
READ (input_unit, NML = PERSON_LIST)
```

   the input might be:

```
&PERSON_LIST PRESIDENT%LEN=4,
             PRESIDENT%NAME="LAMB",
             VICE_PRES%LEN=6,
             VICE_PRES%NAME="MARTIN"/
```

4. If there are fewer values in the expanded sequence than array elements or structure components, null values are supplied for the missing values.

5. If a slash occurs in the input (except within a character value or a namelist comment), it is as if null values were supplied for the remaining variables, and the

namelist input data transfer is terminated. Any remaining characters after the slash within the current record are ignored.

6. An integer value is interpreted as if the data edit descriptor were Iw for a suitable value of w.
7. Binary, octal, and hexadecimal values cannot appear.
8. A complex value consists of a pair of parentheses surrounding the real and imaginary parts, separated by a CS symbol. Blanks may appear before and after these values. The end of record may occur after the real part, or before the imaginary part.
9. A logical value must not contain slashes, commas, semicolons, equals, or blanks as part of the optional characters after the .T, .F, T, or F.
10. A character value may contain slashes, commas, semicolons, or blanks as part of the string. Within the string, apostrophes or quotes must be doubled if they are the same as the delimiting marks. A character value may be continued on as many records as necessary until the matching delimiter is reached. The end-of-record in a continued value has no effect on the value. If the value does not contain the same number of characters as the variable, it is as if the value were assigned to the variable using normal assignment. On namelist input, the DELIM mode is ignored; character input must be delimited.

10.11.1.3 Separators

The name=value pairs and values in an array list are separated by value separators that are of the form:

- a CS symbol, optionally preceded or followed by contiguous blanks
- a slash, optionally preceded or followed by contiguous blanks
- one or more contiguous blanks

Blanks on either side of the = in a name=value are allowed and have no effect.

10.11.1.4 Namelist Comments

A namelist comment may appear as the first nonblank character following a value separator. It begins with an exclamation mark (!); note that an exclamation mark in a character value does not begin a comment. A comment extends to the end of the namelist input record. A slash within the comment does not end the input record; the comment itself ends the input record. A namelist comment may start in the first nonblank position of a namelist input record; in this case, the input record consists only of the comment. Namelist comments are ignored.

10.11.1.5 Blanks

Blanks are part of the value separator unless they appear

1. in a character value
2. before or after the parts of a complex value
3. before an equal sign
4. after an equal sign, unless they are followed immediately by a CS symbol or slash
5. before the ampersand indicating the namelist group name and after the namelist group name

10.11.1.6 Use of Namelist Input

Namelist input requires the namelist group name, preceded by an ampersand, to be on the first nonblank record read by the namelist READ statement unless the record is a namelist comment record.

Example of namelist input:

```
REAL A (3), B (3)
CHARACTER (LEN = 3) CHAR
COMPLEX X
LOGICAL LL
NAMELIST / TOKEN / I, A, CHAR, X, LL, B
READ (*, NML = TOKEN)
```

If the input records are:

```
&TOKEN A(1:2) = 2*1.0  CHAR = "NOP" B = ,3.13,,
       X = (2.4,0.0)  LL = T /
```

results of the READ statement are:

Name	Value
I	Unchanged
A (1)	1.0
A (2)	1.0
A (3)	Unchanged
B (1)	Unchanged
B (2)	3.13
B (3)	Unchanged

Name	Value
CHAR	"NOP"
X	(2.4, 0.0)
LL	True

10.11.2 Namelist Output

With a few exceptions, the form of namelist output is the same as that required for namelist input. The names in the output will be in uppercase. The processor has the freedom to select reasonable output forms. With the exception of undelimited character values, value separators may be blanks, CS symbols, or a combination of blanks and CS symbols. A new record may begin anywhere, except within a name or value, unless the value is a character string or a complex value; a record may begin anywhere within a character string, or may begin before or after the CS symbol, or left or right parenthesis of a complex value. A blank may occur anywhere, except in a name or a noncharacter value. The only blanks that may occur in a character value are those that are in the character string; no additional blanks may be added.

10.11.2.1 Form of Namelist Output

A number of rules, similar to those for list-directed formatting, apply for namelist output.

Rules and restrictions:

1. Namelist output consists of a series of records. The first nonblank record begins with an ampersand, followed by the namelist group name, followed by a sequence of name=value pairs, one pair for each variable name in the namelist group object list of the NAMELIST statement and ends with a slash. Execution of the statement terminates when the namelist group object list is exhausted.

2. A logical value is either T or F.

3. An integer value is one that would be produced by an Iw edit descriptor using a suitable value of w..

4. For real output, the rules for list-directed output are followed using reasonable values for the w, e, and d that appear in real data edit descriptors and are appropriate for the output value.

5. The parts of a complex value are enclosed in parentheses and are separated by a CS symbol. An end of record may occur after the CS symbol only if the value is longer than the entire record. Blanks may be embedded after the CS symbol and before the end of the record.

6. Character values follow the rules in 10.10.2 for list-directed output. As with list-directed output, undelimited namelist output might not be usable as namelist input.

7. Repeat factors of the form $r*c$ are allowed, but not required, on output for successive identical values.

8. An output record will not contain null values.

9. Each record begins with a blank unless it is a continuation of a delimited character string.

10. No values are written for zero-sized arrays.

11. For zero-length character strings, the name is written, followed by an equal sign, followed by a zero-length character string, and followed by a value separator or slash.

12. If defined output is involved, the output form is not specified and might not be usable as namelist input.

Example:

```
NAMELIST / CALC / DEPTH, PRESSURE
DIMENSION DEPTH (3), PRESSURE (3)
WRITE (output_unit, NML = CALC)

&CALC DEPTH(1) = 1.2, DEPTH(2) = 2.2, DEPTH(3) = 3.2,
      PRESSURE = 3.0, 3.1, 3.2 /
```

10.11.2.2 DELIM Specifier for Character Values

Character values are delimited with the specified delimiter, either quote or apostrophe.

Example: for LEFT and RIGHT with values SOUTH and NORTH, the program with the DELIM specifier of QUOTE for unit 10:

```
CHARACTER (5) LEFT, RIGHT
NAMELIST / TURN / LEFT, RIGHT
WRITE (10, NML = TURN)
```

produces the output record:

```
&TURN  LEFT = "SOUTH",  RIGHT = "NORTH" /
```

Note that if the DELIM specifier is NONE in an OPEN statement, namelist output will not be usable as namelist input if character values are transferred, because namelist input of character values requires delimited character values.

11 Program Units

- A **Program Unit** is a sequence of Fortran statements; it may be a main program, a module, an external subprogram, or a block data program unit.
- A **Program** consists of program units and possibly program elements written in languages other than Fortran. There must be one main program.
- A **Main Program** contains the first construct that is executed; it may or may not be a Fortran main program.
- A **Module** is a nonexecutable program unit used to collect related declarations, type definitions, procedure definitions, and procedure interfaces.
- A **USE statement** accesses this information outside the module.
- A **Subprogram** defines a procedure, which may be either a function or a subroutine. A subprogram can be an external program unit or can be contained within a program unit.
- A **Block Data** is used to initialize variables in named common blocks.

A program unit is a main program, module, external subprogram, or block data. This chapter describes each of these and also introduces subprograms, which can be program units or can be contained within program units. It also describes the closely related concept of use association.

Procedures are discussed in 12.

Chapter 15 explains the interoperation of Fortran with other languages.

11.1 Overview

Each program unit is a collection of constructs and statements. The heading statement identifies the kind of program unit; it is optional in a main program. An ending statement marks the end of the unit. The categories of program units are:

main program (11.2)
module (11.3)
external subprogram (11.4)
block data (11.5)

Program execution begins with the first executable construct in the main program. Chapter 2 explains the high-level syntax of Fortran. It shows how statements can be combined to form a program unit.

J.C. Adams et al., *The Fortran 2003 Handbook*,
DOI: 10.1007/978-1-84628-746-6_11, © Springer-Verlag London Limited 2009

The module program unit helps organize elements of the program. A module itself is not executable but contains data declarations, derived-type definitions, procedure interface information, and subprogram definitions.

A subprogram may be a program unit used to define an external procedure, but it also may be a module subprogram or internal to either a main program, a module subprogram, or an external subprogram.

Block data program units are also nonexecutable and are used only to specify initial values for variables in named common blocks. With the addition of modules to Fortran, block data program units are no longer needed because modules can provide global data initializations.

The basic structure of all program units and subprograms is the same:

```
[ heading ]
    [ specification-part ]
    [ execution-part ]
[ CONTAINS
    subprogram
    [ subprogram ] . . . ]
ending
```

The subprogram part consists of the CONTAINS statement and the subprograms.

Rules and restrictions:

1. The heading establishes whether this is a program, module, function, subroutine, or block data. It also establishes its name. The form of the heading is different for the different categories.

2. The heading is optional for a main program.

3. An execution part is prohibited for a module.

4. An execution part and a subprogram part are prohibited for a block data.

5. A subprogram part is prohibited for an internal subprogram (12.4.3).

6. If the PROGRAM, MODULE, FUNCTION, SUBROUTINE, or BLOCK DATA keyword appears in the ending, the same keyword must appear in the heading, if there is one. If the name of the program unit appears in the ending, the same name must appear in the heading.

There are additional rules for each category.

The subprograms in a module define module procedures; the subprograms in a main program, module subprogram, or external subprogram define internal procedures.

11.1.1 The Specification Part

The purpose of the specification part is to describe the nature of the environment—types and attributes of variables, initial values, procedure interfaces, etc. The form of the specification part is given in 2.5.2.

11.1.2 The Execution Part

The execution part is a sequence of executable constructs, FORMAT statements, DATA statements, and ENTRY statements as described in 2.5.3. A DATA statement in the execution part is obsolescent.

11.1.3 The Subprogram Part

The CONTAINS statement and one or more subprograms comprise the subprogram part. An internal subprogram must not have a subprogram part. Internal subprograms are described in 12.4.3.

11.1.4 Example Program

The Fortran program in Figure 11-1 contains four program units: a main program, a module, and two external subroutine subprograms.

```
PROGRAM DRIVER
   USE STOCK_ROOM
   . . .
   CALL MECHANIC(TUNEUP)
   . . .
END PROGRAM DRIVER
```

```
MODULE STOCK_ROOM
   . . .
END MODULE STOCK_ROOM
```

```
SUBROUTINE PARTS &
     (PART, MODEL, YEAR)
   USE STOCK_ROOM
   . . .
END SUBROUTINE PARTS
```

```
SUBROUTINE MECHANIC (SERVICE)
   USE STOCK_ROOM
   . . .
   CALL PARTS (PLUGS, "CRX",1992)
   . . .
END SUBROUTINE MECHANIC
```

Figure 11-1 Four program units

Module STOCK_ROOM contains data and procedure information used by the main program and subroutines MECHANIC and PARTS. The main program DRIVER invokes the task represented by subroutine MECHANIC.

The USE statements in the example access the information in the module STOCK_ROOM.

11.2 Fortran Main Program

A main program may be in another language. Most rules in this section apply to a main program written in Fortran.

Execution of the program begins with the first executable construct in the main program. Execution may terminate for several reasons; executing the END statement in the main program causes normal termination (2.3.1).

The form of a main program (R1101) is:

```
[ PROGRAM program-name ]
    [ specification-part ]
    [ execution-part ]
    [ CONTAINS
       internal-subprogram
       [ internal-subprogram ] . . . ]
END [ PROGRAM [ program-name ] ]
```

The form of a main program is a special case of the general form for a program unit as given in 11.1.

Rules and restrictions:

1. There must be exactly one main program in an executable program; therefore, if the main program is written in another language, no Fortran main program may appear.
2. The PROGRAM statement is optional in a main program.
3. Main programs have no provisions for dummy arguments. However, the use of the intrinsic subroutines GET_COMMAND and GET_COMMAND_ARGUMENT provides similar functionality.
4. OPTIONAL and INTENT attributes or statements must not appear in the specification part of a main program; they are applicable only to dummy arguments.
5. The accessibility specifications, PUBLIC and PRIVATE, must not appear in a main program; they are applicable only within modules.
6. An automatic object (2.2.4) must not appear in a main program.
7. The SAVE attribute or statement may appear, but it has no effect in a main program.
8. The main program must not be referenced anywhere within the program—that is, the main program must not be recursive (either directly or indirectly).

9. The main program must not contain a RETURN or ENTRY statement (but an internal procedure in a main program may have RETURN statements).

The simplest of all programs is:

```
END
```

Of course, this is not a very interesting program! A more interesting simple program is:

```
program sine_calculation
   print *,  "The sine of 0.5 is", sin(0.5)
end program sine_calculation
```

The name of the main program is global to the program (16.1.1). There are no uses for the main program name except in the END statement, but it might be used by other operating systems software, such as a linker.

11.3 Modules

The module program unit packages data specifications and procedure specifications. The name of a module is global to the program.

Fortran defines some intrinsic modules (13.6.1, 14.3, and 15.3) and allows user-defined modules. Procedures and types defined in an intrinsic module are not intrinsic. This means, for example, that they are available only by use association (11.3.8), type specifiers have the form of derived-type specifiers, and there are subtle differences in generic procedure resolution (12.8.1).

Module program units serve the following specific needs for Fortran:

1. They provide a reliable mechanism for specifying global data, including variables, type definitions, and procedure interfaces.
2. They facilitate information hiding.
3. They facilitate the implementation of object-oriented concepts.
4. They reduce argument mismatch errors because they provide explicit interfaces (12.5.1).

Anything required by other program units may be packaged in a module and made available where needed. A module is not itself executable, although the module procedures it contains are executable. A module may use any number of other modules as long as the access path does not lead back to itself.

11.3.1 The Form of a Module

The form of a module (R1104) is:

```
MODULE module-name
   [ specification-part ]
```

```
      [ CONTAINS
          module-subprogram
          [ module-subprogram ] . . . ]
    END [ MODULE [ module-name ] ]
```

The form of a module is a special case of the general form for a program unit as given in 11.1.

Rules and restrictions:

1. The module subprograms may contain internal subprograms.
2. If a procedure declared in the scoping unit of a module has an implicit interface (12.5.1), it must be given the EXTERNAL attribute in that scoping unit; if it is a function, its type and type parameters must be declared explicitly in a type declaration statement in that specification part.
3. If an intrinsic procedure is declared in the scoping unit of a module, it must be given the INTRINSIC attribute explicitly or be used as an intrinsic procedure in that scoping unit.

Example:

```
module fish_or_fowl
   public :: fish, fowl, sqrt
   integer, external :: fish ! fish is a function because it has a type
   external :: fowl ! fowl is a subroutine because it has no type
   real :: sqrt    ! sqrt is a variable because it does not have
                   ! the EXTERNAL or INTRINSIC attribute
end module fish_or_fowl
```

11.3.2 The Specification Part

The form of the specification part (R204) of a module is similar to that for other program units. The following rules and restrictions apply to the specification part of a module; the specification parts of the module procedures, however, have the same rules as those for external procedures.

Rules and restrictions:

1. PUBLIC and PRIVATE attributes and statements are allowed. The only place they are allowed is the specification part of a module.
2. OPTIONAL or INTENT attributes or statements are not allowed.
3. ENTRY statements are not allowed.
4. FORMAT statements are not allowed.
5. Automatic objects are not allowed.
6. Statement function statements are not allowed.

7. If an object appears that has a type with default initialization (4.4.9) and the object has neither the ALLOCATABLE nor POINTER attribute, it must have the SAVE attribute.

The SAVE attribute and statement may be used in the specification part of a module to ensure that module data object values and status are retained. Without SAVE, module data objects remain defined as long as any program unit using the module has initiated, but not yet completed, execution. However, when all such program units become inactive, any data objects in the module not having the SAVE attribute become undefined. SAVE can be used to specify that module objects continue to be defined under these conditions.

The following is an example of a simple module for providing global data:

```
module t_ford
   save
   integer  ::  a, ka
   real     ::  x = 7.14
   real     ::  y(10,10), z(20,20)
end module t_ford
```

This module declares three scalar variables and two arrays. X is given an initial value. These five variables selectively can be made available outside the module.

11.3.3 Module Subprograms

Module subprograms define module procedures, which are discussed in 12.4.2.

11.3.4 Identifiers in a Module

The following may be defined, declared, or specified in a module, and may be public or private. If public, they may be accessed by a USE statement outside the module, and any public entity, except an assignment interface, may be renamed (11.3.7.1) in the using program unit.

1. declared variables
2. named constants
3. derived-types
4. procedure interfaces
5. abstract interfaces
6. module, intrinsic, and external procedures
7. generic identifiers
8. namelist groups

Note that this list does not contain the implicit type rules of the module; these are not accessible via a USE statement.

An entity in a module accessed by use association from another module is treated as an identifier in the module.

Common blocks may be placed in modules. The names of common blocks cannot be accessed, but the names of the variables can be accessed and renamed.

11.3.5 Accessibility

Each identifier in a module has the PUBLIC or PRIVATE attribute (5.8.1), which determines the accessibility of that name in a program unit using the module. A private identifier is not accessible (that is, is hidden) from program units using the module. A public identifier is accessible. Accessibility is described in 5.8.1.

PUBLIC and PRIVATE attributes are specified by the module writer, and the module user has no say in these decisions. However, both the ONLY option on the USE statement and the renaming provisions give the module user additional forms of information hiding and environment tailoring. Between PUBLIC and PRIVATE accessibility and the USE...ONLY feature (11.3.7.3), the module facilities provide considerable flexibility for program design that effectively employs information hiding.

11.3.6 The PROTECTED Attribute

A PROTECTED entity may be referenced (if it is public), but its value must not be changed outside the module. Unlike accessibility, the PROTECTED attribute applies to the entity, not to its identifier, so it is not permitted to modify a protected entity from outside the module, even by means of an alias. The PROTECTED attribute is described in 5.8.2.

11.3.7 The USE Statement

A scoping unit may use the specifications and definitions in a module by referencing (using) the module. This is accomplished with a USE statement. Such access causes an association between named objects in the module and the using scoping unit and is called use association (16.2.1.2). USE statements must be the first statements in a specification part (2.5.2).

11.3.7.1 Form of the USE Statement

The general form of the USE statement (R1109) is:

```
USE [ [ , module-nature ] : : ] module-name [ , rename-list ]
USE [ [ , module-nature ] : : ] module-name , ONLY : [ only-list ]
```

where the module nature is either INTRINSIC or NON_INTRINSIC. Each item in a rename list (R1111) has one of the forms:

```
[ local-name => ] module-entity-name
[ OPERATOR ( local-defined-operator ) => ] OPERATOR ( module-defined-operator )
```

The only list is described in 11.3.7.3.

Rules and restrictions:

1. If the module nature is INTRINSIC, the module name must be the name of an intrinsic module.
2. If the module nature is NON_INTRINSIC, the module name must be the name of a nonintrinsic module.
3. USE statements in a scoping unit must not access an intrinsic module and a nonintrinsic module of the same name.
4. If no module nature appears and the module name is the name of both an intrinsic and a nonintrinsic module, the nonintrinsic module is accessed.

The next two sections discuss accessing some or all of the public entities in a module. This is illustrated in Figure 11-2.

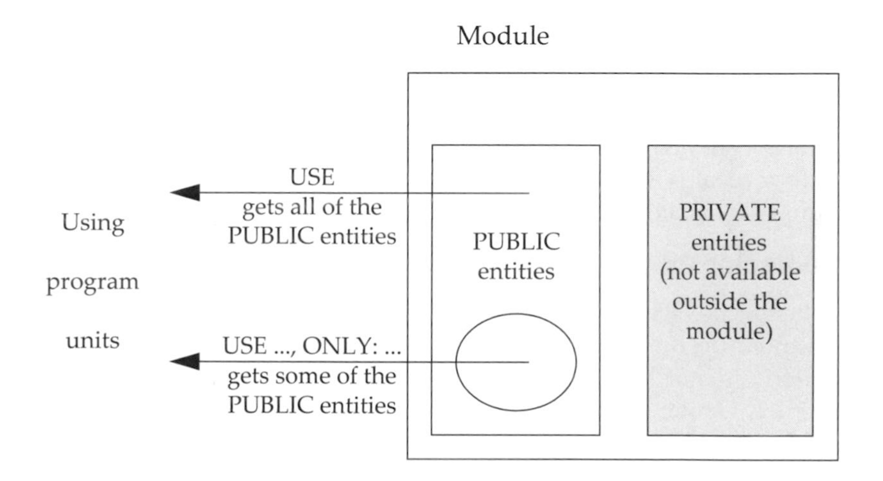

Figure 11-2 Use association of module entities

11.3.7.2 Accessing All Public Entities in a Module

The USE statement without the ONLY option gives the scoping unit access to all public entities in the module.

The optional rename list allows any of the public entities in the module to be renamed to avoid name conflicts or to blend with the naming conventions in the using program unit.

Rules and restrictions:

1. Each module entity and module defined operator must be a public entity in the module.

2. The module defined operator must not be a generic binding (4.4.11.2).

3. Each local defined operator must have the form of a defined operator (3.2.4).

Because there is only one symbol for assignment, it cannot be renamed.

A module entity name or module defined operator may appear more than once in one or more USE statements referencing the same module. As a result, it is possible for one entity to be accessible by more than one local identifier.

Examples of accessing all public entities:

```
USE FOURIER
USE S_LIB, PRESSURE => X_PRES
USE M_LIB, OPERATOR(.matrixmultiply.) => OPERATOR(.mm.)
```

With the USE statements in this example, all public entities in the respective modules are made accessible. In the case of FOURIER, the names are those specified in the module. In the case of S_LIB, the entity named X_PRES is renamed PRESSURE in the program unit using the module. The other entities accessed from S_LIB have the same name in the using program unit as in the module. Note the similarity between the rename syntax and pointer assignment (the only difference is that a rename is part of a statement, not a complete statement itself); this is because the local name is conceptually similar to a local pointer to the module entity. In the case of M_LIB, the module operator .mm. is renamed to .matrixmultiply.

11.3.7.3 Accessing Only Some of the Public Entities

Restricting the entities accessed from a module is accomplished with the ONLY form of the USE statement (R1109). In this case, the using program unit has access only to those entities explicitly identified in the ONLY clause of the USE statement. All items in this list must identify public entities in the module. As with the unrestricted form of the USE statement, named accessed entities may be renamed for local purposes. A rename is described in the previous section. The possible forms for each item in the only list (R1112) are:

> generic-name
> OPERATOR (module-defined-operator)
> ASSIGNMENT (=)
> dtio-generic-spec
> module-entity-name
> rename

Rules and restrictions:

1. Each generic specification and module name must be a public entity in the module.

2. A generic specification must not identify a generic binding (4.4.11.2).

3. A module defined operator must not identify a generic binding (4.4.11.2).

The above constraints do not prevent accessing a generic specification that is declared by an interface block, even if a generic binding has the same generic specification.

Examples of accessing some of the public entities:

```
USE MTD, ONLY :  X, Y, OPERATOR( .ROTATE. )
USE MONTHS, ONLY : January => Jan, May, June => Jun
```

In the case of MTD, only X, Y, and the defined operator **.**ROTATE**.** are accessed, with no renaming. In the case of MONTHS, only Jan, May, and Jun are accessed from the module. Jan is renamed January and Jun is renamed June.

11.3.7.4 Name Conflicts When Using Modules

There are two ways in which potential name conflicts can be avoided when using modules:

1. A public entity in a module might have the same name as a local entity in the using program. In this case, rename the module entity in the USE statement.
2. Two modules being used may each have a public entity with the same name. Such a name conflict is allowed if and only if that name is never referenced in the using program. If a name is to be referenced in the using program, potential conflicts involving that name must be prevented via the rename or ONLY facilities of the USE statement. This is the case even if the using program is another module.

   For example:

```
module BLUE
   integer  A, B, C
end module BLUE

module GREEN
   use BLUE, only : AX => A
   real  B, C
end module GREEN

PROGRAM RED
   USE BLUE
   USE GREEN, BX => B
      . . .
END PROGRAM RED
```

In the program RED, integer A is accessed as AX or A; integer B is accessed as B and real B is accessed as BX; C must not be referenced, because there is a name conflict.

11.3.8 Use Association

The USE statement gives a program unit access to other entities not defined or specified locally within the using program. As mentioned earlier, the association between a module entity and a local entity in the using program unit is termed **use association**, which is described in 16.2.1.2.

An entity accessed by use association has the attributes specified in the module, except that it may have a different accessibility attribute and it may have the ASYNCHRONOUS or VOLATILE attribute even if the module entity does not.

A local identifier in a USE statement must not appear in any other nonexecutable statement that would specify any of its attributes, except that it may appear in a PUBLIC or PRIVATE statement and it may be given the ASYNCHRONOUS or VOLATILE attribute.

11.3.9 Typical Applications of Modules

A number of different Fortran applications are easier to write and understand using modules. Modules provide a way of packaging:

1. global data, including data structures and common blocks
2. user-defined types
3. user-defined operators
4. data abstraction
5. procedure libraries

These uses for modules are summarized in the following sections.

11.3.9.1 Global Data

A module provides an easy way of making type definitions and data declarations global in a program. Notice that COMMON is not used in the example below, although it could have been. Data in a module does not have an implied storage association or an assumption of any form of sequence or any order of appearance, unless it is a sequence structure or in a common block. Global data in a module may be of any type or combination of types.

Another advantage of using modules rather than COMMON is more attributes, such as ALLOCATABLE, may be specified in a module.

Example:

```
MODULE MODELS
   COMPLEX                  :: GTX (100, 6)
   REAL, PROTECTED          :: X (100)
   REAL, SAVE, ALLOCATABLE :: Y (:), Z (:, :)
   INTEGER                  :: CRX, GT, MR2
END MODULE
```

There are alternative ways to "use" this module. For example,

```
USE MODELS
```

makes all the variables (and their attributes) of the module available.

```
USE MODELS, ONLY : X, Y
```

makes only the variables named X and Y and their attributes available.

```
USE MODELS, T => Z
```

makes the data object named Z available, but it is renamed to T. In addition, it makes the other public entities of the module MODELS available with the same names they have in the module.

One way of packaging common blocks is by putting them in a module. For example:

```
MODULE LATITUDE
   COMMON . . .
   COMMON . . .
   COMMON / BLOCK1 / . . .
END MODULE

SUBROUTINE ROUTE
USE LATITUDE
   . . .
END SUBROUTINE

PROGRAM NAVIGATE
USE LATITUDE
   . . .
END PROGRAM
```

The USE statements in this example make all of the variables in the common blocks in the module available. This technique minimizes errors in transcription and omission when the common blocks are used in many routines in the program.

The variables in a common block can be accessed through a module in one subprogram and declared directly in a common block in another. This may be useful as a transition stage between using common blocks and using modules.

11.3.9.2 User-Defined Types

A derived type defined in a module is a user-defined type that can be made accessible outside the module. The same type definition can be referenced via a USE statement by more than one program unit. This is important, for example, in order to allow the type of an actual argument and a dummy argument to match.

Example:

```
module new_type
   type :: tax_payer
      integer :: ssn
      character(len=20) :: name
   end type tax_payer
end module new_type
```

The module new_type contains the definition of a new type called tax_payer. Procedures using the module new_type may declare objects of type tax_payer.

11.3.9.3 User-Defined Operators

An interface block may declare new operators or give additional meanings to the intrinsic ones, such as +, **.**EQ**.**, **.**OR**.**, and //. In the following example, the addition operator (+) is extended, using a module procedure, to type MATRIX. The external function MATRIX_INVERSE defines a desired operation and the interface is used to indicate that it implements the operator **.**INVERSE**.** The operators **.**INVERSE**.** and + then may be used in an expression with operands of type MATRIX. The module supplies the function MATRIX_SUM, but INVERSE is an external function.

```
MODULE MATRIX_OPS
   INTERFACE OPERATOR (+)
      PROCEDURE MATRIX_SUM
   END INTERFACE

   INTERFACE OPERATOR (.INVERSE.)
      FUNCTION MATRIX_INVERSE (MATRIX_1)
         . . .
      END FUNCTION MATRIX_INVERSE
   END INTERFACE
      . . .
   CONTAINS
   FUNCTION MATRIX_SUM (X, Y)
      TYPE(MATRIX), INTENT(IN) :: X, Y
      TYPE(MATRIX) :: MATRIX_SUM
      MATRIX_SUM % VALUES = X % VALUES + Y % VALUES
   END FUNCTION MATRIX_SUM
END MODULE MATRX_OPS
```

An example of its use with variable A and B of type MATRIX might be

```
B + (.INVERSE. A)
```

11.3.9.4 Data Abstraction

Type definitions and operations may be packaged together in a module to facilitate data abstraction. Program units using this module will have the convenience of a new data type specific to a particular application. A simple example might be:

```
MODULE POLAR_COORDINATES

   TYPE POLAR
      PRIVATE
      REAL RHO, THETA
   END TYPE POLAR

   INTERFACE OPERATOR (*)
      MODULE PROCEDURE POLAR_MULT
   END INTERFACE

CONTAINS
   FUNCTION POLAR_MULT (P1, P2)
      TYPE (POLAR), INTENT(IN) :: P1, P2
      TYPE (POLAR) POLAR_MULT
      POLAR_MULT = POLAR (P1 % RHO * P2 % RHO,  &
                          P1 % THETA + P2 % THETA)
   END FUNCTION POLAR_MULT
      . . .
END MODULE POLAR_COORDINATES
```

In the function POLAR_MULT, the structure constructor POLAR computes a value that represents the result of multiplication of two arguments in polar coordinates. Any program unit using the module POLAR_COORDINATES has access to both the type POLAR and the extended intrinsic operator * for polar multiplication.

11.3.9.5 Procedure Libraries

An obvious way to build a procedure library is to put procedures and their related data in a module. In some cases, instead of putting a collection of procedures in a module, a module may be used to collect interface blocks for related procedures.

```
MODULE ENG_LIBRARY
   INTERFACE
      FUNCTION FOURIER (X, Y)
         . . .
      END
      SUBROUTINE INPUT (A, B, C, L)
         . . .
      END SUBROUTINE INPUT
   END INTERFACE
END MODULE ENG_LIBRARY
```

A program or subprogram that uses the module can call the procedures with optional and keyword arguments (for example) and argument matching can be checked because the interfaces for the procedures are explicit. This scheme looks to the programmer much like one that puts the procedures themselves in a module. A USE statement gives access to all the procedure interfaces. Thus, this scheme has most of the advantages of putting the procedures in a module and it also allows separate compilation of the individual procedures. One disadvantage of this method is that the procedures cannot use host association (16.2.1.3) to share data and type definitions.

A third way to build a procedure library is to reference several modules in one module.

```
module MATH_LIB
   use PDE_LIB
   use DIFFEQ_LIB
   . . .
end module MATH_LIB
```

11.4 External Subprograms

An external program unit is simply an external subprogram. It is global to the Fortran program; the procedure it defines may be referenced or called anywhere. External subprograms are described in 12.4.1.

11.5 Block Data Program Units

A block data program unit initializes data values in named common blocks. The block data program unit contains only data specifications and initial data values. There are no executable statements in a block data program unit and the block data program unit can be referenced only in EXTERNAL statements in other program units; its only purpose is to initialize data. The module facility is a natural extension to the very limited block data facility, making block data program units superfluous. The form of the block data program unit is standardized; however, the mechanisms for incorporating block data into the program are processor dependent. Therefore, the use of replacements, such as modules, is recommended.

A block data program unit (R1116) has the following form:

```
BLOCK DATA [ block-data-name ]
   [ specification-part ]
END [ BLOCK DATA [ block-data-name ] ]
```

The form of a block data program unit is a special case of the general from for a program unit as given in 11.1.

An example of a block data program unit is:

```
BLOCK DATA SUMMER
   COMMON / BLOCK_2 / X, Y
   DATA X / 1.0 /,  Y / 0.0 /
END BLOCK DATA SUMMER
```

The name SUMMER appears on the BLOCK DATA statement and the END statement. X and Y are initialized in a DATA statement; both variables are in named common block BLOCK_2.

Rules and restrictions:

1. There must be at most one block data program unit without a name.
2. The specification part may contain only the following statements.

ASYNCHRONOUS*	PARAMETER*
BIND	POINTER*
COMMON	SAVE*
DATA	TARGET*
DIMENSION*	USE
EQUIVALENCE	VOLATILE*
IMPLICIT	derived-type definition
INTRINSIC*	type declaration statement

3. The attribute specification list of a type declaration statement may contain only the attributes marked with * above in (2).
4. A type declaration statement may contain initialization.
5. BIND can refer only to common block names, not variables.
6. A block data program unit may initialize objects in more than one named common block.
7. It is not necessary to initialize an entire common block.
8. A common block must be completely specified, if any object in it is initialized.
9. A given named common block may appear in only one block data program unit.
10. Only an object in a named common block may be initialized.

12 Using Procedures

- A **Procedure** is a **subroutine** or **function**.
- A **Defined Operator**, **Defined Assignment,** and **Defined Input/Output** use procedures to allow user extension of the language in a natural way.
- A **Generic Procedure** allows actual arguments to determine which procedure from a specified group is selected for execution.
- A **Pure Procedure** has limitations on side effects so that it may be used in parallel computations and in specification expressions.
- An **Elemental Procedure** is defined with scalar arguments but may be called with array actual arguments.
- A **Recursive Procedure** is one that can directly or indirectly invoke itself.
- An **Optional Argument** is one that may be omitted.
- An **Argument Keyword** provides order independence of arguments and facilitates the omission of optional arguments.
- An **Explicit Interface** is required for many of the above features. The interface of an intrinsic, module, or internal procedures is always explicit. An explicit interface can be created for an external procedure.
- The **ENTRY Statement** may be used to define an additional procedure in a subroutine or function subprogram.

A procedure encapsulates a sequence of related computations.

A **procedure reference** is the appearance of program syntax that can cause the procedure to be invoked (executed). When a procedure is invoked, execution of the program or procedure making the reference is suspended while the procedure is executed. When execution of the procedure is completed, execution of the invoking program or procedure is resumed.

A **subroutine** is the fundamental form of a procedure. The basic form of subroutine reference is a CALL statement; other forms of reference are described in 12.1.2. When it is invoked, the computations of the subroutine are executed and control then returns to the invoking program or procedure. A subroutine can include any kind of computation. Invocation of a subroutine will generally have effects such as modifying the values of variables or performing input/output.

A **function** is much like a subroutine with the extra feature of returning a result value, which can be used as a part of an expression. A function reference is in an expression and is either a primary or a defined operator. The main purpose of a function is to compute its result value.

J.C. Adams et al., *The Fortran 2003 Handbook*,
DOI: 10.1007/978-1-84628-746-6_12, © Springer-Verlag London Limited 2009

A procedure can be defined intrinsically, by Fortran code, or by means other than Fortran.

Intrinsically defined procedures include the standard intrinsic procedures described in 13 and the procedures in the standard intrinsic modules described in 14 and 15.5.4. Compilers may define intrinsic procedures and modules in addition to those specified by the standard, but code that uses them is nonportable.

A subprogram is a Fortran source code construct that defines one or more procedures. It is either a subroutine subprogram or a function subprogram. The Fortran standard distinguishes between a subprogram, which is a chunk of source code, and a procedure, which is the abstract executable entity defined by the source code. Common parlance uses the word subroutine or function to refer either to the procedure or the source code for the procedure. For most purposes, the distinction is of little importance and just adds an extra level of complexity to the terminology. Partly adopting the common parlance, this book refers to a subroutine or function definition instead of a subprogram in most places. The main place where this terminology falls short is in the discussion of the ENTRY statement, which allows a single subprogram to define multiple procedures.

A procedure can be written in a language other than Fortran (C, for example). As long as all references to such a procedure are consistent in terms of the properties of the interface to this procedure, the referencing program remains standard conforming. It might, however, not be portable, because the non-Fortran procedure or its method of argument communication might differ across implementations.

If the non-Fortran procedure is written in C, portability is facilitated by using the BIND(C) attribute, as described in 15.6.1. This also applies to a procedure written in some other language, provided that it "looks like" a C function—that is, if it is capable of being invoked as though it were written in C; for example, this applies to a C++ function that has the "extern C" attribute.

12.1 Subroutines

A Fortran procedure definition for a subroutine is a subroutine subprogram. When a subroutine is invoked other than via an alternate entry (12.4.5), its execution begins with the first executable construct in the subprogram. Data objects, procedures, and other entities may be communicated to and from the subroutine through argument association, host association, use association, or common storage association.

12.1.1 Subroutine Subprograms

A subroutine subprogram (R1221) has the form of a subprogram as described in 11.1, with a heading of the form

```
[ subroutine-prefix ] SUBROUTINE subroutine-name &
      [ ( [ dummy-argument-list ] ) [ subroutine-suffix ] ]
```

A subroutine prefix consists of one or more of:

```
RECURSIVE
PURE
ELEMENTAL
```

A subroutine suffix is:

```
[ BIND ( C [ , NAME= scalar-character-initialization-expression ] ) ] ]
```

A dummy argument is either a dummy argument name or an asterisk (*), where the asterisk designates an alternate return. When a subroutine is invoked, the dummy arguments become associated as specified by the actual arguments in the subroutine reference (12.6).

The BIND attribute for procedures is described in 15.6.1.

Examples of SUBROUTINE statements are:

```
subroutine initialize_database
SUBROUTINE CAMP (SITE)
SUBROUTINE TASK ()
SUBROUTINE LIGHT (INTENSITY, M, *)
PURE RECURSIVE SUBROUTINE YKTE (Y, KE)
subroutine send_buffer(buffer, buffer_len, status) bind(c)
```

An example of a subroutine subprogram is:

```
SUBROUTINE TROUT (STREAM, FLY)
   CHARACTER (10) STREAM
   OPTIONAL FLY
   STREAM = . . .
      . . .
END SUBROUTINE TROUT
```

Rules and restrictions:

1. No prefix may appear more than once in a given SUBROUTINE statement.

2. If the subroutine is recursive, that is, it calls itself either directly or indirectly, the prefix RECURSIVE is required in the SUBROUTINE statement.

3. A subroutine must not be both recursive and elemental. (See additional notes for pure and elemental procedures in 12.7.1 and 12.7.2.)

4. The * dummy argument is an obsolescent feature (12.6.9).

5. Each named dummy argument is a local entity of the subroutine. The attributes of the entity can be specified by separate statements in the specification part of the subprogram.

12.1.2 Subroutine References

The basic form of a reference to a subroutine is a CALL statement. The other forms of reference can be described in terms of equivalent CALL statements; the other forms are defined assignment, defined input/output, and finalizers. A subroutine reference specifies the subroutine and actual arguments to be used.

The form of the CALL statement (R1218) is:

CALL procedure-designator [([actual-argument-specification-list])]

where a procedure designator (R1219) is one of:

procedure-name (procedure name, possibly a dummy or pointer)
scalar-variable % procedure-component-name (procedure pointer component)
designator % binding-name (type-bound reference)

A CALL statement using the designator%binding form of a subroutine designator references a subroutine determined by the dynamic type of the designator. This is referred to as a type-bound reference. The example in 4.4.11.1 illustrates type-bound subroutine references.

An actual argument specification (R1220) has the form:

[keyword =] actual-argument

where a keyword is a dummy argument name in the accessible procedure interface and each actual argument (R1221) is one of:

expression
variable
procedure-name
scalar-variable % procedure-component-name
* label (an alternate return specifier

Each actual argument must have a corresponding dummy argument. The correspondence is, as described in 12.6.1, by its position in the argument list or the name of its keyword.

Argument association is described in 12.6.

Rules and restrictions:

1. The procedure designator must designate a subroutine.
2. The binding name must be the name of a procedure binding for the declared type of the designator.
3. The label in the alternate return specifier must be a branch target in the same scoping unit as the reference.

Examples of subroutine references:

```
CALL TEST (X = 1.1, Y = 4.4)  ! SUBROUTINE TEST (Y, X)
CALL TYR (2.0*A, *99)         ! SUBROUTINE TYR (R, *)
   . . .
99 . . .  ! Error recovery
   . . .
```

Keyword arguments are used for X and Y in the first CALL statement; therefore, the order of the actual arguments does not matter. In the second example, an alternate return to statement 99 in the calling program unit is the last argument.

Another way to reference a subroutine is with defined assignment (7.5.3, 12.5.4.3). A defined assignment is an assignment statement that is implemented by a user-written subroutine instead of intrinsically by the compiler. Defined assignment requires an ASSIGNMENT interface to specify which subroutines implement which assignments. A defined assignment reference has the form of an assignment statement:

$$arg_1 = arg_2$$

and is equivalent to a CALL statement with an actual argument list of $(arg_1, (arg_2))$.

The types, kinds, and rank of the arguments select the particular defined assignment subroutine, as described in 12.8.

Example of defined assignment:

```
MODULE POLAR_COORDINATES

   TYPE POLAR
      REAL  ::  RHO, THETA
   END TYPE POLAR

   INTERFACE ASSIGNMENT (=)
      MODULE PROCEDURE ASSIGN_POLAR_TO_COMPLEX
   END INTERFACE

      . . .

   SUBROUTINE ASSIGN_POLAR_TO_COMPLEX (C, P)
      COMPLEX, INTENT(OUT)        ::  C
      TYPE (POLAR), INTENT(IN)    ::  P
      C = CMPLX (P%RHO * COS (P%THETA),  &
                 P%RHO * SIN (P%THETA))
   END SUBROUTINE ASSIGN_POLAR_TO_COMPLEX

END MODULE POLAR_COORDINATES
```

```
USE POLAR_COORDINATES
COMPLEX ::  CARTESIAN
   . . .
CARTESIAN  =  POLAR (R, PI/6)
```

This last assignment is equivalent to the subroutine call:

```
CALL ASSIGN_POLAR_TO_COMPLEX (CARTESIAN, (POLAR(R,PI/6)))
```

The structure constructor POLAR constructs a value of type POLAR from R and PI/6 and assigns this value to CARTESIAN according to the computations specified in the subroutine.

A third way to reference a subroutine is through defined input/output. As with defined assignment, an interface is required to specify which subroutines implement defined input/output. The reference then takes the form of a READ, WRITE, or PRINT statement. The details of defined input/output are described in 12.5.4.4 and 9.5.1.4.

The final way to invoke a subroutine is through finalization. Finalization requires that the subroutine be specified in a final binding in a type definition. Finalization does not have a specific syntactic form of reference. Instead, finalization occurs when a finalizable entity goes out of existence, which can happen in multiple ways. The details of finalization are described in 4.4.11.3.

12.2 Functions

A function is similar to a subroutine, with the additional feature of returning a result value that is used as a primary in an expression. Either a subroutine or a function can return values through variables in its arguments or through other forms of association, but the result value of a function is special in that it is directly used in evaluating an expression. Thus a function reference appears as part of a statement, in contrast to a subroutine call, which is always the whole of a statement. A Fortran procedure definition for a function is a function subprogram or a statement function. Statement functions are obsolescent and are described in 12.4.4.

12.2.1 Function Subprograms

A function subprogram has the form of a subprogram as described in 11.1, with a heading of the form:

```
[ function-prefix ] FUNCTION function-name ( [ dummy-argument-name-list ] ) &
    [ function-suffix ]
```

A function prefix (R1227), if present, consists of one or more of:

```
type-spec
RECURSIVE
PURE
ELEMENTAL
```

which is the same list as that for subroutine prefix, with the addition of type specifier.

A function suffix (R1229) consists of zero or more of:

```
RESULT ( result-name )
BIND ( C [ , NAME= scalar-character-initialization-expression ] )
```

Note that, unlike in a SUBROUTINE statement, the parentheses are required in a FUNCTION statement even when there are no dummy arguments. A function dummy argument cannot be an asterisk because a function cannot have an alternate return; after a function returns, evaluation of the expression referencing it continues. When a function is invoked, the dummy arguments become associated as specified by the actual arguments in the function reference (12.6).

The BIND attribute for procedures is described in 15.6.1.

Example function statements are:

```
FUNCTION HOSPITAL (PILLS)
REAL FUNCTION LASER (BEAM)
FUNCTION HOLD (ME, YOU) RESULT (GOOD)
RECURSIVE CHARACTER(10) FUNCTION POLICE (STATION) RESULT (ARREST)
function handler() bind(c)
```

Rules and restrictions:

1. The type of the function result may be specified in the function statement or in a type declaration statement, but not both. If the type is not explicitly specified in one of these ways, the default typing rules apply.
2. If the function result is an array, a pointer, or an allocatable, these attributes must be specified in the specification part. There is no syntax for specifying these attributes in the FUNCTION statement.
3. No prefix or suffix may appear more than once in a given FUNCTION statement.
4. If the function is recursive, that is, it invokes itself either directly or indirectly, the prefix RECURSIVE is required in the FUNCTION statement.
5. A function must not be both recursive and elemental. (See additional rules for pure and elemental procedures in 12.7.1 and 12.7.2.)
6. Each dummy argument is a local entity of the function. The attributes of the entity can be specified by separate statements in the specification part of the subprogram.
7. If there is no result suffix, the function name is also the name of the result variable. In this case, all appearances of the function name in the execution part refer to the result variable rather than the function.
8. If there is a result suffix, it specifies the name of the result variable; the name must be different from the function name. In this case, all appearances of the function name in the execution part refer to the function itself rather than the result variable.

9. If there is a result suffix, the result name rather than the function name must be used in any specification statements that specify attributes of the result.
10. If the result of a function is not a pointer, its value must be defined when execution of the function completes.
11. If the result is a pointer, its association status must be defined when execution of the function completes.

Note that, in the case of direct recursion, both the RECURSIVE keyword and the RESULT suffix must be specified; this is the only case in which the RESULT suffix is required.

12.2.2 Result Variable

Every function has a result variable. By default, the result variable has the same name as the function. The RESULT suffix is used to specify a different name for the result variable. The result variable may be declared, defined, and referenced as an ordinary variable. The value of the function is the value that the result variable has when execution of the function completes.

Why is it sometimes necessary to distinguish between the function name and the result name? If the function result is an array and the function is directly recursive, a recursive reference to the function may be indistinguishable from a reference to the array result. This ambiguity is resolved by providing one name for the result value and a different name for recursive references. For example, if F is a recursive function that returns a rank-one array of reals and has a single integer argument, in the statement

```
A = F(K)
```

F(K) could be interpreted as either a reference to the Kth element of the array result of F or a recursive call to F with actual argument K. If there were no result suffix, this would be interpreted as a reference to the array element, leaving no syntax to express direct recursion.

The following is a simple example of a recursive function REVERSE, which reverses the words in a given phrase.

```
PROGRAM TEST_FLIPPED
PRINT *, REVERSE ("I have flipped for you.")
CONTAINS
```

```
RECURSIVE FUNCTION REVERSE (PHRASE) RESULT (FLIPPED)
   CHARACTER (*)             PHRASE
   CHARACTER (LEN(PHRASE)) FLIPPED
   L = LEN_TRIM (PHRASE)
   N = INDEX (PHRASE(1:L), " ", BACK=.TRUE.)
   IF (N == 0) THEN
      FLIPPED = PHRASE
   ELSE
      FLIPPED = PHRASE (N+1:L) // " " // REVERSE (PHRASE (1:N-1))
   END IF
END FUNCTION REVERSE
END PROGRAM TEST_FLIPPED
```

Note that in this example there is a dynamic function result length, which requires an explicit interface (12.5.1.2(6)).

12.2.3 Function References

A function reference always appears as a primary in an expression. It most commonly has the form of a function designator followed by a parenthesized actual argument list. It can alternatively have the form of a defined unary or binary operator. Invoking the function returns a value which is then used in the expression. For example, in the expression

```
A + F(B)
```

where F is a function of one argument that delivers a numeric result, this result becomes the value of the right-hand operand of the addition operation.

The basic form of a function reference (R1217) is:

procedure-designator ([actual-argument-specification-list])

where a procedure designator and actual argument specification list have the same form as in a CALL statement, except that the procedure designator must designate a function and the actual argument list for a function must not contain an alternate return. This is because a function dummy argument cannot be an alternate return. The rules and restrictions for actual and dummy arguments are listed in 12.6.

Examples of function references are:

```
PRINT *, TIME (TODAYS_DATE)          ! FUNCTION TIME (DATE)

Y = 2.3 * CAPS (size=4*12, kk=K)    ! FUNCTION CAPS (SIZE,KK)
```

The other form of reference is with a defined operation (12.5.4.2). A number of arithmetic, logical, relational, and character operators are predefined in Fortran; these are called **intrinsic operators**. These operators may be given additional meanings, and new operators may be defined. Functions define these operations; interface blocks or generic bindings connect them with operator symbols. A function may be invoked by using one of its defined operators in an expression.

The following is an example of a function reference via a defined operator:

```
INTERFACE  OPERATOR (.BETA.)
    FUNCTION BETA_OP (A, B)
          . . .    ! attributes of BETA_OP, A, and B
                   ! (including intent IN for A and B)
    END FUNCTION
END INTERFACE
   . . .
PRINT *, X .BETA. Y
```

The presence of .BETA. in the expression in the PRINT statement invokes the function BETA_OP, with X as the first actual argument and Y as the second actual argument. The function value is returned as the value of X .BETA. Y in the expression.

12.2.4 Function Side Effects

The main purpose of a function is to compute its result value. Like a subroutine, invocation of a function can also have other effects such as modifying the values of variables or performing input/output. Any such other effect that outlasts a single invocation of the function is called a function side effect. Comparable effects in a subroutine are not usually called side effects because they are the primary purpose of a subroutine invocation.

Examples of function side effects include modifying the value of an argument or other variable visible outside of the function. Modifying the value of a variable local to the function is also a side effect if that variable is saved between invocations. Modifying the value of an unsaved local variable is not a side effect because it does not outlast a single invocation. Input/output on an external file is a side effect.

Some function side effects are disallowed. The most notable case relates to the fact that a function invocation occurs as part of a statement instead of being a separate statement like a subroutine call. A function invocation is not allowed to affect the execution of other parts of the statement (for instance, other function invocations in the same statement). If such intra-statement effects were allowed, they would have significant impact on optimization, in addition to being confusing. There are particular prohibitions against intra-statement side effects in input/output, ALLOCATE, and DEALLOCATE statements.

Even in cases where function side effects are allowed, a program is prohibited from depending on them. If this prohibition is taken to its extreme, there are very few cases where functions with side effects can be safely used. However, compilers do not tend to take such extremes; there are many programs that rely on function side effects and that work well in practice. Problems with function side effects become more likely with highly aggressive optimization.

Suppose, by way of example, that get_value is a function that reads from a file, sets its argument to the value read, and returns an error code as its function result value. This is typical of some coding styles. If this function is invoked twice with code like

```
j = get_value(x)
j = get_value(y)
```

an aggressive optimizer could conceivably optimize away the first invocation, giving results other than intended. This example does not violate any of the prohibitions against effects on other parts of the same statement because the two function invocations are in separate statements. However, if x or y is subsequently referenced, this would violate the prohibition against depending on a function side effect.

For another example, suppose that my_random is a random number generator that has a seed as its argument and returns a pseudorandom number as its function result. The value of the seed is modified by the function. If this function is invoked twice in a statement as in

```
x = my_random(seed) + my_random(seed)
```

an aggressive optimizer could conceivably evaluate this as

```
x = 2 * my_random(seed)
```

which would not have the intended statistical properties. Optimizations like this have been known to cause problems for actual programs with actual compilers. This example violates the prohibition against affecting other parts of the same statement because both references to the function affect each other by changing the value of seed. This problem is why the intrinsic random_number procedure is a subroutine instead of a function.

The authors generally recommend that function side effects be avoided, even where the standard allows them. They introduce potential sources of error, including the possibility of accidentally using the functions in disallowed ways. A function with a side effect can be recast as a subroutine with an additional argument. C interoperation is, however, a major exception to this recommendation. All C procedures are functions and the use of function side effects is ubiquitous in C.

12.3 RETURN Statement

A RETURN statement terminates execution of an instance of a procedure and returns control to the invoking program or procedure. In the case of a recursive invocation, the invoking procedure might be a different instance of the same procedure. A RETURN statement is allowed only in a subprogram, which may be either a subroutine or function subprogram.

The form of the RETURN statement (R1236) is:

```
RETURN [ scalar-integer-expression ]
```

The scalar integer expression option is applicable only to subroutines and is used in conjunction with alternate returns. If the expression is omitted from a return statement in a subroutine, control returns to the statement following the subroutine invocation. A return statement in a function always returns to the evaluation of the expression in which the function invocation appeared.

Executing the END statement of a subprogram has the same effect as executing a RETURN statement that has no scalar integer expression. It is purely a style choice as to whether to put an explicit RETURN statement instead of this implicit one. One must use a RETURN statement in order to return from places other than at the end of the subprogram or to specify an alternate return.

If the scalar integer expression appears, it must be of type integer. If its value is in the range 1:*n* where *n* is the number of asterisks in the dummy argument list, it selects which alternate return, counting from left to right in the actual argument list, is to be used for this particular return from the procedure. If the value is outside that range, it is as if the expression were omitted. The effect of an alternate return is the same as

```
CALL SUBR(..., IRET)
GO TO ( label-list ), IRET
```

where inside SUBR, the dummy argument corresponding to IRET is assigned the integer expression alternate return value prior to returning from SUBR. There are better ways to achieve the functionality of alternate return, such as with a CASE construct controlled by a return code with an appropriate mnemonic value; for this reason alternate return is an obsolescent feature.

12.4 Procedure Definition

A procedure can be defined intrinsically, by Fortran code, or by means other than Fortran. A procedure defined by Fortran code is further classified according to its means of definition as an external procedure, module procedure, internal procedure, or statement function.

In most cases, a procedure is defined by a subprogram. The context of the subprogram determines the classification of the procedure.

12.4.1 External Procedures

An external subprogram is a stand-alone subprogram that is not part of any other program unit. It defines an external procedure.

An external procedure may share information, such as data and procedures via argument lists, modules, and common blocks, but otherwise it does not share information with any other program unit. It may be developed and compiled completely independently of other procedures.

Alternatively, an external procedure can be defined by means other than Fortran. The C interoperability features described in 15.6.1 facilitate the use of such procedures.

12.4.2 Module Procedures

A module subprogram is one that appears in the module subprogram part of a module and is not internal to another module subprogram. It defines a module procedure.

A module procedure inherits the module's environment via host association. The accessibility of a module procedure outside of the module can be controlled with the PUBLIC or PRIVATE attribute.

The syntax of a module subprogram has a restriction—its END statement must include the SUBROUTINE or FUNCTION keyword.

Some module procedures are defined by intrinsic modules (14.3, 15.5.4).

12.4.3 Internal Procedures

An internal subprogram is one that appears in the internal subprogram part of some other subprogram or main program, which is called its **host**. The host can be the main program, a module procedure, or an external procedure. Notably, the host cannot be an internal subprogram; this restriction is to facilitate simplicity of implementation. An internal subprogram defines an internal procedure.

An internal procedure can be referenced only from within its host; this includes any internal procedures of the host. An internal procedure inherits its host's environment via host association. An internal procedure can be a subroutine or function, regardless of whether its host is a subroutine, function, or main program.

The syntax of an internal subprogram has the following restrictions in addition to the general requirements of a subprogram.

Rules and restrictions:

1. Its END statement must include the SUBROUTINE or FUNCTION keyword.
2. It must not have an internal subprogram part.
3. It must not have any ENTRY statements.
4. An internal procedure cannot be passed as an actual argument or used as the target of a procedure pointer. This restriction avoids complications in the definition and implementation of recursive procedures.

Although the context of an internal subprogram looks much like that of a module subprogram, in that both of them follow a CONTAINS statement, they are not the same thing. Thus, these restrictions do not apply to a module subprogram.

12.4.4 Statement Functions

A statement function (R1238) is a function whose definition consists of a single Fortran statement in the specification part of a main program or subprogram. The subprogram may be an external, module, or internal one. There is no subroutine counterpart to a statement function.

A statement function has similarities to an internal procedure in that it is defined within some other subprogram or main program, although the placement is different. The restriction to a single statement is a significant limitation and has several consequences.

The form of a statement function statement, which defines a statement function, is:

```
function-name ( [ dummy-argument-name-list ] ) = scalar-expression
```

Statement functions are obsolescent. A statement function statement can be replaced (except within an internal procedure) with the following equivalent three-line internal function definition

```
FUNCTION function-name ( [ dummy-argument-name-list ] )
function-name=scalar-expression
END FUNCTION
```

providing the function and its arguments have the same types in both the statement function and the internal function.

Because a statement function is only a single statement, there is no place for declarations inside of the statement function. Consequently, any declarations for entities referenced in the statement function have to be outside of the statement function, in its host procedure, even if the only reference to the entity is in the statement function. This is most notable for type declarations of the dummy arguments and of the statement function. To explicitly declare the types of the dummy arguments or of the statement function (as is recommended practice, and as is required if implicit none is in effect), those declarations must be outside of the statement function. This is particularly strange for the dummy arguments because a statement function dummy argument has a scope limited to the statement function, so the type declaration is actually outside of the scope of the entity it declares. Because the scope of a statement function dummy argument is limited, the host scoping unit that the statement function is in may have a different entity with the same name. In that case, a single type declaration declares the types of two different entities. Although this confusing situation is allowed, it is not recommended practice.

The syntax of a statement function statement is similar to that of an assignment statement for an array element. Although they are technically distinguishable in proper context, their similarity often results in user confusions. Such confusion is exacerbated when there are errors in the context. A missing dimension attribute can cause a compiler to interpret an assignment statement as a statement function statement; this usually causes compilation errors, but the error messages can be quite misleading.

Several of the following restrictions on statement functions seem arbitrary and nonintuitive if considered out of context. The important context here is that statement functions were made obsolescent as of Fortran 90. It was decided not to enhance the feature at the same time as it was being made obsolescent. Therefore, many features new to Fortran 90 are not allowed in statement functions. The rules make a little more sense if statement functions are considered as a compatibility feature for old code, rather than a feature that would be used in new code.

Rules and restrictions:

1. The function and all the dummy arguments are scalar.

2. The expression must not use any defined operations. This excludes both defined operators and user-defined extensions of intrinsic operators.

3. Each primary in the expression must be a constant, a variable, a function reference, or a parenthesized expression that meets the same requirements as a statement function expression. This surprisingly limited list excludes array constructors, structure constructors, type parameter inquiries, and subobjects of constants.

4. If the expression references a function, neither the function nor the reference must require an explicit interface unless the function is an intrinsic. A referenced intrinsic function must have a scalar result and must not be transformational. If a function argument in the expression is an array, it must be an array name; this rules out array components, slices, and expressions.

5. A statement function is defined in the specification part of a program or subprogram. Any statement function referenced in the expression must have been defined earlier in the same specification part, and hence a statement function cannot be recursive (either directly or indirectly).

6. If a named constant is used in the expression, it must have been declared earlier in the specification part or made available by use or host association.

7. If an array element is used in the expression, the array must have been declared as an array earlier in the specification part or made available by use or host association.

8. The appearance of any entity in the expression that has not previously been typed explicitly constitutes an implicit type declaration; any subsequent explicit type declaration for that entity must be consistent with the implicit type. This same rule applies to the statement function name.

9. If the statement function is in a contained scoping unit and has the same name as an entity in the host scoping unit, the type of the statement function must be declared in a type declaration statement prior to the statement function statement. This rule avoids a possible syntax ambiguity if the entity in the host is an array, but the rule applies regardless of whether the particular case would be ambiguous or not.

10. Statement function dummy arguments have a scope of the statement function statement.

11. No function reference in the expression may change the value of any dummy argument of the statement function.

12. A statement function must not be used as an actual argument.

13. The interface of a statement function is always implicit, not explicit; see 12.5.1 for a detailed discussion of implicit versus explicit interfaces.

14. A statement function is referenced in the same manner as any other function, except that because statement function interfaces are implicit, the keyword form of actual arguments is not allowed; the argument association rules are the same.

15. The type and type parameters of the expression do not have to be the same as those of the statement function, but they must be compatible with intrinsic assignment. The result of a statement function is obtained by evaluating the expression and then converting it to the type and type parameters of the statement function.

The following are examples of statement functions:

```
CHARACTER(5)  ZIP_5              !Notice these are scalar
CHARACTER(10) ZIP_CODE           !  character strings.
ZIP_5(ZIP_CODE) = ZIP_CODE(1:5)

INTEGER TO_POST, MOVE
TO_POST(MOVE) = MOD(MOVE, 10)

REAL FAST_ABS
COMPLEX Z
FAST_ABS(Z) = ABS(REAL(Z)) + ABS(AIMAG(Z))
```

12.4.5 Alternate Entries

Normally a subprogram defines a single procedure (not counting internal procedures or statement functions, which can be used only within the subprogram). In fact, this one-to-one correspondence is so ubiquitous that it is common to forget about the distinction between a procedure and a subprogram. A subprogram is a chunk of source code; a procedure is the conceptual set of actions defined by the subprogram. The subroutine or function statement at the beginning of the subprogram defines the name of the procedure and its dummy arguments. If a subprogram has ENTRY statements, it defines multiple procedures, one for the function or subroutine statement and one for each ENTRY statement. Each ENTRY statement specifies a procedure name, a set of dummy arguments, and a place for execution of the procedure to begin; these are often referred to as alternate entries.

The procedure specified by an ENTRY statement shares the same data environment as the procedure specified by the subroutine or function statement, except for the dummy arguments.

When a procedure specified by an ENTRY statement is invoked, execution begins with the first executable construct after the ENTRY statement; this might or might not be the same as the first executable construct in the subprogram.

The procedure specified by an ENTRY statement is either an external or module procedure, depending on whether it is in an external or module subprogram; ENTRY statements are not allowed in internal subprograms. The procedure specified by an ENTRY statement is either a subroutine or function depending on whether it is in a subroutine or function subprogram. Such a procedure is referenced just like any other external or module procedure; there is no way for a referencing scoping unit to tell the difference. The only difference is in how the procedure is defined—not in how it is referenced.

An ENTRY statement (R1235) has the form:

```
ENTRY entry-name [ ( [ dummy-argument-list ] ) [ entry-suffix ] ]
```

where an entry suffix is one of

subroutine-suffix
function-suffix

depending on whether it is in a subroutine or function subprogram.

Examples of ENTRY statements are:

```
ENTRY FAST(CAR, TIRES)

ENTRY LYING(X,Y) RESULT(DOWN)
```

The attributes of the dummy arguments and result variable are described in the specification part of the subprogram. The dummy argument list of an ENTRY statement must meet the same requirements as for a SUBROUTINE or FUNCTION statement, as appropriate. For example, the dummy argument list in a function subprogram must not have an alternate return specifier. Additionally, all of the entries in a function subprogram must be compatible as described in rules 5 and 6 below.

If the RESULT suffix is omitted from an ENTRY statement in a function subprogram, it is as though it were specified with the same name as the ENTRY name.

Note the absence of a prefix from the form of the ENTRY statement. The procedure specified by an ENTRY statement is recursive, pure, or elemental according to whether the procedure defined by the FUNCTION or SUBROUTINE statement has those properties; it is neither necessary nor allowed to respecify those on the ENTRY statement. The type of the result variable for an ENTRY statement in a function is specified either implicitly or by a separate type declaration statement.

One common application is for an alternate entry to initialize the shared data environment. Another application is to share code. The example below illustrates a typical code-sharing scenario. Because there is no RETURN statement before the last ENTRY, execution of that starts at the first ENTRY continues at that point and shares the remaining code with execution that starts at the last ENTRY.

```
SUBROUTINE name-1 ( dummy-argument-list-1 )
      . . .
   RETURN
   ENTRY name-2 ( dummy-argument-list-2 )
      . . .
   !  This falls through past the next ENTRY statement.
   ENTRY name-3 ( dummy-argument-list-3 )
      . . .
   RETURN
END
```

Rules and restrictions:

1. An ENTRY statement may appear only in an external or module subprogram; an internal subprogram must not contain an ENTRY statement.

2. An ENTRY statement must not appear in an executable construct or a nonblock DO loop.

3. The entry names must be different from one another and from the original subprogram name.

4. If a result name is specified, it must not be the same as any entry name or the function name in the subprogram. If a result name is specified, the entry name must not appear in any specification statement in the subprogram; any attribute specification must use the result name instead.

5. If all of the entry result variables in a function subprogram have the same type, type parameters, and shape as the function result variable, all of them are aliases for the same variable. In this case there is no restriction on the nature of the result. For example, the result could be of derived type, either scalar or array, and could have the pointer attribute.

6. If any of the result variables in a function subprogram do not have the same type, type parameters, and shape as the function result variable, they must all be nonpointer, nonallocatable scalars and must be of type default integer, default real, default logical, double precision real, or default complex. In this case, all the result variables are storage associated with each other.

7. A dummy argument of an ENTRY statement must not appear before that ENTRY statement either in an executable statement or in the expression of a statement function statement, unless it is also a dummy argument of a preceding ENTRY, FUNCTION, or SUBROUTINE statement, or the statement function statement. This restriction is a little strange in that it depends on the physical order of statements in the source code, but not on the logical order of execution.

8. If a dummy argument, or an object declared with a specification expression that depends on the dummy argument, appears in an executable statement, then that statement may be executed only if the dummy argument appears in the dummy argument list of the referenced procedure.

The order, number, types, type parameters, and names of the dummy arguments in an ENTRY statement may differ from those in the FUNCTION or SUBROUTINE statement or any other ENTRY statement in that subprogram. Note, however, that all of the entry result variables of a function subprogram must be as described in items 5 and 6 above.

The interface to a procedure defined by an ENTRY statement in an external subprogram may be made explicit in another scoping unit by supplying an interface body for it in a procedure interface block. In this case the procedure heading for the interface body is not an ENTRY statement, but must be a FUNCTION or SUBROUTINE statement, as appropriate.

Although the ENTRY statement is not formally designated as obsolescent in the standard, similar ends can be achieved in a more structured way by using module procedures or internal procedures, with the shared data environment being in the host scoping unit. Those other approaches allow the data environment to be divided into portions that are shared and other portions that are not. Alternate entries allow no such division and consequently are prone to additional errors. Therefore, alternate entries are deprecated by many practitioners.

12.5 Procedure Declaration

A procedure is usually referred to from a different scoping unit than the one that defines it. Therefore, the referring scoping unit has no direct information about the properties of the procedure, or even that it is a procedure. Such information is necessary for implementation. This section discusses the ways in which such information can be provided to a referring scoping unit. The necessary information is called the interface of the procedure; it can be either explicit or implicit, with implications elaborated below.

Information about a procedure can be declared using an interface block, type declaration statement (5.1), EXTERNAL statement (5.10.1), INTRINSIC statement (5.10.2), or PROCEDURE declaration statement (5.11). The declarations can be in the referencing scoping unit or made accessible there by host or use association. This section provides an overview of the use of these procedure-related declarations. There are three forms of interface blocks—specific, abstract, and generic—described in subsections below; the details of the other relevant statements are described in 5.

If a scoping unit refers to an internal procedure or recursively refers to itself, then the full source code for the procedure is at hand and no further information is needed. If a scoping unit refers to a module procedure, then all the necessary information is automatically made available, either via a USE statement or because the scoping unit is in the module. In all of these cases, the interface is explicit in the referring procedure.

If a scoping unit refers to a statement function, then the full source code for the statement function is also at hand and no further information is needed. In this obsolescent case, the interface is implicit for historical reasons, even though everything about the function is evident. There is no way to make the interface of a statement function explicit.

If a scoping unit refers to an intrinsic procedure, then the compiler inherently has all the information about that intrinsic procedure except possibly the fact that an intrinsic procedure is intended; if that fact is specified, the details of the procedure are then known by the compiler. The fact that an intrinsic procedure is intended can be explicitly declared by an INTRINSIC statement or the INTRINSIC attribute in a type declaration statement. In most cases, it is evident from the syntax that a procedure reference is intended. Explicit declaration of the INTRINSIC attribute is optional in those cases. A type declaration for an intrinsic function is also allowed, but it achieves nothing useful; in particular, it does not disallow generic references that return types other than the declared one.

If a scoping unit references an external or dummy procedure, then the compiler does not necessarily have any information about that procedure, which might not be written yet. In most cases, the fact that it is a procedure is evident from context. In those cases no explicit declarations are required if there is no intrinsic procedure of the same name and if nothing about the references requires an explicit interface. The fact that a nonintrinsic procedure is intended can be explicitly declared by an EXTERNAL statement or the EXTERNAL attribute in a type declaration statement. The type of an external or dummy function may be declared explicitly in a type declaration statement. All these forms of declaration make it explicit that the interface is implicit (an unfortunate confusion of terminology).

Alternatively, an external or dummy procedure can be declared by an interface body, which provides an explicit interface, or by a PROCEDURE statement, which can provide either an explicit or implicit interface.

If a scoping unit refers to a procedure pointer, then both the EXTERNAL attribute and the POINTER attribute must be declared explicitly. This can be done with combinations of interface bodies and type declaration, EXTERNAL, POINTER, and PROCEDURE statements. For a procedure pointer component, the declaration is in the type definition. A procedure pointer can have either an implicit or explicit interface.

If a scoping unit refers to a nonintrinsic generic procedure, there must be an interface block to specify that it is generic and to detail the specific procedures that are part of the generic procedure. Defined operations, defined assignment, and defined input/output are always generic and this requirement therefore applies to them.

If a scoping unit refers to a procedure type binding, that type binding is declared in the type definition.

12.5.1 Implicit and Explicit Interfaces

The interface of a procedure is the information necessary for the compiler to implement invocation of the procedure. The information that constitutes the interface is described in 12.5.1.1. If this information is provided to the compiler in a scoping unit, the interface of the procedure is explicit in that scoping unit. If this information is not provided, it must be inferred from the form of the procedure reference; in that case, the interface is implicit.

Many features of procedures require an explicit interface; the information necessary to implement the features cannot be deduced solely from the form of the procedure reference. The situations that require an explicit interface are described in 12.5.1.2.

Even in cases that do not require an explicit interface, having one is a significant help in catching errors. The explicit interface allows the compiler to check that the procedure reference is consistent with the interface; an inconsistency should cause a compilation error message. With an implicit interface, the compiler infers the interface from the form of the reference, so there is no opportunity for independent verification; an inconsistency between the reference and the procedure will often go undetected and result in code that does not work correctly. Some compilers can detect errors in implicit interfaces in some situations (for example, when a procedure and the reference to it are in the same source file), but this cannot be relied on portably. Errors in procedure references are very common sources of bugs in code that uses implicit interfaces.

12.5.1.1 Interface Properties

The properties that constitute a procedure interface are

1. Whether it is a function or a subroutine

2. Whether it is pure or elemental

3. Whether it has the BIND attribute

4. The characteristics of the dummy arguments and result variable

5. The names of the dummy arguments

6. The name, generic identifiers, and binding label of the procedure

Items 1-4 constitute the characteristics of a procedure; there are contexts where only the characteristics are important. Items 1-5 constitute an abstract interface, which might apply to multiple procedures. The addition of item 6 makes the interface unique to a specific procedure.

One characteristic of a dummy argument is its classification as a dummy data argument, dummy procedure, or alternate return.

Additional characteristics of a dummy data argument or result variable are

1. Its type, type parameters, rank, and shape

2. Whether it is polymorphic

3. All of the remaining attributes that could apply to a dummy data argument (ALLOCATABLE, ASYNCHRONOUS, INTENT, OPTIONAL, POINTER, TARGET, VALUE, and VOLATILE)

4. The form of any dynamic dependence in specification expressions for type parameters or bounds

5. Which type parameters, shapes, sizes, or lengths are deferred or assumed

Additional characteristics of a dummy procedure argument are

1. Whether its interface is implicit or explicit

2. Its OPTIONAL or POINTER attributes

3. Its characteristics as a procedure, but only if its interface is explicit

It is notable that the characteristics of a dummy procedure with implicit interface do not even include whether it is a function or subroutine.

A dummy asterisk argument has no additional characteristics.

12.5.1.2 Where an Explicit Interface is Required

Most of the features that were introduced after Fortran 77 and are related to procedure invocation require explicit interfaces.

In particular, the following forms of reference require an explicit interface:

1. keyword actual argument

2. generic name

3. defined operator
4. defined assignment
5. reference in a context that requires it to be pure.

If a procedure has any of the following, any reference to the procedure requires an explicit interface:

1. dummy argument with the ALLOCATABLE, ASYNCHRONOUS, OPTIONAL, POINTER, TARGET, VALUE, or VOLATILE attribute
2. assumed-shape dummy argument
3. dummy argument of parameterized derived type
4. polymorphic dummy argument
5. function result that is an array, pointer, or allocatable
6. function result with a length type parameter value that is determined dynamically
7. elemental attribute
8. BIND attribute

Some of these situations are elaborated on below.

An explicit interface is required for an optional dummy argument or keyword actual argument so that the proper correspondence between actual and dummy arguments can be established.

An assumed-shape dummy argument is generally implemented quite differently from an explicit-shape or assumed-size one. For compatibility with Fortran 77, almost all implementations of explicit-shape or assumed-size dummy arguments expect the invoking procedure to pass the address of a contiguous block of memory where the array is stored. If the array to be passed is a discontiguous slice, the compiler must make a temporary contiguous copy of the slice and pass the address of this copy; after the procedure returns, data from the temporary array is copied back to the original locations. These copy operations can cause a significant performance penalty in some cases; in extreme cases, the whole of a large array might be copied in and out in order to operate on a single element. In contrast, an assumed-shape dummy array is generally implemented in a way that allows a discontiguous slice to be handled directly without copying; this requires that the invoking procedure pass information about how the actual argument elements are stored; this information is often referred to as a dope vector or descriptor, but it is handled internally by the compiler, without the user needing to know about its details. When compiling a procedure that passes an array as an actual argument, the compiler needs to know whether to pass a dope vector or just a storage address. An implicit interface implies that the array will be passed in a way that is compatible with an explicit-shape or assumed-size dummy; if the dummy is assumed shape, there needs to be an explicit interface so that the compiler knows to pass the array in a way that is compatible with that.

If an actual argument is a pointer, then what is passed might be either the pointer itself or the target. Which one is passed depends upon whether or not the dummy argument is a pointer. The explicit interface provides the required information. The default for implicit interfaces is to presume that the dummy argument is not a pointer; thus the target is passed.

For a generic procedure reference there are multiple specific procedures that can be referenced using the same generic name. The referencing routine must be able to figure out which specific procedure is intended; the process of determining the correct specific procedure is called disambiguation. Disambiguation is based on comparing the actual argument list with the multiple dummy argument lists to find one that matches. Generic procedures, including the use of interface blocks for configuring generic names, are discussed in detail in 12.5.4.

User operators and defined assignments are essentially special syntax for some forms of generic procedure references. They require explicit interfaces for exactly the same reasons as other generic procedure references. These topics are treated in detail in 12.5.4.2 and 12.5.4.3.

A reference to an elemental procedure generally requires different implementation from an identical-appearing reference to a nonelemental procedure. An explicit interface is required for the elemental case so that the appropriate code is generated.

The requirements on pure procedures are designed to allow compile-time verification of purity: if a procedure is used in a context that requires purity (such as in a FORALL construct or a specification expression), the interface must be explicit so that the compiler can verify that the interface meets this requirement.

12.5.2 Interface Bodies

An interface body defines an explicit interface for a procedure that does not otherwise have one in the current scoping unit. It is important to understand, however, that an interface body is not the only way for a procedure to have an explicit interface. It is a common error to equate having an explicit interface with having an interface body. For example, a module procedure has an explicit interface wherever it is accessible. It is not necessary, or even allowed, to provide an interface body for a module procedure.

An interface body can appear in any of the three forms of interface block (12.5.3, 12.5.4, 12.5.5). It can never appear outside of an interface block. In isolation, the form of an interface body is indistinguishable from that of a subprogram with no execution part or internal subprogram part; it is distinguished only by the context of being in an interface block.

The form of an interface body (R1205) is either:

```
subroutine-statement
    [ specification-part ]
END [ SUBROUTINE [ subroutine-name ] ]
```

or

```
function-statement
[ specification-part ]
END [ FUNCTION [ function-name ] ]
```

The name of the interface body is the subroutine name or function name.

Rules and restrictions:

1. The interface body specifies all the properties of an interface.

2. An interface body also may specify attributes of entities that are not part of the interface; such specifications have no effect. This provision is to facilitate copying source code from the specification part of a subprogram into an interface body. Such copied code may include declarations of local variables that are not relevant to the interface.

3. An interface body must not contain an ENTRY, DATA, FORMAT, or statement function statement, even though those statements are allowed in the specification part of a subprogram.

4. An entry interface may be specified by using the entry name as the function or subroutine name in an interface body. The property of being an entry is not part of an interface.

5. An interface body does not inherit implicit type mappings from its host. It does not inherit identifiers via host association, except as described below under the IMPORT statement.

6. An interface body may contain a USE statement to access entities via use association.

An interface body is unique among contained scoping units in that it does not by default inherit anything by host association. The original rationale for this was to facilitate coding an interface body by copying the specification part from an external procedure. Host association from the host of the interface body could cause differences between the properties of the interface body and the original procedure.

However, there are some situations where host association into an interface body is highly desirable. Notable among these is an interface body that is in a module and needs to access identifiers from the module. Such an interface body cannot access the module identifiers via USE association because a module must not use itself; furthermore, some of the identifiers might not be public. The IMPORT statement accommodates these situations.

The form of an IMPORT statement (R1209) is

```
IMPORT [ [ :: ] import-name-list ]
```

Rules and restrictions:

1. An IMPORT statement is allowed only in an interface body.

2. Each import name must be a name of an entity accessible in the host of the interface body. If it is a local entity of the host, it must have been explicitly declared prior to the interface body.

3. An entity named in an IMPORT statement is accessible via host association in the interface body. The name must not appear in the interface body in any of the contexts described in 16.2.1.3 as preventing host association.

4. An IMPORT statement without an import name list makes every host entity accessible by host association, except for those entities whose host association is prevented as described in 16.2.1.3. If the interface body accesses a local entity of the host by host association, that entity must have been explicitly declared prior to the interface body. Note the subtle point that an entity that does not meet the requirement of prior explicit declaration is still said to be "made accessible" even though it cannot actually be accessed, contradicting the normal English meaning of the word "accessible". The only effect of such "accessibility" is to block implicit declaration of an entity of the same name in the interface body.

12.5.3 Specific Interface Blocks

A specific interface block provides explicit specific interfaces for procedures. A specific interface is an interface for a specific procedure. A specific interface block also declares the EXTERNAL attribute for the same procedures. It can be used for specific external procedures, dummy procedures, and procedure pointers. It cannot be used for intrinsic, internal, and module procedures, because these already have explicit interfaces. The form of a specific interface block is:

```
INTERFACE
    [ interface-body ] ...
END INTERFACE
```

If the name of an interface body is not that of a dummy procedure or a pointer, it must be that of an external procedure; the characteristics of the interface must match those of the external procedure, except that a nonpure interface may be specified for a pure procedure. Note that the names of dummy arguments are not characteristics and are not required to match.

A procedure must not have more than one explicit specific interface in the same scoping unit. Also, the EXTERNAL attribute for a procedure must not be declared both by an interface body and by other means in the same scoping unit.

12.5.4 Generic Interface Blocks

A generic interface block specifies a generic interface for a set of procedures. Also, it can optionally specify explicit specific interfaces. Generic interface blocks can be used for generic names, defined operators, defined assignment, and defined input/output. The use of generic interfaces is described in the following subsections. The general form of a generic interface block is:

```
INTERFACE generic-spec
    [ interface-spec ] ...
END INTERFACE [ generic-spec ]
```

where a generic specification (R1207) is one of

```
generic-name
OPERATOR ( defined-operator )
ASSIGNMENT(=)
READ (FORMATTED)
READ (UNFORMATTED)
WRITE (FORMATTED)
WRITE (UNFORMATTED)
```

and an interface specification (R1202) is one of

```
interface-body
procedure-statement
```

The form of a procedure statement (R1206) (not to be confused with a procedure declaration statement) is:

```
[ MODULE ] PROCEDURE procedure-name-list
```

Rules and restrictions:

1. If the END INTERFACE statement has a generic specification, it must be the same as the generic specification in the INTERFACE statement. For this purpose, the defined operators <, <=, >, >=, ==, and /= are considered to be the same as .LT., .LE., .GT., .GE., .EQ., and .NE., respectively.
2. A procedure specified by a PROCEDURE statement must be an accessible procedure with an explicit interface in the current scoping unit. It may be an external procedure, module procedure, procedure pointer, or dummy procedure.
3. If a PROCEDURE statement has the optional MODULE keyword, then the procedures specified by that statement must be module procedures.
4. A procedure specified by an interface body must be an accessible procedure that would not otherwise have an explicit interface in the current scoping unit; the interface body provides an explicit interface for the procedure, which may be an external procedure, procedure pointer, or dummy procedure.
5. A procedure may be part of multiple generics as long as it separately meets the requirements for each one. Because of the previous rule, at most one may be specified by an interface body; the others must use PROCEDURE statements.
6. A procedure must not be specified by either a PROCEDURE statement or an interface body if that procedure is previously specified to be part of the same generic in the current scoping unit.

7. There may be multiple interface blocks for the same generic in the same scoping unit. Furthermore, the same generic identifier may be accessed via host association and via use association from multiple modules. The procedures specified for the generic by all such means are part of the same generic in the current scoping unit.

A procedure that is specified by either a PROCEDURE statement or an interface body becomes part of the generic specified by the INTERFACE statement. Additionally, a procedure specified by an interface body acquires an explicit interface in the same manner as for a specific interface block.

Note that the MODULE keyword in the PROCEDURE statement has no effect except to add a restriction. There is little reason to use this keyword in new code; it is primarily for compatibility with existing code.

There is a subtlety in the interaction of rules 6 and 7 above. Rule 6 prohibits redundant specification of a procedure that is previously specified to be part of the same generic in the same scoping unit. A specification accessed by host or use association counts as a previous specification for this purpose. However, the prohibition of rule 6 applies only to interface blocks in the current scoping unit. There is no prohibition against redundant specification via host and use association. Thus, if two different modules specify the same procedure to be in the same generic, that does not prohibit a scoping unit from accessing the generic from both modules under rule 7. This can be useful in cases where the generics in the modules have partial overlap, as can reasonably happen. The following example illustrates this.

```
module intrinsic_types_module
   interface display
      procedure :: display_integer, display_real
   end interface
      ...
end module

module my_type_1_module
   use intrinsic_types_module
   interface display
      procedure :: display_my_type_1
   end interface
      ...
end module

module my_type_2
   use intrinsic_types_module
   interface display
      procedure :: display_my_type_2
   end interface
      ...
end module
```

```
program main
   use intrinsic_types_module, my_type_1_module, my_type_2_module
     ...
end program
```

The first module defines a generic named display, with specific procedures for displaying some intrinsic types. The next two modules independently extend the generic to add specifics for two derived types. The main program uses all three modules, merging their generics. The redundant specification of the specifics display_integer and display_real is allowed because it is done via use association.

12.5.4.1 Generic Procedure Names

A simple example of a generic procedure name is provided by the intrinsic function INT:

```
INT(R)
INT(D)
```

where R and D are respectively default real and double precision objects. It looks like there is only one procedure involved here (INT), but there are really two. There is a specific procedure lurking around that accepts a default real argument and another one that accepts a double precision argument. Because the purpose of these two procedures is so similar, it is desirable to refer to each of them with the same generic name. This is sometimes referred to as overloading the name. The type of the argument is sufficient to identify which of the specific procedures is involved in a given reference.

The term generic refers to a set of different specific procedures that all have the same generic name. All of the intrinsic procedures are generic. Users can define additional generics or add additional specific procedures to intrinsic generics. The mechanism for this is a generic interface block with a generic specification that is a generic name.

A simple example of such a generic interface block is:

```
INTERFACE INT
MODULE PROCEDURE RATIONAL_TO_INTEGER
END INTERFACE INT
```

which adds a user-defined procedure to the intrinsic generic INT.

A generic name may be the same as one of the specific names of the procedures included in the generic set, but the authors recommend against this practice as potentially confusing.

A generic name may be the same as the name of a derived type. In this case, the specific procedures must be functions. A reference to such a generic can appear identical in form to a structure constructor for the type (4.4.15), in which case the generic essentially overrides the structure constructor. A form is interpreted as a structure constructor only if it does not resolve to a generic function reference.

If two specific procedures are part of the same generic, the interfaces to those procedures must be sufficiently different so that it can be uniquely determined which specific procedure is referenced by any particular reference to the generic. This uniqueness must hold for all possible forms of invocation, including positional and keyword actual argument specification and omitted optional arguments. Such uniqueness is required even if the program does not reference the generic using all the forms. The requirements are on the construction of the generic in the first place; once a valid generic is constructed, there are no possible forms of reference that are ambiguous. The limitations are conservative in that there are cases that are disallowed even though it can be shown that they cannot possibly cause ambiguity. Thus, in order to determine whether two procedures are sufficiently different to be allowed in the same generic, one must check the particular limitations specified in the standard; it is not sufficient to prove uniqueness by some other reasoning.

The essence of the standard's limitations is uniqueness with respect to type, kind, and rank patterns under both positional or keyword references. The precise rules are as follows. Two specific procedures that have the same generic name in a scoping unit must be distinguishable as specified by these rules, regardless of how they acquired the generic names; for example, the procedures might have been specified in the same generic interface block in the scoping unit in question, or they might have been given the same generic name in two different modules, both of which are accessed by the scoping unit in question.

The rules for distinguishability of specific procedures are built on the definition of distinguishability of dummy arguments. Two dummy arguments are distinguishable if neither is a subroutine and they differ in at least one of the following ways:

1. Neither is type-compatible (5.2) with the other. In the simple nonpolymorphic case, this just means that the two types are different.

2. They differ in the value of a kind type parameter.

3. They differ in rank.

Two specific procedures with the same generic name in a scoping unit must differ in at least one of the following ways:

1. Passed-object dummy. Both have passed-object dummy arguments (4.4.8), and those are distinguishable.

2. Argument count. The idea of this way of distinguishing is that one has more nonoptional dummy arguments of a given pattern than the other has possible matches for. Only dummy data objects are considered for this rule; also any passed-object dummy argument is ignored. For the nonpolymorphic case, the pattern in question consists of the type, kind parameter values, and rank.

   The polymorphic case is more complicated. A dummy argument of one procedure is selected. A count is made of the number of nonoptional dummy arguments in that procedure that are type compatible with the selected dummy argument and have matching kind type parameter values and rank; this count includes the selected dummy argument if it is nonoptional. Then a count is made of the number of arguments in the other procedure that are distinguishable with the selected dummy argument. If there is any dummy argument for which the first count is greater than the second, the procedures differ sufficiently.

3. Position and name. This criterion is the most complicated, but is also the most common. For this criterion, one of the procedures must have both a dummy argument that disambiguates by position and a dummy argument that disambiguates by name. Furthermore, the one that disambiguates by position must either be the same as the one that disambiguates by name or be prior to it.

   An argument disambiguates by position if it is nonoptional and either the argument in the corresponding position of the other procedure is distinguishable or there is no such corresponding position. Any passed-object dummy arguments are ignored for the purpose of evaluating this position correspondence.

   An argument disambiguates by name if it is nonoptional and either the argument of that name in the other procedure is distinguishable or there is no argument of the same name.

For an example, consider a simple two-argument subroutine G (P, Q) with generic name G, dummy argument names P and Q, and neither argument optional. A reference to G, with actual arguments X and Y, could take any of the following four forms:

```
CALLG(X,Y)
CALLG(X,Q=Y)
CALLG(P=X,Q=Y)
CALLG(Q=Y,P=X)
```

What subroutine H could be added to the generic set with G? The first of the above four calls rules out any two-argument H whose first argument has the same type, kind, and rank as the P argument of G and whose second argument has the same type, kind, and rank as the Q argument of G. The third and fourth of these four calls rules out any subroutine H of the form H (Q, P), whose first argument is named Q and has the same type, kind, and tank as the Q (second) argument of G and whose second argument is named P and has the same type, kind, and rank as the P (first) argument of G. The reason for this last case is that a reference to H in which all the actual arguments had keywords would look exactly like a call to G; such a reference would not be uniquely resolvable to either a call to G only or to H only. Any other H could be included in the generic set with G.

Giving a generic name to a procedure provides an additional name by which the procedure can be referenced, but it does not remove any other names. A specific procedure can be referenced by multiple generic names and also by its specific name. However, it is possible for a procedure to be accessible only by a generic name in some scoping units; this can be achieved by making a module in which the specific name is private, but a generic name is public.

12.5.4.2 Defined Operations

An operator (7.1.1.1) can be viewed as a special syntax for special cases of function references. An operator takes one or two values and computes a result that is then used in an expression. A function can do that and more. Thus, a language could easily be defined without operators. One could write something like negate(mult(add(a,b),c)) instead of –(a+b)*c, but the form using operators is simpler to work with and is far more Fortran-like. Just as generic functions are useful, so are generic operators. In fact, all operators are generic; there is no concept of a specific operator. Generic operators are composed from specific functions.

The intrinsic operators come in two forms: symbols such as + and //, and forms with dots and letter sequences such as .NOT. and .EQ. Users can define additional generic operators or add additional specific procedures to intrinsic generic operators. Additional user-defined operators must be in the form with dots and letter sequences; there are no user-defined symbols, but the user can add specific procedures for the intrinsic symbols.

There are two mechanisms for specifying a defined operation: a generic binding (4.4.11.2) or a generic interface block with a generic specification

```
OPERATOR ( defined-operator )
```

The two mechanisms can be intermixed. Both mechanisms follow the same rules except for scoping. A generic interface bock has local scope, most commonly in a module. A generic binding for a type is accessible in any scoping unit where an entity of the type is accessible.

An operator is either unary or binary. A unary operator operates on a single value, which follows the operator in an expression. A unary operator is sometimes called a prefix operator because it precedes its operand. The intrinsic operator .NOT. is an example of a unary operator. An example of a user-defined unary operator is the .INVERSE. operator in 11.3.9.3. A binary operator operates on two values, which precede and follow it in an expression. A binary operator is sometimes called an infix operator because it is in between its operands. The intrinsic operators * and .OR. are examples of binary operators. Some operators have both unary and binary forms, distinguished by parsing the expression they are in. The intrinsic operator – is an example of this; it is a unary operator in the expression –x, but a binary one in x–y.

A specific procedure specified for a defined operator must meet several requirements in order to fit the ways in which operators are used:

1. The procedure must be a function. The function result is the result of the operation.

2. The function must have either one or two dummy arguments.

3. If the function has one dummy argument, it is a specific for a unary operator. The single operand of the unary operator is the actual argument for the dummy.

4. If the function has two dummy arguments, it is a specific for a binary operator. The leftmost operand of the operator is the actual argument corresponding to the first dummy argument; the rightmost operand corresponds to the second dummy argument.

5. If the operator is an intrinsic one, the number of dummy arguments must be consistent with the intrinsic usage. That is, if the operator is an intrinsic unary operator, there must be exactly one dummy argument; if the operator is an intrinsic binary operator, there must be two dummy arguments; if the operator is an intrinsic one with both unary and binary forms, there may be either one or two dummy arguments.

6. All dummy arguments must be nonoptional data objects and have INTENT (IN).

7. The function result must not have assumed length because there is no way to specify the length in the referencing scope as is required for assumed-length functions (4.3.5.1(4)(d)).

8. If the operator is an intrinsic one, the function dummy arguments must not be compatible with the rules for an intrinsically defined operation. There must be a difference in type, kind, or rank.

9. A defined operator identifier must have no more than 63 letters and must not be .TRUE. or .FALSE. Those identifiers are effectively reserved.

Rule 8 above means, for example, that "+" cannot be specified in an operator interface for a function with a scalar integer argument and a scalar real argument, because "+" already has a meaning for this type, kind, and rank pattern. Specifying such an operator extension would mean that I+R, where I is a scalar integer and R is a scalar real, would be ambiguous between the intrinsic meaning and the extended meaning. The user is not allowed to override the intrinsic meaning.

This rule reflects a difference in philosophy on extending intrinsic procedures versus extending intrinsic operators. A user may override a specific intrinsic procedure, but may not override a specific intrinsic operator. This is because the intrinsic operators are relatively few in number and are considered fundamental to the language. A programmer can reasonably be expected to know all of the intrinsic operators so that he can avoid conflicts with them when selecting names for user-defined operators. It would likely cause great confusion and maintenance problems if a user redefined something as fundamental as addition of default integers; if some special treatment of such addition is needed, it is always possible to express it as a function, which would make its special nature more obvious.

On the other hand, the number of intrinsic procedures is relatively large and reasonably likely to increase with subsequent standards. It would be slightly inconvenient to a programmer to make sure that he selects generic names that did not conflict with any intrinsic generics, but there is no way for a programmer to make sure that he selects generic names that will not conflict with any future intrinsic generics. The potential for confusion caused by overriding intrinsic generic procedures is also lower than that caused by overriding intrinsic operators.

The syntax for referencing functions via operators is more restrictive than the general syntax for function reference. In particular, there are no optional arguments or keyword forms to deal with. Therefore, the restrictions needed to ensure uniqueness of generic operator resolution are simpler than the corresponding restrictions for generic names. For each operator, no two specific procedures may have the same argument type, kind, and rank pattern on the basis of argument position, but there are no requirements relating to dummy argument names.

An example of a generic interface block for an operator is:

```
INTERFACE OPERATOR(+)
   FUNCTION INTEGER_PLUS_INTERVAL(X, Y)
      USE INTERVAL_ARITHMETIC
      TYPE(INTERVAL) :: INTEGER_PLUS_INTERVAL
      INTEGER, INTENT(IN) :: X
      TYPE(INTERVAL), INTENT(IN) :: Y
   END FUNCTION INTEGER_PLUS_INTERVAL
   PROCEDURE RATIONAL_ADD
END INTERFACE
```

which extends the "+" operator with two functions, a function INTEGER_PLUS_INTERVAL which presumably computes an appropriate value for the sum of an integer value and something called an "interval", and a function RATIONAL_ADD which probably computes the sum of two "rational numbers". Both functions now can be referenced in the form A+B, where A and B are the two actual arguments. An example defining a new operator, rather than extending an existing operator, is:

```
INTERFACE OPERATOR(.INVERSETIMES.)
   PROCEDURE MATRIX_INVERSE_TIMES
END INTERFACE (.INVERSETIMES.)
```

Now the inverse of matrix A can be multiplied by B using the expression A .INVERSETIMES. B, which produces $A^{-1}B$, and in effect solves the system of linear equations, $Ax = B$ for x.

A function with an operator interface may be referenced with the operator form, but it also may be referenced via the traditional functional form using the specific function name.

12.5.4.3 Defined Assignments

As with defined operators, an assignment can be viewed as a special syntax for a procedure reference. An assignment assigns a value to the variable; the value assigned depends on the value of an expression. A subroutine can do that; one could write something like "call assign(x,(y))" instead of x=y. Both forms are allowed. There is only one assignment symbol, as opposed to the multiplicity of operators.

Assignment is not always the trivial matter that might be implied by its terminology. The value assigned depends on the value of the expression, but is not necessarily the same as the value of the expression. For example, in assigning a real expression to an integer variable, the value assigned is obtained by truncation and conversion; in assigning a scalar real expression to a real array variable, the value assigned is obtained by replicating copies of the scalar value. The specifics of the assignment depend on the types, kinds, and ranks of the expression and variable; this ought to bring generics to mind.

A programmer can specify additional cases of assignment to be implemented by specific procedures. There are two mechanisms for specifying defined assignment: a generic binding (4.4.11.2) or a generic interface block with the generic specification

```
ASSIGNMENT (=)
```

As with defined operations, the two methods follow the same rules except for scope.

A specific procedure specified for a defined assignment must meet the following requirements in order to be consistent with the ways in which assignment is used:

1. The procedure must be a subroutine.
2. It must have exactly two arguments, both of which must be nonoptional.
3. The first argument must have either INTENT (OUT) or INTENT (INOUT). The variable in the assignment is the actual argument corresponding to this dummy argument.
4. The second argument must have INTENT (IN). The expression in the assignment is the actual argument corresponding to this dummy argument.
5. The dummy arguments must not be compatible with the rules for intrinsic assignment of intrinsic types (7.5.2). If both arguments are of intrinsic type, then their types, kinds, or ranks must be incompatible with intrinsic assignment.

Rule 5 above means, for example, that the user cannot define assignment of a scalar default integer expression to a rank 2 default real array variable; that assignment is defined intrinsically and cannot be overridden. However, the user could define assignment of a rank 2 default integer expression to a scalar default real variable, or assignment of a scalar default integer to a scalar default character; neither of these assignments are defined intrinsically. The user may define assignments involving derived types, even where there is a conflicting intrinsic assignment; this overrides the intrinsic assignment for assignment statements.

The restrictions needed to ensure uniqueness of generic assignment resolution are the same as those for generic operators. No two specific procedures may have the same argument type, kind, and rank pattern on the basis of argument position, but there are no requirements relating to dummy argument names.

An example of an assignment interface block is:

```
INTERFACE ASSIGNMENT(=)
   SUBROUTINE ASSIGN_STRING_TO_CHARACTER(C,S)
      USE STRING_DATA
      CHARACTER(*), INTENT(OUT) :: C
      TYPE(STRING), INTENT(IN) :: S
   END SUBROUTINE ASSIGN_STRING_TO_CHARACTER
   PROCEDURE INTEGER_TO_RATIONAL
END INTERFACE ASSIGNMENT(=)
```

This interface block allows ASSIGN_STRING_TO_CHARACTER (which extracts the character value) to be referenced in the form:

```
C = S
```

In addition, INTEGER_TO_RATIONAL may be referenced in the form

```
R = K
```

where R is of derived type RATIONAL and K is an integer. The purpose of INTEGER_TO_RATIONAL presumably is to convert an integer value into the appropriate RATIONAL form.

A subroutine with an assignment interface may be referenced with the assignment statement form, but it also may be referenced via the traditional CALL statement form using the specific subroutine name.

12.5.4.4 Defined Input/Output

Defined input/output is described in detail in 9.5.1.4. It is similar to defined operators and defined assignment in that it involves user-defined specific procedures for a generic whose use is tied to language features. In the case of defined input/output, the language feature in question is input/output.

The mechanisms for specifying defined input/output are a generic binding (4.4.11.2) or a generic interface block with a generic specification that is READ(FORMATTED), READ(UNFORMATTED), WRITE(FORMATTED), or WRITE(UNFORMATTED).

The requirements for the specific procedures are detailed in 9.5.1.4. One aspect of the requirements worth restating is that the DTV argument of the specific procedures must be of derived type; defined input/output is not allowed for intrinsic types. The standard defines input/output for intrinsic types and the user may not override the intrinsic definitions. The standard also defines input/output for derived types that meet some restrictions (9.4.4), but the user can override those definitions. For other derived types, there is no definition of input/output unless the user provides one.

The uniqueness requirement for defined input/output is based only on the type and kind of the DTV argument; the rank of DTV is always zero, arrays being handled as described in 9.4.4. For each of the four input/output generics, no two specific procedures may have the same type and kind for the dtv argument. The other arguments are all specified by the standard to be the same for each specific procedure.

Examples of defined input/output are in 9.5.1.4.

As with other generic procedures, a procedure with a defined input/output interface may also be referenced via the traditional CALL statement form using the specific subroutine name. For the most part, defined input/output does little more than provide an alternate syntax for functionality that is also otherwise available. There is one exception, which relates to input/output of multiple entities in the same record. Defined input/output allows a single input/output statement to reference multiple specific procedures, one for each effective list item, while processing a single record. For unformatted input/output, if the multiple procedures are referenced via CALL statements, they will necessarily process separate records.

12.5.5 Abstract Interface Blocks

An abstract interface block defines abstract interfaces. An **abstract interface** is an interface that is defined on its own, independent of any particular procedure. It gives a name to an interface so that the same interface can be subsequently specified for multiple procedures without needing to duplicate the same lines of source code. It is analogous in this sense to a type definition. It is particularly useful in declarations of procedure pointers. The form of an abstract interface block is:

```
ABSTRACT INTERFACE
  [ interface-body ] ...
END INTERFACE
```

The keyword ABSTRACT is the only difference between this and the form of a specific interface block.

The name of an interface body in an abstract interface block is the name of the abstract interface rather than the name of any procedure.

The following is an example of using an abstract interface to declare two procedure pointers.

```
abstract interface
   function f(x)
      real, intent(in) :: x
      real :: f
   end function f
end interface
procedure(f), pointer :: p, q
```

12.6 Argument Association

An actual argument is specified in a procedure reference. An actual argument can be a variable, an expression, a procedure designator, or an alternate return specifier. Different references to the same procedure can use different actual arguments.

A dummy argument has no utility on its own. It is like a spirit, which must be linked to a physical body to have concrete existence. When the procedure is referenced, an actual argument gives concrete meaning to the dummy. Within the procedure, the dummy argument stands for the actual argument in some sense. Thus, during execution of a procedure reference, the appropriate connection must be established between the actual arguments specified in the reference and the dummy arguments defined within the procedure. This connection is called **argument association**.

Argument association causes the dummy argument to act like a temporary local name for the actual argument. There is a close parallel between argument association and pointer association—in many respects a dummy argument is like a pointer, and an actual argument is like a target. A dummy argument provides a name which can be used to refer to its corresponding actual argument much in the same way that a pointer provides a name which can be used to refer to its target.

In some cases, a dummy argument is implemented with separate storage all of its own; data is copied from and to the actual argument as appropriate. In other cases, a dummy argument is implemented as a reference to the storage location of the actual argument. The standard intentionally defines argument association in such a way that multiple implementation mechanisms are allowed. Different compilers may make different implementation choices for the same source code. The restrictions in 12.6.10 are largely to ensure that standard-conforming code does not depend on which implementation method the compiler selects.

The general model of the standard is that the actual and dummy arguments are separate entities that are related by argument association. The relationship might be very "close", as in when they are different names for the same storage, or it might be more distant, as when they have separate storage.

In most cases the actual argument is associated with the corresponding dummy argument as described above. There are two exceptions, where the entity associated with the dummy argument is one other than the actual argument itself. The first exception is where a pointer actual argument corresponds to a nonpointer dummy argument; in this case, the entity associated with the dummy argument is the target of the pointer—not the pointer itself. The second exception is where a dummy argument has the VALUE attribute; in this case, the entity associated with the dummy argument is a copy of the actual argument.

The standard is imprecise in its terminology in that it uses the term "actual argument" both for the entity that appears in the procedure reference and for the entity that is associated with the dummy argument. In most cases these two entities are the same, and in many other cases the distinction has no important consequences, but the ambiguous terminology has generated questions of interpretation of a few special situations. In particular, if the actual argument is specified by an expression, it is always classified as not definable; this is so even if the expression is a function reference that returns a pointer to a definable target.

A variable is distinguished from a general expression in the context of an actual argument in that a variable is definable, whereas other forms of expressions are not. This affects the interpretation of the restrictions related to the INTENT attribute (5.9.1). For example, the actual argument associated with an INTENT (OUT) dummy must be definable and thus cannot be an expression other than a variable. Note that (A), where A is a variable, is not a variable, but a more general expression.

12.6.1 Argument Correspondence

In general there can be multiple dummy arguments in a procedure and multiple actual arguments in a reference to the procedure. A correspondence must be established between the actual and dummy arguments. There are three ways of establishing such a correspondence: positional, keyword, or passed object; mixtures of these three ways can also be used.

The simplest and most common way of establishing argument correspondence is positional. Positional argument correspondence is illustrated in Figure 12-1. The form of

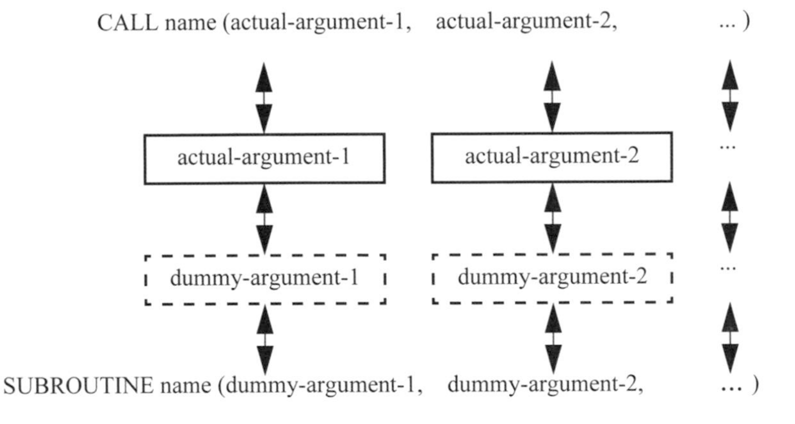

Figure 12-1 Actual and dummy argument lists

a positional argument list in a reference is a comma-separated sequence of actual arguments. The form of the argument list in a procedure definition is a comma-separated list of dummy arguments. The arguments are numbered from left to right. The first dummy argument corresponds to the first actual argument, the second dummy argument corresponds to the second actual argument, and so on.

The actual argument list must not be longer than the dummy argument list. If the actual argument list is shorter, it corresponds to the first part of the dummy argument list; the remaining dummy arguments have no corresponding actual arguments and must be declared optional. Positional argument correspondence can be used for any procedure.

In Figure 12-1 the actual arguments are shown as solid boxes because these represent concrete data values, variables, or other actual argument forms. The dummy arguments are shown as dashed boxes to indicate that they are virtual. The dummy arguments represent "empty names" until they become associated with actual arguments.

Keyword argument correspondence provides an order-independent way of constructing an actual argument list. The name of the dummy argument is used as a keyword to specify which dummy argument corresponds to the actual argument. The form of such a specification is:

dummy-argument-name = actual-argument

This form may be used to establish the correspondence of an actual argument with any named dummy argument. It cannot be used with an alternate return because the corresponding dummy argument has no name. In an actual argument list using keywords, the arguments may be in any order.

Keyword actual argument specification can aid readability by decreasing the need to remember the precise sequence of dummy arguments in dummy argument lists. The order-independence also facilitates more flexible use of optional arguments by not requiring omitted arguments to be at the end of the list. These functionalities, and the forms that they take, are modeled after keyword specifiers in input/output statements.

Keyword argument correspondence can be used only where the interface of the procedure is explicit; the explicit interface is needed to provide the dummy argument names. If the explicit interface is provided by an interface body, the keywords are the dummy argument names specified in the interface body, which can differ from those specified in the procedure subprogram.

A procedure reference may use a mixture of positional and keyword correspondence, specifying keywords for some actual arguments but not for others. In this case, all the actual arguments without keywords must appear prior to those with keywords in the actual argument list. After the appearance of the first keyword actual argument, all subsequent actual arguments in the list must use keywords. Examples are:

```
CALL GO(X, HT=40)

CALL TELL(XYLOPHONE, NET=10, QP=PI/6)
```

The actual arguments without keywords correspond positionally to the first arguments in the dummy argument list. The actual arguments with keywords correspond by keyword and may be in any order, except for the requirement that they must be after the positional arguments.

The third way of establishing argument correspondence involves a passed-object dummy argument. A passed-object dummy argument applies only to references to a procedure pointer component or a procedure type binding. The actual argument corresponding to the passed-object dummy argument is not explicitly written in the actual argument list, but is taken from the form of the reference as described in detail in 4.4.8. A procedure reference can have at most one passed object. Any other actual arguments must be specified using positional and/or keyword forms. The passed-object dummy argument, having been separately handled, is omitted from the dummy argument list used for positional and keyword correspondence.

Regardless of the means used to establish argument correspondence, each reference to a procedure must have one and only one actual argument corresponding to each nonoptional dummy argument and at most one actual argument corresponding to each optional dummy argument.

12.6.2 Optional Arguments

A dummy argument may be specified to be optional. This means that an actual argument need not be supplied for it in a particular invocation. An optional dummy argument is so specified by giving it the OPTIONAL attribute (5.9.3). For a particular invocation of a procedure, an optional dummy argument is said to be present if it is associated with an actual argument.

There are two ways for a procedure invocation to result in a dummy argument that is not present. The first way is for the procedure reference to have no actual argument that corresponds to the dummy argument. In a positional actual argument list, optional arguments at the end of the dummy argument list can be omitted by supplying an actual argument list that is shorter than the dummy argument list; with keyword actual arguments, an optional argument is omitted by not using its argument name as a keyword.

The second way to have a dummy argument that is not present involves multiple levels of procedure references with optional arguments, as illustrated in the following example:

```
CALL TELL(1.3, T=F(K))
   ...
SUBROUTINE TELL(X, N, T)
   OPTIONAL N, T
      ...
   CALL WILLIAM(X, N, T)
END

SUBROUTINE WILLIAM(XX, NN, TT)
OPTIONAL NN,TT
   ...
END
```

In subroutine TELL, the dummy arguments X and T are present for the invocation shown because they have corresponding actual arguments; the dummy argument N has no corresponding actual argument and so is not present. In subroutine WILLIAM, all three dummy arguments have corresponding actual arguments. However, the actual argument corresponding to NN is N, which is, as just noted, not present for this invocation. In this situation, NN is also not present. Even though the form of reference to subroutine WILLIAM has an actual argument corresponding to NN, that actual argument is not present; nowhere is there anything concrete for NN to be associated with. If N is present in a different invocation of TELL, then NN is also present in the resulting invocation of WILLIAM.

During execution of a procedure with an optional dummy argument, it is often necessary to know whether the dummy argument is associated with an actual argument for the particular invocation. The PRESENT intrinsic function answers that question. It is an inquiry function and has one argument, which must be an optional dummy argument in the procedure. The result is a default logical, which has the value true if the dummy argument is associated with an actual argument and false otherwise. Note that the PRESENT intrinsic function must not be used on a nonoptional dummy argument; one might imagine that this would always return a result of true, but instead it is disallowed. An example of the use of the PRESENT intrinsic is:

```
IF (PRESENT(NUM_CHAR)) THEN
   ! Processing if an actual argument has been
   !    supplied for optional dummy argument NUM_CHAR
   USABLE_NUM_CHAR = NUM_CHAR
ELSE
   ! Processing if nothing is supplied for NUM_CHAR
   USABLE_NUM_CHAR = DEFAULT_NUM_CHAR
END IF
```

This illustrates how the PRESENT function can be used to control the processing in the procedure as is appropriate depending on the presence or absence of an optional argument.

An optional dummy argument that is not present is subject to several restrictions. The restrictions mostly rule out things that would make no sense for a nonpresent dummy; they could be summarized loosely by saying that you cannot do much in that case. The programmer is responsible for following these restrictions; typically, but not necessarily, the portions of the code that would violate these restrictions are bypassed depending on the result of the PRESENT intrinsic. The restrictions that apply to a nonpresent dummy argument are:

Rules and restrictions:

1. It must not be referenced or defined.
2. If it is of a type with default initialization, the initialization does not occur.
3. It must not be the target of a pointer assignment.
4. If it is a procedure, it must not be invoked.
5. It must not be supplied as an actual argument corresponding to a nonoptional dummy argument, except in a reference to the PRESENT intrinsic function, a reference to the NULL intrinsic function that meets the requirements of 7.4.1(1)(g), or a specification inquiry that meets the requirements of 7.4.1(7-8).
6. A subobject of it must not be supplied as an actual argument.
7. It must not be used to determine the shape of an elemental procedure reference.
8. If it is a pointer, it must not be allocated, deallocated, nullified, pointer-assigned, or passed as an actual argument to a nonpointer dummy argument.
9. If it is allocatable, it must not be allocated, deallocated, or passed as an actual argument to a nonallocatable dummy argument.
10. If it has length type parameters, they must not be inquired about.
11. It must not be used as a selector in a SELECT TYPE or ASSOCIATE statement.

There are a few fairly common things that can be done with an optional dummy argument that is not present. It can be used as an actual argument corresponding to an optional dummy argument. It can also be used as an actual argument to some intrinsic functions as described in the exceptions in rule 5 above.

If a procedure has any optional arguments, then the interface of the procedure must be explicit in any scoping unit where the procedure is referenced. This requirement holds regardless of whether or not any arguments are omitted from the actual argument list. It is possible for optional arguments to need special treatment by an implementation; the explicit interface provides the information necessary to control such special treatment.

12.6.3 Type and Type Parameters

A dummy data argument, like any data object, has a type and a set (possibly empty) of type parameters. An actual argument corresponding to a dummy data argument must be a data entity. The type and type parameters of the actual and dummy arguments must agree as described below.

The dummy argument must be type compatible (5.2) with the actual argument. This type compatibility requirement can be separated into four cases. In the simplest and by far most common case, the dummy and actual arguments are both nonpolymorphic, in which case they must be of the same type. See 4.4.10 for a description of exactly what it means to be of the same type, which is not always as obvious as one might think. For a nonpolymorphic dummy argument with a polymorphic actual argument, the declared type of the actual argument must be the same as the type of the dummy argument; the dynamic type of the actual argument can then be an extension of the type of the dummy, in which case the dummy argument essentially operates on the nonextended part of the actual argument. For a limited polymorphic dummy argument, the actual argument must have a declared type that is either the same as or an extension of the declared type of the dummy argument; this ensures that the dynamic type of the actual argument will be one that is allowed as a dynamic type of the dummy argument. For an unlimited polymorphic dummy argument, the actual argument can be of any type.

If a dummy argument is a pointer or allocatable, then the requirements for type agreement are more stringent. In either of these cases, the declared type of the actual argument must be the same as the declared type of the dummy argument; furthermore, the actual argument must be polymorphic if and only if the dummy argument is.

In most cases, the values of the corresponding type parameters of actual and dummy arguments must agree. An assumed type parameter of a dummy argument gets its value from that of the actual argument and thus always counts as agreeing for this purpose. A pointer or allocatable dummy argument may have deferred type parameters. A deferred type parameter for a dummy argument agrees only with a deferred type parameter for the actual argument. This is because the value of a deferred type parameter may change during execution of the procedure; the actual argument must be capable of accommodating the same change.

The exception to the requirement for type parameter agreement involves the length parameter of character type. The exception is specific to default character kind and C character kind (which might or might not be the same kind as default). If a character dummy argument of either of these kinds is a scalar, an explicit-shape array, or an assumed-size array, then the length parameter of the actual argument need not agree with that of the dummy argument. If the dummy argument is scalar, the actual argument must be at least as long as the dummy; the dummy argument is associated with the leftmost characters of the actual. The array case is more complicated and is discussed in the next section.

For default character kind, this exception is largely historical. This exception was extended to C character kind to facilitate interoperation with C as illustrated in 15.5.3. Other character kinds might have representations that are incompatible with an exception like this; a character string of length n might not have the same representation as an array of n characters of length 1. Therefore, the length parameters for other character kinds must agree.

In most situations, it is strongly recommended that character dummy arguments not be declared with explicit length; use assumed length or deferred length instead. If an explicit-length character dummy is used, using a shorter actual argument is an error.

12.6.4 Array Arguments

There are several significantly different means of dealing with array arguments. Elemental procedures have scalar dummy arguments, but allow conformable array actual arguments to be used; they are described in 12.7.2. Pointer and allocatable dummy arguments may be arrays and are described in 12.6.5 and 12.6.7, respectively. The association of nonpointer, nonallocatable array arguments is called sequence association and is based on array element order, as described in 6.6.6.

12.6.4.1 Sequence Association

For a nonelemental procedure, the general rule is that an actual argument and its corresponding dummy argument must either both be scalar or both be arrays. There are two exceptional cases of sequence association, where a scalar actual argument may be treated essentially as though it were an array. There are exceptions described later, but in the usual case, an array actual argument corresponds to an array dummy argument.

In sequence association, an actual argument array is treated as a linear sequence of elements in array element order.

Similarly, the dummy array argument is treated as a linear sequence of elements, in array element order. Each element in the dummy argument sequence is associated with the corresponding element in the actual argument sequence.

In the simplest and most common case, the bounds of the actual and dummy argument arrays are the same. Each element of the dummy argument then corresponds to the element with the same index values in the actual argument. Almost as simple is the case where the shape of the actual and dummy arguments is the same, but the lower bounds differ. In that case, the corresponding index values are offset by the difference in the lower bounds. For example, if an actual array argument is declared as AA(5,5,5) and the corresponding dummy array argument is declared as DA(2:6,1:5,0:4), the difference of the lower bounds is +1 for the first dimension, 0 for the second dimension, and -1 for the third dimension. Thus, the dummy element DA(2,1,0) corresponds to the actual element AA(1,1,1), the dummy element DA(6,5,4) corresponds to the actual element AA(5,5,5), and the dummy element DA(3,3,3) corresponds to the actual element AA(2,3,4).

If the dummy argument is an assumed-shape array, the above simple cases are the only ones that can occur. By definition, an assumed-shape array has the same shape as its corresponding actual array, although the lower bounds may differ. The actual argument corresponding to an assumed-shape dummy must not be assumed size because the needed shape information is not available to be assumed; however, a section of an assumed-size array where the upper bound in the last dimension is specified is allowed because the section has complete shape information.

If the dummy argument is an explicit-shape or assumed-size array, the shape of the actual and dummy arguments can be different. The element correspondence is not necessarily immediately evident without laboriously laying out the element sequences. Such a layout is illustrated in Figure 12-2, in which a three-dimensional actual array is associated with a two-dimensional dummy array. In this example, the actual argument element AA(1,2,2) becomes associated with dummy argument element DA(4,2), for example. The size of the dummy array must not exceed the size of the actual array; any excess elements in the actual argument element sequence, such as AA(1,3,2) and AA(2,3,2) in this example, are not associated with any dummy element. An assumed-size dummy array extends to and "cuts off" at the end of the actual argument array sequence.

Figure 12-2 Example of array element sequence association

If the actual argument is an array section, the actual argument sequence is formed from the section, rather than the original array. For example, if the actual argument is the section A(9:1:-2) then actual argument element sequence is A(9), A(7), A(5), A(3), A(1). If the corresponding dummy array is declared as D(2,2) then A(9) is associated with D(1,1), A(7) is associated with D(2,1), A(5) is associated with D(2,1), and A(3) is associated with D(2,2). A(1) is not associated with a dummy element.

The first exceptional case of sequence association is largely historical, from versions of the language prior to the introduction of array section syntax. The use of this exception is discouraged in new code, where the same ends can usually be achieved with an array section as an actual argument. An actual argument that is an element of an explicit-shape, assumed-size, or allocatable array may be sequence associated with a dummy array that is explicit shape or assumed size. The actual argument element sequence for this case begins with the specified array element and extends to the end of the element sequence of the array. A problematic consequence of this case is that it is not necessarily evident, when compiling the procedure reference, whether an array element is to be interpreted as such an element sequence or as a scalar; that determination depends on the rank of the dummy argument, which might not be available when the reference is being compiled. The necessity for the compiled reference to work in either case places significant restrictions on the implementation.

The second exceptional case of sequence association applies to the default character and C character kinds. As mentioned previously, the actual and dummy arguments for these character kinds may have different character length parameters. If a dummy argument of either of these character kinds is an explicit-shape or assumed-size array, the sequence association is on a character-by-character basis rather than element-by-element. The actual and dummy argument character sequences are formed by concatenating the characters in all the array elements, in array element order; the corresponding characters in these sequences are associated. If the actual and dummy character lengths are the same, this has the same effect as element-by-element association; otherwise, it does not. The number of characters in the dummy character sequence must not exceed the number of characters in the actual character sequence.

Figure 12-3 illustrates character-by-character sequence association. The dummy argument element DA(4) to be associated with the last character of the actual argument element AA(2,1) and the first two characters of actual argument element AA(1,2).

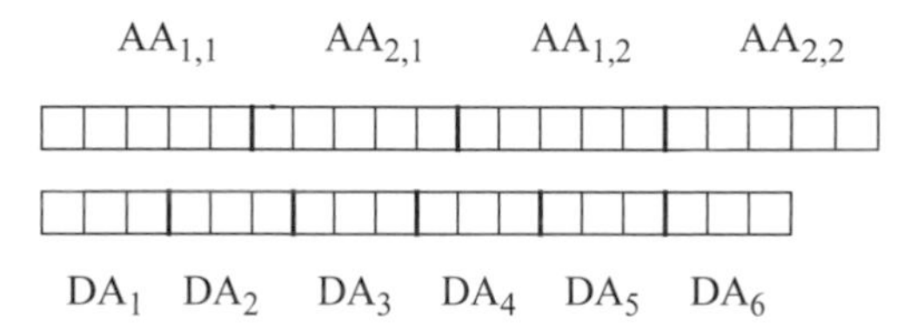

Dummy array: CHARACTER (3) DA(6)

Figure 12-3 Array element sequence association for default characters

A combination of the two special cases applies if the dummy argument is an explicit-shape or assumed-size character array and the actual argument is an element or a substring of an element of an explicit-shape, assumed-size, or allocatable array. The actual character element sequence begins with the first character of the specified element or substring and extends to the last character of the last element in the element sequence.

12.6.4.2 Noncontiguous Arrays

Noncontiguous array actual arguments merit special attention. The main reason for the importance involves interoperability with other languages. Other languages will not generally be able to directly accommodate noncontiguous array arguments. For this purpose, versions of Fortran prior to Fortran 90 also count as "another language" of particular importance. Subsequent Fortran standards introduced the capability of handling noncontiguous arrays, but did so in a way designed to facilitate interoperation with older Fortran compilers. The same provisions also facilitate interoperation with languages such as C.

Because of these interoperability considerations, compilers typically assume that a dummy argument that is explicit shape or assumed size cannot directly handle a noncontiguous array. These dummy argument array forms are the ones that were allowed in Fortran 77. If a noncontiguous actual argument is to be associated with such a dummy argument, the compiler makes a temporary contiguous copy of the array. The dummy argument is then contiguous even though it is associated with a noncontiguous actual argument. If appropriate, data from the contiguous dummy argument is copied back to the noncontiguous actual argument after the invoked procedure returns. The compiler handles this automatically, but there is an important consequence:

There can be considerable performance overhead associated with allocating and deallocating space for a temporary array and copying the data. The resultant performance penalty varies widely. In severe cases, if the array is very large, but the invoked procedure is very small, the performance penalty can be many orders of magnitude. In other cases, the performance overhead can be negligible and might even be outweighed by speed benefits of contiguity.

The exact conditions that trigger making a temporary copy are compiler dependent. However, in practice, there are only a few variations. A significant variation is that some compilers make the full determination at compile time, while others do a run-time test. The compilers that make the determination at compile time often end up making copies of arrays that turn out not to need it because they are actually contiguous, but that contiguity could not be guaranteed from the information available at compile time. Newer compilers are more likely to make a run-time test in such cases, avoiding superfluous copying.

An array is noncontiguous if its elements are stored in memory out of order or with gaps in memory between the elements. Strictly speaking, the standard does not refer to memory storage and does not define a concept of contiguity. However, the requirements of the standard are far simpler to understand when expressed in terms of contiguity instead of in terms of syntax. This is because the requirements were derived from considerations of contiguity, but then expressed in terms of syntax so as to avoid unduly restricting implementations at a low level. In actual implementation, the concept of contiguity described here might not translate into physical contiguity at the hardware level. For example, storage that is contiguous in virtual memory space might not be contiguous in physical memory. Such low-level implementation distinctions are not made by the Fortran standard.

For current purposes, an array component or named array is considered to be contiguous unless it is a pointer or an assumed-shape dummy argument. An array denoted by an array element using the first special case of sequence association described in 12.6.4.1 is also considered to be contiguous. An array section can potentially be noncontiguous. A possibly noncontiguous array section could have any of the following three forms

1. an array reference using a subscript triplet
2. an array reference using a vector subscript
3. a structure component reference in which a part other than the rightmost is an array

An array pointer or assumed-shape dummy argument can potentially be associated with a noncontiguous array section and thus is potentially noncontiguous.

One restriction that applies to the use of an array section as an actual argument, regardless of the nature of the dummy argument, is that an array section generated by a vector subscript is not definable—that is, it must not be assigned a new value by the procedure. The associated dummy argument must not have intent OUT or intent INOUT. If it has unspecified intent, it must not be defined by the procedure. The reason for this is that, with a vector subscript, the same actual array element could be a part of an array section more than once and thus become associated with more than one element of the dummy array. If such an object could be defined, conflicting values might be specified for the same actual array element.

12.6.5 Pointer Arguments

A dummy data or procedure argument may have the POINTER attribute. In this case, the corresponding actual argument must also have the POINTER attribute. As discussed in 12.6.3, the requirements for type and type parameter agreement for pointer dummy arguments are more stringent than for nonpointer ones.

On entry to a procedure, a dummy pointer acquires the same pointer association as the actual argument, unless the dummy has INTENT (OUT), in which case it has an undefined association status. Unless the dummy has INTENT (IN), the association of the dummy pointer may be changed during execution of the procedure; the actual argument acquires the association of the dummy on return from the procedure.

A dummy data pointer argument may be an array. Dummy array pointers use pointer association rather than sequence association. Therefore, the rank of the actual argument corresponding to a dummy pointer must be the same as the rank of the dummy. Furthermore, the bounds of the dummy argument are always the same as those of the actual argument. This is different from an assumed-shape nonpointer dummy argument, which assumes the shape but not the bounds of the actual argument. The following example illustrates this difference.

```
program arrays
   real, target :: t(0:9)
   real, pointer :: p(:)
   p => t
   call sub(p,p)
contains
   subroutine sub(tt, pp)
      real :: tt(:)
      real, pointer :: pp(:)
      ...
   end subroutine
end program
```

The lower and upper bounds of the assumed-shape tt are 1 and 10, while those of the pointer pp are 0 and 9.

The restrictions on dummy argument aliasing (12.6.10) are less stringent for pointer dummy arguments than for nonpointer ones. This is because it is normal for multiple pointers to have the same target; the compiler is expected to deal appropriately with such aliasing, even if this impacts performance.

An actual argument may have the POINTER attribute, regardless of whether the dummy argument does or not. The case where the dummy argument also has the POINTER attribute is discussed above. If the dummy argument does not have the POINTER attribute, then the pointer actual argument must be associated, and its target is associated with the dummy argument using the same rules as if no pointers were involved.

12.6.6 Target Arguments

A dummy argument may have the TARGET attribute, as may an actual argument. A dummy argument and its corresponding actual argument are not required to agree in this attribute; all four combinations are allowed of the dummy argument and its corresponding actual argument having the TARGET attribute or not.

A procedure with a TARGET dummy argument must have an explicit interface where it is referenced. Other than that, the only special considerations for TARGET arguments relate to the status of pointers that are associated with them. Consider each of the four combinations of attributes.

If neither the actual argument nor the dummy argument have the TARGET attribute, no pointers can be associated with either one and there is therefore nothing to be said on this topic.

If the dummy argument has the TARGET attribute and the actual argument does not, pointers might become associated with the dummy argument during execution of the procedure. Such pointers are not associated with the actual argument—only with the dummy argument. Because the actual argument does not have the TARGET attribute, no pointer can ever be pointer associated with it. When the procedure returns, any pointers that are associated with the dummy argument become undefined. This is comparable to any other case where a pointer's target goes away other than by deallocating the pointer.

If the actual argument has the TARGET attribute and the dummy argument does not, no pointers can be associated with the dummy argument. The association of pointers with the actual argument is not affected by the procedure invocation or return. The only variant of this combination worthy of note involves a second level of procedure reference as in the following example.

```
module m
  integer, pointer :: p
contains
  subroutine sub_1
    integer, target :: x
    p => x
    call sub_2(x)
  end subroutine sub_1
  subroutine sub_2(y)
    integer :: y
    call sub_3(y)
  end subroutine sub_2
  subroutine sub_3(z)
    integer, target :: z
      ...
  end subroutine sub_3
end module m
```

The pointer p is associated with the actual argument x in the call to sub_2. Because the dummy argument y does not have the TARGET attribute, p is not associated with y. However, y is an actual argument corresponding to the TARGET dummy argument z in sub_3. Thus there is an indirect relationship between the TARGET variables x in sub_1 and z in sub_3. In this example, it is processor dependent whether or not p is associated with z on entry to sub_3. A strict reading suggests that on a processor where p becomes associated with z on entry to sub_3, p would then become undefined on return from sub_3 because y does not have the target attribute. This scenario is confusing at best; the safest portable practice would be treat p as though it became undefined on any processor.

The combination where both the actual and dummy arguments have the TARGET attribute is the most complicated in that the behavior also depends on several other factors. To simplify the presentation, the condition that both the dummy and actual arguments have the TARGET attribute is implied throughout the rest of this section rather than being repeated multiple times.

The simple cases of this combination act somewhat as though the actual and dummy arguments are the same target. Any pointers associated with the actual argument become associated with the dummy argument on entry to the procedure. Any pointer that is associated with the dummy argument when the procedure returns remains associated with the actual argument unless the pointer is a nonsaved local or module variable that becomes undefined on the return. These rules apply except in the following cases.

If the dummy argument has the VALUE attribute, the argument association is between the dummy argument and a temporary copy of the actual argument. That temporary copy does not exist except during the execution of the procedure. Thus, any pointers associated with the dummy argument on return become undefined.

If the actual argument is an array section with a vector subscript, it cannot have a pointer associated with it, even if the array has the TARGET attribute. Thus, this case acts as though the actual argument did not have the TARGET attribute.

If the dummy argument is an explicit-shape or assumed-size array, and the preceding paragraph does not apply, the behavior is largely processor dependent. This allows for the possibility that the processor might implement copy-in/copy-out in such cases. In particular, it is processor dependent whether pointers associated with the actual argument become associated with the dummy argument on entry. Likewise, the association status of pointers associated with the dummy argument on return is processor dependent.

12.6.7 Allocatable Arguments

A dummy data argument may have the ALLOCATABLE attribute. In this case, the corresponding actual argument also must have the ALLOCATABLE attribute. As discussed in 12.6.3, the requirements for type and type parameter agreement for allocatable dummy arguments are more stringent than for nonallocatable ones.

On entry to a procedure, a dummy allocatable acquires the same allocation status, bounds, and value as the actual argument, unless the dummy has INTENT (OUT), in which case it becomes unallocated. Unless the dummy has INTENT (IN), the allocation of the dummy argument may be changed during execution of the procedure; the actual argument acquires the allocation status, bounds, and value of the dummy on return from the procedure.

The rank of the actual argument corresponding to a dummy allocatable must be the same as the rank of the dummy. Furthermore, the bounds of the dummy argument are always the same as those of the actual argument. Allocatable dummy arguments do not use sequence association.

An actual argument may have the ALLOCATABLE attribute, regardless of whether the dummy argument does or not. The case where the dummy argument also has the ALLOCATABLE attribute is discussed above. If the dummy argument does not have the ALLOCATABLE attribute, the allocatable actual argument must be allocated and it is associated with the dummy argument using the same rules as if no allocatables were involved.

12.6.8 Procedure Arguments

A dummy procedure argument may be a pointer or a nonpointer. A dummy procedure must be specified to be a procedure by referencing it as a procedure or by declaring it with an EXTERNAL statement, a PROCEDURE statement, or an interface body.

If it is a pointer, the general rules for pointer arguments apply. The actual argument must be a procedure pointer, a reference to a function that returns a procedure pointer, or a reference to the NULL intrinsic function.

If it is not a pointer, the associated actual argument must be an external, module, intrinsic, or dummy procedure.

Rules and restrictions:

1. The actual argument must not be an internal procedure or a statement function.
2. The actual argument must not be a generic procedure. If there is a specific procedure with the same name as a generic procedure, that name designates the specific procedure when used as an actual argument.
3. The actual argument must not be an intrinsic procedure other than the specific intrinsic functions listed without an asterisk in 13.4.
4. The actual argument must not be a nonintrinsic elemental procedure.
5. If the interface of the dummy procedure is explicit, the associated actual procedure must be consistent with this interface as described in 12.5.3.
6. If the dummy procedure is typed, referenced as a function, or has an explicit function interface, the actual argument must be a function.
7. If the dummy procedure is referenced as a subroutine or has an explicit subroutine interface, the actual argument must be a subroutine.
8. A procedure actual argument must be pure if it is in a reference to a pure procedure.

A dummy procedure argument may be optional. A dummy procedure that is not a pointer must not have the INTENT attribute.

12.6.9 Alternate Returns

An alternate return is neither a data argument nor a procedure argument. It may appear only in a subroutine argument list. It is used to specify a return different than the normal execution upon completion of the subroutine. As mentioned in 12.3, there are usually superior ways of achieving the desired control, and therefore the alternate return is an obsolescent feature. It could be removed from the next revision of the Fortran standard.

There may be any number of alternate returns in a subroutine argument list, and they may appear at any position in the list. In the dummy argument list each alternate return is designated by an asterisk. For example, the following dummy argument list for subroutine CALC_2 has two alternate return indicators, in the second and fifth argument positions.

```
SUBROUTINE CALC_2(A,*,P,Q,*)
```

An alternate return dummy argument has no name. Consequently, it cannot be optional and cannot use the keyword method of establishing argument correspondence.

An actual argument associated with an alternate return dummy argument must be an asterisk followed by the label of a branch target in the scope of the invoking scoping unit. It specifies the return point for the corresponding alternate return. For example, the following is a valid reference to CALC_2:

```
CALL CALC_2(X,*330,Y,Z,*200)
```

provided the statements labeled 200 and 330 are valid branch targets in the invoking scoping unit. The statement having the label 330 is the return point for the first alternate return, and the statement having the label 200 is the return point for the second alternate return.

An alternate return is taken when a RETURN statement (12.3) with an optional integer expression is executed. The integer expression value selects which of the alternate returns, counting from left to right, is to be utilized. Using the above example call:

```
RETURN 2    ! returns to statement 200
            !   in the calling program
RETURN (1) ! returns to statement 330
RETURN      ! normal return from the call
```

12.6.10 Argument Aliasing

A dummy data object is aliased if any part of its actual argument can be accessed via a name other than the dummy argument name while the procedure is executing. This can happen in the following ways:

1. The actual argument or any subobject of it is associated with some other actual argument or a subobject of it in the same invocation. The simplest case of this is that the same actual argument appears twice in the actual argument list. Other cases involve one actual argument being a subobject of or associated with the other.
2. The actual argument is a variable accessed from common or a module, and the same common or module variable is also accessed separately in the procedure or any lower-level procedures that it invokes.
3. The actual argument is a variable that is also accessible via host association in the procedure, or the actual argument is associated with or a subobject of such a variable.

Unrestricted dummy argument aliasing could potentially cause performance problems. This is because the aliasing is not necessarily evident while compiling the procedure. Therefore, the compiler would have to generate code that worked correctly in the presence of aliasing. This would inhibit many basic forms of optimization such as storing values in registers. Also, ambiguities of meaning could arise with mixtures of argument aliasing and copy-in/copy-out implementations of argument association; if two copies of a variable acquired different values, the results would depend on the order of the copying. The standard deliberately leaves the implementation details of argument association to the processor so as to facilitate optimization.

Any definition or undefinition of the value of a dummy argument during execution of the procedure must be done through the dummy argument rather than through an alias, except in the cases detailed below. With the same exceptions, if the value becomes defined or undefined, it may not be referenced through an alias during execution of the procedure, either before or after the definition or undefinition.

The following are the exceptions to the restrictions on definition and undefinition of values:

1. The restrictions do not apply to a pointer dummy argument. Aliasing of pointer targets is normal and the compiler is expected to deal with it correctly.

2. The restrictions do not apply if both the dummy and actual arguments have the TARGET attribute, the dummy argument is scalar or assumed shape, and the actual argument is not an array section with a vector subscript. In essence, the TARGET attribute disallows argument copying in most cases; although the standard says it indirectly, that is the model behind this exception. Argument copying is expected in typical implementations of vector subscripts and some cases of explicit-shape and assumed-size dummy arguments. This relates to the material in 12.6.6.

If a dummy argument or a subobject of it is allocatable, any allocation or deallocation during execution of the procedure must be done through the dummy argument rather than through an alias. If it is allocated or deallocated, it may not be referenced through an alias during execution of the procedure, either before or after the allocation or deallocation.

The restrictions described here do not disallow dummy argument aliasing. They just restrict the behavior that is allowed when there is aliasing. In order to determine whether a program violates these restrictions, it is necessary to examine both the procedure and its invocation. Examining the invocation is necessary to see whether any aliasing occurs. Examining the procedure is necessary to see whether the restrictions on aliased arguments are violated.

These restrictions have a subtle point relating to parenthesized expressions as actual arguments. The following two sample invocations have an important difference.

```
call sub(x, x)      !-- Aliased
call sub(x, (x))    !-- Not aliased
```

With the first of these sample calls, the two dummy arguments are trivially aliased; with the second sample, they are not. This is because the (x) in the second sample is not the variable x, but rather is an expression that references x. The expression is evaluated prior to invoking the subroutine and subsequent changes to the value of x do not change the previously evaluated expression.

12.7 Special Categories of Procedures

This section describes special categories of procedures.

12.7.1 Pure Procedures

A pure procedure is one that may be used in contexts where some side effects would be particularly problematic. A pure procedure definition must meet a stringent set of constraints that allow the limitations on side effects to be verified at compile time. All of the functions defined in the standard are pure, as are some of the subroutines (13.3, 14.3, 15.3).

There are two basic situations where pure procedures are required—parallel processing and specification expressions. If a procedure is referenced in a context that conceptually allows multiple instances of the procedure to be executing in parallel, then some side effects from those multiple instances could conflict and cause nondeterministic behavior; to prevent this, such a procedure is required to be pure. This applies to the FORALL statement and construct; it also applies to elemental references, as described in 12.7.2. If a procedure is referenced in a specification expression, side effects could cause complications; therefore, such a procedure is required to be pure. Secondarily, a procedure is required to be pure if it is referenced in the body of a pure procedure or is passed as an actual argument to a pure procedure.

The essence of the restrictions on a pure procedure is that it must not perform any input/output to external files and must not modify the values of any variables accessible outside of the procedure except that a pure subroutine (but not a function) can modify its actual arguments. A pure procedure cannot save the values of local variables. Each invocation of a pure procedure is independent of any other. The only useful effect of a pure procedure is the computation of its result value if it is a function, or modifications of its arguments if it is a subroutine.

The prefix specification PURE in a FUNCTION or SUBROUTINE statement specifies that the procedure is pure. The prefix specification ELEMENTAL also implies purity, which can be confirmed by explicit declaration. A statement function is pure if it references only pure functions.

If a procedure other than a statement function is used in a context that requires it to be pure, its interface must be explicit and must specify that the procedure is pure.

Rules and restrictions:

1. A nonpointer dummy data argument of a pure procedure must have its intent specified; if the procedure is a function, the specification must be INTENT (IN).

2. A local variable of a pure procedure must not have the SAVE attribute, either explicitly or implicitly. Note that this means that local variables in pure procedures cannot have explicit initialization, because explicit initialization implicitly specifies the SAVE attribute.

3. A procedure referenced in a pure procedure must be pure. A dummy or internal procedure in a pure procedure must be pure, regardless of whether it is referenced. Note that procedure references include defined assignments, defined operations, defined input/output and finalization.

4. A pure procedure must not contain a STOP statement, an INQUIRE statement, or an input/output statement, that specifies an external file.

5. In a pure procedure, any variable that is in common, is host associated, is use associated, has INTENT (IN), or is storage associated with any such variable, must not be used in the following contexts:

    a. in a variable definition context (16.3.1)

b. as the data target in a pointer assignment statement

c. as a pointer component of a structure constructor

d. as the right-hand side of an intrinsic assignment statement in which the left-hand side has a pointer component (at any level)

e. as the actual argument corresponding to a pointer dummy argument

Example:

```
PURE FUNCTION DRIVEN_SNOW(CRYSTAL)
   INTEGER DRIVEN_SNOW
   INTEGER, INTENT(IN) :: CRYSTAL
      . . .
END FUNCTION DRIVEN_SNOW
```

12.7.2 Elemental Procedures

An elemental procedure is one that is defined with scalar dummy arguments, but may be referenced with actual arguments that are of any rank, provided that the actual arguments are conformable.

When an elemental procedure is invoked with array arguments, the effect is as if the procedure had been invoked multiple times, with scalar arguments corresponding to the individual elements of the actual argument arrays. If any of the original actual arguments are scalar, those scalars are used as is for each of the invocations. A single elemental procedure substitutes for writing multiple nonelemental procedures—one for each rank.

Several of the procedures defined by the standard are elemental (13.3, 14.3). The prefix specification ELEMENTAL in a FUNCTION or SUBROUTINE statement specifies that the procedure is elemental.

An elemental procedure must be pure because the multiple invocations implied in the array case are conceptually done in parallel. The ELEMENTAL prefix implies purity, which may be confirmed by a PURE prefix. All the rules for pure procedures apply. An elemental procedure also must meet the additional requirements described below.

An elemental subprogram is defined with all of its dummy arguments scalar; an elemental function is also defined with a scalar result. The expressive power of elemental procedures comes from the provision that the actual arguments may in general be of any rank, as long as all of the actual arguments in a given invocation of an elemental procedure are conformable. There are two exceptions to the allowance of array actual arguments. First, the KIND actual argument of a standard intrinsic elemental must be scalar. Second, if any of the actual arguments in an elemental subroutine reference is an array, each actual argument corresponding to an INTENT (OUT) or INTENT (INOUT) dummy must be an array.

The interface of an elemental procedure must be explicit in any scope in which it is referenced.

Rules and restrictions:

1. An elemental procedure must not be recursive.
2. A dummy argument of an elemental procedure must be a nonpointer, nonallocatable, scalar data object.
3. The result variable of an elemental function must be scalar and must not be a pointer or allocatable.
4. A dummy argument of an elemental procedure must not be used in a specification expression except as an argument to one of the intrinsic functions BIT_SIZE, KIND, LEN, or a numeric inquiry function. This rather strange restriction is alleged to facilitate optimization, but since the same ends can be achieved with more complicated syntax, it is not clear that the restriction actually achieves anything. The workarounds are to either use an allocatable variable or write and use a lower-level pure function.

Example:

```
ELEMENTAL FUNCTION VIP_CALC(X, Y)
   REAL VIP_CALC
   REAL, INTENT(IN) :: X, Y
      . . .
END FUNCTION VIP_CALC
   . . .
   Q = VIP_CALC(1.0, 2.0)
     !  A reference to VIP_CALC with scalar arguments
   A(1:N) = VIP_CALC(A1(1:N), 3.0)
     !  A reference with conformable arguments
```

12.7.3 Recursive Procedures

A **recursive procedure** is one that directly or indirectly invokes itself. A procedure must be declared recursive explicitly if it is ever invoked recursively.

Because a recursive procedure can have multiple instances simultaneously active, an additional definition of what it means for an entity to be local to the procedure is required. If a local entity has the SAVE attribute (either explicitly or implicitly), then all instances of the procedure share a single copy of that entity. If a local entity does not have the SAVE attribute, then each instance of the procedure has a separate copy of that entity; changing the value of an unsaved variable in one instance has no effect on the value in other instances.

12.7.4 Procedure Pointers

A procedure pointer is a pointer whose target must be a procedure. When the pointer is associated with a target, the procedure pointer can be referenced, which causes the target procedure to be invoked. A procedure pointer can be either a named entity or a structure component.

A named procedure pointer is declared by specifying both the EXTERNAL and POINTER attributes. The EXTERNAL attribute can be specified using an EXTERNAL statement, a PROCEDURE statement, or the EXTERNAL attribute in a type declaration statement. The POINTER attribute can be specified using a POINTER statement or the POINTER attribute in a PROCEDURE or type declaration statement.

A procedure pointer structure component is declared as described in 4.4.7.

Although a procedure pointer has much in common with a data pointer, there is a subtle difference in terminology: a procedure pointer is not a variable. The association of a procedure pointer can vary during execution, just like the association of a data pointer. However, the Fortran definition of a variable requires that a variable be able to have a value. A procedure does not have a value.

12.8 Resolving Procedure References

Procedure resolution is the process of determining what specific procedure is invoked by a particular reference. Procedure resolution depends on the form of reference and on the names, types, kinds, and ranks of the actual arguments. If a program has a procedure reference that does not resolve to any specific procedure, the program is invalid.

For a procedure reference in the form of an operator, assignment, defined input/output, or type-bound reference, it is generally straightforward to resolve what specific procedure is invoked. The only cases where more than one specific procedure could apply are resolved as follows:

1. If both an elemental and a nonelemental specific procedure could be compatible with the reference, the reference is to the nonelemental one. The nonelemental reference could be thought of as a more precise match.

2. If both an intrinsic and a defined derived-type assignment could apply, the reference is to the defined one. This is reasonably obvious in that otherwise such a defined assignment could never be referenced.

3. A type-bound reference to a generic procedure is resolved to a type-bound specific reference based on the generic bindings in the declared type of the data reference.

4. A type-bound reference to a specific procedure is resolved based on the specific binding in the dynamic type of the data reference.

For a procedure reference by name, the large majority of the cases are trivial because there is only one compatible procedure. However, a single name can possibly identify a specific procedure, a generic procedure, and an intrinsic procedure, all in the same scoping unit. Most such cases can be resolved using the following simplified guidelines:

1. In the absence of any contrary declarations, a procedure reference using the name of an intrinsic procedure is a reference to the intrinsic. A type declaration does not count as conflicting with an intrinsic function, even if the intrinsic function has no specific function with a result of that type.

2. If both an intrinsic and a user-specified generic could be compatible with the reference, the reference is to the user-specified generic.

3. If both an elemental and a nonelemental specific procedure could be compatible with the reference, the reference is to the nonelemental one.

The precise rules depend on whether or not the name is established to be generic.

12.8.1 Resolving Generic Name References

A name is established to be generic in a scoping unit if the scoping unit

1. has a generic interface block for that name,

2. has an INTRINSIC specification for a generic intrinsic of that name, or

3. accesses a generic of that name via USE or host association.

The name might also be the name of a specific procedure in the same scoping unit, but the generic resolution procedure is followed regardless. If a name is established to be generic, candidate procedures are checked in the following order, with the resolution being to the first procedure that is consistent with the reference.

1. A nonelemental specific procedure of the generic.

2. An elemental specific procedure of the generic.

3. An intrinsic procedure.

4. A generic reference in the host scoping unit, checked in this same order.

The distinguishability restrictions on generic procedure names (12.5.4.1) ensure that no more than one specific procedure can be consistent at each step.

The following example illustrates application of these rules:

```
INTERFACE GEN
   ELEMENTAL SUBROUTINE SP_E(X)
      REAL, INTENT(INOUT) :: X
   END SUBROUTINE
   SUBROUTINE SP_A(A)
      REAL, INTENT(OUT) :: A(:)
   END SUBROUTINE
END INTERFACE
```

```
REAL B(10), C(10,10)
CALL GEN(B)            ! Resolves to SP_A
```

12.8.2 Resolving Specific Name References

If a name is not established to be generic in a scoping unit, its resolution is as follows. The resolution is explicitly established if the name is one of the following; no more than one of these can apply in a scoping unit.

1. A dummy argument of the scoping unit.
2. A procedure defined by a subprogram or statement function in the scoping unit.
3. A procedure accessed by USE or host association.
4. An intrinsic procedure, if the name is explicitly given the INTRINSIC attribute.
5. An external procedure, if the name is explicitly given the EXTERNAL attribute and is not a dummy argument.

   If none of the above applies to the name, then

1. If it is a function reference using the name of an intrinsic function, or a subroutine reference using the name of an intrinsic subroutine, the reference is to that intrinsic.
2. Otherwise, the reference is to an external procedure.

The procedure arguments play no role in the resolution of a nongeneric reference. Once the steps above resolve to a procedure, that completes the resolution. If the resolution is to a procedure whose arguments are not consistent with the reference, then the program is invalid. For example, consider the following program.

```
program wrong
   write (*,*) cos('zero') !-- Invalid.
end program wrong
```

This program is invalid because the resolution is to the intrinsic function cos, which has no specific function with a character argument. Even if there were an appropriate external function named cos, the resolution would not be to that external function.

12.9 Procedure Properties

Table 12-1 summarizes the properties of Fortran procedures.

Table 12-1 Summary of Fortran procedure properties

	Type of procedure				
Property of procedure	**External**	**Intrinsic**	**Module**	**Internal**	**Statement function**
Dummy arguments may be optional	Yes	Yes	Yes	Yes	No
Reference may use keywords	Yes	Yes	Yes	Yes	No
Reference may be recursive	Yes	N/A	Yes	Yes	No
Definition may have CONTAINS	Yes	N/A	Yes	No	No
May be passed as an actual argument	Yes	Yes	Yes	No	No
May be described by an interface body	Yes	No	No	No	No
Interface automatically explicit	No	Yes	Yes	Yes	No
May be referenced elementally	Yes	Yes	Yes	Yes	No
May be used to define operators	Yes	No	Yes	No	No
May be a specific in a generic	Yes	No	Yes	No	No
Definition may contain ENTRY statements	Yes	N/A	Yes	No	N/A

13 Intrinsic Procedures and Modules

- **A Standard Intrinsic Procedure** is an intrinsic procedure defined by the standard. Other intrinsic procedures may be specified by the processor. There are elemental procedures, inquiry functions, and transformational functions. All intrinsic functions are pure.
- **A Standard Intrinsic Module** is a module defined by the standard; it may be required or may be optional. The standard intrinsic modules are a Fortran environment module, three IEEE arithmetic modules, and a C interoperability module.
- An **Elemental Intrinsic Procedure** is one that is defined with scalar arguments and may be called with conformable arguments.
- An **Inquiry Intrinsic Function** is one whose result depends on the properties, other than value, of its arguments. It returns information about the status, nature, or attributes of its arguments.
- A **Transformational Intrinsic Function** is one that is neither an elemental nor inquiry function. One argument is usually an array but the result value is not related to the arguments in the same way as with elemental procedures.
- A **Generic Intrinsic Procedure** may be used to reference one of a group of intrinsic procedures. The particular procedure selected is determined by the argument attributes in the reference. Some generic intrinsic procedures have named specific procedures.
- There is a **Representation Model** for each of the real and integer number systems and for bits. Some of the intrinsic functions compute values based on how data is represented with respect to these models. The physical representation is not required to match the models exactly.

Procedures (functions and subroutines) that are part of the Fortran processor and are not module procedures are called **intrinsic**; the standard specifies a collection of intrinsic procedures which are described in this chapter. Examples of standard intrinsic procedures are COS, SUM, RANDOM_NUMBER, and SHAPE.

There may be nonstandard intrinsic procedures that are supplied by a particular Fortran processor but such procedures are not portable in the sense that they might not be in all Fortran processors.

Intrinsic procedures are always "there", and may be invoked from any program unit. However, a user-written function or subroutine with the same name as an intrinsic function or subroutine might override the intrinsic procedure (12.5.4.2, 12.8).

J.C. Adams et al., *The Fortran 2003 Handbook*,
DOI: 10.1007/978-1-84628-746-6_13,

All of the intrinsic procedures are described in 13.3 and each of these procedures is described in detail in A.

The Fortran standard defines three categories of intrinsic modules (13.6). The first is the Fortran environment intrinsic module ISO_FORTRAN_ENV; it is required and specifies some aspects of the Fortran environment such as input/output logical unit numbers and the size of the numeric storage unit in bits. It defines named constants for these values.

A second category consists of three intrinsic modules which specify named constants, derived types, exceptions, arithmetic modes, and module procedures to support exception handling and arithmetic; these modules are compatible with the binary arithmetic international floating-point standard [13]. Exceptions and arithmetic modes are manipulated (signaled, recognized, enabled, and disabled) by calls to procedures provided by these modules. These intrinsic modules are optional for a standard conforming processor and have the names IEEE_EXCEPTIONS, IEEE_ARITHMETIC, and IEEE_FEATURES. They are described in 14.3.

A third category is the intrinsic module ISO_C_BINDING; it provides named constants, derived types, and module procedures to support references to C programs from Fortran programs and vice versa. This intrinsic module is required and is described in 15.3.

The required intrinsic modules are always available, and are accessible in any scoping unit via a USE statement referencing the name of the intrinsic module. The module procedures in all these intrinsic modules are generic, have no specific names, and therefore cannot be passed as actual arguments to dummy procedures. The module functions have INTENT (IN) arguments only and are pure, with some being elemental. Module procedures in intrinsic modules are not intrinsic procedures. This affects generic resolution (12.8.1) and specification functions (7.4.2.2).

A user-defined module may have the same name as an intrinsic module, but to refer to the intrinsic module, the USE statement must specify the INTRINSIC nature; the program unit may specify the NON_INTRINSIC nature to refer to the user-defined module but access to it is provided by default if it is accessible. Always specifying the module nature (11.3.7.1) facilitates portability and may prevent conflicts with future standards.

13.1 Properties of Intrinsic Procedures

Intrinsic procedures are "predefined" by the processor, but otherwise conform to the principles and rules for procedures as described in 12. In particular, intrinsic procedures are invoked in the same way as other procedures (12.1.2 and 12.2.3) and employ the same argument association mechanisms (12.6). An intrinsic procedure's interface is explicit. The functionality and partial interfaces to the standard intrinsic procedures, including the argument keywords (dummy argument names) and argument optionality, are described in 13.3, 13.4, and 13.5; the complete interfaces are provided in A.

Generic Intrinsic Procedures. The intrinsic procedures listed in 13.3 are generic. Some intrinsic procedures have two or more underlying specific intrinsic procedures, which are listed in Table 13-1. The intrinsic functions LLT, LLE, LGT, and LGE each have one underlying named specific function. Many, such as ALL, EPSILON, or SUM and all intrinsic subroutines, have no named underlying specific procedures.

Specific Intrinsic Procedures. For historical reasons some intrinsic procedures have specific names related to the type of the argument. In most cases (except those marked with an asterisk * in Table 13-1), such a procedure may be used as an actual argument associated with a dummy procedure argument. For example, CSQRT is a specific intrinsic procedure for computing the complex square root of a default complex argument. Note that a specific intrinsic function may have the same name as the generic intrinsic function; for example, SQRT is both the generic intrinsic function and the specific intrinsic function for a default real argument. Thus, when SQRT is used as an actual argument, is associated with a dummy procedure argument, and is given the attribute INTRINSIC, the specific function SQRT is being passed; there are other ways to have the specific function SQRT passed (12.6.8).

All specific intrinsic procedures have INTENT (IN) dummy arguments. All specific intrinsic functions return results with a type that the corresponding generic function would have if called with the same argument types, except the specific intrinsic functions AMAX0, AMIN0, MAX1, and MIN1; the differences are specified in Table 13-1.

Elemental Intrinsic Procedures. Many of the intrinsic functions and one intrinsic subroutine (MVBITS) are elemental. This extends those intrinsic procedures to array arguments and array results in a natural way.

Transformational Intrinsic Functions. A transformational intrinsic function is one that is not elemental and not an inquiry function. Most of the transformational functions have an array argument (for example, the SUM function) that is treated as a whole and not elementally. The transformational categorization does not apply to subroutines.

Inquiry Intrinsic Functions. An inquiry intrinsic function returns information about the status, nature, or attributes of its arguments. Unless stated otherwise, these arguments need not have defined values. They may be unallocated allocatables or pointers that are not associated. They may be array arguments whose shape is not defined.

A **Pure Intrinsic Procedure** is an intrinsic procedure that is pure as defined in 12.7.1. All intrinsic functions are pure, but only one of the intrinsic subroutines (MVBITS) is pure.

Keyword and Optional Arguments. Intrinsic procedure references may use keyword arguments, as described in 12.6.1. Some arguments of intrinsic procedures are optional (12.6.2), and the use of keywords makes possible the omission of actual arguments, corresponding to optional dummy arguments. For example, in

```
CALL RANDOM_SEED (PUT=SEED_VALUE)
```

the keyword form shown must be used because the optional first argument SIZE is omitted.

Intrinsic Argument Intent. The intent of the dummy arguments of each intrinsic subroutine is specified in the description of the subroutine in A. The nonpointer dummy arguments of intrinsic functions have INTENT (IN); the pointer dummy arguments never change either their associated actual arguments or their targets.

13.2 Representation Models

Some of the intrinsic functions compute values related to how data is represented. These values are based upon and determined by the underlying **representation model**. There are three such models: the **bit model**, the **integer number system model**, and the **real number system model**.

These models, and the corresponding functions, return values related to the models, allowing development of robust and portable code. For example, by obtaining information about the spacing of real numbers, the convergence of a numerical algorithm can be controlled so that maximum accuracy may be achieved while attaining convergence.

In a given implementation the model parameters are chosen to match the implementation as closely as possible, but an exact match is not required and the model does not impose any particular arithmetic on the implementation.

13.2.1 The Bit Model

The bit model interprets a nonnegative scalar data object of type integer as a sequence of binary digits (bits), based upon the model

$$\sum_{k=0}^{z-1} w_k 2^k$$

where z is the number of bits and each w_k has a bit value of 0 or 1. The bits are numbered from right to left beginning with 0.

The bit computation functions in 13.3.4.3 are based upon the bit model. The model deals only with nonnegative integers interpreted through these functions and the MVBITS subroutine, and it is not necessarily related to the implementation of the integer data type. However, the bit model is identical to the integer model (13.2.2) for $w_{z-1} = 0$.

The interpretation of an integer with the first bit set is processor dependent; this is typically a negative value. Such a negative value may be the result of assigning a negative integer value to a variable, but it also may be created by shifting an integer left so that its first bit becomes 1. The interpretation of such an entity as an integer is not provided by this bit model. However, the bit model does predict the value as a string of bits. For example, suppose on a processor with $z = 32$ an integer entity with positive value 2^{30} is left shifted one position using the intrinsic function ISHIFT. The interpretation of this value as an integer is not specified by the model even though the model predicts the result is a string of 32 bits, all of which are zero except the first one which is one.

The BOZ literal constants (4.3.1.4) that are interpreted as integers are not necessarily related to this model, regardless of the issue of negative integers.

A common model for processors where integers occupy 32 bits is for the value of z to be 32.

13.2.2 The Integer Number System Model

The integer number system is modeled by

$$i = s \sum_{k=0}^{q-1} w_k r^k$$

where

- i is the integer value
- s is the sign (+1 or –1)
- r is the base and is an integer greater than 1
- q is the number of digits and is an integer greater than 0
- w_k is the kth digit and is an integer $0 \le w_k < r$

A common model for processors that use 4 bytes for integers is: $q = 31$ and $r = 2$.

13.2.3 The Real Number System Model

The real number system is modeled by

$$x = s b^e \sum_{k=1}^{p} f_k b^{-k}$$

where

- x is the real value
- s is the sign (+1 or –1)
- b is the base (real radix) and is an integer greater than 1
- e is an integer between some minimum and maximum value
- p is the number of mantissa digits and is an integer greater than 1
- f_k is the kth digit and is an integer $0 \le f_k < b$, but f_1 may be zero only if e and all the f_k are zero

One common implementation is the IEEE binary floating-point standard [13], which has single precision model numbers with:

$b = 2$
$p = 24$

$-125 \le e \le 128$

This IEEE standard does not represent f_1, which is presumed to be 1. Thus, the mantissa, including its sign, can be represented in 24 bits. The exponent, including its sign, takes 8 bits, for a total of 32 bits in the single precision representation. What normally would be an exponent value of –127 or –126 is not included in the exponent range; rather, IEEE uses these cases to identify the real value zero (the one case in which f_1 is 0), out-of-range values (infinities), or NaNs (not a number—illegal values). It should be noted that the specific model described above where $b = 2$, $p = 24$, $e_{max} = 128$, and $e_{min} = -125$ is not the specific model used in the Fortran 2003 standard where $b = 2$, $p = 24$, $e_{min} = -126$, and $e_{max} = 127$.

The numeric inquiry and manipulation functions return information about the real number system model pertaining to an implementation. The IEEE intrinsic module procedures permit a more precise manipulation of floating-point values that are of IEEE (arithmetic) type (14.3.3). On the other hand, the numeric computation and manipulation intrinsic functions assume or in some cases specify that the results are processor dependent when the result value would exceed the range of floating-point representable values.

13.3 Intrinsic Procedures

The intrinsic procedures are grouped into subsections. Each subsection summaries the classification of its intrinsic procedures and summarizes the functionality of each of the procedures. The intent of the arguments of the functions is not stated for any of the functions except PRESENT; for all but PRESENT, the intent is INTENT (IN). For PRESENT and all intrinsic subroutines, the intent of each dummy argument of the intrinsic subroutines is given specifically.

Generic intrinsic subroutines that assign values to arguments of type character do so in accordance with the rules of intrinsic assignment (7.5.2); this includes truncation or blank padding as appropriate.

Intrinsic procedure argument keyword names (dummy argument names) are made as consistent as possible. Dummy arguments that play a similar or identical role have the same name. For example, these include such keyword dummy argument names as KIND, DIM, MASK, and BACK.

KIND is an argument to specify the kind type parameter of the function result. The KIND actual argument must be a scalar integer initialization expression, even in elemental references; it is usually optional. It is used in conversion functions (13.3.3); it is also used in functions, such as SIZE and LEN, whose results might not fit into a default integer when applied to large arrays or large character strings.

DIM is used mostly in the array reduction functions (13.3.4.4) and in some of the other array functions (13.3.1.4), to specify which dimension of the array is involved, if not the whole array. DIM is a scalar integer and usually is optional.

MASK, when used in an array reduction function, selects elements of the array that are to be involved in the operation. For example, in the function SUM, any element of the array that is to be included in the sum of the elements can be selected by use of an appropriate mask. The MASK must be conformable with the array it is masking; it usually is an optional argument.

BACK is an optional logical argument used in several of the character intrinsic functions. For example, if BACK=.TRUE. in the INDEX function, the search is for the rightmost occurrence of the target rather than the leftmost.

The permitted argument types for each procedure are specified in the description of each procedure in A. For many of these procedures, the permitted kinds are not specified explicitly, following the convention that if the kinds are not specified, all available kinds for a given type are permitted. The possible kinds for some intrinsic procedures are restricted, and this is stated explicitly. For example, the arguments for intrinsic function DPROD are restricted to default real kind.

The types and kinds of the results are always explicitly stated in A. In some cases they are the same as one of the arguments, sometimes they are specified by a KIND argument, and sometimes they are of default kind.

13.3.1 Inquiry Functions

The inquiry functions, rather than performing some computation with their arguments, return information concerning the status or nature of the argument; the returned value is independent of the value of the argument so that the actual argument of a reference to such a function need not be defined. Note however that the pointer association status of arguments used in certain inquiry intrinsic functions, namely ASSOCIATED, EXTENDS_TYPE_OF, and SAME_TYPE_AS, must be defined when referenced.

The inquiry functions are nonelemental. The inquiry function COMMAND_ARGUMENT_COUNT is described in 13.3.5.3.

13.3.1.1 Character and Bit Inquiry Functions

The LEN intrinsic function returns the length type parameter (number of characters) of the character string argument as defined in 7.2.2.1. The BIT_SIZE function returns the number of bits z provided by the bit model (13.2.1). The NEW_LINE function returns the new line character as specified in A.

Function	**Value returned**
BIT_SIZE	Number of bits in the bit model
LEN	Length of a character string argument
NEW_LINE	New line character for the character kind of the argument

13.3.1.2 Kind Functions

The KIND inquiry function returns the kind type parameter of its argument, which may be of any intrinsic type. Somewhat related to the KIND function, but providing a complementary functionality, are three transformational functions, SELECTED_REAL_KIND, SELECTED_INT_KIND, and SELECTED_CHAR_KIND. The values of the arguments to these functions must be defined. Also see 14.3.3.7 for a similar transformational function IEEE_SELECTED_REAL_KIND from the IEEE arithmetic module.

Function	Value returned
KIND	Kind parameter
SELECTED_CHAR_KIND	Kind parameter of a specified character set
SELECTED_INT_KIND	Kind parameter of an integer data type, specified by a minimum decimal range
SELECTED_REAL_KIND	Kind parameter of a real data type, specified by a minimum decimal precision and/or exponent range

13.3.1.3 Numeric Inquiry Functions

The environmental intrinsic inquiry functions together describe the numerical environment in terms of the integer model (13.2.2) and real model (13.2.3).

Function	Value returned
DIGITS	Number of model digits in a model number
EPSILON	Value that is small relative to 1 for a real value
HUGE	Largest number in the real or integer model
MAXEXPONENT	Maximum value of the model exponent
MINEXPONENT	Minimum value of the model exponent
PRECISION	Decimal precision of a model number
RADIX	Base of a model number
RANGE	Decimal exponent range of a model number
TINY	Smallest positive number in the real model

13.3.1.4 Array Inquiry Functions

Many of the intrinsic functions are related to arrays. A subset of these functions, called array inquiry functions, allow certain properties of an array to be queried dynamically. These properties include shape, extents, and size of arrays as well as the allocation status of an allocatable array.

Function	Value returned
ALLOCATED	Allocation status of the argument
LBOUND	Lower bound(s) of an array or a dimension of an array
SHAPE	Number of elements in each dimension of an array
SIZE	Number of elements of an array or a dimension of an array
UBOUND	Upper bound(s) of an array or a dimension of an array

13.3.1.5 Inquiry of Dynamic Properties

Several of the intrinsic functions permit inquiry about dynamic properties of their arguments. They include functions to inquire whether:

- an allocatable object is allocated.
- a pointer object is associated and with what it is associated.
- an argument is present or not.

- a polymorphic object has the same type as another object or has a type that is an extension of the type of another object.

Function	Value returned
ALLOCATED	Allocation status of the argument
ASSOCIATED	Tests the association status of a pointer or its association with a specific target
EXTENDS_TYPE_OF	True if the dynamic type of the first argument is an extension type (4.4.12) of the dynamic type of the second argument
PRESENT	True if an actual argument of a procedure is present
SAME_TYPE_AS	True if two objects are of the same dynamic type

13.3.2 Numeric Manipulation Functions

The numeric manipulation functions manipulate parts of the floating-point representation, primarily relative to the real model (13.2.3). In contrast, the NEAREST function returns a result in terms of the machine representation.

The numeric manipulation functions are elemental.

Function	Value returned
EXPONENT	Exponent of a real value
FRACTION	Fractional part of a real value
NEAREST	Nearest machine-representable number in a given direction
RRSPACING	Reciprocal of model relative spacing near a specified value
SCALE	Value scaled by a power of the radix
SET_EXPONENT	Value with its exponent set to a specified value
SPACING	Model absolute spacing near a specified value

13.3.3 Conversion Functions

Fortran has a number of intrinsic functions to transfer or convert data values from one type and kind combination to another combination; most of these have been in Fortran for a long time, although the optional KIND argument is relatively new.

The conversion functions ACHAR and IACHAR (and the character computation functions LLT, LLE, LGT, and LGE) use the collating sequence specified in ISO/IEC 646:1991 (International Reference Version).

Where the functions are defined, ACHAR and IACHAR and CHAR and ICHAR are inverses of one another. That is, for an example using the ASCII character functions, for values of C where IACHAR (C, KIND (I)) is defined, ACHAR (IACHAR (C, KIND (I)), KIND (C)) = C where I is of integer type with any integer kind. Similarly, for values of I where ACHAR(I, KIND (C)) is defined, IACHAR (ACHAR (I, KIND (C)), KIND (I)) = I where C is of character type with any character kind.

Where the functions are defined, ACHAR and IACHAR and CHAR and ICHAR are inverses of one another. That is, for example using the ASCII character functions, where C is of default character kind with values C for which IACHAR (C) is defined,

ACHAR (IACHAR (C)) = C. Similarly, for I of default integer kind with values for which ACHAR(I) is defined, IACHAR (ACHAR (I)) = I.

All conversion functions below are elemental.

Function	Value returned
ACHAR	Character in a specified position of the ASCII character set
AIMAG	Imaginary part of a complex value
AINT	Real value truncated to a whole number
ANINT	Real value rounded to the nearest whole number
CHAR	Character in a specified position of a character set
CMPLX	Complex value
CONJG	Complex conjugate of a complex value
DBLE	Double precision value
IACHAR	Position of a specified character in the ASCII character set
ICHAR	Position of a specified character in a character set
INT	Truncated integer value
LOGICAL	Logical value
NINT	Real value rounded to the nearest integer
REAL	Real value

13.3.3.1 NULL and Transfer Procedures

The transformational function NULL is used in pointer assignment contexts and returns a disassociated pointer (null pointer) or unallocated allocatable object.

The transformational function TRANSFER and pure subroutine MOVE_ALLOC transfer data without changing any bits. The elemental subroutine MVBITS transfers a sequence of bits from one integer to another.

Procedure	Operation
MOVE_ALLOC	Transfer an allocation from one object to another of the same type
MVBITS	Copies a sequence of bits from one integer to another
NULL	A disassociated pointer or unallocated allocatable component of a structure constructor
TRANSFER	Value transferred from an object to the result without conversion

13.3.4 Computation Procedures

The computation intrinsic procedures perform computational operations, delivering the results as function results or INTENT (OUT) subroutine arguments.

13.3.4.1 Numeric Computation Procedures

The numeric computations include trigonometric, logarithmic, exponential, differences, products, maxima, minima, remainder, square root, absolute value operations, and some matrix operations.

All of the numeric computation functions are elemental except DOT_PRODUCT, and MATMUL, which are transformational.

Function	Value returned
ABS	Absolute value
ACOS	Arc cosine
ASIN	Arc sine
ATAN	Arc tangent
ATAN2	Angle in radians of a complex value X+Y*i*
CEILING	Smallest whole number greater than or equal to a value
COS	Cosine
COSH	Hyperbolic cosine
DIM	Difference of two values, if positive, or otherwise zero
DOT_PRODUCT	Dot product of two rank-one arrays
DPROD	Double precision product of two single precision values
EXP	Natural exponential
FLOOR	Greatest integer less than or equal to a value
LOG	Natural logarithm
LOG10	Logarithm to the base 10
MATMUL	Matrix multiplication
MAX	Maximum of specified values
MIN	Minimum of specified values
MOD	Remainder function, having the sign of the first argument
MODULO	Remainder function, having the sign of the second argument
SIGN	Value with a specified sign
SIN	Sine
SINH	Hyperbolic sine
SQRT	Square root
TAN	Tangent
TANH	Hyperbolic tangent

The subroutines RANDOM_NUMBER and RANDOM_SEED provide pseudorandom number sequences and a means to specify and control them. They are not elemental, and not pure.

Subroutine	Operation
RANDOM_NUMBER	Generate pseudorandom scalar or array of real type
RANDOM_SEED	Retrieve or set the seed of the pseudorandom number generator

These procedures to create pseudorandom numbers can be implemented in many ways. One of these ways creates an internal seed for the pseudorandom number generator from the value specified in the PUT argument. An immediately subsequent reference to RANDOM_SEED may return a value in the GET argument that is different than that specified earlier. If these values do differ, use of either in the PUT argument of a reference to RANDOM_SEED will generate the same sequence of pseudorandom numbers from calls to RANDOM_NUMBER. For example:

```
call  RANDOM_SEED (PUT=SEED1)
```

```
call  RANDOM_SEED (GET=SEED2)
```

SEED2 might not equal SEED1. But in the code segment:

```
call  RANDOM_SEED (PUT=SEED1)
call  RANDOM_SEED (GET=SEED2)
call  RANDOM_NUMBER (X1)
call  RANDOM_SEED (PUT=SEED2)
call  RANDOM_NUMBER (X2)
```

X1 must equal X2.

One typical implementation permitted by the standard is for the processor to initialize the pseudorandom number seed to the same value at the beginning of execution. This will produce the same sequence of random numbers each time the program is executed. To avoid this behavior, the values from the real time clock can be used to specify an initial seed that is likely different each time the program is executed.

A second typical implementation permitted by the standard is for the processor to initialize the pseudorandom number seed to a different value at the beginning of each execution. This will produce a different sequence of random numbers each time the program is run. For debugging code, this may not be desirable; a solution is to retrieve the seed when execution begins and record it so that the same seed can be used to initialize the pseudorandom number sequence when program is rerun.

13.3.4.2 Character Computation Functions

The character manipulation functions perform operations on character strings such as adjustment, searching, scanning, indexing, trimming, comparing, maxima, minima, and replication.

All of the character computation procedures, except LEN, REPEAT, and TRIM, are elemental functions. The procedures REPEAT and TRIM are transformational functions. LEN is an inquiry function.

Function	**Value returned**
ADJUSTL	Leading blanks removed and placed on the right
ADJUSTR	Trailing blanks removed and placed on the left
INDEX	Location of a given substring in a character string
LEN	Length of a character string
LEN_TRIM	Length of a string after trailing blanks have been removed
LGE	Greater than or equal to comparison based on the ASCII collating sequence
LGT	Greater than comparison based on the ASCII collating sequence
LLE	Less than or equal to comparison based on the ASCII collating sequence
LLT	Less than comparison based on the ASCII collating sequence
MAX	Maximum of specified values
MIN	Minimum of specified values
REPEAT	Concatenation of several copies of a character string
SCAN	Position in a string of any one of a given set of characters

TRIM	String without trailing blanks
VERIFY	Position in a string of a character that is not one of a given set

13.3.4.3 Bit Computation Procedures

The bit computation procedures perform disjunction, conjunction, exclusive disjunction, bit setting and clearing, and bit shifting and bit moving in terms of the bit model specified in 13.2.1.

Most of the bit computation procedures are elemental functions. All of their arguments are of integer type. BIT_SIZE is an inquiry function. MVBITS is an elemental subroutine. All of their arguments are of integer type. BIT_SIZE is an inquiry function. MVBITS is an elemental subroutine. Because of the dependence on the number of bits in the bit model, the bit manipulation procedures may yield nonportable results.

The BOZ literal constants are not directly permitted as arguments to these functions. However, as illustrated in the examples for some of these procedures in A, the function INT with a BOZ literal constant argument is permitted and provides this functionality.

Function	**Value returned**
BIT_SIZE	Number of bits in the bit model
BTEST	Test of the bit value in a specified position
IAND	Logical AND of two integers
IBCLR	Value with a specified bit set to zero
IBITS	Specified bits extracted from an integer value
IBSET	Value with a specified bit set to one
IEOR	Logical exclusive-OR of two integers
IOR	Logical inclusive-OR of two integers
ISHFT	Logical end-off shift of an integer
ISHFTC	Logical circular shift in a field of an integer
NOT	Logical complement of an integer

Subroutine	**Operation**
MVBITS	Copies a sequence of bits from one integer to another

13.3.4.4 Array Functions

Array functions provide array operations such as sum, product, conjunction, disjunction, counting, shifting, dot product, matrix multiplication, matrix transposition, packing, unpacking, merging, reshaping, maxima, and minima. All are transformational except MERGE, which is elemental.

The reduction functions ALL, ANY, COUNT, MAXVAL, MINVAL, PRODUCT, and SUM reduce an argument array in one of two senses; either

- all or selected array elements are reflected in (reduced to) a scalar result, or
- the reduction takes place along a dimension specified by an optional argument DIM.

In the latter case the function result is in general an array whose rank is one less than that of the argument array; it is a scalar if the array argument is of rank one. Because the rank is reduced when DIM is present, the actual argument for DIM must not be an optional dummy argument; otherwise the rank of the result could not be determined and fixed at compile time because the presence or absence of DIM would depend on the characteristics of the calling procedure.

The argument DIM that specifies the reduction along a particular dimension satisfies the same requirements for each of these reduction functions. Namely, DIM is an integer of any kind in the range [1, *n*] where *n* is the rank of the array argument (ARRAY for the functions MAXVAL, MINVAL, SUM, and PRODUCT and MASK for the functions ALL, ANY, and COUNT). For two-dimensional arrays, the reduction operation is performed down the columns if DIM=1 or across the rows if DIM=2. For example, if ARRAY has a shape of [5 10], SUM (ARRAY, DIM=1) produces a ten element result array, each element being the sum of the five corresponding column elements in ARRAY. Similarly, MAX (ARRAY, DIM=2) produces a five element result array, each being the maximum of the corresponding ten element row. For higher dimensional arrays, the arrays can be thought of as a collection of pencils and the reduction is performed down each hyperpencil. For example, an array of shape [3 4 5] can be reduced to a result array of shape [4 5], [3 5], or [3 4], depending on the value of DIM being 1, 2, or 3, respectively. All of the reductions functions, when DIM is specified, perform their operations in this way.

The array over which the reduction is performed is the first argument. These functions require this array and the masking array, if present, to be conformable. The descriptions of the results treat these arrays as if the lower bound in each dimension of each array is one. These arrays may be of size zero and the result for each function when these arrays are of size zero is specified in the description of the function.

The array construction functions MERGE, PACK, RESHAPE, SPREAD, and UNPACK construct new array values from the elements of existing arrays.

The array manipulation functions CSHIFT, EOSHIFT, and TRANSPOSE rearrange elements of an array. Although they have an optional DIM argument, the DIM argument does not change the rank of the result from that of the actual argument and so is not restricted from being an optional dummy argument.

The location functions MAXLOC and MINLOC locate the maximum and minimum values in the array or along a specified dimension.

Function	**Value returned**
ALL	True if all array elements are true
ANY	True if any array elements are true
COUNT	Number of true array elements
CSHIFT	Circular shift of the elements of an array
DOT_PRODUCT	Dot product of two rank-one arrays
EOSHIFT	End-off shift of the elements of an array
MATMUL	Matrix multiplication
MAXLOC	Location of the first maximum element of an array
MAXVAL	Maximum value of array elements

MERGE	Selection of values under control of a mask
MINLOC	Location of the first minimum element of an array
MINVAL	Minimum value of array elements
PACK	Masked array packed into a vector
PRODUCT	Product of array elements
RESHAPE	Rank-one array reshaped to an array of a specified shape
SPREAD	Array replicated by adding a dimension
SUM	Sum of array elements
TRANSPOSE	Matrix transpose
UNPACK	Array unpacked from a vector under mask control

13.3.5 System Environment Procedures

Several intrinsic procedures provide information about the environment in which a Fortran program is executing.

13.3.5.1 Time and Date Subroutines

The time and date subroutines are nonelemental and nonpure. The definitions of the CPU_TIME and SYSTEM_CLOCK subroutines are imprecise because time measurements are typically processor dependent; in general, the intention is that CPU_TIME measures processor time to run a program and often does not include time consumed by the processor for tasks other than the particular program. The procedure SYSTEM_CLOCK measures elapsed time in units and typically measures the wall-clock time taken to run the particular program.

Subroutine	Operation
CPU_TIME	Obtain the processor time
DATE_AND_TIME	Obtain date and time information in various formats
SYSTEM_CLOCK	Obtain data from the system clock

13.3.5.2 Testing Input/Output Status

These functions test input/output status values for end-of-file and end-of-record conditions:

Function	Value returned
IS_IOSTAT_END	True if a value indicates an end-of-file IOSTAT condition
IS_IOSTAT_EOR	True if a value indicates an end-of-record IOSTAT condition

These functions are elemental and thus pure.

13.3.5.3 Command Line Manipulation Procedures

The command line manipulation procedures are used to inquire about the environment that invoked the program. The procedure COMMAND_ARGUMENT_ COUNT is an inquiry function whereas the procedures GET_COMMAND, GET_COMMAND_

ARGUMENT, and GET_ENVIRONMENT_VARIABLE are subroutines. These procedures depend upon the aspects of the operating system that invoked the program and that are nonstandard and thus are processor dependent.

Procedure	**Operation**
COMMAND_ARGUMENT_COUNT	Number of command line arguments
GET_COMMAND	Obtain the entire command initiating the program
GET_COMMAND_ARGUMENT	Obtain a specified command argument
GET_ENVIRONMENT_VARIABLE	Obtain the value of a system environment variable

These subroutines are not elemental.

13.4 Specific Names for Generic Intrinsic Procedures

Some of the intrinsic functions have specific names for specific argument types. These functions may be invoked with the generic name or with the specific name for the appropriate argument. A generic procedures must not be passed as an actual argument; a specific procedure may be, except for ones marked with an asterisk in Table 13-1.

Table 13-1 List of intrinsic procedures with specific names

Generic name	Specific name and arguments	Specific argument types
ABS	ABS (A)	Default real
	CABS (A)	Default complex
	DABS (A)	Double precision real
	IABS (A)	Default integer
ACOS	ACOS (X)	Default real
	DACOS (X)	Double precision real
AIMAG	AIMAG (Z)	Default complex
AINT	AINT (A)	Default real
	DINT (A)	Double precision real
ANINT	ANINT (A)	Default real
	DNINT (A)	Double precision real
ASIN	ASIN (X)	Default real
	DSIN (X)	Double precision real
ATAN	ATAN (A)	Default real
	DATAN (A)	Double precision real
ATAN2	ATAN2 (A)	Default real
	DATAN2 (A)	Double precision real
CHAR	* CHAR (I)	Default integer

Table 13-1 *(Continued)* List of intrinsic procedures with specific names

Generic name	Specific name and arguments	Specific argument types
COS	COS (X)	Default real
	CCOS (X)	Default complex
	DCOS (X)	Double precision real
CONJG	CONJG (X)	Default complex
COSH	COSH (X)	Default real
	DCOSH (X)	Double precision real
DIM	DIM (X,Y)	Default real
	IDIM (X,Y)	Default integer
	DDIM (X,Y)	Double precision real
DPROD	DPROD (X,Y)	Default real
EXP	EXP (X)	Default real
	CEXP (X)	Default complex
	DEXP (X)	Double precision real
ICHAR	* ICHAR (C)	Default character
INDEX	INDEX (STRING, SUBSTRING)	Default character
INT	* INT (A)	Default real
	* IFIX (A)	Default real
	* IDINT (A)	Double precision real
LEN	LEN (STRING)	Default character
LGE	* LGE (STRING_A, STRING_B)	Default character
LGT	* LGT (STRING_A, STRING_B)	Default character
LLE	* LLE (STRING_A, STRING_B)	Default character
LLT	* LLT (STRING_A, STRING_B)	Default character
LOG	ALOG (X)	Default real
	CLOG (X)	Default complex
	DLOG (X)	Double precision real
LOG10	ALOG10 (X)	Default real
	DLOG10 (X)	Double precision real
MAX	* MAX0 (A1, A2, A3, ...)	Default integer
	* AMAX1 (A1, A2, A3, ...)	Default real
	* DMAX1 (A1, A2, A3, ...)	Double precision real
	* MAX1 (A1, A2, A3, ...)	Default real
	* AMAX0 (A1, A2, A3, ...)	Default integer
MIN	* MIN0 (A1, A2, A3, ...)	Default integer
	* AMIN1 (A1, A2, A3, ...)	Default real
	* DMIN1 (A1, A2, A3, ...)	Double precision real
	* MIN1 (A1, A2, A3, ...)	Default real
	* AMIN0 (A1, A2, A3, ...)	Default integer

Table 13-1 *(Continued)* List of intrinsic procedures with specific names

Generic name	Specific name and arguments	Specific argument types
MOD	MOD (A, P)	Default integer
	AMOD (A, P)	Default real
	DMOD (A, P)	Double precision real
NINT	NINT (A)	Default real
	IDNINT (A)	Double precision real
REAL	* REAL (A)	Default integer
	* FLOAT (A)	Default integer
	* SNGL (A)	Double precision real
SIGN	SIGN (A, B)	Default real
	DSIGN (A, B)	Double precision real
	ISIGN (A, B)	Default integer
SIN	SIN (X)	Default real
	CSIN (X)	Default complex
	DSIN (X)	Double precision real
SINH	SINH (X)	Default real
	DSINH (X)	Double precision real
SQRT	SQRT (X)	Default real
	CSQRT (X)	Default complex
	DSQRT (X)	Double precision real
TAN	TAN (X)	Default real
	DTAN (X)	Double precision real
TANH	TANH (X)	Default real
	DTANH (X)	Double precision real

Note:

MAX1 is equivalent to INT (MAX (. . .))

AMAX0 is equivalent to REAL (MAX (. . .))

MIN1 is equivalent to INT (MIN (. . .))

AMIN0 is equivalent to REAL (MIN (. . .))

13.5 Alphabetical List of All Intrinsic Procedures

Table 13-2 lists the intrinsic procedures. The argument names shown are the keywords for keyword argument calls. All of the optional arguments are noted as such. These procedures are described in detail, in alphabetical order, in A.

13.6 Standard Intrinsic Modules

The standard specifies several intrinsic modules, called **standard intrinsic modules**, some of which are required, namely, the modules ISO_FORTRAN_ENV and ISO_C_BINDING, and some of which are optional, namely the modules

Table 13-2 List of intrinsic procedures and arguments

Procedure and arguments	Optional arguments
ABS (A)	
ACHAR (I, KIND)	KIND
ACOS (X)	
ADJUSTL (STRING)	
ADJUSTR (STRING)	
AIMAG (Z)	
AINT (A, KIND)	KIND
ALL (MASK, DIM)	DIM
ALLOCATED (ARRAY)	
ALLOCATED (SCALAR)	
ANINT (A, KIND)	KIND
ANY (MASK, DIM)	DIM
ASIN (X)	
ASSOCIATED (POINTER, TARGET)	TARGET
ATAN (X)	
ATAN2 (Y, X)	
BIT_SIZE (I)	
BTEST (I, POS)	
CEILING (A, KIND)	KIND
CHAR (I, KIND)	KIND
CMPLX (X, Y, KIND)	Y, KIND
COMMAND_ARGUMENT_COUNT ()	
CONJG (Z)	
COS (X)	
COSH (X)	
COUNT (MASK, DIM, KIND)	DIM, KIND
CPU_TIME (TIME)	
CSHIFT (ARRAY, SHIFT, DIM)	DIM
DATE_AND_TIME (DATE, TIME, ZONE, VALUES)	DATE, TIME, ZONE, VALUES
DBLE (A)	
DIGITS (X)	
DIM (X, Y)	
DOT_PRODUCT (VECTOR_A, VECTOR_B)	
DPROD (X, Y)	

Table 13-2 *(Continued)* List of intrinsic procedures and arguments

Procedure and arguments	Optional arguments
EOSHIFT (ARRAY, SHIFT, BOUNDARY, DIM)	BOUNDARY, DIM
EPSILON (X)	
EXP (X)	
EXPONENT (X)	
EXTENDS_TYPE_OF (A, MOLD)	
FLOOR (A, KIND)	KIND
FRACTION (X)	
GET_COMMAND (COMMAND, LENGTH, STATUS)	COMMAND, LENGTH, STATUS
GET_COMMAND_ARGUMENT (NUMBER, VALUE, LENGTH, STATUS)	VALUE, LENGTH, STATUS
GET_ENVIRONMENT_VARIABLE (NAME, VALUE, LENGTH, STATUS, TRIM_NAME)	VALUE, LENGTH, STATUS, TRIM_NAME
HUGE (X)	
IACHAR (C, KIND)	KIND
IAND (I, J)	
IBCLR (I, POS)	
IBITS (I, POS, LEN)	
IBSET (I, POS)	
ICHAR (C, KIND)	KIND
IEOR (I, J)	
INDEX (STRING, SUBSTRING, BACK, KIND)	BACK, KIND
INT (A, KIND)	KIND
IOR (I, J)	
ISHFT (I, SHIFT)	
ISHFTC (I, SHIFT, SIZE)	SIZE
IS_IOSTAT_END (I)	
IS_IOSTAT_EOR (I)	
KIND (X)	
LBOUND (ARRAY, DIM, KIND)	DIM, KIND
LEN (STRING, KIND)	KIND
LEN_TRIM (STRING, KIND)	KIND
LGE (STRING_A, STRING_B)	
LGT (STRING_A, STRING_B)	
LLE (STRING_A, STRING_B)	

Table 13-2 *(Continued)* List of intrinsic procedures and arguments

Procedure and arguments	Optional arguments
LLT (STRING_A, STRING_B)	
LOG (X)	
LOG10 (X)	
LOGICAL (L, KIND)	KIND
MATMUL (MATRIX_A, MATRIX_B)	
MAX (A1, A2, A3, ...)	A3, ...
MAXEXPONENT (X)	
MAXLOC (ARRAY, DIM, MASK, KIND)	MASK, KIND
MAXLOC (ARRAY, MASK, KIND)	MASK, KIND
MAXVAL (ARRAY, DIM, MASK)	MASK
MAXVAL (ARRAY, MASK)	MASK
MERGE (TSOURCE, FSOURCE, MASK)	
MIN (A1, A2, A3, ...)	A3, ...
MINEXPONENT (X)	
MINLOC (ARRAY, DIM, MASK, KIND)	MASK, KIND
MINLOC (ARRAY, MASK, KIND)	MASK, KIND
MINVAL (ARRAY, DIM, MASK)	MASK
MINVAL (ARRAY, MASK)	MASK
MOD (A, P)	
MODULO (A, P)	
MOVE_ALLOC (FROM, TO)	
MVBITS (FROM, FROMPOS, LEN, TO, TOPOS)	
NEAREST (X, S)	
NEW_LINE (A)	
NINT (A, KIND)	KIND
NOT (I)	
NULL (MOLD)	MOLD
PACK (ARRAY, MASK, VECTOR)	VECTOR
PRECISION (X)	
PRESENT (A)	
PRODUCT (ARRAY, DIM, MASK)	MASK
PRODUCT (ARRAY, MASK)	MASK
RADIX (X)	
RANDOM_NUMBER (HARVEST)	

Table 13-2 *(Continued)* List of intrinsic procedures and arguments

Procedure and arguments	Optional arguments
RANDOM_SEED (SIZE, PUT, GET)	SIZE, PUT, GET
RANGE (X)	
REAL (A, KIND)	KIND
REPEAT (STRING, NCOPIES)	
RESHAPE (SOURCE, SHAPE, PAD, ORDER)	PAD, ORDER
RRSPACING (X)	
SAME_TYPE_AS (A, B)	
SCALE (X, I)	
SCAN (STRING, SET, BACK, KIND)	BACK, KIND
SELECTED_CHAR_KIND (NAME)	
SELECTED_INT_KIND (R)	
SELECTED_REAL_KIND (P, R)	P, R
SET_EXPONENT (X, I)	
SHAPE (SOURCE, KIND)	KIND
SIGN (A, B)	
SIN (X)	
SINH (X)	
SIZE (ARRAY, DIM, KIND)	DIM, KIND
SPACING (X)	
SPREAD (SOURCE, DIM, NCOPIES)	
SQRT (X)	
SUM (ARRAY, DIM, MASK)	MASK
SUM (ARRAY, MASK)	MASK
SYSTEM_CLOCK (COUNT, COUNT_RATE, COUNT_MAX)	COUNT, COUNT_RATE, COUNT_MAX
TAN (X)	
TANH (X)	
TINY (X)	
TRANSFER (SOURCE, MOLD, SIZE)	SIZE
TRANSPOSE (MATRIX)	
TRIM (STRING)	
UBOUND (ARRAY, DIM, KIND)	DIM, KIND
UNPACK (VECTOR, MASK, FIELD)	
VERIFY (STRING, SET, BACK, KIND)	BACK, KIND

IEEE_EXCEPTIONS, IEEE_ARITHMETIC, and IEEE_FEATURES. These intrinsic modules define module procedures, derived types, and named constants. All of the module functions of these intrinsic modules are pure, but only some of the module subroutines of these modules are pure. The Fortran environment module ISO_FORTRAN_ENV is described in this chapter whereas the IEEE modules and interoperability module are described in 14 and 15, respectively.

13.6.1 The Fortran Environment Module

The Fortran environment module ISO_FORTRAN_ENV specifies named constants only. The named constants are all of default integer type; their values specify quantities related to the Fortran environment. The names and meanings of the named constants are provided in Table 13-3.

Table 13-3 The ISO_FORTRAN_ENV Module Constants

Name	Value
INPUT_UNIT	The unit number of the processor-dependent preconnected external unit identified by an asterisk in a READ statement (9.1.6.2)
OUTPUT_UNIT	The unit number of the processor-dependent preconnected external unit identified by an asterisk in a WRITE statement (9.1.6.2)
ERROR_UNIT	The unit number of the processor-dependent preconnected external unit used for error reporting
NUMERIC_STORAGE_SIZE	The number of bits in a numeric storage unit (4.3.1.1)
CHARACTER_STORAGE_SIZE	The number of bits in a character storage unit (4.3.5.1)
FILE_STORAGE_SIZE	The number of bits in a file storage unit (9.1)
IOSTAT_END	The value of the IOSTAT specifier (9.2.3) in an input statement when an end-of-file condition occurs and no error condition occurs (9.6)
IOSTAT_EOR	The value of the IOSTAT specifier (9.2.3) in an input statement when an end-of-record condition occurs and no error condition occurs (9.6)

The values of the INPUT_UNIT, OUTPUT_UNIT, and ERROR_UNIT may be negative, but will not be −1. The value of ERROR_UNIT may be equal to the value of OUTPUT_UNIT. The values of IOSTAT_END and IOSTAT_EOR must be unequal and negative.

14 IEEE Exceptions and Arithmetic

- Three **IEEE Intrinsic Modules** are defined by the Fortran standard to support the IEEE-style arithmetic and exceptions. Which intrinsic modules are provided is processor-dependent; whether some or all of the IEEE features of a particular module are provided is also processor-dependent. The three intrinsic modules IEEE_FEATURES, IEEE_ARITHMETIC, and IEEE_EXCEPTIONS define the set of specific IEEE features, arithmetic, modes, and exceptions that are supported by the processor. None of the procedures in intrinsic modules are intrinsic procedures.
- **IEEE Arithmetic** is a term used by the Fortran standard to refer to the subset of the features described in the IEEE international standard for floating-point arithmetic [13] that is supported by the IEEE intrinsic modules. For a particular implementation, these modules determine the kinds of the real (or complex) data type that can be used and the details of that support. In effect, the three modules define the term "IEEE Arithmetic" for a particular implementation. The Fortran standard specifies a minimum subset of the IEEE specifications that must be supported before the processor can claim it is "supporting IEEE arithmetic".
- An **IEEE Exception** is one of five anomalies that can occur during a floating-point operation. These exceptions are overflow, divide-by-zero, invalid, underflow, and inexact.
- An **IEEE Exceptional Value** is either a denormalized number, one of the infinities, or one of the NaN (Not-a-Number) values. These values are typically the result of an arithmetic operation that is anomalous or exceptional.

The three IEEE intrinsic modules supply derived types, named constants, and module procedures which support the IEEE exceptions, arithmetic, and procedures, described in [13]. The features of these modules have been provided in Fortran to support the programming of computation that requires detailed control of the arithmetic and careful response to exceptions. With these IEEE features, robust, reliable, accurate, and yet very efficient algorithms can be implemented in a straightforward way.

The features are defined to be consistent with the IEEE floating-point binary standard [13]. This standard specifies the format for floating-point numbers in arithmetic units and storage, the accuracy of the primary arithmetic operations, the arithmetic exceptions that can be raised by these operations, and the rounding, underflow, and halting modes that processors must follow to be compliant.

On the other hand, the IEEE features supported by these three intrinsic modules may be difficult or nearly impossible to implement with sufficient efficiency to satisfy the requirements of a Fortran implementation on a particular hardware processor. Consequently, the IEEE modules and the features and details of each module are, in

J.C. Adams et al., *The Fortran 2003 Handbook*,
DOI: 10.1007/978-1-84628-746-6_14,

many cases, optional to the Fortran processor. The possible options are constrained by the Fortran standard; the intention is to encourage the implementation of as much as possible of the arithmetic and exceptions support, following the intent and spirit of the IEEE floating-point standards to an extent that is feasible and reasonable. This is why these Fortran intrinsic modules are said to support the IEEE style of floating-point arithmetic and exceptions rather than to require IEEE arithmetic and exceptions.

However, these modules, and thus the Fortran standard, also permit a conforming processor to provide non-IEEE kinds of real and complex type that might provide support for only some of these features. For example, a processor might support most of the IEEE features for a particular kind but not support the IEEE square root function or the divide exception for that kind, say K. It would be nonconforming for the IEEE module procedure IEEE_SUPPORT_DATATYPE (1.0_K) to return true when it does not meet all the requirements.

14.1 Terms and Concepts

The following terms and concepts apply to each kind of real supported as an IEEE real. The particular forms or representations of the term or concept depend on the kind of real being used.

Not-a-Number (NaN) is a value that has no numerical significance. It usually is the result of an operation where no mathematical result is appropriate. For example, the result of 0.0/0.0 or 0.0 raised to the power 0.0 has no mathematical value and thus results in a NaN. A NaN may also be generated by other means, such as by the procedure IEEE_VALUE. A NaN is either a signaling NaN or a quiet NaN. A signaling NaN signals an invalid operation whenever it is used as an operand; a quiet NaN propagates through almost every arithmetic operation without signaling an exception.

A normal number is a value that is neither a denormalized number, an infinity, nor a NaN. Numbers, such as `+0.0`, `-0.0`, `1.0`, and `-1.0`, are examples of normal numbers. A number such as `3.1` is a normal number, even though it cannot be represented exactly in binary IEEE format.

A signed zero is an IEEE real value that behaves like zero in all but a few special cases. Some of the special cases include division of a nonzero finite divisor by a signed zero, the intrinsic functions SIGN, ATAN2, and SQRT, and the IEEE module function IEEE_COPY_SIGN.

Infinity (Inf) represents the mathematical concept of infinity, but also represents a value so large in magnitude that it cannot be represented as a precise processor value. It can be signed. It is created by overflow, a division of a nonzero finite value by a zero value, or the module function IEEE_VALUE. In relational comparisons with real types, positive infinity tests larger than all other real values and negative infinity tests smaller than all real values. An arithmetic operation on infinity is considered to return an exact value and raises no exceptions, except for those operations that are invalid such as infinity minus infinity.

Huge (huge) is the largest positive representable number in magnitude. The notation +huge is the largest positive representable value and, because the IEEE representation is a sign-magnitude representation, -huge is the smallest negative representable value. Because the value of the intrinsic function HUGE is based on formulas related to a "best-fit" model for processor numbers of a particular real kind, the IEEE value +huge might not be the same value as the value returned by the intrinsic HUGE. However, for IEEE kinds of real, they are likely the same values.

Tiny (tiny) is the smallest magnitude positive number representable with full precision. The notation +tiny is the smallest positive representable value with full precision and -tiny is the smallest magnitude negative representable value with full precision. Because the value returned by the intrinsic function TINY is based on formulas related to a "best-fit" model for all (small and large) real processor numbers of a particular real kind, the value +tiny might not be the same value as the value returned by the intrinsic TINY. However, for IEEE kinds of real, they are likely to the same values.

A denormalized number (denorm) is a number between +0.0 and +tiny or -0.0 and -tiny that cannot be represented with full precision of the arithmetic. The smallest denormalized number in magnitude has one bit of precision and the largest denormalized number has one bit of precision less than +tiny. The notation +denorm or -denorm is used to denote any positive or negative nonzero denormalized number, respectively.

An exceptional value is a nonnormal number, that is, an infinity (either sign), a denormalized value (either sign), or a NaN (either signaling or quiet). An infinity can be created by any operation that overflows or by division of a nonzero value by zero. A denormalized value can be created by any arithmetic operation when the underflow mode is gradual. A NaN is created by an invalid operation.

Divide by zero is an exception that occurs when a nonzero dividend is divided by a zero divisor. The result of such a division is a correctly signed infinity.

Underflow is an exception that occurs when the result of an operation is a number smaller in magnitude than the smallest nonzero number that has full precision but is not exactly zero. That is, the result is a nonzero number strictly between -tiny and +tiny.

Overflow is an exception that occurs when the result of an operation is a number so large in magnitude that it is outside the range of representable numbers. That is, the result is a number that exceeds +huge or is less than -huge.

Inexact is an exception that occurs when the exact result of an operation has to have its precision reduced (or rounded) because the exact result is too precise to be represented as a real value. In particular, the inexact exception is raised if the rounded result overflows, or if it underflows and the exact result is not equal to the returned result. As well as floating-point operations, this flag may be raised when a real is converted to an integer and is too large to be represented in the particular kind of integer.

Invalid is an exception that occurs when an arithmetic operation is invalid. An operation is invalid when there is no mathematically meaningful value to provide for the result. Invalid operations include:

- zero divided by zero or infinity divided by infinity;
- zero times infinity;
- magnitude subtraction where both operands are infinite (for example, Inf–Inf);
- the square root of a value that is less than zero
- the IEEE remainder operation where the divisor is zero or the dividend is infinite;
- an arithmetic operation (except exponentiation) with a signaling NaN as an operand; or
- IEEE comparisons, if they are supported.

An IEEE Mode is a state of the arithmetic processor; there is an IEEE mode that specifies how rounding is performed, how underflow is handled, or the behavior or responses when an exception is handled. Modes are described in 14.2.8. Although other modes in the processor are possible, such as input/output modes, the word "mode" in the remainder of this chapter refers to an IEEE mode.

The concept **"Supporting IEEE Arithmetic"** means at least the following subset of features of the IEEE standard [13]:

- the normalized numbers must be those of the IEEE floating-point formats;
- the operations of addition, subtraction, and multiplication with at least one of the rounding modes must be supported;
- the functions IEEE remainder, copy sign, scaling by a power of the exponent, extracting the exponent, next-after value, and unordered must be provided (via the function interfaces described in later sections);
- the inquiry function IEEE_SUPPORT_DIVIDE must be provided; and
- the addition, subtraction, and multiplication operations with normalized operands produce the results specified by the IEEE standard when the result is also normalized and within range.

All the features that constitute the concept "supporting IEEE arithmetic" must be available for a particular kind of real X for the IEEE module inquiry function IEEE_SUPPORT_DATATYPE (X) to return true. Many of the IEEE module procedures must not be invoked with arguments for which IEEE arithmetic is not supported. In all cases, this is a program requirement in the same sense that the intrinsic function SQRT must not be invoked with a negative argument whose kind of real does not support IEEE arithmetic; the processor is not required diagnose such a violation in a program.

Note that the support for NaNs and relational operations are not required in the definition of supporting IEEE arithmetic but the unordered function which indicates whether operands are ordered is required. If a processor's support for NaNs includes the use of NaNs as arguments for the unordered function and IEEE exceptions are supported, but IEEE comparisons are not supported, then the IEEE comparisons can be implemented by the programmer using procedures. Note that the definition of supporting IEEE NaNs below requires IEEE comparisons to be supported as well, specifically with respect to NaNs as operands of relational operators (14.2.4).

Some other IEEE features are optional but support within the IEEE modules is provided for them if they are available; these are:

- denormalized numbers,
- infinities,
- NaNs,
- operations on these special values,
- gradual underflow,
- exceptional results for specific operations such as divide by zero, and
- the availability of an IEEE-conforming intrinsic SQRT function.

These features are, in a sense, extensions to the required features, encouraged by the specification of the IEEE intrinsic modules and may be partially or completely implemented by the processor. An example of partial support is the availability of an inquiry module function that detects whether the intrinsic SQRT function satisfies the accuracy and functionality of the square root operation described in the IEEE standard [13]. In addition, some of the IEEE module procedures have restrictions not specified in the IEEE standard, mainly because some hardware implementations on which Fortran is used cannot support these operations; the expectation is that on most current and future hardware these restrictions are unnecessary; it is likely that most implementations of the IEEE modules will remove these restrictions.

Other features specified in the IEEE standard [13] are not mentioned by the current Fortran standard, for example IEEE traps and trap handlers. However, the features specified by the Fortran standard and the IEEE intrinsic modules do not preclude these features being implemented by a processor as an extension.

The concepts **"Supporting IEEE Denormalized Numbers, Infinities, or NaNs"** means supporting the primary arithmetic (all except exponentiation) operations, relational conditions, and assignments with IEEE denormalized numbers, IEEE infinities, and IEEE NaNs, respectively. That is, support for the generation of these values and their use as operands in all arithmetic operations and arguments to all procedures, and their assignments to variables. In addition, the treatment of these numbers by the intrinsic functions and the IEEE module functions must be consistent with the IEEE standard [13]. If a primary arithmetic operation or intrinsic procedure produces a denormalized value, the underflow and inexact exceptions are signaled if the value is not exact or no exception is signaled if the result is exact. If a primary arithmetic operation or intrinsic function produces a denormalized result and the result is not exact, the underflow and inexact exceptions must be raised.

The concept **"Supporting IEEE Divide"** means supporting a division operation that returns the correctly rounded result, depending on the rounding mode, and that raises the divide-by-zero signal and returns a correctly signed infinity when a finite nonzero value is divided by zero.

The concept **"Supporting IEEE Square Root"** means supporting a square root function that returns the correctly rounded result for nonnegative operands, as specified by the rounding mode, that it returns a positive result for all positive operands (including the result +0.0 for `+0.0` as the argument and +Inf result for `+Inf` as the argument), that it returns –0.0 for an operand of `-0.0`, and that it returns a NaN for any other negative operand with the invalid exception flag signaled. The only other exception that can be raised is inexact if the rounded result is not the exact square root.

The concept **"Supporting IEEE Remainder Operation"** means supporting the remainder operation defined as the exact remainder of x divided by y, provided y is nonzero and x is finite, and otherwise a NaN. When the remainder is zero, it has the sign of x.

The concept **"Supporting IEEE Formatted Input and Output"** means supporting the conversion of a value between an internal IEEE binary floating-point representation and a decimal formatted value as specified by [13]. This means that correctly rounded results are required for numbers with specific ranges for all formatted input/output rounding modes (9.2.4, 10.9.7), and for numbers outside these ranges, the error in the result for nearest rounding mode is at most 0.47 units in the last place, and in all other rounding modes, the error is at most 1.47 units in the last place. The rounding modes set by the IEEE module procedure IEEE_SET_ROUNDING_ MODE have no effect on the input/output conversion modes; they only control rounding during an arithmetic operation. It should be noted that unless the function IEEE_SUPPORT_IO is true, the rounding mode NEAREST might not round the same way when the value is equidistant between two representable or decimal numbers; see the input/output rounding mode NEAREST in 9.2.4.

The concept **"Signaling an Exception or Raising an Exception Flag"** refers to the process of informing a program that an exception has occurred. When an exception such as overflow occurs, the overflow exception is raised. This typically is indicated in some hardware register accessible by special instructions. To detect such an event in a program, an IEEE intrinsic module procedure is referenced which sets a logical variable to true; this is also referred to as "raising an exception flag". From the software point of view, the exception is not raised until it is tested by the IEEE module procedure; from the hardware point of view, the exception is usually raised within a few cycles of the occurrence of the exception.

The concept **"Supporting the IEEE International Standard"** refers to support for a subset of features of the IEEE binary floating-point standard [13] defined by the three IEEE intrinsic modules. The module procedure IEEE_SUPPORT_STANDARD inquiries whether nine specific features are supported—see the description of IEEE_SUPPORT_ STANDARD and Note 5 in Table 14-9.

14.2 IEEE Arithmetic and Exceptions—an Introduction

IEEE arithmetic and exceptions are a collection of the following major items:

- a set of formats (how a floating-point number is represented),
- a set of requirements for the accuracy of the basic floating-point arithmetic operations,
- a specification of supported rounding modes for the basic arithmetic operations,
- a set of exceptions that must be recognized, and
- a prescription on how these exceptions can be handled.

Hardware that is compliant with the IEEE standards has been available to varying degrees for nearly 20 years. Some support for IEEE arithmetic was added in Fortran 95 and more complete support is specified in Fortran 2003.

Before describing the features of the Fortran IEEE intrinsic modules, consider the basic problem that is presented by floating-point arithmetic. To support a wide range of scientific and engineering computations, large number ranges must be provided, say as large in magnitude as 10^{300} and as small in magnitude as 10^{-300}. In order to handle numbers of this size, either 600-700 decimal digits (or 1800-2100 binary digits) must be maintained for all computations (which represents a large cost), or some form of compact representation is needed, such as one that logically partitions a number into parts that contain the exponent to some base, the fraction, and the sign of the number.

Because the range of numbers that can be represented is finite, one has to expect and plan for handling the situation when a result of some computation exceeds the representable range. When a result exceeds the range, an overflow (the result is too large in magnitude) or an underflow (the result is too small in magnitude) occurs. Similarly, because one may require more digits than are available to represent a result exactly, one has to expect the result to be rounded (shortened) to fit into the available space; when this shortening occurs, the result is said to be inexact and an inexact exception is raised. This rounding or shortening of the result is a consequence of the limited precision of the representation. Thus, when a range or precision limit is exceeded, a floating-point exception occurs, which, in these cases, is an overflow, underflow or inexact exception.

The IEEE standard specifies what are the limits for each floating-point container (storage word or intermediate storage), what happens when the limits are exceeded, and what tools (functions and modes) are available to make these limits more acceptable and to diagnose unavoidable arithmetic problems in computations. That is, the IEEE standard allows programmers to perform robust, efficient, and accurate computations where possible, despite these limits.

When the processor raises an exception, the programmer has the option of having the program halt or continue execution. This capability is one of the computational modes supported by the IEEE modules and is introduced in 14.2.8. However, the Fortran standard does not specify traps and procedures to handle exceptions as described in the IEEE standards.

14.2.1 Floating-Point Formats

To understand the details of IEEE arithmetic and exceptions, consider first the representation of IEEE floating-point numbers for single precision values, often referred to as the IEEE 32-bit single precision format. A word of 32 bits is divided into 3 parts; a sign, an exponent part, and a fractional part. This is shown in Figure 14-1.

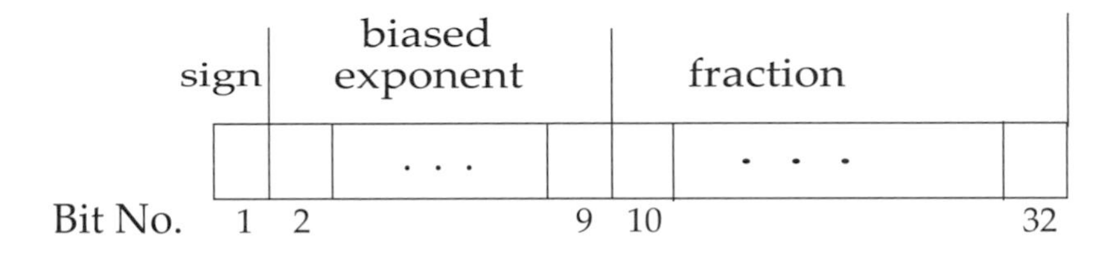

Figure 14-1 Format for a 32-bit IEEE single precision word

The sign field is 1 bit wide and is either a 0 or 1, indicating a positive or negative number, respectively. The exponent field is an unsigned integer of 8 bits, capable of storing integers from 0 to 255. This field represents the exponent as a power of 2, but is stored with a bias of 127; that is, if the true exponent of the number is 0 (that is, a number such as 2^0), it is stored as the integer 127 or a binary 0 followed by 7 binary 1's. The stored value in the exponent field is called the **biased exponent**. The remaining 23 bits hold the fractional part with the binary point at the extreme left of the fraction field (before the first binary bit of the fractional part).

But there is one more subtlety. Floating-point numbers are either zero or nonzero. For all nonzero numbers, the fractional part must be nonzero and therefore has a first nonzero bit (which of course is always 1 because there is no other nonzero bit value). The representation is normalized (that is, the exponent adjusted) so that the first nonzero bit is the first bit before the binary point and the remaining bits represent the fraction. But now all normalized numbers are 1 in this first bit so why store it? Let it be implied—it is always 1. Thus, while there are physically 23 fractional binary bits, they represent a 24-bit mantissa; the first bit is always 1 and is not physically stored. Thus, a floating-point nonzero number is represented as: $s \times 2^e \times (1 + f)$ where f is the factional part, e is the exponent, called the **unbiased exponent**, and s is the sign. Thus, the number 1.0 is represented as the bit pattern

0 01111111 00000000000000000000000

where the sign is 0 (positive), the biased exponent is 127 + 0 (that is, it represents 2^0) and the fraction is zero with an implied bit representing a binary 1 digit before the binary point. (The blanks in the above display separate the sign part, exponent part, and fractional part from one another.) The number 2 is represented by the bit pattern

0 10000000 00000000000000000000000

and the number $2 + 2^{-22}$ is represented by the bit pattern

0 10000000 00000000000000000000001

Notice that the biased exponent exceeds 127 for all numbers 2 or larger in magnitude.

Because of the existence of the implied bit which is always 1, the value zero is represented by a reserved exponent, which is selected to be zero, with all bits in the fraction set to 0. Thus, zero is represented as:

0 00000000 00000000000000000000000

And, of course, there can be a negative zero now, namely:

1 00000000 00000000000000000000000

Denormalized numbers, if they are allowed, are represented with the same biased exponent as zero, namely 00000000, but have at least one binary 1 digit in the fractional part. They may have an unbiased exponent that is in the range [–149, –127], depending on the value of the denormalized number. For example:

0 00000000 10000000000000000000000

is a denormalized number that has the value tiny/2.0 or 2^{-127}.

14.2.2 Floating-Point Exceptions

Before going on to the other special values, consider the anomalous results that can occur when computing with a format such as this. The numbers that are representable by such a format are depicted as vertical bars or tick marks in Figure 14-2:

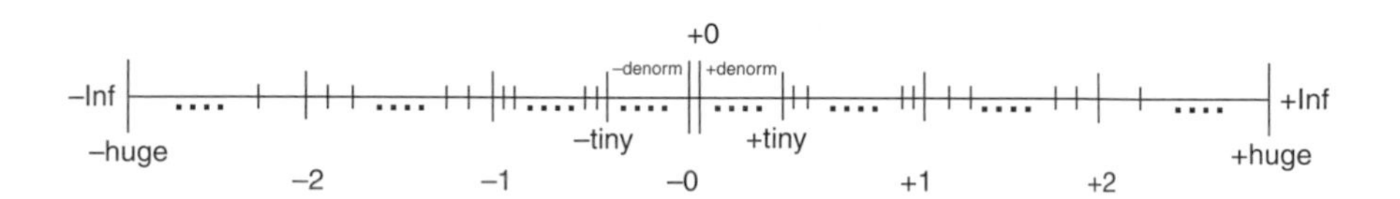

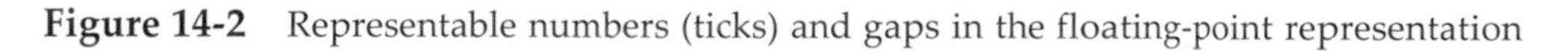
Figure 14-2 Representable numbers (ticks) and gaps in the floating-point representation

Notice that:

1. At the extreme left there is the region minus infinity (-Inf); -Inf denotes any number so large in magnitude that it cannot be represented (there is no room in the finite representation and the limits of the representation are exceeded; that is, the exponent is too large, greater than 127 in the 32-bit format). In the above 32-bit format, -Inf is represented as a value with a negative sign, the largest biased exponent possible and a zero fraction—that is, a number with nine initial 1 bits followed by twenty-three 0 bits as in:

   1 1111111 0000000000000000000000.

2. Towards the left side, there are the negative numbers. Towards the middle of the number system, the numbers are smaller in magnitude and closer together.

3. At the left of the middle, there is a gap, denoted -denorm, between the smallest negative normalized number -tiny and -0—this is the location where the negative denormalized numbers reside. That is, the numbers that are negative and are so small in magnitude that the first 1 bit cannot be placed in the first (implied) position and still have an unbiased exponent in the range [–126, 127].

4. Similarly, to the right of positive zero and the smallest positive normalized number +tiny, there is a gap, denoted +denorm—this is the location where the positive denormalized numbers lie.

5. Similarly, to the right of +tiny, the numbers are distributed all the way to infinity, with the numbers near zero closer together, and the numbers spreading apart as they get larger.

6. At the extreme right, there is the region plus infinity (+Inf) with a large gap between the largest positive number, designated +huge, and +Inf.

7. NaN does not appear in this figure because it is not a number.

It should be noted here that the ticks in Figure 14-2 are representable numbers in the 32-bit IEEE format. Model numbers, as described in 13.2.3, may or may not he the same as representable numbers, and in general are not the same as the IEEE representable numbers. For the IEEE 32-bit and 64-bit formats, this model can be made to include all of the IEEE normal numbers, except negative zero, but there may be some IEEE representable numbers, namely denormalized numbers, that are not model numbers. As pointed out in 14.1, huge and tiny can be the same numbers that are returned by the intrinsic HUGE and TINY (and will likely be) but are not required to be the same in a particular implementation.

Now we are able to describe in detail what exceptions are. Each of the tick marks in the above figure indicates a representable number, that is, a number that can be exactly represented by the format. What happens when the result of a computation creates a value in between these tick marks, which is certainly possible, for example, 1/10? Because of the finite representation, the processor cannot represent such a value and when the arithmetic unit tries to create and store such a value, it creates what is called an exception. There are five types of exceptions; four exceptions correspond to the four types of gaps in the above figure and the fifth exception corresponds to a mathematically invalid operation.

1. If the result of an operation has no mathematical value (such as `0.0/0.0` or $\texttt{0.0}^{\texttt{0.0}}$, or Inf-Inf), an invalid exception occurs—there is no legitimate value to use. If the operation is to complete without halting, the processor will create a NaN and raise the invalid exception.

2. If the generated result is so large in magnitude that it is to the right of +huge or the left of -huge, an overflow exception occurs—that is, there is not enough space in the exponent part of the format to represent a number with this large an exponent. Such a situation could be generated, for example, by the computation huge*huge, which generates a result which is too large to represent. When such an extreme result is generated, the overflow exception is raised. If the operation is to complete without halting, the processor will create a value of +Inf or -Inf, depending on the operator and the signs of its operands.

3. If the generated result is the result of a division of a nonzero value by zero, the result is so large again that it cannot be presented. However, in this case of a large result, a divide-by-zero exception is raised rather than an overflow exception and the result created is either positive or negative infinity, depending on the signs of the numerator and denominator.

4. If the generated result is between -tiny and -0 or between +0 and +tiny (but nonzero), again the number cannot be represented. This time, an underflow exception is raised. Such results may be set to a zero of the appropriate sign; this is called **abrupt underflow** (or the **abrupt underflow mode**). Alternatively, such results can be represented as denormalized numbers, still raising underflow, provided the processor supports such numbers; this is called **gradual underflow** (or the **gradual underflow mod**e).

5. If the generated result is within range but it has a value that causes it to be between the ticks in Figure 14-2 (for example, the computed result is $1.0+2^{-50}$ and the two nearest neighbors or representable numbers are 1.0 and $1.0+2^{-23}$), an inexact exception occurs; that is, because the space to store the fraction is limited to 23 bits in the IEEE single precision format, the number $1.0+2^{-50}$ cannot be represented—it needs 50 bits in the fraction to represent it—so that the number has to be rounded, causing an inexact exception. The inexact exception frequently occurs, because it indicates that a rounding error has occurred in order to store the floating-point result; it is ignored in most computations.

Notice that the above discussion suggests that exceptions occur mostly in cases where the operands of arithmetic operations are normal numbers. Indeed, that is the case; the overflow, underflow, divide-by-zero, and inexact exceptions do not occur when one of the operands is an infinity or a NaN. However, the invalid exception will occur when, for instance, both operands are infinities in a division or a true subtraction. See the IEEE standard [13] for full description of these cases.

Exceptions sound ominous, and certainly were with early architectures. Modern architectures, in many cases, raise these exceptions without skipping a beat (wasting a cycle) but because of parallel and pipelined functional units, cannot stop the computation precisely at the point where the exception occurred. To do so would be prohibitively expensive in terms of cycles used, for example, to back up and redo parts of the computation that had no exception but had to be interrupted. In addition, whereas the initial expectation is that exceptions are rare, the inexact exception probably occurs more often than no exception because it essentially indicates a rounding error has occurred. Thus, because of the cost of interruption and the high frequency of occurrence of at least some of these exceptions, the approach taken by most modern chips is to record their occurrence in a flag and let the computation proceed uninterrupted until the processor, program, and programmer are ready to handle them. This is the preferred approach because it avoids very time-consuming interrupts to the processor. It is encouraged by the IEEE standard. The Fortran intrinsic modules specified below support this approach of delaying the fix-up action upon the occurrence of an exception to a convenient point in the program and not when and where it happens.

The IEEE standard [13] specifies other formats for representing numbers. The other one that is of particular importance in Fortran is the 64-bit IEEE format which is commonly used for the double precision real type. It has all of the same properties as the 32-bit format, except that the biased exponent is 11 bits rather than 8 bits and the fractional part is 52 bits rather than 23 bits. NaNs, infinities, signed zeros, normalized numbers, and denormalized numbers are representable, but there are considerably more NaNs, normalized, and denormalized numbers than in the 32-bit format.

14.2.3 IEEE Arithmetic

Now, with an understanding of the formats and exceptions defined by the IEEE standard, we can now specify what the arithmetic must do. To be fully compliant with the IEEE standards, the result of every operation must be defined as follows:

1. If the operands for an operation are such that the mathematical result is not defined (e.g., 0.0/0.0), the invalid exception must be raised and the result is a NaN.

2. If the result is defined mathematically but cannot be represented because it is too large in magnitude, the overflow exception must be raised and the result is an infinity of the appropriate sign.

3. If the result is defined mathematically but cannot be represented because it is too small in magnitude, either a denormalized or zero result is returned with the underflow exception raised. What the processor does depends on the magnitude of the result and on the floating-point status (14.2.7) flags of the processor which can be set or retrieved using certain module procedures described in 14.3.

4. If the result is valid and within range and if the mathematical result is a representable value, that value becomes the result with no exception raised. If the mathematical result is not representable, a representable result is returned, which is one of the immediate neighbors to the exact result, and the inexact exception is raised. Which neighbor is determined by the rounding mode in effect at the time of the operation. For example, on an IEEE machine using the 32-bit format, if the mathematical result is $1.0+2^{-40}$, the returned result is `1.0` if the rounding mode is round to nearest, round down, or round towards zero and is $1.0+2^{-23}$ if the rounding mode is round up.

14.2.4 Quiet and Signaling NaNs

A NaN is not a number and may be generated when an invalid operation is performed. There are two kinds of NaNs: quiet ones and signaling ones.

For a processor that supports NaNs and for an operation that returns a floating-point result, a quiet NaN propagates through the operation without raising an invalid exception. A signaling NaN, on the other hand in the same situation, signals an invalid exception when used; the result of the operation is a quiet NaN. For relational operations, the result is not real. For the relational operators, if neither operand is a NaN, no exception is signaled; if either one is a NaN (signaling or quiet) and the relational operator is <, <=, >, or >=, an invalid exception is raised. If either operand of a relational operator is a NaN, the result is false for all relational operators except /= and true for /=; the result is false even for equal comparison of the same NaN.

14.2.5 The Programming Approach for the IEEE Standard

The IEEE approach encourages what many feel is an unnatural and rather special programming approach to handling exceptions, namely, let the exceptions occur in large grain computations, and once the computation is complete, inquire about what exceptions occurred and, depending on what occurred, rescale, or use a different algorithm to avoid the erroneous computation, or abort the computation. Using the IEEE support provided by this standard, robust and reliable software can be written to attain very efficient code in the presence of occasionally badly scaled data. The example in Figure 14-4 illustrates this approach.

14.2.6 What If the Processor Is Not Compliant with the IEEE Standard?

A Fortran processor that is running on hardware that is fully compliant with one of the IEEE floating-point standards is expected to, and probably will, provide the IEEE modules described in 14.3. However, because some current hardware is partially compliant with the IEEE standards or because insisting on full compliance incurs unacceptable inefficiencies, it is expected that some Fortran processors will support selected features of the IEEE modules.

What features are actually supported is processor dependent, but to conform with the Fortran standard, the parts of the IEEE modules that are actually implemented are partially constrained (14.6). For example, if the named constant IEEE_DATATYPE is accessible from the IEEE_FEATURES module, the processor must support IEEE arithmetic (as defined in 14.1) for at least one kind of real; other specifications in the Fortran standard require such support for the complex data type of the same kind. As a second example, if either of the modules IEEE_EXCEPTIONS or IEEE_ARITHMETIC is accessible, the overflow and divide-by-zero exceptions for all kinds of real and complex data types must be supported. This accessibility approach means that at least a well written program will generate a compile time error if it tries to use an unsupported feature and will not have undetected run-time errors because an IEEE feature is not supported.

14.2.7 The Processor's Floating-Point Status

The processor's floating-point status is represented by the values of the supported exception flags and modes of the processor. Module subroutines are provided to inquire about the floating-point status (IEEE_GET_STATUS) and to set the status to specific values and modes (IEEE_SET_STATUS). Objects with values representing the floating-point status are of the private type IEEE_STATUS_TYPE.

14.2.8 The Modes of the Floating-Point Processor

There are three processor features that have associated modes, provided the processor can support them.

For the arithmetic rounding feature, there is the mode that determines what rounding algorithm is used when an inexact exception is raised; the possible rounding modes are round to nearest, round up, round down, and round towards zero.

For the underflow feature, there is the underflow mode; the underflow mode is either

- gradual in which case the result of an operation may be a denormalized number that approximates the result or zero when the result is smaller in magnitude than the smallest denormalized number, or
- abrupt in which case the result of an underflowed operation is always zero when an underflow exception occurs.

For exceptions, there is the mode that specifies the processor's action when an exception is raised; either the processor halts, aborting the computation, upon the occurrence of an exception, or it delivers a result, raises the exception flag, and continues execution. What "aborting the computation" really means is processor dependent and unspecified by either the IEEE or Fortran standard.

14.3 Descriptions of the Three Intrinsic Modules

To support IEEE arithmetic and exceptions, Fortran defines three intrinsic modules. The modules are optional and their presence or absence, as well as the particular module entities they define, determines the level of IEEE support on a particular processor.

The first module is the IEEE_FEATURES module; it defines a derived type IEEE_FEATURES_TYPE and named constants of this type. The named constants are individually optional and each represents an IEEE feature; for example, the presence of the named constant IEEE_NAN in this module indicates that the IEEE modules support IEEE NaNs for at least one kind of real. Table 14-2 lists each named constant and the IEEE feature it represents.

The recommended practice is to provide a USE statement with the ONLY option referring to the named constant for a particular feature that is desired; used in this way, the USE statement acts like a compiler command line switch, accessing only the IEEE feature(s) requested, and producing a diagnostic if the requested feature is not supported. In this case, the compiler is expected to create efficient code that implements only the features specified by the named constants and it need not be cognizant of the other features.

Two other intrinsic modules IEEE_EXCEPTIONS and IEEE_ARITHMETIC define the module procedures, named constants, and derived types that give detailed control over the features selected from the IEEE_FEATURES intrinsic module. These modules are described in detail in 14.3.2 and 14.3.3, respectively. The chapter concludes with the discussion of some sample Fortran code illustrating the use of the IEEE arithmetic and exception features.

The module procedures in the modules IEEE_EXCEPTIONS and IEEE_ARITHMETIC have dummy arguments named CLASS, FLAG, FLAG_VALUE, GRADUAL, HALTING, I, P, R, ROUND_VALUE, STATUS_VALUE, X, and Y; in this collection of procedures, these arguments are used consistently and have essentially the same requirements in all procedures. Their common properties are detailed in Table 14-1.

In fact, although these procedures are not intrinsic procedures (that is, there are made accessible only by a USE statement), they are classified as inquiry, transformational, elemental, nonelemental, functions, subroutines, and pure, and have the general properties of these classifications of procedures described in 12.7 and 13.1. In particular, where the module procedure is elemental, the arguments are described as scalars but, because of the elemental nature of the procedure, these procedures may be called with array or scalar arguments in conformance with the usual rules for elemental procedure references (12.7.2). Where an entity is the actual argument of an inquiry function and is the object being inquired about, the corresponding actual argument need not have a defined value and may be a scalar or an array. Otherwise, if a module procedure's argument has the intent IN attribute, it must have a defined value.

The "optional" argument notation {X} (or {,X}) is used for the functions in Tables 14-4 and 14-9. In these cases, the argument X is not optional in the sense of an OPTIONAL argument in 12.6.2 but is a notation to specify the interfaces to two distinct functions; one function has no argument X and the other has a nonoptional argument X. The function with no argument X returns a result indicating a property of all kinds of the real data type and the other function with the argument X returns a result indicating a property of real objects of the specific kind of X. The subtle difference between an optional argument X and the specification of two functions can best be explained by considering the following example:

```
subroutine my_sub(x)
   use ieee_arithmetic
   real, optional :: x
   print *, ieee_support_datatype(x)
   print *, ieee_support_datatype( )
end
```

Because IEEE_SUPPORT_DATATYPE has no optional argument, the two references to IEEE_SUPPORT_DATATYPE are to two different functions; these references do not depend on whether the actual argument corresponding to the dummy argument X is present.

Table 14-1 Properties of the actual arguments of the module procedures, specified by name

Argument name	Description
CLASS	Of type IEEE_CLASS_TYPE with INTENT (IN). It may have any value of this type. It must conform with its companion argument X.
FLAG	Of type IEEE_FLAG_TYPE. It may be a scalar or an array and of INTENT (INTENT (INTENT (INTENT (INTENT (IN) or INTENT (OUT). If it is of INTENT (IN) and a scalar, it may have any value of this type; if it is of INTENT (IN) and an array, each element of the array may have any value of this type, but no two elements may have the same value.
FLAG_VALUE	Of type default logical. It must conform with its companion argument FLAG. It has INTENT (OUT) in the subroutine IEEE_GET_FLAG and INTENT (IN) in the subroutine IEEE_SET_FLAG.
GRADUAL	Of type default logical. It must be a scalar. It has INTENT (OUT) in the subroutine IEEE_GET_UNDERFLOW_MODE and INTENT (IN) in the subroutine IEEE_SET_UNDERFLOW_MODE.
HALTING	Of type default logical. It has INTENT (OUT) in the subroutine IEEE_GET_HALTING_MODE, and INTENT (IN) in the subroutine IEEE_SET_HALTING_MODE and inquiry function IEEE_SUPPORT_HALTING. It must conform with its companion argument FLAG.
I, P, R	Of type integer of any kind with INTENT (IN). I must conform with its companion argument X. P and R must be scalars.
ROUND_VALUE	Of type IEEE_ROUND_TYPE. It must be scalar. It has INTENT (OUT) for subroutine IEEE_GET_ROUNDING_MODE and INTENT (IN) for subroutine IEEE_SET_ROUNDING_MODE.
STATUS_VALUE	Of type IEEE_STATUS_TYPE. It must be scalar. It has INTENT (OUT) for subroutine IEEE_GET_STATUS and INTENT (IN) for subroutine IEEE_SET_STATUS.
X, Y	Of type real of any kind. Each is an INTENT (IN) argument.

Note that the function IEEE_SELECTED_REAL_KIND in Table 14-12 uses the bracket notation, indicating the arguments are optional in the sense of OPTIONAL arguments defined in 12.6.2.

14.3.1 The Intrinsic Module IEEE_FEATURES

The IEEE_FEATURES module defines a derived type IEEE_FEATURES_TYPE with private components and named constants of this type. A particular implementation makes accessible only those named constants corresponding to the IEEE-style features that it supports; the set of named constants that are accessible is processor dependent.

Table 14-2 provides the names and meanings of the named constants used to denote IEEE features supported by the IEEE modules. These constants may be used to assign values to variables of type IEEE_FEATURES_TYPE, but are intended to be used in statements such as USE IEEE_FEATURES: ONLY ... . As such, they behave like command-line switches to specify which IEEE features are expected in the scoping unit that is being compiled. The intention is that the programmer use the ONLY option on the USE statement to insure that only the specified features are used. Where a feature is not supported, the compiler provides a diagnostic or message to this effect.

Table 14-2 The named constants in the module IEEE_FEATURES

Name[1]	Meaning, if accessible
IEEE_DATATYPE	Specifies support for IEEE arithmetic for at least one kind of real.
IEEE_DENORMAL	Specifies support for IEEE denormalized numbers for at least one kind of real.
IEEE_DIVIDE	Specifies support for IEEE divide operation for at least one kind of real.
IEEE_HALTING	Specifies support for IEEE halting when an exception is signaled for at least one kind of real and at least one exception flag.
IEEE_INEXACT_FLAG	Specifies support for the IEEE inexact exception for at least one kind of real.
IEEE_INF	Specifies support for IEEE infinities for at least one kind of real.
IEEE_INVALID_FLAG	Specifies support for the IEEE invalid exception for at least one kind of real.
IEEE_NAN	Specifies support for IEEE NaNs for at least one kind of real.
IEEE_ROUNDING	Specifies support for the four IEEE rounding modes for at least one kind of real.
IEEE_SQRT	Specifies support for the IEEE square root operation for at least one kind of real.
IEEE_UNDERFLOW_FLAG	Specifies support for the IEEE underflow exception for at least one kind of real.

Note 1: All the named constants are scalar of type IEEE_FEATURES_TYPE.

There are no module procedures defined in the IEEE_FEATURES module. However, inquiry module functions in the module IEEE_ARITHMETIC inquire about the extent of support for many of the IEEE features. For example, if the named constant

IEEE_DATATYPE is accessible from module IEEE_FEATURES, the inquiry module function IEEE_SUPPORT_DATATYPE from IEEE_ARITHMETIC must be accessible and return true for at least one kind of real.

14.3.2 The Intrinsic Module IEEE_EXCEPTIONS

The IEEE_EXCEPTIONS module defines the derived types IEEE_FLAG_TYPE and IEEE_STATUS_TYPE with private components and named constants of type IEEE_FLAG_TYPE. The exceptions overflow, divide-by-zero, invalid, underflow, and inexact are represented by named constants of type IEEE_FLAG_TYPE. The type IEEE_STATUS_TYPE permits the declaration of variables used by the intrinsic module subroutines IEEE_GET_STATUS and IEEE_SET_STATUS to retrieve and set the processor's IEEE status (14.2.7).

If either IEEE_EXCEPTIONS or IEEE_ARITHMETIC is accessible, the overflow and divide-by-zero exceptions must be supported for all kinds of real; for those exceptions not supported, their flags are always quiet. The procedure IEEE_SUPPORT _FLAG can be used to inquire about whether theses exceptions are supported.

The constants of Table 14-2 are available even if the corresponding feature is not supported by the processor. These constants are normally used as arguments to certain procedures such as IEEE_GET_FLAG and IEEE_SET_FLAG. No operations, including equality or inequality comparisons, are defined for the types IEEE_FLAG_TYPE and IEEE_STATUS_TYPE.

Table 14-3 The named constants for exceptions in the module IEEE_EXCEPTIONS

Name	Attributes[1]	Meaning
IEEE_DIVIDE_BY_ZERO	scalar	The exception, indicating a division of a nonzero finite value by zero
IEEE_INEXACT	scalar	The exception, indicating an arithmetic operation is inexact
IEEE_INVALID	scalar	The exception, indicating an arithmetic operation is invalid
IEEE_OVERFLOW	scalar	The exception, indicating the magnitude of a result exceeds the processor's upper range limit
IEEE_UNDERFLOW	scalar	The exception, indicating the magnitude of a result is less than tiny but is not zero
IEEE_USUAL	shape (3)	The array of three exceptions [IEEE_OVERFLOW, IEEE_DIVIDE_BY_ZERO, IEEE_INVALID]
IEEE_ALL	shape (5)	The array of five exceptions [IEEE_OVERFLOW, IEEE_DIVIDE_BY_ZERO, IEEE_INVALID, IEEE_UNDERFLOW, IEEE_INEXACT]

Note 1: All the named constants are of type IEEE_FLAG_TYPE.

The IEEE_EXCEPTIONS module also defines module procedures to support inquiries concerning the exceptions, modes, and status of the processor. These module procedures are classified as inquiry functions, or elemental or nonelemental subroutines. Where the procedures are functions returning logical values, the returned type is logical of default kind. All of these procedures are generic (with no specific names), all of the functions are pure, and some of the subroutines are pure as noted in Tables 14-4 to 14-6.

These tables also provide a description of the procedures and their functionality. Braces { } surrounding arguments indicate which arguments are optional in the sense described in 14.3. Details relevant to particular procedures are provided as footnotes in the tables.

Table 14-4 Inquiry functions in the module IEEE_EXCEPTIONS

Function[1] and argument(s)	Value returned (default logical)
IEEE_SUPPORT_FLAG[2] (FLAG {, X})	True, if the processor supports the detection of the exception FLAG for the data type and kind as X (or for all kinds of the real data type, if X is absent), and false, otherwise.
IEEE_SUPPORT_HALTING (FLAG)	True, if the processor supports[3] program control (halting or continuation) for the exception FLAG after the specified exception occurs for those kinds of real that support the exception FLAG, and false, otherwise.

Note 1: All functions in this table are pure.
Note 2: See 14.3 for the meaning of the optional notation {,X} used in this function.
Note 3: "Support" here implies that IEEE_SET_HALTING_MODE (FLAG) may be used to change the halting mode.

Table 14-5 Elemental subroutines in the module IEEE_EXCEPTIONS

Subroutine[1] and arguments	Operation
IEEE_GET_FLAG (FLAG, FLAG_VALUE)	Assigns to FLAG_VALUE the value true if the exception specified by FLAG is signaling and false otherwise.
IEEE_GET_HALTING_MODE (FLAG, HALTING)	Assigns to HALTING the value true if raising the specified exception FLAG will cause halting and false otherwise.

Note 1: All subroutines in this table are pure.

Table 14-6 Nonelemental subroutines in the module IEEE_EXCEPTIONS

Subroutine and argument(s)	Operation
IEEE_GET_STATUS[1] (STATUS_VALUE)	Assigns to STATUS_VALUE the state of the floating-point environment.
IEEE_SET_STATUS (STATUS_VALUE[4])	The state of the floating- point status is restored to that specified by the variable STATUS_VALUE.
IEEE_SET_FLAG (FLAG[2], FLAG_VALUE)	If FLAG is scalar and FLAG_VALUE is true, the exception FLAG is set to signaling; otherwise, it is set to quiet. If FLAG or FLAG_VALUE is an array, they must be conformable.
IEEE_SET_HALTING_MODE[3] (FLAG[2], HALTING)	If FLAG is scalar and HALTING is true, the halting mode is set to cause execution to halt for all kinds of real that support halting when the exception FLAG occurs; otherwise, it is set to allow execution to continue. If FLAG or FLAG_VALUE is an array, they must be conformable.

Note 1: Only the subroutine IEEE_GET_STATUS is pure.
Note 2: If FLAG is an array, no two elements may have the same value.
Note 3: This subroutine may not be invoked for an argument FLAG for which IEEE_SUPPORT_HALTING (FLAG) is false.
Note 4: The value of STATUS_VALUE must have been set by the subroutine IEEE_GET_STATUS.

14.3.2.1 Raising Exceptions from Operations Not To Be Executed

There are cases in the execution sequence where certain code is not to be executed. For example, the execution of if-then-else construct must not execute the else block when the logical expression is true. Similarly, in the WHERE construct, the elements of the expressions on the right hand side of an assignment statement in the where-block where the logical mask is false must not executed. In these and similar cases in the language, any exceptions that would be raised if such code were executed must not be raised. For example:

```
WHERE( A > 0.0 ) x = 1.0/A
```

is not permitted to raise the divide-by-zero exception for elements of A that are zero. This is an issue for optimization because on some architectures it is more efficient to perform the operation that causes an exception and then ignore the exception rather than to prevent the computation from occurring. Processors are allowed to perform unnecessary, or even prohibited, operations as long as they have no effect on the program status.

14.3.3 The Intrinsic Module IEEE_ARITHMETIC

The IEEE_ARITHMETIC module defines derived types, named constants of these types, and operator interfaces for == and /=. The derived types are IEEE_CLASS_TYPE and IEEE_ROUND_ TYPE and have private components.

The IEEE_CLASS_TYPE is used to support the specification and testing of the different classes of floating-point values returned by the basic operations. A class is provided for each of the following values:

- a negative denormalized number,
- a negative infinity,
- a negative normal value,
- a negative zero,
- a positive denormalized value,
- a positive infinity,
- a positive normal value,
- a positive zero,
- a quiet NaN,
- a signaling NaN, and
- a value that is none of the above.

Table 14-7 provides the names and meanings of the named constants used to denote IEEE classes for floating-point values. Any of these constants can be used as an argument to the module function IEEE_VALUE to generate a value of a specified class. The module function IEEE_CLASS (X) returns a value of type IEEE_CLASS_TYPE corresponding to the class of data values that X falls into. Operator interfaces for == and /= are provided in the module to permit comparisons with scalar and array data entities of type IEEE_CLASS_TYPE.

The Fortran standard is unclear whether a signaling NaN returned by the IEEE_VALUE module function is preserved and/or raises an invalid exception when it is assigned to a variable, or when it is an actual argument of a procedure. The IEEE standard specifies that it is implementation dependent whether copying a signaling NaN without a change of format raises an invalid exception. For this reason, use of such a signaling NaN is considered processor dependent.

Also, the module functions IEEE_IS_FINITE, IEEE_IS_NAN, IEEE_IS_NORMAL, and IEEE_IS_NEGATIVE are inquiry functions that return the value true if the argument has a value implied by the name of the function.

Table 14-8 provides the names and meanings of named constants used to designate IEEE rounding modes. Any of these constants can be used as an argument to the procedure IEEE_SET_ROUNDING_MODE to set the rounding mode to the specified mode, to the procedure IEEE_SUPPORT_ROUNDING_MODE to inquire whether the specified rounding mode is possible for a specified data type and kind, and to test the

Table 14-7 The named constants for IEEE classes of floating-point values in the module IEEE_ARITHMETIC

Name[1]	Meaning
IEEE_NEGATIVE_DENORMAL	Negative value that is a denormalized number
IEEE_NEGATIVE_INF	Negative value representing negative infinity
IEEE_NEGATIVE_NORMAL	Negative value that is not a NaN, infinity, nor denormalized value
IEEE_NEGATIVE_ZERO	Zero value that is negative in some operations.
IEEE_POSITIVE_DENORMAL	Positive value that is a denormalized number
IEEE_POSITIVE_INF	Positive value representing positive infinity
IEEE_POSITIVE_NORMAL	Positive value that is not a NaN, infinity, nor denormalized
IEEE_POSITIVE_ZERO	Zero value that is positive in some operations.
IEEE_QUIET_NAN	NaN value that if used in an operation propagates appropriately without raising the IEEE_INVALID exception
IEEE_SIGNALING_NAN	NaN value that if used in an operation will signal the exception IEEE_INVALID
IEEE_OTHER_VALUE	Value that is not one of the other IEEE values specified in this table

Note 1: All the named constants are scalar of type IEEE_CLASS_TYPE.

result of the procedure IEEE_GET_ROUNDING_MODE to determine which rounding mode is active. Operator interfaces for == and /= are provided in the module to permit comparisons with scalar and array data entities of type IEEE_ROUND_TYPE.

The IEEE_ARITHMETIC module also defines other module procedures to support the inquiries concerning the features and modes supported by the processor, kind values supporting IEEE arithmetic, the current modes, and the processor's floating-point status. Also, procedures to manipulate IEEE special values are defined. These module procedures are classified as inquiry functions, or elemental or nonelemental subroutines. Where the procedures are functions returning logical values, the returned type is logical of default kind. All of these procedures are generic (no specific names specified), all of the functions are pure, and two of the four subroutines are pure as noted in Tables 14-9, 14-10, 14-12, and 14-13. These tables provide a brief description of the procedures and their functionality; for more details and comparisons with related intrinsic functions, see 14.3.3.1-14.3.3.7 or B. Braces surrounding arguments indicate which arguments are optional in the sense described in 14.3. Details relevant to particular procedures are provided as footnotes in the tables.

Table 14-8 Named constants for the IEEE rounding modes in the module IEEE_ARITHMETIC

Name[1]	Meaning
IEEE_DOWN	Rounding mode specifying that the exact value is rounded down to the nearest representable number. If the exact value is less than –huge but not infinite, the result is –Inf; if the exact value is greater than +huge but not infinite, the result is +huge.
IEEE_NEAREST	Rounding mode specifying that the exact value is rounded to the nearest representable number. If there are two nearest results, the one whose least significant bit is 0 is returned. For an exact value larger than $2^{emax} \times (2-2^{-p})$ in magnitude using the model parameters in 13.2.3 for IEEE formats, the result is an infinity of the same sign as the exact value.
IEEE_TO_ZERO	Rounding mode specifying that the exact value is rounded towards zero to the nearest representable number. If the magnitude of the exact value is larger than huge, the result is huge with the same sign as the exact value.
IEEE_UP	Rounding mode specifying that the exact value is rounded up to the nearest representable number. If the exact value is greater than +huge, the result is +Inf; if the exact value is less than –huge but not infinite, the result is –huge.
IEEE_OTHER	Rounding mode specifying that the exact value is rounded in a way that does not follow any of the other four IEEE rounding modes in this table.

Note 1: All the named constants are scalar of type IEEE_ROUND_TYPE.

In order to support denormalized values, the processor is likely to support many, if not all, of the IEEE features in Table 14-9; however, it is possible that the underflow control mode is gradual and cannot be set to abrupt in which case IEEE_SUPPORT_UNDERFLOW_CONTROL returns false.

The module functions of Table 14-10 are defined even for arguments that are exceptional values. Where the treatment of such arguments is specific to the function, the special treatment is described in 14-10, its footnotes, or in 14.3.3.1-14.3.3.7.

In general, exceptional values—denormalized numbers, infinities, and NaNs—are permitted as values for X and/or Y. Denormalized numbers are treated like normal arguments with a value that is very small in magnitude and participate in the computation of the module function in a usual mathematical way. For arguments that are infinities, there are two cases to consider: if the infinite argument is considered a valid argument to the function, the operation of the function is considered exact and no overflow or inexact exceptions are raised, even when the infinity may be the result of the function; if the argument is considered an invalid argument, an invalid exception must be raised and the result must be a NaN. For NaNs are arguments, there are again two cases: either the NaNs are quiet or one of the arguments is a signaling NaN. If none are signaling, the result is one of the NaN arguments with no exception raised; if

Table 14-9 Inquiry functions in the module IEEE_ARITHMETIC

Function[1] and argument(s)[2]	Value returned[6]
IEEE_SUPPORT_DATATYPE ({X})	True if the processor supports IEEE arithmetic.
IEEE_SUPPORT_DENORMAL ({X})	True if[3] the processor supports denormalized numbers.
IEEE_SUPPORT_DIVIDE ({X})	True if the processor supports divide with the accuracy specified by the IEEE standard.
IEEE_SUPPORT_INF ({X})	True if the processor supports the IEEE infinities.
IEEE_SUPPORT_IO ({X})	True if the processor supports IEEE base conversion rules during formatted input/output for rounding modes UP, DOWN, ZERO, and NEAREST.
IEEE_SUPPORT_NAN ({X})	True if the processor supports IEEE NaNs.
IEEE_SUPPORT_ROUNDING (ROUND_VALUE {,X})	True if the processor supports[4] a particular rounding mode ROUNDING_VALUE.
IEEE_SUPPORT_SQRT ({X})	True if the processor supports the intrinsic function SQRT satisfies the requirements of IEEE standard.
IEEE_SUPPORT_STANDARD ({X})	True if the IEEE inquiry functions IEEE_SUPPPORT_...[5] return true.
IEEE_SUPPORT_UNDERFLOW_C ONTROL ({X})	True if the processor supports control of the underflow mode. Note that if the inquiry function IEEE_SUPPORT_STANDARD returns true, it does not imply support for underflow control.

Note 1: All functions in this table are pure.
Note 2: See 14.3 for the meaning of the optional notation {X} or {,X} used in these functions.
Note 3: And if IEEE_SUPPORT_DATATYPE (X) is true.
Note 4: "Support" includes the ability to change the rounding mode by a call to IEEE_SET_ROUNDING_MODE (ROUND_VALUE {,X}).
Note 5: The IEEE inquiry functions are: IEEE_SUPPORT_DATATYPE, IEEE_SUPPORT_DENORMAL, IEEE_SUPPORT_DIVIDE, IEEE_SUPPORT_FLAG, IEEE_SUPPORT_HALTING, IEEE_SUPPORT_INF, IEEE_SUPPORT_NAN, IEEE_SUPPORT_ROUNDING, and IEEE_SUPPORT_SQRT. The functions must return true for any exception or rounding mode. Notice that IEEE_SUPPORT_UNDERFLOW_CONTROL is not included in this list.
Note 6: If the argument X appears, the returned value relates only to the real kind that is the same as the kind as X, or otherwise for all real kinds.

one or more is signaling, the result is a signaling NaN. This general behavior is the behavior specified by the IEEE standard [13].

For the IEEE_ARITHMETIC module functions IEEE_CLASS, IEEE_LOGB, IEEE_NEXT_AFTER, IEEE_SCALB, and IEEE_SELECTED_REAL_KIND in Tables 14-10

Table 14-10 Elemental functions in the module IEEE_ARITHMETIC

Function[1] and argument(s)	Value returned
IEEE_CLASS (X)	The IEEE class of a specified value X
IEEE_COPY_SIGN (X, Y)	The value X with the sign of Y
IEEE_IS_FINITE (X)	True if X is finite
IEEE_IS_NAN (X)	True if X is an IEEE NaN
IEEE_IS_NEGATIVE (X)	True if X is negative
IEEE_IS_NORMAL (X)	True if X is normal
IEEE_LOGB (X)	The unbiased IEEE exponent of X
IEEE_NEXT_AFTER (X, Y)	The next representable neighbor of X towards Y
IEEE_REM (X, Y)	The IEEE remainder
IEEE_RINT (X)	The rounded integer value of X
IEEE_SCALB (X, I)	The value $X \times 2^I$
IEEE_UNORDERED (X, Y)	True if the X or Y or both are NaNs
IEEE_VALUE (X, CLASS)	An IEEE value specified by the value of CLASS with the same kind as X

Note 1: All functions in this table are pure. None of these functions may be invoked for arguments X and Y for which IEEE_SUPPORT_DATATYPE (X) or IEEE_SUPPORT_DATATYPE (Y) is false. This is a requirement on a program; it is not a requirement on an implementation to check this restriction.

and 14-12, the details are too lengthy to provide in a tabular format; they are described in 14.3.3.1-14.3.3.7.

Several of the module functions in IEEE_ARITHMETIC have a similar functionality to one or more intrinsic functions. Generally, they are different but for some implementations when the Fortran models are:

For the 32-bit IEEE real data type: $b = 2$, $p = 24$, $e_{min} = -125$, $e_{max} = 128$

For the 64-bit IEEE real data type: $b = 2$, $p = 53$, $e_{min} = -1021$, $e_{max} = 1024$

they may produce the same results. These seven module functions are compared and contrasted with the relevant intrinsic functions below.

One difference that is common to all but one of the module functions is that, if an argument X and/or Y is used in an IEEE module function, it must be one for which IEEE_SUPPORT_DATATYPE(X) and/or IEEE_SUPPORT_DATATYPE(Y) is true. The corresponding intrinsic functions have no such restriction.

14.3.3.1 Module Function IEEE_COPY_SIGN Versus Intrinsic Function SIGN

Each function copies the sign of the second argument to the first argument, maintaining the magnitude of the first argument, except possibly for exceptional argument values as noted below. The intrinsic function SIGN accepts integer arguments of the same kind as well as real arguments of the same kind whereas IEEE_COPY_SIGN accepts only real arguments X and Y, possibly of different kinds. A possible other difference is that the function SIGN is not required to operate on exceptional arguments X and Y whereas the functionality of IEEE_COPY_SIGN is specified for all exceptional arguments; that is, for all arguments X, ABS (X) = IEEE_COPY_SIGN (X, 1.0), even if X is a NaN.

14.3.3.2 Module Function IEEE_LOGB Versus Intrinsic Function EXPONENT

Each function returns an integer related to the exponent of the argument X. The intrinsic function EXPONENT returns a value of type integer and default kind which is the exponent of X to the base b in the form of values used by the real model (13.2.3), and if the base of the Fortran model for values of the kind and type of X is 2, EXPONENT always returns a value one larger than IEEE_LOGB, that is, EXPONENT (X) = IEEE_LOGB (X) + 1 for all nonzero and finite values for X. The module function IEEE_LOGB returns a value of the type and kind of X; its value is the unbiased exponent of X when represented in an IEEE format (14.2.1).

14.3.3.3 Module Function IEEE_NEXT_AFTER Versus Intrinsic Function NEAREST

Each function returns the representable neighbor closest to X in a direction indicated by the second argument but the details are quite different. The module function IEEE_NEXT_AFTER (X, Y) returns X if Y is equal to X; otherwise, the closest representable neighbor to X, but not equal to X, toward Y. The intrinsic function NEAREST (X, S) returns a nearest neighbor to X, never equal to X, to the left if S is negative and nonzero, and to the right if S is positive and nonzero. S is not allowed to be zero. For IEEE_NEXT_AFTER and NEAREST, both arguments have to be real but may be of different kinds. However, for IEEE_NEXT_AFTER, both arguments have to be of a kind such that IEEE_SUPPORT_DATATYPE is true. In addition, for IEEE_NEXT_AFTER, exceptions must be raised when X is huge or -huge and Y is of the same sign as X, or when the result is a denormalized number; see 14.3.3.10 for the details. However, for the intrinsic function NEAREST, such exceptions are not required nor specified, but are allowed.

14.3.3.4 Module Function IEEE_REM Versus Intrinsic Functions MODULO or MOD

All three functions compute a "remainder" upon division but have different definitions for the quotient which in general causes different values for the result. The arguments of IEEE_REM must be of type real and can have different kinds but those of MOD and MODULO must have the same kinds but may be of type integer or real.

The intrinsic function MOD (A, P) for real arguments defines the quotient Q to be INT (A/P), and the result is A – QxP; the intrinsic function MODULO (A, P) for real arguments defines the quotient to be FLOOR(A/P), and the result is A – QxP; the IEEE function IEEE_REM (X, Y) defines the quotient Q to be the integer nearest the exact quotient X/Y and when there are two such nearest integers, the even integer Q where |Q–X/Y|=0.5, and the result is exactly X–QxY. Also, for IEEE_REM, if the result is zero, it will have the sign of X.

Table 14-11 illustrates the differences in the results of these functions when their arguments are real.

Table 14-11 Comparison of results of the remainder functions IEEE_REM, MOD, and MODULO

Arguments		IEEE_REM	MOD	MODULO
5.0	3.0	–1.0	2.0	2.0
5.0	–3.0	–1.0	2.0	–1.0
–5.0	3.0	1.0	–2.0	1.0
–5.0	–3.0	1.0	–2.0	–2.0

14.3.3.5 Module Function IEEE_RINT Versus Intrinsic Functions AINT or ANINT

All three functions "round" a real value to an integer value represented as a real value with the type and kind of the argument. However, they use different rounding rules. IEEE_RINT rounds according to the current IEEE rounding mode and follows IEEE's rules for rounding. The intrinsic function AINT in effect always uses the IEEE rounding mode DOWN. The intrinsic function ANINT always uses the rounding rule COMPATIBLE as specified in 10.5.2.3, which is like the IEEE NEAREST rule except when the value is halfway between two integers in which case the integer furthest away from zero is returned by ANINT.

14.3.3.6 Module Function IEEE_SCALB Versus Intrinsic Function SCALE

Both functions scale their first argument X by a power of the base raised to the second argument; IEEE_SCALB computes $X \times 2^I$ whereas the intrinsic function SCALE computes $X \times b^I$. Where the Fortran model for numbers of the type and kind of X uses b=2, the functions compute the same values except possibly when the result is an exceptional value, but when $b \neq 2$, the results are different; see 14.3.3.11 for the details. For exceptional arguments, the results for the function SCALE are processor-dependent whereas the module function IEEE_SCALB specifies precisely the exception signals and the exceptional returned values. Lastly, the IEEE standard specifies the scaling be performed without forming 2^I explicitly; the scaling is performed by adjusting the biased exponent appropriately.

14.3.3.7 Module Function IEEE_SELECTED_REAL_KIND Versus Intrinsic Function SELECTED_REAL_KIND

Both functions have two optional arguments P and Q and return a kind number of a real data type satisfying the same minimum precision and range requirements. The difference is that the kind number for the IEEE function must be an IEEE real data type (that is, one that supports IEEE arithmetic and thus a kind of real data type for which IEEE_SUPPORT_DATATYPE returns true). Secondly, the intrinsic function SELECTED_REAL_KIND has an additional return value of –4, whereas the IEEE module function IEEE_SELECTED_REAL_KIND is not required to return this value. A returned value of –4 indicates the processor has a different kind of real for the P and R specification separately but not a single kind value that satisfies for the requirements for both precision and range simultaneously.

Table 14-12 Kind transformational function in the module IEEE_ARITHMETIC

Function[1] and arguments	Value returned
IEEE_SELECTED_REAL_KIND ([P, R][2])	Kind type parameter value for an IEEE real data type with given minimum precision P and range R

Note 1: The function IEEE_SELECTED_REAL_KIND is pure.
Note 2: This optional notation means either just P is present, just R is present, or both P and R are present, with the keyword arguments as appropriate. The interface is the same as that for the intrinsic function SELECTED_REAL_KIND.

Table 14-13 Nonelemental subroutines in the module IEEE ARITHMETIC

Subroutine and argument	Operation
IEEE_GET_ROUNDING_MODE[1] (ROUND_VALUE)	Store the current IEEE rounding mode[2] into ROUND_VALUE.
IEEE_SET_ROUNDING_MODE[4] (ROUND_VALUE)	Set the IEEE rounding mode to ROUND_VALUE.
IEEE_GET_UNDERFLOW_MODE[1,3] (GRADUAL)	Set GRADUAL to true if the current IEEE underflow mode is gradual or false if the underflow mode is abrupt.
IEEE_SET_UNDERFLOW_MODE[3] (GRADUAL)	Set the IEEE underflow mode to gradual underflow if GRADUAL is true, or otherwise, set the IEEE underflow to abrupt underflow.

Table 14-13 *(Continued)* Nonelemental subroutines in the module IEEE ARITHMETIC

Subroutine and argument	Operation
Note 1: The subroutines IEEE_GET_ROUNDING_MODE and IEEE_GET_UNDERFLOW_MODE are pure. Note 2: If the current rounding mode is not IEEE_NEAREST, IEEE_UP, IEEE_DOWN, or IEEE_TO_ZERO, the value IEEE_OTHER is returned. Note 3: This subroutine must not be invoked unless IEEE_SUPPORT_UNDERFLOW_CONTROL (X) is true for some X. This is a requirement on a program; it is not a requirement on an implementation to check this restriction. Note 4: The subroutine IEEE_SET_ROUNDING_MODE must not be invoked unless IEEE_SUPPORT_ROUNDING (ROUND_VALUE, X) and IEEE_SUPPORT_DATATYPE (X) are true for some X. This is a requirement on a program; it is not a requirement on an implementation to check this restriction.	

14.3.3.8 Details of Returned Results for Module Function IEEE_CLASS

The result of IEEE_CLASS (X) is the class of X, specified by values of type IEEE_CLASS_TYPE as follows:

Value of X	Result Value
Quiet NaN	IEEE_QUIET_NAN, if IEEE_SUPPORT_NAN (X) is true.
Signaling NaN	IEEE_SIGNALING_NAN, if IEEE_SUPPORT_NAN (X) is true.
+Inf	IEEE_POSITIVE_INF, if IEEE_SUPPORT_INF (X) is true.
–Inf	IEEE_NEGATIVE_INF, if IEEE_SUPPORT_INF (X) is true.
+denorm	IEEE_POSITIVE_DENORMAL if IEEE_SUPPORT_DENORMAL (X) is true.
–denorm	IEEE_NEGATIVE_DENORMAL, if IEEE_SUPPORT_DENORMAL (X) is true.
+normal	IEEE_POSITIVE_NORMAL.
–normal	IEEE_NEGATIVE_NORMAL.
+zero	IEEE_POSITIVE_ZERO.
–zero	IEEE_NEGATIVE_ZERO.

Otherwise, if the value is not one of these values or the appropriate support inquiry module function is false for the type and kind of X, the returned result is IEEE_OTHER_VALUE.

14.3.3.9 Details of Returned Results for Module Function IEEE_LOGB

The Fortran standard additionally specifies the result value of IEEE_LOGB(X) for X==0 to be -inf when IEEE_SUPPORT_INF (X) is true, or –HUGE (X) otherwise, and in addition requires divide-by-zero to signal. The IEEE standard [13] requires that when X is finite and positive, IEEE_SCALB (X, –IEEE_LOGB (X)) be strictly in the open interval (0,2) and that it be less than 1 only when X is denormalized.

14.3.3.10 Details of Returned Results for Module Function IEEE_NEXT_AFTER

The result of IEEE_NEXT_AFTER (X, Y) is the nearest representable IEEE number to X toward Y and so applies to normal and exceptional values. When X=Y, the result is X with no exception raised. If X≠Y but X is zero, the result is the smallest representable number in magnitude toward Y; if denormalized numbers are supported, the result is the smallest denormalized number in magnitude of the sign of Y, and both underflow and inexact exceptions are signaled; if denormalized numbers are not supported, then the result is TINY (X) of the appropriate sign and no signals are raised. If, for large X in magnitude where infinities are supported, the result is infinite, then overflow and inexact exceptions are raised; this can only happen when Y is an infinity. See 14.3.3.3.

14.3.3.11 Details of Returned Results for Module Function IEEE_SCALB

The result of IEEE_SCALB (X, I) is exactly $X \times 2^I$ when the result is a normal number; if the result is too large, an overflow exception occurs and the result is an infinity of the sign of X if IEEE_SUPPORT_INF (X) is true, and SIGN (HUGE (X), X) otherwise; if the result is too small and the result is not exact, the underflow and inexact exceptions are raised, and the result is the nearest representable number to $X \times 2^I$, which maybe a denormalized number if they are supported; if X is an infinity, the result is X, with no exception raised.

14.3.3.12 Details of Returned Results for Module Function IEEE_SELECTED_REAL_KIND

The result of IEEE_SELECTED_REAL_KIND ([P,R]) is an integer specified by the following table:

Value	Meaning
Positive integer	The kind number of a real type that supports IEEE data types.
–1	No kind number is available that supports the minimum precision P as defined by the intrinsic function PRECISION and meets IEEE data type requirements.
–2	No kind number is available that supports the minimum exponent range R as defined by the intrinsic function RANGE and meets the IEEE data type requirements.
–3	No kind number is available that supports either the minimum precision P as defined by the intrinsic function PRECISION or the minimum exponent range R as defined by the intrinsic function RANGE and meets the IEEE data type requirements.

14.4 Initial and Final Status Requirements Entering and Leaving Any Procedure

Recall that the floating-point status of a processor is a combination of the states of the exception flags, the rounding mode, the underflow mode, the halting mode, and possibly other modes and flags determined by the processor. The floating-point status may change as the program executes because of references to various procedures, including the intrinsic procedures, intrinsic operations, format processing, and IEEE module procedures. The Fortran standard specifies what the status is initially, how the status changes on entry to and exit from procedures, and what has to be reported about the final status upon normal termination of the program.

In case the main program or any user-written procedure accesses any IEEE intrinsic modules, Table 14-14 summaries the changes in the floating-point status for the IEEE flags and modes when events, such as exiting and entering a procedure or main program, occurs. In this case, the status of non-IEEE floating-point flags and modes is processor dependent.

Table 14-14 Floating-Point Status on Entry and Exit From Procedures, Accessing Any IEEE Module.

Location in the execution of a program	Exception flags and status	IEEE modes: rounding, underflow, halting
On entry to main program	Quiet	Processor-dependent
On entry to any procedure except intrinsic and special procedures	Set to quiet	No change to mode on entry
On exit from any procedure except intrinsic and special procedures	Restore any flag signaling on entry and add any flags raised	Restore to entry mode
On entry to intrinsic procedures	Set to quiet	No change to mode on entry
On exit from intrinsic procedures	Reset to entry OV, DV, IV flags[1]. May signal OV if result is too large; may signal IV if result is a NaN. Whether the UN and IE flags are set to the entry values, set quiet, or passed on is processor dependent	Restore to entry mode
On exit from special IEEE module procedures:		
IEEE_SET_STATUS	Set to STATUS_VALUE[2]	Set to STATUS_VALUE[2]
IEEE_SET_FLAG	Set to FLAG_VALUE[3]	Unchanged
IEEE_SET_ROUNDING_MODE	Unchanged	Set to ROUND_VALUE[4]
IEEE_SET_UNDERFLOW_MODE	Unchanged	Set to GRADUAL[5]
IEEE_SET_HALTING_MODE	Unchanged	Set to HALTING[6]

Table 14-14 *(Continued)* Floating-Point Status on Entry and Exit From Procedures, Accessing Any IEEE Module.

Location in the execution of a program	Exception flags and status	IEEE modes: rounding, underflow, halting
Note 1: OV, DV, IV, UN and IE are the flags IEEE_OVERFLOW, IEEE_DIVIDE_BY_ZERO, IEEE_INVALID, IEEE_UNDERFLOW, and IEEE_INEXACT, respectively. Note 2: STATUS_VALUE is the STATUS_VALUE argument of the procedure IEEE_SET_STATUS. Note 3: FLAG_VALUE is the FLAG_VALUE argument of the procedure IEEE_SET_FLAG. Note 4: ROUND_VALUE is the ROUND_VALUE argument of the procedure IEEE_SET_ROUNDING_MODE. Note 5: GRADUAL is the GRADUAL argument of the procedure IEEE_SET_UNDERFLOW_MODE. Note 6: HALTING is the HALTING argument of the procedure IEEE_SET_HALTING_MODE.		

The consequences of Table 14-14 are that initially at the start of a main program that accesses an IEEE module, all exception flags are quiet but the rounding, halting, and underflow modes are processor dependent; to make the program portable, the programmer should initially set these modes. As execution continues, flags will signal as a result of intrinsic operations and the invocation of intrinsic procedures. On entry to procedures defined by the programmer, exception flags will be set quiet but all modes will be inherited from the caller. If the procedure needs special modes to compute robustly, the programmer must set them initially to the needed modes; these modes will be reset to the status at entry to the procedure on exit. On the other hand, exception flags are sticky in the sense that both the exception flags on entry and the flags raised during execution of the procedure become the exception flags on exit. If the programmer does not wish to have exception flags raised from the execution of the procedure to be communicated on exit from the procedure, these flags should be cleared explicitly by invoking IEEE_SET_FLAGS before the return.

In case the main program or any user-written procedure accesses no IEEE intrinsic module, the floating-point status is processor dependent when entering or leaving the main program or any procedure, except that any flags raised on entry to a procedure or main program remain signaling on exit. The consequences are that modes and flags are initially processor dependent, and essentially remain so throughout a computation. It is expected although that the exception flags are sticky in the sense that once raised by any procedure, they remain raised at the termination of the program. However, the standard, when IEEE modules are not accessed, essentially treats this whole issue as processor dependent by being silent on what happens.

Upon termination of the program by a STOP statement, the processor is required to indicate which exception flags are signaling by writing to the logical unit specified by the named constant ERROR_UNIT in the ISO_FORTRAN_ENV intrinsic module. To avoid this warning, the programmer should clear all flags before the termination of the program. Whether the exceptions are reported on a normal termination that is not caused by the execution of a STOP statement is processor dependent.

14.5 Interoperability Issues for IEEE Arithmetic and Exceptions

The Fortran standard supports interoperability of C code with Fortran and as such the IEEE floating-point status flags and modes must also be interoperable. However, the requirements are very limited and are listed below, but do not require the C processor to support any of the derived types, modes, or flags described above; the C processor is permitted to support them. The restrictions on interoperability are:

1. The C procedure is not permitted to use the C signal capability to change the handling of an exception that is being handled by the Fortran processor.

2. The C procedure must not alter the floating-point status, other than setting an exception flag to signaling; because of the restriction above, the floating-point status must not be changed by using the C signal capability but may be changed by executing a C operation, such as floating-point multiplication.

3. The values of the floating-point exception flags on entry to the C procedure are processor dependent.

Although stated above in terms of procedures written in C, the same restrictions apply to a procedure defined by any means other than Fortran. In such a case, if a signal capability is available in the other language, it must not meddle with Fortran's floating-point status and modes.

14.6 A Summary of the Optional Features

The three IEEE intrinsic modules are individually optional. One might expect the following combinations:

- No IEEE intrinsic modules are provided.

- The modules IEEE_FEATURES and IEEE_EXCEPTIONS only are provided. For this case, a subset of the named constants of IEEE_FEATURES would be accessible where the subset represents some or all of the features of IEEE exceptions.

- All three IEEE modules are provided with some or all of the named constants of IEEE features accessible.

With the latter two combinations, the named constants in IEEE_FEATURES can and will likely be used as compiler switches in the program to specify the IEEE features needed. In this case, the processor will execute the program only if the features specified in a program are accessible.

From the programmer's point of view, programming in a portable manner in the presence of this apparent optionality could become a nightmare. To ameliorate the situation, the standard's implementation model appears to be that except for IEEE_FEATURES, the IEEE modules must define all the types, named constants, and module procedures specified in the Fortran standard. For example, all of the named constants for the IEEE exceptions will be provided, even though the processor may not support all exceptions; the IEEE_SUPPORT... procedures will return false for those exceptions not supported. Also, all of the module procedures will be provided so that a reference can be present to any of them without compilation or linking failing because of an unresolved symbol. However, a program is non-conforming if it invokes a module procedure when its corresponding support procedure returns false. For example, a condition for invoking the procedure IEEE_IS_NAN (X) is that the procedure IEEE_SUPPORT_NAN (X) return true. If IEEE_SUPPORT_NAN returns false and yet IEEE_IS_NAN is invoked, the program is nonconforming.

To illustrate how a portable program can be written using this model, consider the following program segment. The segment references IEEE_SCALB to perform scaling of the variable x if the module IEEE_ARITHMETIC supports IEEE arithmetic for the kind of x. This scaling by IEEE_SCALB must be performed without forming 2 raised to the power i. On the other hand, if IEEE arithmetic is not supported for the kind of x, the code still needs to scale x. This can be done with the intrinsic function SCALE. However, this intrinsic function potentially forms this power and if i is large enough, scale can overflow computing the wrong result. Consequently, the alternative code scales x twice when i is too large, thus avoiding a potential overflow exception.

```
use IEEE_FEATURES, only: IEEE_DATATYPE
use IEEE_ARITHMETIC, only: IEEE_SCALB, IEEE_SUPPORT_DATATYPE
   . . .
if( IEEE_SUPPORT_DATATYPE(x) )  then
  y = IEEE_SCALB (x, i)
else
  y = x
  if( i > maxexponent(x) )  then
    y = scale( x, maxexponent(x) )
    i = i - maxexponent(x)
   endif
   y = scale (y, i)
 endif
```

The above code avoids an overflow whether or not IEEE arithmetic is supported or whether or not the intrinsic function forms the factor 2^i. Also, in the situation that all module names are defined, the code segment compiles and links whether or not the module function IEEE_SCALB has been implemented to conform with the IEEE standard, provided the modules IEEE_FEATURES and IEEE ARITHMETIC are accessible.

There are, however, other constraints and conditions on the program and processor concerning the accessibility and use of IEEE features. They have been described throughout the previous sections but are summarized here; violation of these constraints and conditions generally results in processor-dependent programs.

- Which IEEE modules are provided is processor dependent.
- Which named constants of the IEEE_FEATURES module are provided is processor dependent.

- If a scoping unit does not have access to any of the IEEE modules, the level of support of IEEE features is processor dependent.
- If a scoping unit accesses IEEE_EXCEPTIONS or IEEE_ARITHMETIC without accessing IEEE_FEATURES, the supported subset of IEEE features is processor dependent.
- If IEEE_EXCEPTIONS or IEEE_ARITHMETIC is accessible, the overflow and divide-by-zero exception will be supported for all kinds of real and complex. The invalid, underflow, and inexact exceptions are optional; the module procedure IEEE_SUPPORT_FLAG returns true for those exceptions that are supported. Also, the accessibility of the IEEE_FEATURES's named constants IEEE_INEXACT_FLAG, IEEE_INVALID_FLAG, and IEEE_UNDERFLOW influences the extent of the support of these exceptions.
- Control of the halting mode after a particular exception is processor dependent. If IEEE_HALTING is accessible, halting control must be supported for at least one exception; an exception FLAG supports halting control if IEEE_SUPPORT_HALTING (FLAG) returns true. The halting mode can be determined by IEEE_GET_HALTING_MODE and altered with the procedure IEEE_SET_HALTING_MODE or IEEE_SET_STATUS.
- If IEEE_UNDERFLOW_FLAG of IEEE_FEATURES is accessible within a scoping unit, the underflow exception must be supported for at least one kind of real within that scoping unit, and the procedure IEEE_SUPPORT_FLAG (IEEE_UNDERFLOW, X) must return true for a variable X of a supported kind. Similarly, for the inexact or invalid exception, if the corresponding named constant IEEE_INEXACT_FLAG or IEEE_INVALID_FLAG is accessible, the corresponding exception must be supported for at least one kind of real and the corresponding reference to IEEE_SUPPORT_FLAG must return true. The signaling of a particular exception can be determined by IEEE_GET_FLAG and altered with the procedure IEEE_SET_FLAG or IEEE_SET_STATUS.
- If IEEE_DATATYPE of IEEE_FEATURES is accessible within a scoping unit, IEEE arithmetic must be supported for at least one kind of real within that scoping unit, and the procedure IEEE_SUPPORT_DATATYPE (X) must return true for a variable X of that supported kind. Similarly, for IEEE_DENORMAL, IEEE_DIVIDE, IEEE_INF, IEEE_NAN, IEEE_ROUNDING, and IEEE_SQRT of IEEE_FEATURES, the processor must support the feature for at least one kind of real, and the corresponding inquiry function must return true for each kind of real supported. In addition, if IEEE_ROUNDING is accessible, the rounding modes IEEE_NEAREST, IEEE_TO_ZERO, IEEE_UP, and IEEE_DOWN must be supported.
- The ability to alter the rounding mode is processor dependent. The rounding mode can be altered to a particular rounding mode for a particular kind if IEEE_SUPPORT_ROUNDING returns true for that rounding mode and kind, using

the procedures IEEE_SET_ROUNDING_MODE or IEEE_SET_STATUS; the current rounding mode can be determined with the procedure IEEE_GET_ROUNDING_MODE.

- The availability of rounding for base conversion in formatted input/output (9.2.4, 10.9.7) is processor dependent. The inquiry function IEEE_SUPPORT_IO returns true for a particular kind of real if such conversion is available.

- The ability to alter the underflow mode is processor dependent. If the processor supports the control of the underflow mode for a particular real data type, IEEE_SUPPORT_UNDERFLOW_CONTROL (X) must return true for a variable X of that supported kind. The underflow mode may be determined with the procedure IEEE_GET_UNDERFLOW_MODE and set with the procedure IEEE_SET_UNDERFLOW_MODE or IEEE_SET_STATUS.

- The ability to determine or generate exceptional values is processor dependent. The inquiry function IEEE_SUPPORT_DENORMAL, IEEE_SUPPORT_INF, or IEEE_SUPPORT_NAN is true, respectively, if denormalized numbers, infinities, or NaNs are supported. The functions IEEE_IS_FINITE, IEEE_IS_NAN, IEEE_IS_NEGATIVE, and IEEE_IS_NORMAL can be used to determine the particular exceptional value and its sign, if appropriate.

- Whether IEEE arithmetic or the IEEE standard is supported for any kind is processor dependent. If the processor supports either of these concepts, the corresponding inquiry function IEEE_SUPPORT_DATATYPE or IEEE_SUPPORT_STANDARD for a particular real kind returns true.

- In a sequence of code with no invocations of IEEE_GET_FLAG, IEEE_SET_FLAG, IEEE_GET_STATUS, IEEE_SET_STATUS, or IEEE_SET_HALTING_MODE, if there is no variable that depends on an operation in which an exception occurred, the signaling of that exception is processor dependent. For example, in the code segment:

```
X = sqrt(Z)
X = 3.0
```

  the raising of the invalid exception when Z is negative and nonzero is processor dependent because the value of X is 3.0 at the end of the code segment whether or not the invalid exception is raised.

14.7 Examples of the Use of IEEE Features, Arithmetic, and Exceptions Modules

This section discusses two examples of the use of the IEEE intrinsic modules. First, the simple example program below is briefly discussed in the next subsection. Secondly, a more thorough and significant use of these intrinsic modules is presented that provides a robust and accurate computation of dot product, despite the fact that overflow, underflow, inexact, and invalid exceptions may occur in a naive version of this computation.

14.7.1 Discussion of the Simple Example

The program example in Figure 14-3 is a portable program to determine whether the multiplication x*y overflows. It is somewhat more complicated than necessary in order to illustrate various features of the IEEE intrinsic modules.

```
program  Has_overflow_occurred
  ! This simple program illustrates the detection of overflow.
  use, intrinsic :: IEEE_FEATURES, only : IEEE_DATATYPE
  use, intrinsic :: IEEE_ARITHMETIC, only : IEEE_SUPPORT_DATATYPE
  use, intrinsic :: IEEE_EXCEPTIONS, only : IEEE_OVERFLOW, &
                    IEEE_GET_FLAG, IEEE_SET_HALTING_MODE
  real :: x, y, z;   logical :: overflow_flag, IEEE_SUPPORT_HALTING

  ! Is there IEEE arithmetic support for entities of the
  ! type and kind of x and is halting supported for overflow?
  if(IEEE_SUPPORT_DATATYPE(x) .and.  &
     IEEE_SUPPORT_HALTING(IEEE_OVERFLOW))  then
    ! There is support for IEEE exceptions, including overflow.
    ! It is unnecessary to set overflow to nonsignaling
    ! as the processor is required to set the initial overflow
    ! flag to quiet at the start of execution of a program.
    ! Read in two values and check for overflow of their product.
    read *, x, y
    ! Set the halting mode to continue execution after overflow.
    call  IEEE_SET_HALTING_MODE( IEEE_OVERFLOW, .false. )
    ! Multiply x and y and check for overflow.
    z = x * y
    call  IEEE_GET_FLAG( IEEE_OVERFLOW, overflow_flag )
    if( overflow_flag )  then
      print "(a/3G16.8)", "An overflow occurred in the product" // &
            " x*y: x, y and x*y:", x, y, z
    else
      print "(a/3G16.8)", "No overflow occurred in the product" // &
            " x*y: x, y and x*y:", x, y, z
    endif
  else
    print *, "There is no arithmetic or exceptions support for" // &
             " detecting overflow"
  endif

end program  Has_overflow_occurred
```

Figure 14-3 A simple Fortran program to test for an overflow exception

The USE statements at the beginning of the program are used as compiler switches to ensure that the needed IEEE features are available. For example, if any of the three intrinsic modules or named constant IEEE_DATATYPE are not accessible, compilation of this program will abort.

The program first checks that there is IEEE support for the data type of X, which, in this case is of type default real and checks that there is support for the halting mode for the overflow exception. If the result of this inquiry function is true, then the signaling of overflow, when it occurs, is supported for the default real type. Next, the program sets the halting mode for the overflow exception to continue execution after the occurrence of any overflow by calling the module subroutine IEEE_SET_HALTING_MODE. The operation x*y is next performed and assigned to z. After this operation, the status of the overflow exception flag is examined using the intrinsic subroutine IEEE_GET_FLAG. Finally, a message is printed indicating whether or not overflow has occurred and the program terminates.

14.7.2 Computing a Dot Product Carefully

A more realistic use of the IEEE modules is to create a robust, efficient implementation of the dot product of two vectors, despite the scaling of the two vectors. It is actually a difficult problem to compute the correct answer in general, as will soon be appreciated.

To begin the discussion, consider the computing the sum:

$$\sum_{k=1}^{n} x_k y_k$$

Mathematically, this function is computing the product of the norms of the two vectors times the angle between the vectors; that is:

$$\|x\| \|y\| \cos \langle x, y \rangle$$

From this mathematical formulation, one can see that the result can be very small or very large, depending on the norms of x and y and the cosine of the angle between them, which of course can be small when the vectors are orthogonal or nearly so. On the other hand, the sum formulation can exhibit several computational anomalies, such as overflow (products or sums too large), underflow (products too small or subtraction causing a complete loss of accuracy, resulting in underflow), or, even worse, invalid computations (the product of two elements can overflow to positive infinity, making the partial sum infinite, followed by the product of the next two elements being negative infinity, causing the next partial sum to be invalid—not a number is the result of +Inf-Inf). Checking for all of these conditions when they happen in this application (and more so in a real application) is a nightmare.

Consider instead the following approach. Ignoring all exceptions, compute the dot product using the sum formulation. Check to see if the overflow flag raised. If it has been raised, scale the vectors x and y so that their largest element in magnitude is near one. Recompute the sum as before, but this time with the scaled vectors x and y; no overflow or invalid operations can occur. Because the computed result can overflow or underflow, return the result as an integer power of two and a significant part. Figure 14-4 gives a listing of the subroutine that performs this computation as a module subroutine mult in the module dot.

Instead of computing it twice, why not just scale the vectors x and y in the first place? The reason is that this can be inefficient for data that is well scaled. The strategy above makes the computation costly only when the vectors are badly scaled; for the expected usual case where the vectors are nicely scaled, the computation is quick and efficient.

```
module dot

  ! Module for dot product of two real arrays of rank 1.

  use, intrinsic :: ieee_features, only : &
          IEEE_DATATYPE, IEEE_HALTING
  use, intrinsic :: ieee_arithmetic, only : ieee_get_flag,   &
          ieee_set_flag, ieee_set_halting_mode, ieee_logb,   &
          ieee_scalb, IEEE_OVERFLOW, IEEE_ALL

  private
  logical, public :: matrix_error = .false.
  public  mult

  contains

    subroutine mult( a, b, x, i )
      real, intent(in) :: a(:),b(:)
      real, intent(out) :: x
      integer, intent(out) :: i
      ! Local variables.
      integer  exp_a, exp_b
      real  max_a, max_b
      logical overflow
      intrinsic  abs, maxval, size, sum

      if( size(a)/=size(b) )  then
        matrix_error = .true.
        return
      endif

      ! The processor ensures that flags, particularly
      ! IEEE_OVERFLOW, are quiet on entry to a procedure.
      ! Set the halting mode for all exceptions to continue
      ! execution. Assume the calling program has checked that IEEE
      ! arithmetic and exceptions are supported for the default
      ! real type.
```

```
      call  ieee_set_halting_mode( IEEE_ALL, .false. )
      x = sum( a*b )
      i = 0
      call ieee_get_flag( IEEE_OVERFLOW, overflow )

      if( overflow )  then

        ! An overflow has occurred.  Clear all exception flags.

        call ieee_set_flag( IEEE_ALL, .false. )

        ! Scale x and y so that the element of maximum magnitude is
        ! near 1.

        max_a = maxval( abs(a) )
        max_b = maxval( abs(b) )
        exp_a = ieee_logb( max_a )
        exp_b = ieee_logb( max_b )
        x = sum( ieee_scalb( a, -exp_a )*ieee_scalb( b, -exp_b ) )
        i = exp_a + exp_b
      endif
    end subroutine mult
end module dot
```

Figure 14-4 A robust, efficient subroutine to compute a scaled dot product

15 Interoperability with C

- A **Companion Processor** is a mechanism for defining procedures and using data. It is most often a C compiler. The provisions for interoperation are described in terms of C.
- **Interoperability** is the property of being usable with both the Fortran processor and a companion processor.
- A **Binding Label** is a global identifier which bridges the gap in syntax between Fortran names and C identifiers. A Fortran procedure, module variable, or common block can be given a binding label, possibly different from its Fortran name.
- The **BIND Attribute** is used to specify interoperability of a derived type, procedure, procedure interface, or common block. It is also used to specify a binding label for a procedure, module variable, or common block.

Interoperability allows processing to be done using a mixture of languages. Fortran supports the following aspects of interoperability with C:

1. Procedures. A Fortran program can incorporate functions written in C. The Fortran code can invoke those C functions; conversely, the C functions can invoke Fortran procedures.
2. Data. The Fortran and C portions of a Fortran program can communicate data by argument passing or global data.
3. Files. Separate Fortran and C programs can communicate by means of files.

This chapter covers interoperability of procedures and data in a program that has a mixture of Fortran and C code. Interoperability of files is facilitated by stream access, discussion of which is integrated into the chapters on input/output.

A Fortran and a C entity are interoperable with each other if they have corresponding properties. The details of how their properties must correspond depend on whether the entities are types, data objects, or procedures. In some contexts, the interoperability of an entity in one language is discussed, regardless of whether there is a corresponding entity that it is interoperable with. An entity is interoperable in this sense if it has the properties necessary for interoperation.

The BIND attribute has multiple roles. For a derived type, the BIND attribute specifies interoperability. For a procedure, procedure interface, or common block, the BIND attribute specifies interoperability and a binding label. For a variable, the BIND attribute specifies a binding label; a variable without the BIND attribute may be interoperable, but it does not have a binding label.

J.C. Adams et al., *The Fortran 2003 Handbook*,
DOI: 10.1007/978-1-84628-746-6_15, © Springer-Verlag London Limited 2009

15.1 Companion Processors

The interoperability features of Fortran are designed around interoperation with C, but can indirectly support other languages as well. Each processor has one or more companion processors with which it can interoperate. The particular set of companion processors and the means of selecting among them are processor-dependent.

In order to be constructively useful, a companion processor must be capable of working with data and procedures that can be described in terms of C. The obvious case of a candidate companion processor is a C compiler. Compilers for other languages can also be companion processors if they are capable of interoperating with C. It might be said that C serves as the "lingua Franca" of programming languages.

For example, C++ compilers can use the "extern C" attribute to interoperate with C functions. Therefore, interoperation between Fortran and C++ can be achieved through the common ground of C. Likewise, Ada compilers can interoperate with C and thus could conceivably serve as companion processors. One simple case that is easy to overlook is that the companion processor could be a Fortran compiler. The interoperability facilities of the language could be used to facilitate interoperation of multiple Fortran compilers.

As a trivial case, a Fortran compiler could serve as its own companion processor; this trivial case allows a vendor to meet the requirement of supporting one or more companion processors even in an environment where there might be no other processors. Although the interoperability features have little direct relevance to such an environment, their support facilitates portability of code that might be used in other environments.

15.2 Binding Labels

Fortran and C have different rules on the validity and uniqueness of identifiers. In particular, C identifiers are case sensitive, but Fortran identifiers are not. The concept of a binding label bridges this gap and provides a means for Fortran code to refer to C identifiers. A binding label follows the rules for C identifiers.

A binding label is the global identifier by which a Fortran variable, common block, or procedure is known to the C compiler. If an entity has a binding label, it can be referred to by that identifier in C code. A variable or procedure that has no binding label has no global identifier in C, but can still be interoperable and used in ways that do not require a global identifier. For example, a global identifier is not needed for a procedure argument.

The BIND attribute specification for a variable, common block, procedure, or procedure interface confers a binding label, either explicitly or implicitly. If the BIND attribute specification includes a NAME specifier, then the binding label is the character value so specified, with any leading or trailing blanks removed; the case of the letters is significant in a binding label specified this way. If there is no NAME specifier, the binding label is the name of the variable, common block, procedure, or procedure interface, with all letters in lower case.

However, the character value in a NAME specifier may be zero-length or entirely blank, in which case there is no binding label. This allows multiple instances of such NAME specifications, which otherwise would be disallowed by the prohibition, described below, against duplication of binding labels.

This special rule allows the specification of a procedure that is interoperable, but has no binding label. This is useful for a procedure that is passed as an actual argument to a C function. Such a procedure must be interoperable, but there might be no need for it to have a global identifier. The BIND attribute specification for a procedure normally specifies both interoperability and a binding label. For consistency, the special rule for the blank case also applies to the BIND attribute specification of a variable or common block, but it is not obvious that it serves any useful function in those cases.

Binding labels are global identifiers (16.1.1) of a program and are required to be unique. Having two Fortran variables or procedures with the same binding label does not establish a linkage between those entities; instead, it is disallowed. If multiple scoping units have declarations of the same COMMON block, all the declarations must specify the same binding label or all of them must have no binding label. A binding label may be the same as another global identifier such as an external procedure name; it is likely to be common for an interoperable external procedure to have a binding label the same as its procedure name.

15.3 The ISO_C_BINDING Intrinsic Module

The ISO_C_BINDING intrinsic module includes named constants, derived types, and module procedures to support interoperability. Most of the named constants designate interoperable kind values for Fortran intrinsic types; these are described in 15.4.1. Other named constants designate character values that correspond to C characters with special semantics such as the new line character; these are described in 15.5.3. The derived types, module procedures, and remaining named constants provide support for interoperation with C pointers, as described in 15.5.4.

The standard permits this module to have other public entities. However, a standard-conforming program must not use any such processor-dependent entities.

15.4 Interoperability of Types

Interoperability of data depends on the more abstract matter of interoperability of types and type parameters. C has no concept like a Fortran type parameter; rather, the notion of a C type corresponds to that of a Fortran type with a particular set of type parameter values. For example, Fortran has a single integer type with a type parameter, which distinguishes multiple representations; C, on the other hand, has a different type for each integer representation.

A C type is interoperable with a Fortran type with a particular set of type parameter values if the C and Fortran objects take the same amount of space and if the same representations have corresponding meanings. These principles apply to both kind and length type parameters.

Each language has some type constructs that have no direct correspondences in the other language. For example, there is no Fortran type that corresponds to a C union type or a C struct type with a bit field; there is no C type that corresponds to a Fortran type with an allocatable component. Types using these constructs are not interoperable.

15.4.1 Intrinsic Types

Table 15-1 shows the correspondence between C types and intrinsic Fortran types and type parameters. A C type in the first column corresponds to the Fortran type specifiers in the same row of the second column. The names used for the kind values in the second column are all named integer constants from the ISO_C_BINDING intrinsic module. For example, C type int corresponds to the Fortran type integer with kind C_INT.

The C types given in the above table are standard C types. Any unqualified C type that is compatible with one of these types also corresponds to the same Fortran type and type parameter values. For example, a C type derived from int via a C typedef corresponds to INTEGER(C_INT). A Fortran type with a particular kind value may correspond to more than one C type.

Fortran has no unsigned types. C has unsigned integer types, each of which corresponds to the same Fortran integer kind as the signed C integer of the same size. The C standard requires unsigned integers to have the same representation as nonnegative integers.

Note that the C char kind corresponds to a Fortran character with length 1. There is a further requirement that this length be specified by an initialization expression or by omission. Fortran characters with lengths other than 1 are not directly interoperable, but there are special provisions for using them in conjunction with C (15.5.3).

The values of the three kind type parameters for complex are required to be the same as the corresponding values for real.

All the named constants in the above table are public entities in the ISO_C_BINDING module, but this does not imply that all these C types will be interoperable. If the value of one of these named constants is positive, it will be a valid kind value for the intrinsic type; the corresponding C type is interoperable with the Fortran intrinsic type of that kind. If the value of one of these named constants is negative, there is no interoperable Fortran kind for the corresponding C type. The named constant C_INT is required to be positive; all the others may be negative.

The particular negative value indicates the reason why there is no interoperable Fortran kind for a particular C type. Table 15-2 provides a description of the meaning of the various negative values.

15.4.2 C Enum Types

Fortran provides a facility for interoperation with C Enum types. This facility is described in 4.6.

Table 15-1 Interoperable intrinsic types

C Type	Fortran Type and Type Parameter Values
int	INTEGER(C_INT)
short int	INTEGER(C_SHORT)
long int	INTEGER(C_LONG)
long long int	INTEGER(C_LONG_LONG)
signed char	INTEGER(C_SIGNED_CHAR)
unsigned char	INTEGER(C_SIGNED_CHAR)
size_t	INTEGER(C_SIZE_T)
int8_t	INTEGER(C_INT8_T)
int16_t	INTEGER(C_INT16_T)
int32_t	INTEGER(C_INT32_T)
int64_t	INTEGER(C_INT64_T)
int_least8_t	INTEGER(C_INT_LEAST8_T)
int_least16_t	INTEGER(C_INT_LEAST16_T)
int_least32_t	INTEGER(C_INT_LEAST32_T)
int_least64_t	INTEGER(C_INT_LEAST64_T)
int_fast8_t	INTEGER(C_INT_FAST8_T)
int_fast16_t	INTEGER(C_INT_FAST16_T)
int_fast32_t	INTEGER(C_INT_FAST32_T)
int_fast64_t	INTEGER(C_INT_FAST64_T)
intmax_t	INTEGER(C_INTMAX_T)
intptr_t	INTEGER(C_INTPTR_T)
float	REAL(C_FLOAT)
double	REAL(C_DOUBLE)
long double	REAL(C_LONG_DOUBLE)
float _Complex	COMPLEX(C_FLOAT_COMPLEX)
double _Complex	COMPLEX(C_DOUBLE_COMPLEX)
long double _Complex	COMPLEX(C_LONG_DOUBLE_COMPLEX)
_Bool	LOGICAL(C_BOOL)
char	CHARACTER(1, C_CHAR)

15.4.3 C Pointer Types

C pointers and Fortran pointers are different in several fundamental ways, two of which have implications for interoperability. One is that a Fortran pointer is of the same type as its target, while a C pointer is a separate type. The other is that a Fortran array pointer includes information about shape and bounds, which means that its representation is not directly compatible with the simple address form of a C pointer.

The ISO_C_BINDING module includes the public derived types C_PTR and C_FUNPTR. The type C_PTR is interoperable with any C data pointer type; C_FUNPTR is interoperable with any C function pointer type.

Table 15-2 The meaning of negative values

For Intrinsic Types	Value	Meaning
INTEGER	–1	The C type exists but there is no Fortran kind that interoperates with it.
	–2	The C type is not defined by the C processor.
REAL or COMPLEX	–1	The C type exists but there is no Fortran kind that has the same precision.
	–2	The C type exists but there is no Fortran kind that has the same exponent range.
	–3	The C type exists but there is no Fortran kind that has the same precision and exponent range.
	–4	There is no interoperable Fortran kind for some reason other than precision or range requirements.
LOGICAL	–1	There is no Fortran kind that interoperates with the C type.
CHARACTER	–1	There is no Fortran kind that interoperates with the C type.

All components of these derived types are private; user programs cannot directly manipulate the components of objects of these types.

A single Fortran type interoperates with all C data pointer types; this implies that the C compiler must have the same representation for all data pointer types. Likewise, the C compiler must have the same representation for all function pointer types. The C standard does not require this, but all current C compilers meet the restrictions.

An entity of type C_PTR cannot be dereferenced directly in Fortran. Instead, the features discussed in 15.5.4 can be used to achieve these ends.

15.4.4 Derived Types

A Fortran derived type is defined to be interoperable if it has the BIND(C) attribute. An interoperable derived type is nonextensible (4.4.12) by definition. The following restrictions apply to a type with the BIND(C) attribute.

1. It must not have the SEQUENCE attribute. BIND(C) types have properties similar to those of SEQUENCE types—so much so that BIND(C) types might be thought of as a category of SEQUENCE types, but they are not categorized that way in the standard.
2. It must not have the EXTENDS attribute.
3. It must not have type parameters.
4. It must not have procedure bindings.

5. Each component must be a nonpointer, nonallocatable data component with interoperable type and type parameters.

The restriction against Fortran pointer components might seem severe at first glance, but recall that the Fortran types C_PTR and C_FUNPTR are interoperable with C pointer types. An interoperable derived type must not have Fortran pointer components, but it may have components of type C_PTR and C_FUNPTR.

A Fortran derived type is interoperable with a C struct type if:

1. the Fortran type is interoperable,
2. the Fortran type and the C type have the same number of components,
3. the type and type parameters of each Fortran component are interoperable with the type of the corresponding C component,
4. each scalar Fortran component corresponds to a scalar C component, and
5. each array Fortran component corresponds to an array C component and the shape matches as described in 15.5.2.

The last two rules above are not specified explicitly in the standard, but they are necessary. It is likely that they were so obvious that the need to specify them was overlooked.

The component correspondence in these rules is according to position in the type declarations.

The component names of the Fortran and C types are not required to agree. This avoids problems relating to the difference in case sensitivity in the two languages. Component names of the Fortran type may be private. If so, the PRIVATE attribute has no effect on C.

The following Fortran derived type and C struct type are interoperable with each other.

```
type, bind(c) :: real_node_type
  real(c_float) :: data
  type(c_ptr) :: next_node
end type

typedef struct {
  float data;
  nodeType *nextNode;
} nodeType
```

15.5 Interoperation of Data

A Fortran variable is interoperable if it has interoperable type and type parameters and also meets other conditions detailed below.

It is notable that having the BIND attribute is not one of those conditions. There is no need to explicitly specify that a variable is interoperable; if it has the needed properties then it is interoperable without further specification. The only purpose of the BIND attribute for variable is to give it a binding label (15.2). A binding label is needed for communication with C via a global variable, but it is not needed for communication via procedure arguments.

An interoperable Fortran variable cannot be polymorphic because extensible types cannot be interoperable.

15.5.1 Scalar Variables

A scalar Fortran variable is interoperable if it has interoperable type and type parameters and is neither allocatable nor a pointer. An interoperable scalar Fortran variable is interoperable with a scalar C variable if the type and type parameters of the Fortran variable are interoperable with the type of the C variable.

15.5.2 Array Variables

An array Fortran variable is interoperable if it has interoperable type and type parameters, is neither allocatable nor a pointer, and has either explicit shape or assumed size. The prohibition against being allocatable or a pointer is redundant with the requirement for explicit shape or assumed size; it is included in this definition for parallelism with the scalar case.

The requirement for explicit shape or assumed size rules out assumed-shape and deferred-shape arrays, plus all array sections. This requirement ensures contiguity in memory, as required by C. However, one should not misinterpret this to mean that an array passed as an actual argument to a C function must be explicit shape or assumed size. See 15.7 for an explanation of why such an interpretation would be incorrect.

An interoperable Fortran array of rank 1 is interoperable with a C array if the types of their elements are interoperable and their sizes match. For this purpose, the size of a rank 1 Fortran assumed-size array matches that of a C array of unspecified size. However, C has no zero-sized arrays, so a Fortran array of size zero is not interoperable with any C array.

C arrays of rank greater than 1 are formally constructed as arrays whose elements are arrays. However, this formalism obscures more than it helps the description of interoperability. Instead, consider such a C array to be described by the type of the final elements and the sizes of each dimension. An interoperable multi-dimensional Fortran array is interoperable with such a C array if the types of their elements are interoperable and the size of each dimension of the Fortran array (also known as the extent) matches the size of the corresponding dimension of the C array. The matching of sizes is as described above for the rank 1 case, except that only the last dimension of an assumed-size array follows the assumed-size rule. The dimensions of the Fortran array correspond to those of the C array in reverse order.

For example, arrays declared as follows in Fortran and C are interoperable.

Fortran	C
`real(c_float) :: x(5,6,7)`	`float y[7][6][5];`
`real(c_double) :: xx(-10:10)`	`double yy[21];`
`integer(c_int) :: i(8,9,*)`	`int i[][9][8];`

The double example illustrates that the Fortran lower bounds do not directly matter except as they are used to compute the extents. Recall that C arrays always have lower bounds of zero. Thus, the double C array in this example has elements numbered 0 through 20, which correspond to the Fortran elements numbered –10 through 10. Similar index correspondence rules apply to each dimension of a multi-dimensional array.

A common idiom in C is to use a rank-1 array of pointers to represent a two-dimensional array. This construct is not interoperable with a rank-2 Fortran array. Instead, it can interoperate with an array of type C_PTR.

15.5.3 Character Data

A single C character is interoperable with a single Fortran character of length 1. However, C has nothing that directly corresponds to Fortran character strings of lengths other than 1; such a Fortran character string is not interoperable. A character string in C is represented as an array of character elements; this is interoperable with a Fortran array of single-character elements, but not with a scalar Fortran string.

It is usually most natural for multi-character Fortran data to use scalar strings instead of arrays. For example, it would be untenably awkward to have to write the array ['Y'_c_char, 'e'_c_char, 's'_c_char] instead of the more natural 'Yes'_c_char.

Because character data is so basic and ubiquitous, a special-case feature was introduced to alleviate the awkwardness of the differences between C and Fortran in this regard. As detailed in 12.6.4.1, argument association is character-by-character sequence association if the dummy argument is an explicit-shape or assumed-size array of the C character kind. This special rule overrides the normal distinction between arrays and strings; it allows a Fortran string to be passed as an actual argument to a C function that expects an array of characters. An example of the application of this rule is shown in 15.7.

To further facilitate portable interoperation with C character data, the ISO_C_BINDING module includes several named character constants with values corresponding to characters with special meaning in C. Table 15-3 lists these named constants. The specifications of the values depend on whether or not there is a Fortran character kind interoperable with C char. Normally there is such a Fortran character kind, in which case these named constants are of that kind and have values specified in terms of C as shown in the last column of the table. If there is no such Fortran character kind, these named constants are of default character kind and have values specified in terms of the character intrinsic functions as shown in the next-to-last column.

15.5.4 Pointers

Fortran pointers and C pointers are not directly interoperable with each other. A Fortran array pointer must include information about the bounds and strides of the array; this invariably forces the physical representation of a Fortran array pointer to be differ-

Table 15-3 Named constants for special C characters

Named Constant	C Character	Character Value	
		C_CHAR=–1	C_CHAR>0
C_NULL_CHAR	null character	CHAR(0)	'\0'
C_ALERT	alert	ACHAR(7)	'\a'
C_BACKSPACE	backspace	ACHAR(8)	'\b'
C_FORM_FEED	form feed	ACHAR(12)	'\f'
C_NEW_LINE	new line	ACHAR(10)	'\n'
C_CARRIAGE_RETURN	carriage return	ACHAR(13)	'\r'
C_HORIZONTAL_TAB	horizontal tab	ACHAR(9)	'\t'
C_VERTICAL_TAB	vertical tab	ACHAR(11)	'\v'

ent from the simple address form of a C pointer. Fortran scalar pointers are often implemented with a representation like that of a C pointer, but such similarity is not guaranteed by the standard.

Instead, a C pointer type is interoperable with one of the Fortran types C_PTR or C_FUNPTR, as described above. The ISO_C_BINDING module has procedures for conversion between the Fortran and C pointer forms. The module functions C_LOC and C_FUNLOC convert Fortran pointers into C ones or directly generate C pointers to specified Fortran targets. The module subroutines C_F_POINTER and C_F_PROCPOINTER convert C pointers into Fortran ones. Additionally, the module function C_ASSOCIATED provides functionality corresponding to that of the ASSOCIATED intrinsic for Fortran pointers. The named constants C_NULL_PTR and C_NULL_FUNPTR provide convenient shorthand forms for a C null data pointer and function pointer, respectively.

C_LOC (X) Inquiry Function

A C data pointer to the argument.

X A suitable target for a C data pointer. X must either have the TARGET attribute or be an associated pointer. If X is an array, it must have nonzero size and either be an interoperable variable or an allocated allocatable variable with interoperable type and type parameters. X must not be polymorphic. X must not have a length type parameter unless it is of type character with length 1.

Result Characteristics. Scalar of type C_PTR.

Result Value. The result is a C pointer to the argument or, if the argument is a pointer, to the target of that pointer. If the target of the pointer is not interoperable, the resulting C pointer cannot be dereferenced in C, but it can be converted to a Fortran pointer by C_F_POINTER.

C_FUNLOC (X) Inquiry Function

A C function pointer to the argument.

X An interoperable procedure or an associated pointer to one.

Result Characteristics. Scalar of type C_FUNPTR.

Result Value. The result is a C pointer to the argument.

C_F_POINTER (CPTR, FPTR [, SHAPE]) Subroutine

Converts a C data pointer to a Fortran data pointer.

CPTR A scalar INTENT (IN) argument of type C_PTR. It must either point to an interoperable data entity or be the result of a reference to C_LOC. Note that, in either case, it cannot point to a Fortran variable that does not have the TARGET attribute. The target of CPTR must not have been deallocated and must not have become undefined as the result of termination of a procedure.

FPTR An INTENT (OUT) pointer suitable for pointing to the target of CPTR. If CPTR points to an interoperable data entity, FPTR must have type and type parameters interoperable with the type of the C entity. If CPTR is the result of a reference to C_LOC with a noninteroperable target, FPTR must be a nonpolymorphic scalar pointer with the same type and type parameters as the target. FPTR will be set to point to the specified target. If FPTR is an array, its lower bounds will be set to 1 and its shape will be as specified by the SHAPE argument.

SHAPE An INTENT (IN) rank-1 integer array. It must be present if and only if FPTR is an array. Its size must equal the rank of FPTR. It specifies the shape of FPTR.

C_F_PROCPOINTER (CPTR, FPTR) Subroutine

Converts a C function pointer to a Fortran procedure pointer.

CPTR A scalar INTENT (IN) argument of type C_FUNPTR. It must point to an interoperable procedure.

FPTR An INTENT (OUT) procedure pointer suitable for pointing to the target of CPTR. The interface of FPTR must be interoperable with a C prototype describing the procedure. FPTR will be set to point to the specified target.

C_ASSOCIATED (C_PTR_1 [, C_PTR_2]) Inquiry Function

Association status of a C pointer or whether two C pointers are associated with the same target.

C_PTR_1 A scalar of type C_PTR or C_FUNPTR.

C_PTR_2 An optional scalar of the same type as C_PTR_1

Result Characteristics. Default logical scalar.

Result Value. If C_PTR_2 is absent, the result is true if C_PTR_1 is not a C null pointer. If C_PTR_2 is present, the result is true if both arguments point to the same target and are not C null pointers.

15.5.5 Global Data

Global data in C is stored in a variable with external linkage, which roughly means either a variable declared as extern or a file scope variable declared without an explicit storage class. Such a C global variable can be linked to either a Fortran common block or a Fortran module variable. Although the syntax used for linking a module variable would also appear to fit and to make sense for a local variable in a procedure, that is not allowed.

Linkage between a C variable and a Fortran variable or common block is a form of association. Modifications to the values of the C variable cause corresponding modifications to the values of the linked Fortran entity, and vice versa. In practice, this is implemented by assigning them to the same memory locations; other implementations are allowed in theory, but are unlikely to happen.

The BIND attribute for a variable or common block specifies a binding label either explicitly or implicitly. The forms for specifying the BIND attribute for a variable or common block are described in 5.1 and 5.8.3.

If there is a C variable with external linkage that has an identifier the same as the binding label of the Fortran variable or common block, that C variable is linked to the Fortran variable or common block. A binding label must be valid as the form of a C identifier, but it is not required that there be such a C variable with that identifier; if there is none, the Fortran variable or common block stands on its own.

Both C and Fortran provide ways to specify initialization for a variable. If a Fortran variable or common block is linked to a C variable, initialization may be specified by Fortran, by C, or by neither, but it must not be specified by both.

If a Fortran variable is linked to a C variable, they must be interoperable with each other.

If a Fortran common block is linked to a C variable, one of the following must hold:

1. The common block contains only a single variable, which is interoperable with the C variable.

2. The C variable is a structure and each of the variables in the common block is interoperable with the corresponding component of the C structure.

   The following is a simple example of linkage between a C and Fortran variable.

```
integer(c_int), bind(c) :: status = 0
```

```
int status;
```

In this example, the Fortran declaration must be in the specification part of a module and the C declaration must be outside of any functions. The Fortran declaration initializes the variable, so the C declaration must not.

Linkage to a C variable with a mixed- or upper-case name requires that the name be explicitly specified with a NAME specifier in the BIND attribute as in the following example.

```
real(c_double), bind(c,name='ElapsedTime') :: ElapsedTime

double ElapsedTime;
```

Even though the declaration of the Fortran variable in this example uses the same mixed case as the C declaration, the case in the Fortran declaration is insignificant. Without the explicit NAME specifier, the binding label of the Fortran variable would be all lower case. The linkage to this C variable could alternatively be done with a common block instead of a module variable. The Fortran declarations for such a common block might look like:

```
common /ElapsedTime/ ElapsedTime
real(c_double) :: ElapsedTime
bind(c,name='ElapsedTime') :: /ElapsedTime/
```

In this example, the name of the single variable in the common block is the same as the name of the common block. Such duplication of the name of a common block and a variable is allowed but is not required. It is sometimes considered poor style, but is fairly common in cases like this where the common block has only a single variable. The default binding label comes from the name of the common block rather than that of the variable, although in this example the default binding label is not used.

15.6 Interoperation of Procedures

An interoperable Fortran procedure is potentially invocable from C. An interoperable C function is potentially invocable from Fortran. Such inter-language invocation requires interoperability between Fortran interfaces and C prototypes.

A C function that is not interoperable cannot be invoked directly from Fortran, but it can be invoked indirectly by invoking an interoperable C function which in turn invokes the non-interoperable one. Conversely, a noninteroperable Fortran procedure can be invoked indirectly from C by means of an interoperable intermediary procedure.

15.6.1 Interoperable Fortran Procedures and Interfaces

The BIND attribute specification in the SUBROUTINE or FUNCTION statement of a subprogram or in an ENTRY statement specifies that the procedure named in that statement is interoperable.

The BIND attribute specification in the SUBROUTINE or FUNCTION statement of an interface body specifies that the interface is interoperable.

Rules and restrictions:

1. An interoperable procedure must not be defined by an internal subprogram; it must be defined by an external or module subprogram.

2. A Fortran reference to an interoperable procedure requires an explicit interface. It is possible that the implementation mechanisms for interoperable procedures might be different from those for non-interoperable ones. The explicit interface provides the information necessary to guarantee that the appropriate mechanisms are used.

3. An interoperable procedure or interface must not be elemental.

4. A dummy argument of an interoperable procedure or interface must not be optional and must not be an alternate return.

5. Each dummy argument of an interoperable procedure or interface must be an interoperable variable or an interoperable procedure. This restriction indirectly prohibits several things. For example it implies that a dummy argument must not be a pointer, an allocatable object, an assumed-shape array, or a character string with length other than 1. See 15.5.3 for more on the character length issue.

6. The result variable of an interoperable function procedure or interface must be an interoperable scalar variable.

7. A binding label (15.2) must not be specified for a dummy procedure or an abstract interface. A binding label would make no sense in these cases because a binding label specifies a particular procedure. For a dummy procedure, the particular procedure involved is specified by the actual argument, not by the dummy. The whole point of an abstract interface is that it is an interface that is defined on its own without being tied to a particular procedure.

8. A Fortran procedure must not be invoked as a C signal handler, even if it has an appropriate interface.

15.6.2 Interoperability of Fortran Interfaces and C Prototypes

Fortran procedure interfaces are comparable to C function prototypes. A particular Fortran procedure interface is interoperable with a particular C function prototype if the following conditions hold.

1. The Fortran interface must be interoperable; that is it must have the BIND attribute. This condition implies several others, as described in the preceding section.

2. If the Fortran interface describes a subroutine, the C prototype must have a result type of void.

3. If the Fortran interface describes a function, the result variable of the function must be interoperable with the result of the C prototype.

4. The number of Fortran dummy arguments must equal the number of C formal parameters.

5. Each Fortran dummy argument that has the VALUE attribute (5.9.2) must be interoperable with the corresponding C formal parameter.

6. Each Fortran dummy argument without the VALUE attribute must correspond to a C formal parameter that is a pointer; the Fortran dummy argument must be interoperable with a target of the C pointer.

The correspondence between Fortran dummy arguments and C formal parameters is positional; the names do not matter. The syntax for invoking a C procedure may use keyword forms of argument specification; the keyword names for such usage are those of the Fortran interface—not the C function.

The standard separately specifies that the C function prototype must not have variable arguments denoted by an ellipsis, although that condition seems implied by the others. In any case, it is notable that there is no support for interoperation of a procedure with a variable number of arguments. The closest available approximation to that functionality would be to have a single argument that is an array of C pointers as in the following example.

```
type(c_ptr), allocatable :: args(:)
   . . .
allocate(args(3))
args(1) = c_loc(something)
args(2) = c_loc(something_else)
args(3) = c_null_ptr
call sub(args)
```

15.6.3 Interoperable C Functions

A C function is interoperable if it has external linkage and can be described by a C prototype that is interoperable with some Fortran interface. This definition also applies to procedures in companion processor languages other than C. For example, a C++ function must be describable by a C prototype in order to be interoperable.

15.6.4 Restrictions on C Functions

The definition of an interoperable C function focuses on the issues directly involved in invoking the function and passing arguments to it. For the most part, once a C function is invoked, its internal operations are determined by the C compiler or other companion processor. The Fortran standard has little to say about the matter. However, the Fortran standard places several restrictions on what a C function may do internally.

These restrictions apply to any C function invoked during execution of a Fortran program, even if the C function is not interoperable.

1. The C setjmp and longjmp functions have semantics that do not mix well with Fortran. These functions may be invoked internally within a C function, but only subject to the following conditions, which avoid needing special treatment by the

Fortran compiler. Neither setjmp nor longjmp may be invoked directly from Fortran. If a C function invokes either setjmp or longjmp, that C function must not also invoke a Fortran procedure, either directly or indirectly. In essence, these two restrictions require isolation of the effects of setjmp and longjmp to portions of the C code that do not involve Fortran.

2. A C function must not use the C signal function to change the handling of any exception that is being handled by Fortran. Because it is not specified what exceptions are handled by the Fortran compiler, this restriction essentially prohibits portable programs from using the C signal to change the handling of any exceptions.

3. A C function may cause a floating-point exception flag (14.4) to be set to signaling. A flag that is signaling on entry to the C function must also be signaling on return. The values of the flags on entry to a C function invoked from Fortran are processor dependent.

4. If Fortran and C code both have the same external file open at the same time, the results are compiler-dependent. The two compilers are not required to co-ordinate their input/output handling of a single file. Such mixed-language access might work as expected with some compiler combinations, but is not portable. It is advisable to have all input/output operations for a single file done in a single language at a time. There is no problem with having input/output in both Fortran and C as long as each file is handled by one language at a time; any Fortran connection to the file should be closed before opening the file in C, and vise versa.

In addition to these specific restrictions, there is a general principle which is not explicitly stated as an issue of interoperation. The Fortran standard has many restrictions that are stated in sufficiently general terms that they apply without regard to the language in which a particular procedure is written. This implies restrictions on what may be done in a C function in a Fortran program. In many such cases, violations of these restrictions are likely to go undetected by the compiler, but a program with such violations is nonstandard. In short, although the inner workings of a C function are defined by C, its externally visible effects on Fortran entities must meet the same requirements as if the C function were a Fortran procedure.

The following are some notable cases of such implied restrictions.

1. A dummy argument that has INTENT (IN), or whose actual argument is not definable, must not be modified while its procedure is executing. That restriction applies to any means of modification, whether in Fortran or in C.

2. There are restrictions on modifying dummy arguments that are aliases. Those restrictions also apply to modification done in C.

3. A C function must not use pointer arithmetic to access memory outside of the bounds of its arguments.

15.6.5 Connecting Fortran Procedures and C Functions

Invocation of a Fortran procedure or a C function from the other language requires a means to identify the intended procedure or function. There are two means of accomplishing such identification: directly using binding labels or indirectly using procedure pointers.

The binding label of a C function with external linkage is the C identifier of the function. The binding label of an interoperable Fortran procedure is given, either implicitly or explicitly, by the BIND attribute specification in the subroutine or function statement (12.1.1, 12.2.1).

An interoperable C function may be referred to directly in Fortran if it has been given an explicit interface with the same binding label. The Fortran interface must be interoperable with the C function prototype.

If a Fortran procedure has a binding label, the procedure may be referred to directly in C code using that binding label as its C identifier. Such a Fortran procedure may also be referred to from other Fortran code. Having a BIND attribute is one of the conditions that requires an explicit interface for references from Fortran code (12.5.1.2).

Pointers provide an indirect means of cross-language identification. A datum of type C_FUNPTR can point to a Fortran procedure or C function. Such a datum can be communicated between Fortran and C via argument passing or as global data. The target of such a pointer need not have a binding label.

15.7 Examples of Interoperation

It is important to understand that the definitions of interoperability do not stand alone, but integrate with the rest of the language. A Fortran procedure reference to a C function still follows all the rules for a Fortran procedure reference. A narrow focusing solely on the requirements of interoperability misses the bigger picture.

For example, an interoperable Fortran interface must not have a dummy argument that is allocatable. If one looks too narrowly, this might sound like a restriction against passing an allocatable array as an argument to an interoperable procedure. But that conclusion would be incorrect because the restriction against allocatable objects applies to the dummy arguments of an interoperable procedure. The normal rules for the association of actual and dummy arguments still apply. Those rules allow an allocated allocatable object to be the actual argument for a nonallocatable dummy argument. Likewise, the restrictions against assumed-shape and pointer dummy arguments of interoperable procedures do not mean that an actual argument must not be an assumed-shape array or a pointer.

A particularly important case of this principle relates to character strings. A Fortran character string with a length other than 1 is not interoperable. A C character string is done as an array and is thus interoperable with a Fortran array of characters rather than with a Fortran character string. It would be unacceptably awkward if this implied that Fortran character strings could not be passed to C functions. However, the rules for argument association include a special case specifically to allow this. As described in 12.6.3, for characters of kind c_char, a scalar character string actual argument of length *n* may be associated with a dummy argument that is an array of size *n* as in the following example.

```
use :: iso_c_binding, only :: c_char, c_null_char
interface
  subroutine copy (in, out)
    import c_char
    character(kind=c_char) :: in(*), out(*)
  end subroutine copy
end interface

character(len=11 ,kind=c_char) :: &
     digit_string = c_char_'0123456789' // c_null_char
character(kind=c_char) :: digit_array(11)

call copy(digit_string, digit_array)
```

A C function described by the prototype

```
void copy(char in[], char out[]);
```

would be interoperable with the interface in this example.

The following example illustrates interoperation of procedure pointers. One common use of procedure pointers is for callback procedures in event-driven graphic user interfaces. The definition of a graphic element might include a procedure that will be called when that graphic element is selected interactively. A simple example is a button with a procedure to be called when the button is clicked on.

```
module m
  interface
    subroutine define_button(text, text_len, proc) bind(c)
      use iso_c_binding
      character, intent(in) :: text(text_len)
      integer, value :: text_len
      type(c_funptr) :: proc
    end subroutine define_button
    subroutine poll_for_events(status) bind(c)
      integer, intent(out) :: status
    end subroutine poll_for_events
  end interface
contains
  subroutine graphic_interface
    type(c_funptr) :: proc
    integer :: status
    proc = c_funloc(do_button)
    call define_button('Push me', 7, proc)
    ...!-- Define other graphic elements.
```

```
      event_loop: do
        call poll_for_events(status)
        if (status/=0) exit event_loop
      end do graphics_loop
    end subroutine graphic_interface
    subroutine do_button() bind(c)
      write(*,*) 'The button was pushed.'
      return
    end subroutine do_button
  end module m
```

The procedures define_button and poll_for_events are presumed to be part of a graphics interface library.

16 Scope, Association, and Definition

- The **Scope** of an identifier is the part of the program in which the identifier is known and accessible. A scope may encompass an entire program or be limited to a part of a single statement; usually it is something in between. Often it is desirable to limit the scope of an identifier to avoid name conflicts.
- **Association** provides communication among entities in the same or different scopes. In a sense it performs a function opposite to limiting the scope of an identifier. It allows an entity to be identified by different names in a scope or by the same or different names in different scopes. There are various forms: name association, pointer association, storage association, and inheritance association. Name association includes argument association, use association, host association, linkage association, and construct association.
- **Definition** is the state of having a value. An entity may be initialized. Various events may occur during execution, such as assignment or input, that result in entities acquiring or changing values. Other events such as an error condition or execution of a RETURN or DEALLOCATE statement may cause an entity to become undefined. An undefined entity must not be referenced.

The topics of scope, association, and definition are related. Scope specifies the part of a program where an identifier is known and accessible. Association is the pathway along which entities in the same or different scopes communicate. Definition characterizes the ways in which variables are given values. Its opposite, undefinition, characterizes the way values become unpredictable.

Various aspects of scope, association, and definition have already been discussed throughout the earlier chapters of this handbook. If simple programming disciplines are followed, correct programs can be created without concern for the subtle issues related to these topics, but there are situations in which it is necessary to know the details, particularly when modifying or maintaining programs. This chapter provides many of those details and indicates where the others may be located in the earlier chapters.

16.1 Scope

The **scope** of an identifier is that part of a Fortran program in which that identifier has a given meaning and can be used, defined, or referenced. With the possibility that different parts of a program are developed by different programmers, it is reasonable to allow and expect that something named X in one module or external subprogram, for example, has nothing to do with something named X in another program unit. This permits different programmers to work independently. The concepts of scope and

J.C. Adams et al., *The Fortran 2003 Handbook*,
DOI: 10.1007/978-1-84628-746-6_16,

classes of identifiers are some of the mechanisms that support this capability. Of course, some identifiers will have the scope of the entire program so that, in a large project with nested scopes, the usage of identifiers of global entities must be managed carefully. The scope of an identifier may be

- a program
- a scoping unit
- a construct
- a statement
- a part of a statement

A global identifier has the scope of a program; a local identifier has the scope of a scoping unit; the identifier of a construct entity has the scope of a construct; and the identifier of a statement entity has the scope of a statement or part of a statement. An entity may be identified by

- a name (3.2.2)
- a statement label (3.2.5)
- an external input/output unit number (9.1.6)
- an identifier of a pending data transfer operation (9.7)
- a generic specification (12.5.4)
- a binding label (15.2)

Note that Fortran keywords are not reserved and may be used as identifiers, with the exception of some type names as noted in 4.2.1.

16.1.1 Global Identifiers

The global entities of a program are

- program units
- external procedures
- common blocks
- entities with binding labels
- external input/output units
- pending data transfer operations

Each of these has a global identifier. Although an external subprogram is one form of program unit, it may specify more than one external procedure. The names of external procedures specified by ENTRY statements are global names. That is why both program units and external procedures appear in the list above.

Within a program:

- The name of a program unit, external procedure, or common block must not be the same as the name of any other such entity, except that an intrinsic module may have the same name as a program unit, external procedure, or common block.
- An entity of the program must not be identified by more than one binding label.
- The binding label of an entity must not be the same as the binding label of another entity.
- Ignoring differences in case, a binding label must not be the same as the name of another global entity unless the other global entity is an intrinsic module. A binding label may be the same as the name of a global entity that identifies the same entity.

Note that module procedure names are not global. One module may contain a module procedure that has the same name as a module procedure in a different module. The same is true for internal procedures in different hosts. An implementation may choose to assign global designators to these entities; if so, it is the responsibility of the processor to ensure that none of these are the same as those of external procedures, other global entities, or each other. This might be done by including in each such designator added characters that are not allowed in a Fortran name or by using such characters together with the global name of the program unit in which the entity appears.

16.1.2 Local Identifiers

A local identifier has the scope of a scoping unit. The scoping units are:

- a derived-type definition
- a procedure interface body, excluding any derived-type definitions and procedure interface bodies contained within it
- a program unit or subprogram, excluding derived-type definitions, procedure interface bodies, and subprograms contained within it

To visualize the concept of scope and scoping units, consider Figure 16-1. The outer rectangle bounds the pieces of an executable Fortran program; it is not a scoping unit but could be said to represent global scope. Within the executable program four other rectangles depict program units. One is the main program, two others are external subprogram units, and the fourth one is a module program unit.

All four of these program unit rectangles represent scoping units, excluding any rectangles within them. The main program in this example encloses no rectangle and so is an integral scoping unit without holes. External subprogram A has two internal subprograms within it, and therefore subprogram A's scoping unit is this rectangle, excluding internal subprograms B and C. External subprogram D has an interface block in it and no internal subprograms. Its scoping unit is subprogram D, excluding the in-

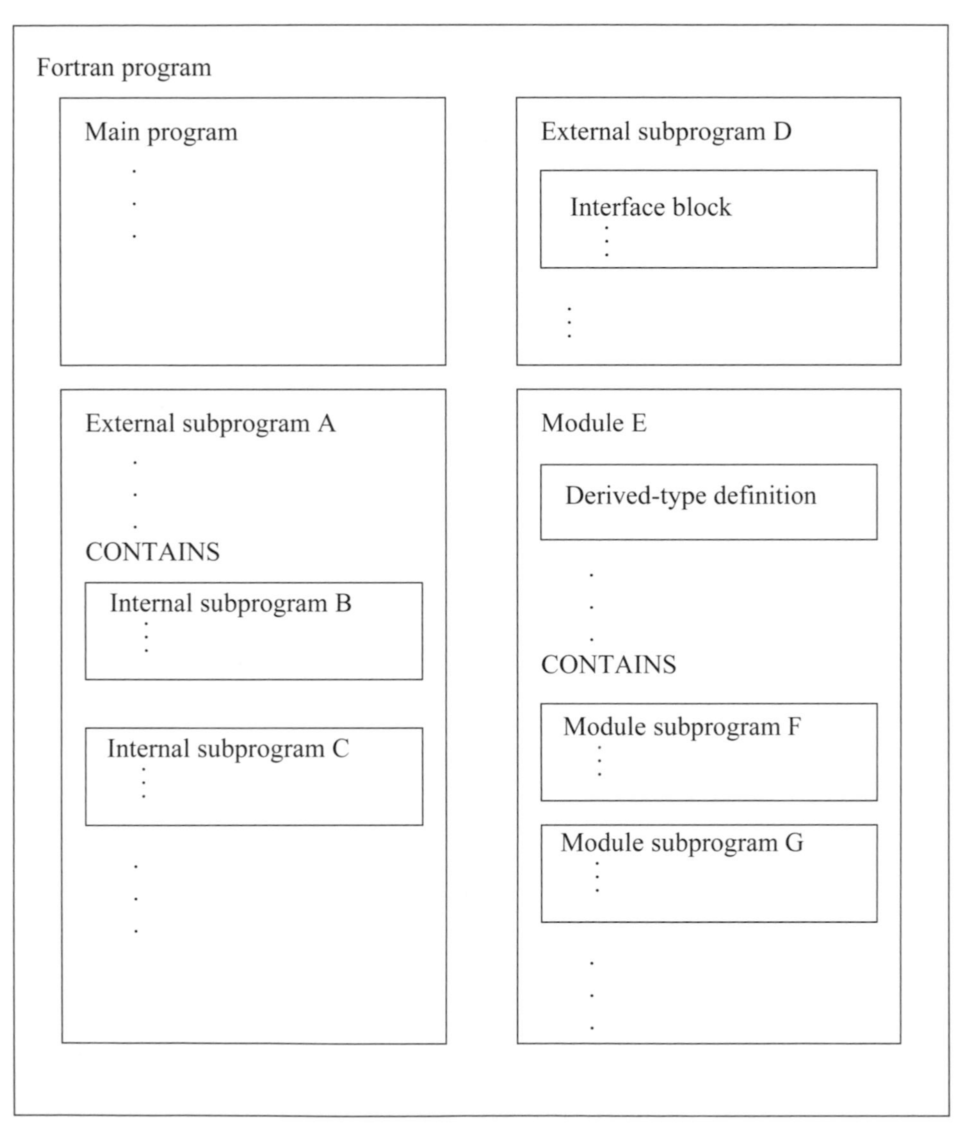

Figure 16-1 Scoping units denoted as boxes

terface block. Module E has a derived-type definition and two module subprograms within it. Its scoping unit is similarly the module program unit, excluding the derived-type definition and the module subprograms.

In addition, the interface block, the derived-type definition, and each of the internal and module subprograms are scoping units. In this example, these latter scoping units have no holes, as they do not themselves contain internal scoping units although, in general, the subprograms and the interface block could do so.

16.1.2.1 Classes of Local Identifiers

Within a scoping unit, there are three classes of local identifiers. The entities whose identifiers fall in each of the classes are listed below:

Class 1 Identifiers

named constants, named constructs, statement functions, internal procedures, module procedures, dummy procedures, intrinsic procedures, abstract interfaces, generic interfaces, derived types, namelist groups, external procedures accessed via USE, statement labels, and named variables that are not statement or construct entities

Class 2 Identifiers

type parameters, components, and type-bound procedure bindings of a derived type—there is a separate class for each derived type, which means that two different derived types can have the same identifiers

Class 3 Identifiers

argument keywords—there is a separate class of names for each procedure with an explicit interface, which means that two different procedures can have the same argument keywords

Rules and restrictions:

1. A local identifier of one class may be the same as a local identifier of another class.
2. A local identifier of one class must not be the same as another local identifier of the same class with one exception for generic names as explained below.

16.1.2.1.1 Class 1 Identifiers

Most local identifiers fall into Class 1. There are several special considerations for entities of this class.

A generic name may be the same as the name of a specific procedure. A generic name may also be the name of a derived type. A derived-type name is used in a structure constructor to construct values of the type, but the form of a structure constructor and the form of a function reference are identical. The form

```
name ( . . . )
```

is interpreted as a generic function reference if possible; it is interpreted as a structure constructor only if it cannot be interpreted as a generic function reference.

A Class 1 local identifier must not be the same as a global identifier if the global identifier is used in the scoping unit unless the global identifier either

- appears only in a USE statement that renames it

- is an external function (or entry) name appearing in the function definition to specify a function result
- is an external procedure name that is also a generic name
- is a common block name

The name of a common block must not be the same as the name of a constant or an intrinsic procedure referenced in the scoping unit. If the common block name is the same as another local identifier, any appearance of the name in a context other than as a common block in a COMMON or SAVE statement is an appearance of the local identifier. Note that the name of a module or internal procedure (which is a local identifier of Class 1) may be the same as the name of a common block (which is a global identifier).

Within the scope of a definition of a subprogram, the subprogram name (or entry name) may appear if the procedure it defines is invoked recursively. If the subprogram is an external subprogram, the appearance of the name in this context is a global reference. If the subprogram is an internal or module subprogram, this is a local reference in the scoping unit of its host.

16.1.2.1.2 Class 2 Identifiers

Entities that have the scope of a derived-type definition are its parameters, its components, and its type-bound procedures.

Outside the type definition (4.4.2), a type parameter name may appear only as a keyword in a derived-type specification or as a type parameter inquiry (6.3).

Outside the type definition, a component name may appear only in a designator for the component or as a keyword in a structure constructor for an object of the type.

Outside the type definition, the binding name (4.4.11) of a type bound procedure may appear only in a procedure reference.

A component name or binding name may appear only where the name is accessible (4.4.5).

16.1.2.1.3 Class 3 Identifiers

A dummy argument name may be used as a keyword in a reference to a procedure. Such a name that is specified in an internal procedure, module procedure, or interface body has the scope of the host of the procedure or interface body. It may appear only as a keyword in a reference to the procedure for which it is a dummy argument. The same is true for an argument name of an intrinsic procedure; it has the scope of the scoping unit in which the intrinsic procedure is referenced. If a procedure or interface body is accessible by use or host association in a scoping unit, a dummy argument used as a keyword is permitted in a procedure reference in that scoping unit.

16.1.2.2 Resolution of Generic Identifiers

In most scoping units, generic identifiers (including names, operators, and the assignment symbol) are specified in interface blocks accessible to the scoping unit. In a type definition, they are the generic bindings for that type including any from a parent type.

If a generic procedure is accessed from a module, the rules for resolution apply to all the specific names even if some of them are inaccessible by their specific names. If a generic procedure reference applies to both a specific procedure from an interface and an accessible generic intrinsic procedure, it is the specific procedure from the interface that is referenced. If a generic name is the same as the name of a generic intrinsic procedure, the generic intrinsic procedure is not accessible if the procedures in the interface and the intrinsic procedure are not all functions or not all subroutines. The resolution of procedure references is covered in 12.8.

Resolution depends on the ability to distinguish among the dummy arguments of procedure references. Two arguments are **distinguishable** if neither is a subroutine and they differ in type, kind, or rank; otherwise, they are compatible and do not contribute to resolution. The detailed rules for distinguishability may be found in 12.5.4.1.

16.1.3 Statement and Construct Identifiers

The following identifiers have the scope of a statement or part of a statement and are known as statement entities:

- a DO variable in a DATA statement or array constructor
- an index variable in a FORALL statement
- a dummy argument in a statement function

The following identifiers have the scope of a construct and are known as construct entities

- an index variable in a FORALL construct
- an associate name in a SELECT TYPE or ASSOCIATE construct

The DO variable in an implied-do loop in a DATA statement or array constructor has the scope of the implied-do loop; that is, its scope is part of a statement. On the other hand, the DO variable of an implied-do loop in an input/output list has the scope of the subprogram in which the input/output statement appears.

Both the DO variables in DATA statements and array constructors and the index variables in FORALL statements and constructs must be of type integer; they have no other attributes. If the scoping unit that includes the DATA statement, array constructor, FORALL statement or construct also includes an IMPLICIT NONE statement or there are contrary implicit typing rules in effect, the statement and construct entities must be explicitly declared in that scoping unit.

A statement function dummy argument is a scalar with the type and type parameters it would have if it were a variable in the scoping unit of the statement function; it has no other attributes.

Within the single block of an ASSOCIATE construct and each separate block of a SELECT TYPE construct, the entity identified by an associate name has the declared type, dynamic type, type parameters, rank, and bounds of the selector; it has the scope of the block.

Rules and restrictions:

1. Within the scope of a statement entity, no other statement entity may have the same name.

2. If a global or local identifier in the scoping unit of a statement has the same name as a statement entity in that statement, the name in the scope of the statement entity is interpreted as the statement entity. Elsewhere it is interpreted as the global or local identifier.

3. If a global or local identifier in the scoping unit of a FORALL construct has the same name as an index variable in that construct, the name in the scope of the FORALL construct is interpreted as the index variable. Elsewhere it is interpreted as the global or local identifier.

4. In rules 2 and 3, the global or local identifier must not have the same name as the statement entity or index variable unless it is the name of a common block or a scalar variable.

5. The name of an index variable of a nested FORALL statement or construct must not be the same as the name of an index variable of an outer FORALL construct.

6. If a global or local name in the scoping unit of a SELECT TYPE or ASSOCIATE construct is the same as an associate name, the name in the scope of the construct is interpreted as that of the associate name. Elsewhere it is interpreted as the global or local identifier.

16.2 Association

Association is the mechanism used to establish a relationship between entities and identifiers. It permits an entity to be identified by different names in the same scoping unit or by the same or different names in different scoping units. Storage association may cause different entities to share the same storage. There are four forms of association:

1. Name association involves the use of names to establish an association.

2. Pointer association allows dynamic association of names within a scoping unit and is essentially an aliasing mechanism.

3. Storage association involves the use of storage sequences to establish an association between data objects. The association may be between two objects in the same scoping unit (EQUIVALENCE) or in different scoping units (COMMON).

4. Inheritance association is the association between the inherited components and the parent component in an extended type.

16.2.1 Name Association

There are five forms of name association: argument, use, host, linkage, and construct.

16.2.1.1 Argument Association

Argument association is explained in detail in 12.6. It establishes a correspondence between the actual argument in the scoping unit containing the procedure reference and the dummy argument in the scoping unit defining the procedure. An actual argument other than an alternate return may be the name of a variable or procedure, a designator or subobject, or an arbitrary expression. The dummy argument name is used in the procedure definition to refer to the actual argument.

The association of an array actual argument, that is not a pointer or allocatable array, may be based on array element order; this form of argument association is called **sequence association** (12.6.4.1). Sequence association is also used for arguments of default character or C character kind. It is the form of association used for interoperability (15.5.3).

When execution of a procedure terminates, the actual and dummy argument association terminates. A dummy argument of the procedure often will be associated with a different actual argument in a subsequent execution of the procedure.

16.2.1.2 Use Association

Use association causes an association between entities in the scoping unit of a module and the scoping unit containing a USE statement referring to the module. It provides access to entities specified in the module. The default situation is that all public entities in the module are accessed by the name used in the module, but entities can be renamed selectively in the USE statement and excluded with the ONLY option. When an entity is renamed by a USE statement, the original name in the module can be used as a local name for a different entity in the scoping unit containing the USE statement. There would be no name conflict. Use association is explained in 11.3.8.

16.2.1.3 Host Association

Host association permits entities in a host scoping unit to be accessible in an internal subprogram, module subprogram, or derived-type definition. An interface body has access via host association only to those entities named in IMPORT statements in the interface body. Unlike use association, with host association there is no mechanism for renaming entities. The accessed entities are known by the same name and have the same attributes as in the host, except that the VOLATILE or ASYNCHRONOUS attribute may be added to the attributes of an accessed entity that, otherwise, would not possess the attribute. Furthermore, a generic entity accessed from the host may be extended in the local scope. The entities that can be accessed by host association are named data objects, derived types, abstract interfaces, procedures, generic identifiers, and namelist groups.

The program unit containing an internal subprogram is called the **host** of the internal subprogram. The program unit (which must be a module) containing a module subprogram is called the **host** of the module subprogram. Because the internal (or module) subprogram also has a local data environment, rules are needed to determine whether a given reference inside that subprogram identifies a host entity or one local to the subprogram.

Rules and restrictions:

1. A name (of a variable or other identifiable object) is local if it is declared explicitly (other than in an ASYNCHRONOUS or VOLATILE statement) in the contained subprogram, regardless of any declarations in the host. A dummy argument in a contained subprogram is an explicit local declaration. Variables accessed by use association are considered to be explicitly declared in the contained scoping unit.

2. An entity not declared explicitly in a contained subprogram is nevertheless local (via implicit declaration) if and only if it is neither explicitly nor implicitly declared in the host nor accessed by use association.

3. If it is not local based on rules 1 and 2 above, the entity is host associated.

4. If a derived-type name in a host is inaccessible, data entities of that type or subobjects of such data entities still can be accessible.

5. If a host entity is inaccessible only because a local variable with the same name is wholly or partially initialized in a DATA statement, the local variable must not be referenced or defined prior to the DATA statement.

6. Local identifiers of a subprogram are not accessible to its host

7. The type of the function name (entry name or statement function name) in an internal or module function is determined by the explicit declaration, if any, or the implicit type rules of the function. The type for such a function is also the type in the host scoping unit.

In a language in which the attributes of all entities must be declared explicitly, the typing rules are very simple. In such languages, local declarations typically override host declarations, and any host declarations not overridden are available in the contained subprogram. Fundamentally these are the rules used in Fortran, and this clean situation can be simulated by using IMPLICIT NONE in both the host and the contained procedure; IMPLICIT NONE forces explicit declaration of all entities, and is highly recommended. In this case, each variable declared in the host is available in the host and in each contained subprogram in which it is not overridden by a local declaration; a variable declared in a contained subprogram is local to that subprogram and locally replaces any host variable with that name.

However, Fortran allows implicit declarations—use of an entity name in the execution part without an explicit declaration of that name in the specification part—and that complicates the situation. For example, suppose the variable TOTAL is referenced in an internal subprogram, and neither the internal subprogram nor its host explicitly declares TOTAL. Is TOTAL a host or local entity? Or worse, suppose that TOTAL is used in two internal subprograms in the same host, without declaration anywhere. Is there one TOTAL in the host or are there two local TOTALs? The possibilities are shown in Figure 16-2.

If TOTAL is referenced in the host, it becomes declared implicitly there and is therefore a host entity. In this case, any internal subprogram use of TOTAL accesses the

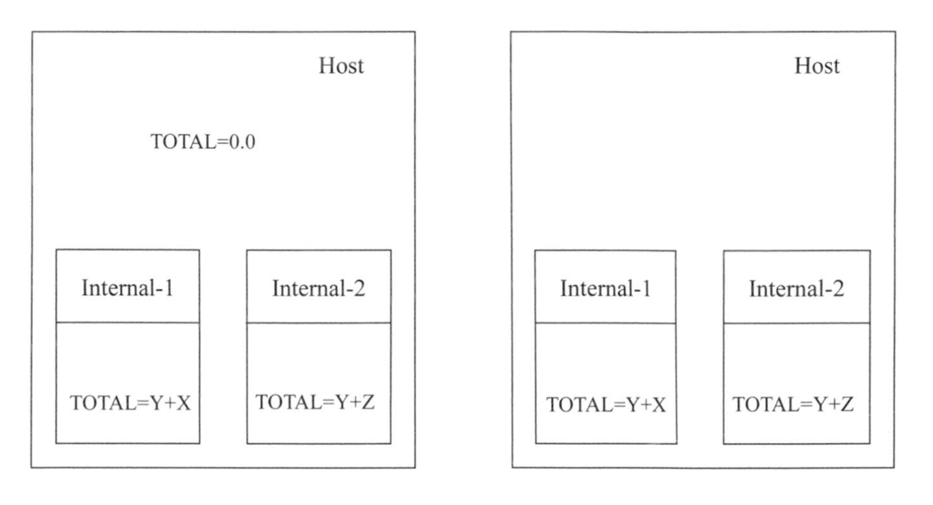

Figure 16-2 Is there one TOTAL in the host or two local TOTALs?

host entity. The situation is the same (TOTAL is a host entity) if it is declared but not referenced in the host and not declared in the internal subprogram. If TOTAL is declared in an internal subprogram, then TOTAL is local regardless of whether TOTAL is declared or referenced in the host.

Implicit declarations are governed by the implicit typing rules and the use of the IMPLICIT statement. The rules governing implicit typing are given in 5.3. The rules governing implicit typing within a contained subprogram are explained below. A particular example is detailed in Figure 16-3.

host implicit typing rules	=	host default implicit rules
	+	host IMPLICIT statements
contained procedure typing rules	=	host implicit typing rules
	+	contained procedure IMPLICIT statements

A particularly interesting case of the host associated implicit rules is when the host has IMPLICIT NONE. With IMPLICIT NONE, no other implicit statements are allowed in that scoping unit, and explicit typing is required for all data objects in the host. IMPLICIT NONE is therefore the default in the contained subprogram, although this may be modified by IMPLICIT statements actually in the contained subprogram. This can result in some of the letters designating implicit types in the contained subprogram and some not. For example, suppose that the host has IMPLICIT NONE and the contained subprogram has the following IMPLICIT statements:

```
IMPLICIT COMPLEX(C,Z)
IMPLICIT LOGICAL(J-L)
```

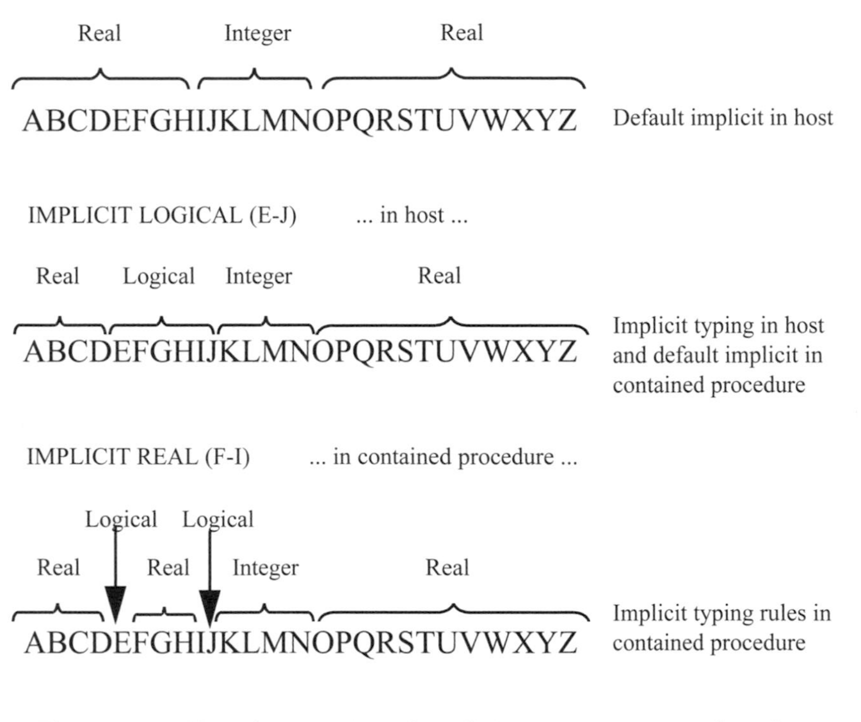

Figure 16-3 How the mapping of implicit typing progresses from host to contained procedure

Then data objects in the contained subprogram with names starting with C or Z may be declared implicitly of type complex; data objects with names starting with J, K, or L may be declared implicitly of type logical. IMPLICIT NONE continues to apply to letters A–B, D–I, and M–Y, and data object names beginning with these letters must be explicitly declared.

16.2.1.4 Linkage Association

Linkage association is between a module variable with the BIND attribute and the C variable with which it interoperates, or between a Fortran common block and the C variable with which it interoperates (15.5). Such an association is in effect throughout the execution of the program.

16.2.1.5 Construct Association

Construct association is effective only within the blocks of the SELECT TYPE and ASSOCIATE constructs. Execution of a SELECT TYPE statement establishes an association between the selector and the associate name of the construct. Execution of an ASSOCI-

ATE statement establishes an association between each selector and the corresponding associate name of the construct. The association is established before execution of any block within these constructs and is not affected by any subsequent changes to variables that were used in subscripts or substring ranges in the selector. The association persists throughout the execution of the executed block. Within the block, each selector may be accessed by the associated associate name. Upon termination of the construct, the association is terminated.

Rules and restrictions:

1. If the selector is allocatable, it must be allocated. The associate name is associated with the data object and does not have the ALLOCATABLE attribute.

2. If the selector has the POINTER attribute, it must be associated. The associate name is associated with the target of the pointer and does not have the POINTER attribute.

3. If the selector is a variable other than an array section having a vector subscript, the association is with the data object specified by the selector; otherwise, the association is with the value of the selector expression.

16.2.2 Pointer Association

Pointer association occurs when a variable with the POINTER attribute becomes associated with a target. A reference to the pointer name is then a reference to the target. A pointer is associated by pointer assignment, intrinsic assignment to an object of derived type with a pointer component, or allocation. If the pointer has deferred type parameters or shape, they are assumed from the target. If the pointer is polymorphic, its dynamic type is the dynamic type of the target. A pointer has both an association status and a definition status.

16.2.2.1 Pointer Association Status

The association status of a pointer is either defined or undefined. If the status is defined, it is either associated or disassociated. Unless a pointer is initialized (explicitly or by default), it has an initial association status of undefined. A pointer may be initialized to have an initial status of disassociated. During execution of a program, the association status may change. Whatever its association status, a pointer may always be nullified (disassociated), allocated, or pointer assigned. If the association status is disassociated or undefined, the pointer must not be referenced or deallocated. A pointer with an undefined association status must not be supplied as an argument to the ASSOCIATED intrinsic function.

There are situations where the target of a pointer may cease to exist, leaving the pointer "dangling". This can happen in several ways: a pointer specified in an outer scope may become associated in a subprogram with a nonsaved local entity that disappears when the subprogram completes execution; or, a pointer may be associated with a whole or part of a previously allocated object that is deallocated. Such dangling pointers have an undefined association status. They must not be used until their status is reestablished. A processor is not required to detect when a target ceases to exist.

16.2.2.1.1 Events that Cause a Pointer to Become Associated

1. successful execution of an ALLOCATE statement (which results in the pointer being allocated)
2. pointer assignment to a target that is either associated or has the TARGET attribute. (If the target is allocatable, it must be allocated.)

16.2.2.1.2 Events that Cause a Pointer to Become Disassociated

1. The pointer is nullified (6.7.2).
2. The pointer is deallocated (6.7.3.2).
3. The pointer is pointer-assigned (7.5.5) to a disassociated pointer.
4. The pointer is an ultimate component of an object of a type for which default initialization is specified for the component and
   a. the object is allocated.
   b. a procedure is invoked with the object as an actual argument corresponding to a nonpointer nonallocatable dummy argument with INTENT (OUT).
   c. a procedure is invoked with the object as a nonsaved nonpointer nonallocatable local object that is not accessed by use or host association.

16.2.2.1.3 Events that Cause the Association Status of a Pointer to Become Undefined

1. pointer assignment to a target that has an undefined association status
2. deallocation of the target of the pointer other than through the pointer
3. execution of MOVE_ALLOC (the allocation transfer procedure) with the pointer associated with the argument FROM and an object without the TARGET attribute associated with the argument TO
4. execution of a RETURN or END statement that causes the pointer's target to become undefined
5. execution of a RETURN or END statement that terminates a procedure defined by a subprogram in which the pointer was defined or accessed unless the pointer
   a. has the SAVE attribute
   b. is in blank common
   c. is in a named common block that appears in at least one other scoping unit that is in execution

d. is in the scoping unit of a module if the module also is accessed by another scoping unit that is in execution

e. is accessed by host association, or

f. is the return value of a function declared to have the POINTER attribute

6. invocation of a procedure with a pointer actual argument corresponding to a pointer dummy argument with INTENT (OUT)

7. invocation of a procedure with an actual argument that has an ultimate component for which default initialization is not specified corresponding to a dummy argument with INTENT (OUT)

16.2.2.1.4 Other Events that Change the Association Status of a Pointer

The effects on the association status of a pointer that becomes associated with another pointer by argument association, construct association, or host association are specified in 16.2.5. While two pointers are name associated, storage associated, or inheritance associated and the association status of one of the pointers changes, the status of the other changes accordingly. If a pointer has the VOLATILE attribute, its status may be changed by a means not specified by the program.

16.2.2.2 Pointer Definition Status

An associated pointer may be defined (have a value) or undefined depending on whether its target is defined or undefined. When a pointer is allocated, it becomes associated but undefined unless there is a SOURCE option (6.7.1) in the ALLOCATE statement for the pointer. When a pointer is pointer assigned, its association and definition status become those of the data or procedure target.

16.2.3 Storage Association

To explain the effects of some Fortran statements, such as COMMON and EQUIVALENCE, the concepts of storage and storage association are needed. These concepts are based on an abstract model of a simple, arbitrarily large computer memory, consisting of sequential memory locations.

Storage is a sequence (2.4) whose elements are called **storage units**. There are several categories of storage: numeric storage, character storage, file storage (9.1.2), and unspecified storage. The elements of numeric storage are called **numeric storage units** and similarly for the other categories of storage.

16.2.3.1 Storage Sequence

A **storage sequence** is a contiguous subsequence of storage units. It can be specified by its initial point and its size. A storage sequence may have size zero.

When storage association applies, each Fortran data object occupies a storage sequence, whose size is defined as follows.

- A nonpointer, nonallocatable scalar data object of type default integer, default real, or default logical occupies one numeric storage unit; that is, its storage sequence is size one. Default complex and double precision real objects occupy two contiguous numeric storage units. For default complex, the storage unit for the real part precedes the storage unit for the imaginary part.

- A nonpointer scalar default character object of length *len* occupies *len* contiguous character storage units.
- A nonpointer scalar character object with the C character kind(15.5.3) of length *len* occupies *len* contiguous unspecified storage units.
- A nonpointer scalar object of sequence type (4.4.10) with no type parameters occupies a sequence of storage sequences corresponding to the sequence of its ultimate components.
- A nonpointer scalar object of any type not specified above occupies a single unspecified storage unit that is different for each type and each set of type parameters and different from the unspecified unit for a character of C character kind.
- A nonpointer contiguous array occupies a sequence of contiguous storage sequences, one for each array element, in array element order (6.6.6).
- A pointer occupies a single unspecified storage unit that is different from that of any nonpointer object and is different for each combination of type, type parameters, and rank.
- Each common block has a storage sequence as described in 5.14.2.
- Each ENTRY statement in a function subprogram specifies a result that has a storage sequence as described in 12.4.5(6).
- EQUIVALENCE statements create storage sequences from the storage units of the objects making up the equivalence lists.

16.2.3.2 Association of Storage Sequences

Two storage sequences are **associated** if the initial point i of either sequence satisfies the relationship $j \leq i < j + size$, where j and $size$ are the initial point and the size of the other sequence. They are **totally associated** if they are the same storage sequences; otherwise, they are **partially associated**. For nonzero-sized storage sequences, this means that the sequences share at least one storage unit.

Two data objects are **associated**, **totally associated**, or **partially associated** if their storage sequences are associated, totally associated, or partially associated.

Partial association applies to character data and also to other composite objects in COMMON and EQUIVALENCE statements. It may exist only between

1. objects of default character or character sequence type
2. an object of default complex, double precision real, or numeric sequence type and an object of one of the default types (integer, real, logical, complex), or double precision real, or numeric sequence type

For noncharacter entities, partial association may occur only through the use of COMMON, EQUIVALENCE, or ENTRY statements. For character entities, partial association may occur only through argument association or the use of COMMON or EQUIVALENCE statements.

The effect of certain Fortran statements is described in terms of storage association. For example

```
EQUIVALENCE (X,Y)
```

causes the initial points of the storage sequences of X and Y to be the same.

In turn, the concept of storage association is used to describe the effects of defining or undefining one of the associated entities. Thus, in the example, assigning a value to X will have an effect on Y as described in 16.3.2(13) and 16.3.3(1).

If a storage unit in a storage sequence has default initialization, that storage unit must be of the same type and have the same default initialization in all associated storage sequences.

16.2.3.3 Examples of Storage Association

First, a simple example involving numeric storage units is:

```
COMPLEX :: C
REAL :: X(0:5)
EQUIVALENCE (C, X(3))
```

The storage sequence occupied by C consists of two numeric storage units, one for the real part and one for the imaginary part.

Cr	Ci

The storage sequence occupied by X consists of six numeric storage units, one for each element of the array.

X(0)	X(1)	X(2)	X(3)	X(4)	X(5)

The EQUIVALENCE statement indicates that X(3) and the real part of C occupy the same storage unit, creating the following association; items above and below each other are storage associated.

X(0)	X(1)	X(2)	X(3)	X(4)	X(5)
			Cr	Ci	

Next, consider an example with character data. If two character objects are to become associated, they must have the same kind type parameter.

```
CHARACTER(KIND=GREEK) :: A(2,2)*2 ,B(2)*3, C*5
EQUIVALENCE (A(2,1)(1:1), B(1)(2:3), C(3:5))
```

A, B, and C occupy unspecified character storage sequences of size 8, 6, and 5 respectively, and the EQUIVALENCE statement sets up the following associations.

A(1,1)(1:1)	A(1,1)(2:2)	A(2,1)(1:1)	A(2,1)(2:2)	A(1,2)1:1)	A(1,2)(2:2)	A(2,2)(1:1)	A(2,2)(2:2)
	B(1)(1:1)	B(1)(2:2)	B(1)(3:3)	B(2)(1:1)	B(2)(2:2)	B(2)(3:3)	
C(1:1)	C(2:2)	C(3:3)	C(4:4)	C(5:5)			

16.2.4 Inheritance Association

Inheritance association is between the inherited components and the corresponding components of the parent in an extended type (4.4.12). It is not affected by the accessibility of the inherited components.

16.2.5 Establishing Associations

When two entities become associated by argument, construct, or host association, the **associating entity** acquires some of the characteristics of the **pre-existing entity**. For argument association, the associating entity is the dummy argument, and the pre-existing entity is the actual argument. For construct association, the associating entity is identified by the associate name, and the pre-existing entity is the selector. For host association, the associating entity is the entity in the contained scoping unit, and the pre-existing entity is in the host. If the host is a recursive procedure, the pre-existing entity that participates in the association is the one from the innermost instance that invoked, directly or indirectly, the contained procedure.

The following rules apply for an association established by argument, construct, or host association:

1. If the associating entity is neither a pointer nor allocatable, its definition status and value (if it is defined) become the same as those of the pre-existing entity. If the entities are arrays and the association is not argument association, the bounds of the associating entity become the same as those of the pre-existing entity.

2. If the associating entity has the POINTER attribute, the pointer association status becomes the same as that of the pre-existing entity. If the pre-existing entity has an association status of associated, the associating entity becomes pointer associated with the same target, and, if they are arrays, the bounds of the associating entity become the same as those of the pre-existing entity.

3. If the associating entity has the ALLOCATABLE attribute, the allocation status becomes the same as that of the pre-existing entity. If the pre-existing entity is allocated, the definition status, value (if it is defined), values of any deferred type parameters, and bounds (if it is an array) become the same as those of the pre-existing entity. If the associating entity is polymorphic and the pre-existing entity is

allocated, the dynamic type of the associating entity becomes the same as that of the pre-existing entity.

16.3 Definition

During the execution of a program, variables are **defined** or **undefined**. If a variable is defined, it has a value. If a variable is undefined, it either does not have a value or its value is unpredictable and thus not portable. Because a processor cannot always check for undefined conditions, or it might be too costly to do so, the responsibility for avoiding the use of undefined values rests with the programmer.

An array is defined if and only if all of its elements are defined. A complex or character scalar object is defined if and only if all of its subobjects are defined. An object of user-defined type is defined if and only if all of its nonpointer components are defined. When it is necessary to refer to the components of an object as well as the components of any subobject of the object (6.5), the term **subcomponents** is used. (This term is slightly misleading as it includes all components—the top-level components as well as the components of components.)

Some variables are always defined; they are zero-sized arrays and zero-sized character strings. All other variables are initially undefined unless they are initialized. They may be initialized in DATA statements, in type declaration statements, by default initialization, or by means other than Fortran. For example, a variable may be declared in a module, but acquire an initial value from a C code. As execution proceeds, events may occur that cause a variable to become defined or undefined (see the somewhat lengthy lists below). An undefined variable must not be referenced.

16.3.1 Variable Definition Contexts

A variable may appear in contexts where it must be capable of being defined or undefined; these are called **variable definition contexts** and are listed below:

1. the variable in an assignment statement
2. an input item in a READ statement
3. a DO variable in a DO statement or an implied do variable in an input/output statement
4. a variable name in a NAMELIST statement
5. an internal file variable in a WRITE statement
6. an ID, IOMSG, IOSTAT, or SIZE specifier in an input/output statement
7. a definable variable in an INQUIRE statement
8. an allocate object in an ALLOCATE statement and a STAT variable, or ERRMSG variable in an ALLOCATE or DEALLOCATE statement

9. an actual argument in a reference to a procedure with an explicit interface if the associated dummy argument has the INTENT (OUT) or INTENT (INOUT) attribute
10. a variable that is the selector in a SELECT TYPE or ASSOCIATE construct if the associate name of that construct appears in a variable definition context
11. an object in a DEALLOCATE statement
12. a pointer in a NULLIFY statement
13. a data or procedure pointer on the left side of a pointer assignment statement

The last three are contexts in which variables may become undefined. Some variables are prohibited from appearing in a variable definition context (5.9.1, 12.7.1).

16.3.2 Events that Cause Variables to Become Defined

Variables become defined as follows:

1. Execution of an intrinsic assignment statement other than a masked array assignment or FORALL assignment causes the variable on the left of the equal sign to become defined. Execution of a defined assignment statement may cause all or part of the variable that precedes the equal sign to become defined.
2. Execution of a masked array assignment or FORALL assignment may cause some or all of the array elements in the assignment statement to become defined.
3. As execution of an input statement proceeds, each variable that is assigned a value from the input file becomes defined at the time that data is transferred to it. Execution of a WRITE statement whose unit specifier identifies an internal file causes each record that is written to become defined. (But see 16.3.3(4).)
4. Execution of a DO statement causes the DO variable, if any, to become defined.
5. In a FORALL construct or statement, the index name becomes defined when the index name value set is evaluated.
6. Beginning execution of the action specified by an implied-do in a synchronous input/output statement causes the implied-do variable to become defined. (But see 16.3.3(5).)
7. A reference to a procedure causes the entire dummy argument data object to become defined if the dummy argument does not have the INTENT (OUT) attribute and the entire associated actual argument is defined. If only a subobject of the actual argument is defined, only the corresponding subobject of the associated dummy argument is defined.
8. Execution of an input/output statement containing an IOSTAT specifier causes the specified integer variable to become defined.

9. Execution of a synchronous READ statement containing a SIZE specifier causes the specified integer variable to become defined.

10. Execution of a wait operation corresponding to an asynchronous input statement containing a SIZE specifier causes the specified integer variable to become defined.

11. Execution of an INQUIRE statement causes any variable that is assigned a value during the execution of the statement to become defined if no error condition exists.

12. If an error, end-of-file, or end-of-record condition occurs during execution of an input/output statement that has an IOMSG specifier, the specified variable becomes defined.

13. When a character storage unit becomes defined, all associated character data objects become defined.

14. When a numeric storage unit becomes defined, all associated storage units of the same type become defined.

15. When an entity of double precision real type becomes defined, all totally associated entities of double precision real type become defined.

16. When an unspecified storage unit becomes defined, all associated unspecified storage units become defined.

17. When a default complex entity becomes defined, a default real entity associated with a part of the complex entity becomes defined.

18. When both parts of a default complex entity become defined as a result of partially associated default real or default complex entities becoming defined, the default complex entity becomes defined.

19. When all components of a structure of numeric sequence type or character sequence type become defined as a result of associated objects becoming defined, the structure becomes defined.

20. Execution of an ALLOCATE or DEALLOCATE statement with a STAT specifier causes the variable specified by the STAT specifier to become defined.

21. If an error condition occurs during execution of an ALLOCATE or DEALLOCATE statement with an ERRMSG specifier, the specified variable becomes defined.

22. Allocation of an object that has a nonpointer default initialized subcomponent causes that component to become defined.

23. Execution of a pointer assignment statement that associates a pointer with a target that is defined causes the pointer to become defined.

24. Allocation of a zero-sized array causes the array to become defined.

25. Invocation of a procedure causes each automatic object of zero size in that procedure to become defined.

26. Invocation of a procedure that contains a nonsaved nonpointer nonallocatable local object causes all nonpointer default-initialized subcomponents of the object to become defined.

27. Invocation of a procedure that has a nonpointer nonallocatable INTENT (OUT) dummy argument causes all nonpointer default-initialized subcomponents of the dummy argument to become defined.

28. Invocation of a nonpointer function of a derived type causes all nonpointer default-initialized subcomponents of the function result to become defined.

29. An object with the VOLATILE attribute that is changed by a means not specified by the program becomes defined (5.7.6).

16.3.3 Events that Cause Variables to Become Undefined

Variables become undefined as follows:

1. When a variable of a given type becomes defined, all associated variables of different type become undefined. However, when a variable of type default real is associated with a part of a variable of type default complex, the complex variable does not become undefined when the real variable becomes defined and the real variable does not become undefined when the complex variable becomes defined. When part of a variable of type default complex is associated with a different part of a variable of type default complex, definition of one does not cause the other to become undefined. When variables of type default complex are associated, definition of one does not cause the other to become undefined

2. If the evaluation of a function may cause a variable to become defined, and if a reference to the function appears in an expression in which the value of the function is not needed to determine the value of the expression, the variable becomes undefined when the expression is evaluated.

3. When execution of an instance of a subprogram completes,

   a. its nonsaved local variables become undefined.

   b. nonsaved variables in a named common block that appears in the subprogram become undefined if they have been defined or redefined, unless another active scoping unit is referencing the common block.

   c. nonsaved finalizable local variables of a module may be finalized at the option of the processor if no other active scoping unit is referencing the module. In either case they become undefined.

   d. nonsaved nonfinalizable local variables of a module become undefined unless another active scoping unit is referencing the module. Note that a module is in use whenever any of its module procedures are active even if no other active scoping units reference the module. This can happen for a type-bound procedure, or a procedure invoked via a procedure pointer or a companion processor.

4. When an error condition or end-of-file condition occurs during execution of an input statement, all of the variables specified by the input list or namelist group of the statement become undefined.

5. When an error condition, end-of-file condition, or end-of-record condition occurs during execution of an input/output statement, all of the implied-do variables in the statement, if any, become undefined.

6. Execution of a defined assignment statement may cause all or part of the variable on the left the equal sign to be undefined.

7. Execution of a direct access input statement that specifies a record that has not been written previously causes all of the variables specified by the input list of the statement to become undefined.

8. Execution of an INQUIRE statement may cause the variables specified in the NAME, RECL, and NEXTREC specifiers to become undefined.

9. When a character storage unit becomes undefined, all associated character storage units become undefined.

10. When a numeric storage unit becomes undefined, all associated numeric storage units become undefined unless the undefinition is a result of defining an associated numeric storage unit of a different type (see Item 1 above).

11. When an entity of double precision real type becomes undefined, all totally associated entities of double precision real type become undefined.

12. When an unspecified storage unit becomes undefined, all associated unspecified storage units become undefined.

13. When an allocatable entity is deallocated, it becomes undefined.

14. When the allocation transfer procedure MOVE_ALLOC is executed and causes the allocation status of an allocatable entity to become deallocated, the entity becomes undefined.

15. Successful allocation of a nonzero-sized object for which default initialization has not been specified for an subcomponent causes the subcomponent to become undefined unless there is a SOURCE option (6.7.1) in the ALLOCATE statement for the object.

16. Execution of an INQUIRE statement causes all inquire specifier variables to become undefined if an error condition exists, except for the variable in an IOSTAT or IOMSG specifier.

17. When a procedure is invoked:

    a. An optional dummy argument that is not associated with an actual argument becomes undefined.

    b. A dummy argument with INTENT (OUT) becomes undefined, except for those nonpointer subcomponents of the argument for which default initialization is specified.

    c. An actual argument associated with a dummy argument with INTENT (OUT) becomes undefined, except for those nonpointer subcomponents of the argument for which default initialization is specified.

    d. A subobject of a dummy argument that does not have INTENT (OUT) specified becomes undefined if the corresponding subobject of the actual argument is undefined.

    e. The result variable of a function becomes undefined except for those nonpointer subcomponents of the result for which default initialization is specified.

18. When the association status of a pointer becomes undefined or disassociated, the pointer becomes undefined.

19. When the execution of a FORALL construct or statement completes, the index names become undefined. Note that if there is a variable in the program unit with the same name as an index name, it retains its value.

20. Execution of an asynchronous READ statement causes all of the variables specified by the input list or SIZE specifier to become undefined. Execution of an asynchronous namelist READ statement causes any variable in the namelist group to become undefined if that variable will subsequently be defined during the execution of the READ statement or the corresponding WAIT operation.

21. When execution of a RETURN or END statement causes a variable to become undefined, any variable of type C_PTR becomes undefined if its value is the C address of any part of the variable that becomes undefined.

22. When a variable with the TARGET attribute is deallocated, any variable of type C_PTR becomes undefined if its value is the C address of any part of the variable that is deallocated.

23. An object with the VOLATILE attribute may become undefined by a means not specified by the program (5.7.6).

A Standard Intrinsic Procedures

This appendix contains detailed specifications of the standard intrinsic procedures. The title of each description gives the name of the procedure and the names of its dummy arguments, with an indication of which arguments are optional. On the right side of each title line, the classification of each procedure is given; it indicates whether the procedure is an inquiry, transformational or elemental function, or an elemental or nonelemental subroutine.

Whether an actual argument is optional is indicated by braces { } in the section titles; the braces indicate that each argument within the braces is optional and may appear as an argument with or without any of the other optional arguments. The arguments within the braces follow all the rules for optional arguments (12.6.2) and their positional order is the order given in the braces. Some procedures such as SELECTED_REAL_KIND and RANDOM_SEED have additional constraints on the optional arguments; such constraints are not implied by the braces notation but are given explicitly in the description.

The text following each title follows several conventions. For functions, a description of the requirements for types and kinds of each argument are given. If the specification for an argument specifies a type but not a kind, any acceptable kind parameter value for that type is permitted. However, for inquiry functions, the arguments need not have defined values, pointer arguments need not have a defined association status, and allocatable arguments need not be allocated. For all functions, all arguments have intent IN but this intent is not stated explicitly. When a function returns an array, the array has lower bounds of 1 and strides of 1 in each dimension. In some of the descriptions of the result values and examples, these result values are described in terms of elements of the function's dummy arguments which have lower bounds of 1 and strides of 1.

For subroutines, the intent of each dummy argument is specified in the description; if those arguments have intent OUT or INOUT, the value returned by the subroutine is specified in the description of the argument.

For all elemental procedures, the type and possibly the kind of the dummy arguments are specified; whether the argument is scalar or array follows the rules for elemental procedure references (12.7.2), except for the argument named KIND which is always a scalar and specified as such.

Examples of references to each of the procedures are provided and use the following conventions:

1. The default real type has seven decimal digits of precision.
2. The mathematical notation $[a, b]$ or (a, b) is used for closed or open intervals, respectively; such intervals are used to specify ranges of values, where closed intervals include the end points and open intervals do not.

3. The character *b* indicates a blank in a character string.

4. NaN is the IEEE designation for "not a number".

All real values cannot be represented exactly in any processor; therefore, when the following text says something like "ACOS (0.1) has the value 1.4706289056333", it means that the value is a processor approximation to 1.4706289056333. The Fortran standard does not specify how accurate the approximation must be.

Many of the intrinsic functions have kind parameter arguments, named KIND. For these intrinsic functions are referenced, the actual argument corresponding to these dummy arguments named KIND must be scalar integer initialization expressions with values limited to the kind parameter values supported by the processor for the particular types.

Some of the array intrinsic functions have optional logical mask arguments, that is, actual arguments associated with dummy arguments named MASK that select the elements of one or more array arguments to be operated on. For these functions, the operations are performed only on the elements of other array arguments corresponding to the elements of the mask argument that are true. The elements corresponding to the false values of the mask arguments need not be defined. However, the mask argument affects only the value of the function; it does not affect the evaluation of the arguments that are array expressions, prior to invoking the function. Note that for the MERGE intrinsic function, the argument MASK is not optional; therefore, the preceding discussion does not apply.

ABS (A) Elemental Function

Absolute value.

A Of any integer, real, or complex type.

Result Characteristics. The same as A except that if A is complex, the result is real.

Result Value. The value of the result is $|A|$; if A is of type complex with value $x + yi$, this is equal to $\sqrt{x^2 + y^2}$.

Examples. ABS (–1) has the value 1. ABS ((3.0, 4.0)) has the value 5.0.

ACHAR (I {, KIND}) Elemental Function

Character in a specified position of the ASCII character set.

I Of type integer. Its value must be in the range of the collating values for the appropriate ASCII or processor collating sequence.

KIND Scalar integer initialization expression whose value is a character kind value.

Result Characteristics. Of type character, length 1 with kind KIND if KIND is present, or otherwise, with the kind of default character.

Result Value. The result is the ASCII character whose position in the ASCII collating sequence is I, provided I is in the range [0, 127] and the processor is capable of representing that character in the character type of the result; otherwise, the result is processor dependent. ACHAR(IACHAR(C)) has the value C for any character C representable in the default character type.

Examples. ACHAR (120) is x and ACHAR (120, SELECTED_CHAR_KIND ("ISO_10646")) is x represented as a character of the ISO 10646 character kind.

ACOS (X) — Elemental Function

Arc cosine.

X Of type real such that $|X| \leq 1$.

Result Characteristics. Same as X.

Result Value. The result is the arc cosine of X, expressed in radians, in the range $[0, \pi]$.

Examples. ACOS (1.0) is 0.0; and ACOS (–1.0d0) is π with double precision kind.

ADJUSTL (STRING) — Elemental Function

Leading blanks removed and placed on the right.

STRING Of type character.

Result Characteristics. Of the type and type parameters of STRING.

Result Value. The result is the argument value with leading blanks removed and the same number of trailing blanks inserted.

Example. ADJUSTL ("*bb*input.f90*bbb*") returns the string input.f90*bbbbb*.

ADJUSTR (STRING) — Elemental Function

Trailing blanks removed and placed on the left.

STRING Of type character.

Result Characteristics. Of the type and type parameters of STRING.

Result Value. The result is the argument value, adjusted to the right with all trailing blanks removed and the same number of leading blanks inserted.

Example. ADJUSTR ("*bb*input.f90*bbb*") returns the string *bbbbb*input.f90.

AIMAG (Z) — Elemental Function

Imaginary part of a complex value.

Z Of complex type.

Result Characteristics. Of type real with the kind of Z.

Result Value. The result is the imaginary part of the complex object Z.

Example. AIMAG ((1.0, 2.0)) is 2.0.

AINT (A {, KIND}) Elemental Function

Real value truncated to a whole number.

A Of type real.

KIND Scalar integer initialization expression whose value is a real kind value.

Result Characteristics. Of type real with kind KIND if KIND is present, or otherwise, with the kind of A.

Result Value. The largest whole number whose magnitude is less than or equal to |A| and whose sign is the same as A.

Examples. AINT (3.7) is 3.0; AINT (–3.7, P) is –3.0 with kind P.

ALL (MASK {, DIM}) Transformational Function

True if all array elements are true.

MASK Array of type logical.

DIM Scalar of type integer with a value in the range [1, n] where n is the rank of MASK. The corresponding actual argument must not be an optional dummy argument.

Result Characteristics. Logical of the kind of MASK. If DIM is absent or MASK has rank one, it is a scalar; otherwise, it is a rank n–1 array whose shape is that of MASK with the DIM dimension removed.

Result Value. If DIM is absent, or MASK has rank one, the result is the scalar equal to the conjunction ("and-ing") of all the elements of MASK or true if MASK is zero sized. If DIM is present and MASK has rank $n \geq 2$, the result is a rank n–1 array where the value of the element $(s_1, s_2, ..., s_{DIM-1}, s_{DIM+1}, ..., s_n)$ is the value of ALL (MASK $(s_1, s_2, ..., s_{DIM-1}, :, s_{DIM+1}, ..., s_n)$).

Examples. ALL ([.true., .false.]) is a scalar of default logical type with value false. If B is the array $\begin{bmatrix} 1 & 3 & 5 \\ 2 & 4 & 6 \end{bmatrix}$ and C is the array $\begin{bmatrix} 0 & 3 & 5 \\ 7 & 4 & 8 \end{bmatrix}$, ALL (B/=C,DIM=1) is [true false false], ALL (B/=C,DIM=2) is [false false], and ALL (B/=C) is false.

ALLOCATED (ARRAY) or ALLOCATED (SCALAR) Inquiry Function

Allocation status of the argument.

ARRAY An allocatable array of any type.

SCALAR An allocatable scalar of any type.

Result Characteristics. Scalar of type default logical.

Result Value. The result is true if the argument is allocated and false if unallocated.

ANINT (A {, KIND}) Elemental Function

Real value rounded to the nearest whole number.

A Of type real.

KIND Scalar integer initialization expression whose value is a real kind value.

Result Characteristics. Of type real with kind KIND if KIND is present, or otherwise, with the kind of A.

Result Value. The result is the nearest whole number to A; if there are two such whole numbers, the one of greater magnitude is returned.

Examples. ANINT (3.1) is 3.0 and ANINT (–3.5) is –4.0.

ANY (MASK {, DIM}) Transformational Function

True if any array elements are true.

MASK Array of type logical.

DIM Scalar of type integer with a value in the range [1, n] where n is the rank of MASK. The corresponding actual argument must not be an optional dummy argument.

Result Characteristics. Logical of the kind of MASK. If DIM is absent or MASK has rank one, it is a scalar; otherwise, it is a rank n–1 array whose shape is that of MASK with the DIM dimension removed.

Result Value. If DIM is absent, or MASK has rank one, the result is a scalar equal to the disjunction ("or-ing") of all the elements of MASK or false if MASK is zero sized. If DIM is present and MASK has rank n ≥ 2, the result is a rank n–1 array where the value of the element (s_1, s_2, ..., $s_{\mathrm{DIM}-1}$, $s_{\mathrm{DIM}+1}$, ..., s_n) is the value of ANY (MASK (s_1, s_2, ..., $s_{\mathrm{DIM}-1}$, :, $s_{\mathrm{DIM}+1}$, ..., s_n)).

Examples. ANY ([.true., .false.]) is a scalar of default logical type with value true. If B is the array $\begin{bmatrix} 1 & 3 & 5 \\ 2 & 4 & 6 \end{bmatrix}$ and C is the array $\begin{bmatrix} 0 & 3 & 5 \\ 7 & 4 & 8 \end{bmatrix}$, ALL (B/=C,DIM=1) is [true false false], ALL (B/=C,DIM=2) is [false false], and ALL (B/=C) is true.

ASIN (X) Elemental Function

Arc sine.

X Of type real such that |X| ≤ 1.

Result Characteristics. Same as X.

Result Value. The result is the arc sine of X, expressed in radians, in the range [–π/2, π/2].

Example. ASIN (1.0) is π/2 and ASIN (–1.0d0) is –π/2 with double precision kind.

ASSOCIATED (POINTER{, TARGET}) Inquiry Function

Association status of a pointer or its association with a specific target.

POINTER A pointer of any type. It may be a procedure pointer. It must have a defined association status.

TARGET Any procedure or data target allowed in a pointer assignment statement (7.5.5) of the form POINTER => TARGET. If TARGET is a pointer, its pointer association status must be defined.

Result Characteristics. Scalar of type default logical.

Result Value. If POINTER is disassociated, the result is false regardless of whether TARGET is present. If TARGET is absent, the result is true if the pointer POINTER is associated with a target.

If TARGET is present and:

TARGET	Conditions	Result
procedure	POINTER is associated with TARGET	true
	POINTER is not associated with TARGET	false
procedure pointer	POINTER and TARGET are associated with the same procedure	true
	POINTER and TARGET are not associated with the same procedure or either POINTER or TARGET is disassociated	false
scalar	the target associated with POINTER occupies the same storage units as the scalar target and TARGET is not a zero-sized storage sequence	true
	the target associated with POINTER does not occupy the same storage units, TARGET is a zero-sized storage sequence, or POINTER is disassociated	false
array	TARGET and the target associated with POINTER have the same shape, occupy the same storage units in array element order, and neither is zero-sized nor an array whose elements are zero-sized storage sequences	true
	TARGET and the target associated with POINTER do not have the same shape, do not occupy the same storage units in array element order, or one of them is zero-sized or an array whose elements are zero-sized storage sequences, or POINTER is disassociated	false
scalar pointer	the targets associated with POINTER and TARGET occupy the same storage units, and neither of the targets is a zero-sized storage sequence	true

TARGET	Conditions	Result
	the targets associated with POINTER and TARGET occupy different storage units, either one or both of the targets associated with POINTER or TARGET is a zero-sized storage sequence, or either one or both of POINTER or TARGET is disassociated	false
array pointer	the targets associated with POINTER and TARGET have the same shape, occupy the same storage units in array element order, and neither is zero-sized nor an array whose elements are zero-sized storage sequences	true
	the targets associated with POINTER and TARGET do not have the same shape, occupy different storage units in array element order, or one or both of them is zero-sized or an array whose elements are zero-sized storage sequences, or either one or both of POINTER or TARGET is disassociated	false

Examples. Consider the type declarations:

```
type node_type
   integer :: value
   type(node_type), pointer :: next => null ()
end type
type(node_type) ::  node
```

ASSOCIATED (node%next) is false initially but after the allocation:

```
ALLOCATE (node%next)
```

ASSOCIATED (node%next) is true.

Consider a rank one array B with the TARGET attribute and bounds 1:N, and a pointer B_PTR of rank-one of the same type. After the execution of the statement:

```
B_PTR => B(:N)
```

ASSOCIATED (B_PTR, B) is true whereas after the execution of the statement:

```
B_PTR => B(N:1:-1)
```

ASSOCIATED (B_PTR, B) is false.

ATAN (X) — Elemental Function

Arc tangent.

X Of type real.

Result Characteristics. Same as X.

Result Value. The result is the arc tangent of X, expressed in radians, in the range $[-\pi/2, \pi/2]$.

Examples. ATAN (1.0) is $\pi/4$ and ATAN (–1.0d0) is $-\pi/4$ with double precision kind.

ATAN2 (Y, X) Elemental Function

Angle in radians of a complex value X + Y*i*.

Y, X Of type real but with the same kind. Y and X must not both be zero.

Result Characteristics. Same as Y.

Result Value. The result is the principal value of the arc tangent of the complex value X + Y*i*, expressed in radians, in the range [−π, π]. For special values, the following table specifies the results:

Y	X	ATAN2 (Y, X)
>0	any	>0
>0	=0	π/2
<0	any	<0
<0	=0	−π/2
0[1]	>0	0
0[1]	<0	π
+0	>0	+0
+0	<0	π
−0	>0	−0
−0	<0	−π

Note 1: Denotes the case where the processor cannot distinguish between +0 and −0.

Examples. ATAN2 (1.0, 1.0) is π/4 and ATAN2 (0.0d0, −1.0d0) is π with double precision kind.

BIT_SIZE (I) Inquiry Function

Number of bits in the bit model.

I Scalar or array of type integer.

Result Characteristics. Scalar of the type and kind of I.

Result Value. The result is the number of bits *z* provided by the bit model.

Example. Using the particular bit model for default integer described in 13.2.1, BIT_SIZE (I) where I is of default integer type is 32.

BTEST (I, POS) Elemental Function

Test of the bit value in a specified position.

I	Of type integer.
POS	Of type integer. Its value must be in the range [0, BIT_SIZE (I)–1].

Result Characteristics. Of type default logical.

Result Value. The result is true if the bit position POS of I is 1 and false otherwise.

Example. Using the particular bit model for default integers described in 13.2.1, BTEST (16, 4) is true.

CEILING (A {, KIND}) — Elemental Function

Smallest whole number greater than or equal to a value.

A	Of type real.
KIND	Scalar integer initialization expression whose value is an integer kind value.

Result Characteristics. Of type integer with kind KIND if KIND is present, or otherwise, with the kind of default integer.

Result Value. The result is the smallest whole number greater than or equal to A.

Examples. CEILING (3.1) is 4; CEILING (3.0) is 3; and CEILING (–4.1, P) is –4 of integer kind P.

CHAR (I {, KIND}) — Elemental Function

Character in the specified position of a character set.

I	Of type integer. Its value must be in the range of the collating values for the processor collating sequence.
KIND	Scalar integer initialization expression whose value is a character kind value.

Result Characteristics. Of type character, of length 1, and with the kind KIND, if KIND is present, or otherwise, with the kind of default character.

Result Value. The result is the character whose position is I in the collating sequence for characters of the kind of the result.

Examples. CHAR (100) is the character d if the default character set uses the ASCII collating sequence. CHAR (107, SELECTED_CHAR_KIND ("ISO_10646")) is the character in position 107, namely k, of the ISO 10646 character set, provided the processor supports this ISO character set.

CMPLX (X {, Y, KIND}) — Elemental Function

Complex value.

X, Y — Of type integer, real, or complex, or a BOZ literal constant. Y is optional, need not be of the same type or kind as X but must not be present if X is complex. The actual argument corresponding to Y must not be an optional dummy argument.

KIND — Scalar integer initialization expression whose value is a real kind value.

Result Characteristics. Of type complex of kind KIND, if KIND is present, or otherwise, with kind default real.

Result Value. If Y is absent, the result is the value X if X is complex, or the complex value REAL (X, KIND) + 0*i*, if X is not complex; if Y is present, the value is that of the complex number REAL (X, KIND) + REAL (Y, KIND)*i*. If X or Y is a BOZ literal constant, the argument corresponding to X or Y is treated as if it were a constant of the type and kind of the result whose bit pattern is that given by the BOZ literal constant; the interpretation of the bit pattern is processor dependent.

Examples. CMPLX (10.0) is the complex value 10; CMPLX (10, 1) is the complex value 10 + *i*; CMPLX ((10.0, 2.0), Q) is 10 + 2*i* of real kind Q.

COMMAND_ARGUMENT_COUNT () Inquiry Function

Number of command line arguments.

Result Characteristics. Scalar of type default integer.

Result Value. The result is the number of command arguments used to invoke the execution of the program containing the reference to this function. If the processor does not support command arguments, the result is zero. If the processor has a concept of command name, the count does not include the command name.

Example. Consider the command:

```
prog -c file.f90
```

On many systems, if the program prog executed the statement:

```
number = COMMAND_ARGUMENT_COUNT ()
```

the value of number is 2. It could be 0 or 3 on some systems.

CONJG (Z) Elemental Function

Complex conjugate of a complex value.

Z — Of complex type.

Result Characteristics. Of type complex of the kind of Z.

Result Value. The result is the complex conjugate of Z.

Examples. CONJG ((1.0, –10.0)) is 1 + 10*i*. CONJG ((–2.0, 1.0_Q)) is –2 + *i* of kind Q where Q is a real kind parameter value of a real type with higher precision than the default real type.

COS (X) Elemental Function

Cosine.

X Of type real or complex.

Result Characteristics. Same as X.

Result Value. The result is the cosine of X. If X is real, X is in radians. If X is complex, the real part of X is expressed in radians.

Examples. COS (0.0) is 1; COS (B) where B is $\pi/3$ and of type default real is 0.5; and COS (A) where A is $-\pi/2 - i$ and of type complex with kind Q is the complex value 1.54308 with kind Q.

COSH (X) Elemental Function

Hyperbolic cosine.

X Of type real.

Result Characteristics. Of the type of X.

Result Value. The result is the hyperbolic cosine of X.

Examples. COSH (0.0) is 1 and COSH (–0.5d0) is 1.127626 with double precision kind.

COUNT (MASK {, DIM, KIND}) Transformational Function

Number of true array elements.

MASK	Array of type logical.
DIM	Scalar of type integer with a value in the range [1, n] where n is the rank of MASK. The corresponding actual argument must not be an optional dummy argument.
KIND	Scalar integer initialization expression whose value is an integer kind value.

Result Characteristics. Of type integer with kind KIND if KIND is present, or otherwise, with the kind of default integer. If DIM is absent or MASK has rank one, it is a scalar; otherwise, it is a rank n–1 array whose shape is that of MASK with the DIM dimension removed.

Result Value. If DIM is absent or MASK has rank one, the result is the count of all the true elements of MASK or 0 if MASK is zero sized. If DIM is present and MASK has rank $n \geq 2$, the result is a rank n–1 array; the value of the element (s_1, s_2, ..., s_{DIM-1}, s_{DIM+1}, ..., s_n) is the value of COUNT (MASK (s_1, s_2, ..., s_{DIM-1}, :, s_{DIM+1}, ..., s_n)).

Examples. COUNT ([.true., .false.]) is 1. If B is the array $\begin{bmatrix} 1 & 3 & 5 \\ 2 & 4 & 6 \end{bmatrix}$ and C is the array $\begin{bmatrix} 0 & 3 & 5 \\ 7 & 4 & 8 \end{bmatrix}$, COUNT (B/=C,DIM=1) is [2 0 1], COUNT (B/=C,DIM=2) is [1 2], and COUNT (B/=C) is 3.

CPU_TIME (TIME) Subroutine

Obtain the processor time.

TIME — Scalar of type real with INTENT (OUT). It is assigned the processor time in seconds. If the processor cannot return a meaningful time, TIME is assigned a processor-dependent negative value.

Example. Consider the following code segment:

```
call  CPU_TIME (start_time)
   . . .
call  CPU_TIME (end_time)
elapsed_time = end_time - start_time
```

The first reference to CPU_TIME saves the start time for the code segment and the second reference saves the completion time; the difference in the two times provides an estimate of the execution time, subject to an error dependent on the resolution of the timer used by the procedure CPU_TIME.

CPU time is not consistently available or interpreted on existing processors. The standard leaves the definition of CPU_TIME imprecise, permitting an implementation to define the result for CPU_TIME in a way useful for its uses. For example, when a single result on a parallel machine is not adequate, the processor might choose to return an array of times, one for each processor. Also, the start time is left imprecise so that a particular processor might measure time from midnight or from the beginning of the year. In addition, the time measured might include system overhead and might not measure anything related to "wall clock time".

The purpose of CPU_TIME is to permit the comparison of different algorithms on the same processor or to return a measure of time that determines what parts of a computation are most time consuming and to compare different methods of improving the performance of various segments of software. An implementation of CPU time is thus encouraged to return a result that permits these uses of CPU_TIME.

CSHIFT (ARRAY, SHIFT {, DIM}) Transformational Function

Circular shift of the elements of an array.

ARRAY — Array of any type.

SHIFT Scalar or array of type integer. If ARRAY is of rank 1, it is a scalar; if ARRAY is an array of rank n>1, SHIFT may be an array of rank $n-1$ whose shape is the shape of ARRAY with the DIM dimension removed.

DIM Scalar of type integer with a value in the range [1, n] where n is the rank of ARRAY. If DIM is absent, it is as if it were present with the value 1.

Result Characteristics. Same as ARRAY.

Result Value. If SHIFT or an element of SHIFT is positive, elements of the array are shifted left; if it is negative, elements are shifted right; and, if it is zero, no shift occurs.

If ARRAY is of rank one, the ith element of the result is ARRAY (1+MODULO (i+SHIFT−1, SIZE (ARRAY))); if ARRAY is of rank greater than 1, the section ($s_1, s_2, ..., s_{DIM-1}, :, s_{DIM+1}, ..., s_n$) of the result has the value CSHIFT (ARRAY ($s_1, s_2, ..., s_{DIM-1}, :, s_{DIM+1}, ..., s_n$), sh, 1), where sh is either SHIFT if SHIFT is a scalar or SHIFT ($s_1, s_2, ..., s_{DIM-1}, s_{DIM+1}, ..., s_n$), if SHIFT is an array.

Examples. If V is the array [1 2 3 4 5 6], the effect of shifting V circularly to the left by two positions is achieved by CSHIFT (V, SHIFT = 2) which has the value [3 4 5 6 1 2]; CSHIFT (V, SHIFT = −2) achieves a circular shift to the right by two positions and has the value [5 6 1 2 3 4].

The rows of an array of rank two may all be shifted by the same amount or by different amounts. If M is the array $\begin{bmatrix}1&2&3\\4&5&6\\7&8&9\end{bmatrix}$, the value of CSHIFT (M, SHIFT = −1, DIM = 2) is

$\begin{bmatrix}3&1&2\\6&4&5\\9&7&8\end{bmatrix}$, and the value of CSHIFT (M, SHIFT = [−1, 1, 0], DIM = 2) is $\begin{bmatrix}3&1&2\\5&6&4\\7&8&9\end{bmatrix}$.

DATE_AND_TIME ({DATE, TIME, ZONE, VALUES}) Subroutine

Obtain date and time information in various formats.

DATE Scalar of type default character with INTENT (OUT). If DATE is present, it is assigned the current date in the form of the string *CCYYMMDD* of type default character where the two characters *CC* are the two decimal digit century, the two characters *YY* are the two decimal digit year, the characters *MM* are the two decimal digit month (01-12), and the characters *DD* are the two decimal digit day (01-31) within the month. If the processor has no date available, DATE is assigned blanks.

TIME Scalar of type default character with INTENT (OUT). If TIME is present, it is assigned the current time in the form of a string *hhmmss.sss* of type default character where the two characters *hh* are the two decimal digit hour of the day (00-23), the characters *mm* are the

	two decimal digits for the minute in the hour, the characters *ss.sss* are the seconds *ss* and milliseconds *sss* in the hour. If no time is available, TIME is assigned blanks.
ZONE	A scalar of type default character with INTENT (OUT). If ZONE is present, it is assigned the time zone in the form of a string +*hhmm* or –*hhmm* where the two characters *hh* and *mm* represent the difference between Coordinated Universal Time (UTC) and the current time zone in hours and minutes, respectively. If this information is not available, ZONE is assigned blanks.
VALUES	Rank-one array of type default integer with INTENT (OUT) whose size is at least 8. If VALUES is present, its elements are set to the following values:

Element	**Value**
1	The year (e.g., 2008)
2	The month of the year (e.g., 12)
3	The day of the month (e.g., 15)
4	The zone as a difference from UTC in minutes (e.g., –360)
5	The hour of the day, in the range [0, 23] (e.g., 23)
6	The minutes of the hour, in the range [0, 59] (e.g., 12)
7	The seconds of the minute, in the range [0, 60] (e.g., 59)
8	The milliseconds of the second, in the range [0, 999]

In case any one of these values is not available, it is assigned –HUGE (0). UTC is defined by the ISO standard ISO 8601:1998.

Examples. The code segment:

```
integer :: time(8)
call  date_and_time (values = time)
print "(a, i5, 2i3, i5, 3i3, i4)", "The date and time are:", time
```

produces a date and time such as:

```
The date and time are: 2008  6 25 -420 13 37 48 747
```

DBLE (A) Elemental Function

Double precision value.

A	Of type integer, real, or complex, or a BOZ literal constant.

Result Characteristics. Of type real with double precision kind.

Result Value. The result is REAL (A, KIND (0.0d0)). If A is a BOZ literal constant, the argument corresponding to A is treated as if it were a constant of the type real with double precision kind whose bit pattern is that given by the BOZ literal constant; the interpretation of the bit pattern is processor dependent.

Example. DBLE (1.23) is the default real value 1.23 converted to the real type with double precision kind by extending the default representation to a real representation with double precision kind. In general, it is not equal to 1.23d0.

DIGITS (X) Inquiry Function

Number of model digits in a model number.

X Scalar or array of type integer or real.

Result Characteristics. Scalar of default integer type.

Result Value. If X is of type integer, the result is q where q is a parameter for the integer model (13.2.2) for the kind of X; if X is of type real, the result value is p where p is a parameter of the real model (13.2.3) for the kind of X.

Examples. If q=31 in the integer model (13.2.2), DIGITS (0) has the value 31; if p=24 in the real model (13.2.3), DIGITS (0.0) has the value 24.

DIM (X, Y) Elemental Function

Difference of two values if positive, or zero otherwise.

X, Y Of type integer or real, but both with the same type and the same type parameter.

Result Characteristics. Same as X.

Result Value. The result is the difference X–Y if the difference is positive; otherwise, zero.

Examples. DIM (2, 3) is 0 and DIM (2.2_Q, –3.3_Q) is 5.5 of real kind Q.

DOT_PRODUCT (VECTOR_A, VECTOR_B) Transformational Function

Dot product of two rank-one arrays.

VECTOR_A, VECTOR_B Rank-one arrays of type integer, real, complex, or logical with the same size. They must both be of numeric type or both be of logical type.

Result Characteristics. Scalar with the same type and kind of the expression VECTOR_A*VECTOR_B if the actual arguments are both of numeric type, or of the expression VECTOR_A .AND. VECTOR_B if both actual arguments are of type logical.

Result Value. The result is the dot product of the vectors VECTOR_A and VECTOR_B. If the actual arguments are of numeric type (integer, real, or complex), the result is SUM (VECTOR_A*VECTOR_B). If the actual arguments are of type logical, the result is ANY (VECTOR_A .AND. VECTOR_B). Note that if the rank-one arrays are of zero size, the result is zero if they are of numeric type or false if they are of logical type.

Examples. DOT_PRODUCT ([1.0, 2.0], [2.0, 3.0]) is 8.0; DOT_PRODUCT ([.false., .true.], [.true., .true.]) is true; DOT_PRODUCT (A(1:0), B(10:9)) is 0 with double precision kind where A is of type default real, and B is of type double precision real. DOT_PRODUCT ([(1.0, 2.0), (2.0, 3.0)], [(1.0, 1.0), (1.0, 4.0)]) is 17+4*i*.

DPROD (X, Y) — Elemental Function

Double precision product of two single precision values.

X, Y — Of default real type.

Result Characteristics. Of type real with double precision kind.

Result Value. The result is the product X×Y.

Examples. DPROD (3.0, 2.1) is 6.3 with double precision kind.

EOSHIFT (ARRAY, SHIFT {, BOUNDARY, DIM}) — Transformational Function

End-off shift of the elements of an array.

ARRAY — Array of any type.

SHIFT — Scalar or array of type integer. If ARRAY is rank one, it is a scalar. If ARRAY is an array of rank $n>1$, it may be an array of rank $n-1$ whose shape is the shape of ARRAY with the DIM dimension removed.

BOUNDARY — Of the type and type parameters of ARRAY. It is either a scalar or and array of rank $n-1$ where $n>1$ is the rank of ARRAY and of the shape of ARRAY with the DIM dimension removed. BOUNDARY is an optional dummy argument. It may be absent only if ARRAY is of one of the intrinsic types integer, real, complex, logical or character; if it is absent, it is assumed to have the value 0, 0.0, 0+0*i*, false, or the blank character with the corresponding kind and length parameter of ARRAY, respectively.

DIM — Scalar of type integer with a value in the range [1, n] where n is the rank of ARRAY. If DIM is absent, it is as if it were present with the value 1.

Result Characteristics. Same as ARRAY.

Result Value. If SHIFT or an element of SHIFT is positive, elements of the array are shifted left; if it is negative, elements are shifted right; and, if it is zero, no shift occurs.

The element $(s_1, s_2, ..., s_n)$ of the result is ARRAY $(s_1, s_2, ..., s_{DIM-1}, s_{DIM}+sh, s_{DIM+1}, ..., s_n)$ where *sh* is SHIFT, if SHIFT is a scalar, or SHIFT $(s_1, s_2, ..., s_{DIM-1}, s_{DIM+1}, ..., s_n)$ if SHIFT is an array provided LBOUND (ARRAY,DIM) $\leq s_{DIM} + sh \leq$ UBOUND (ARRAY,DIM) and is otherwise BOUNDARY or BOUNDARY $(s_1, s_2, ..., s_{DIM-1}, s_{DIM+1}, ..., s_n)$ if BOUNDARY is an array.

Examples. If V is the array [1 2 3 4 5 6], the effect of shifting V end-off to the left by 3 positions is achieved by EOSHIFT (V, SHIFT = 3) which has the value [4 5 6 0 0 0]; EOSHIFT (V, SHIFT = –2, BOUNDARY = 99) achieves an end-off shift to the right by two positions with the boundary value of 99 and has the value [99 99 1 2 3 4].

The rows of an array of rank two may all be shifted by the same amount or by different amounts and the boundary elements can be the same or different. If M is the character array $\begin{bmatrix} A & B & C \\ D & E & F \\ G & H & I \end{bmatrix}$, the value of EOSHIFT (M, SHIFT = –1, BOUNDARY = '*', DIM = 2) is

$\begin{bmatrix} * & A & B \\ * & D & E \\ * & G & H \end{bmatrix}$, and the value of EOSHIFT (M, SHIFT = [–1, 1, 0], BOUNDARY = ['*', '/', '?'],

DIM = 2) is $\begin{bmatrix} * & A & B \\ E & F & / \\ G & H & I \end{bmatrix}$.

EPSILON (X) Inquiry Function

Value that is small relative to 1 for a real value.

X Scalar or array of type real.

Result Characteristics. Scalar of the type and kind of X.

Result Value. The result is the number b^{1-p} which is a number that is almost negligible with respect to 1, for real numbers of the type of X (13.2.3).

Examples. If $b = 2$ and $p = 24$ in the real model (13.2.3), EPSILON (0.0) has the value 2^{-23}; if b=2 and p=53 in the double precision model (13.2.3), EPSILON (0.0d0) has the value 2^{-52} with double precision kind.

EXP (X) Elemental Function

Natural exponential.

X Of type real or complex.

Result Characteristics. Same as X.

Result Value. The result is e^X. If X is complex, the imaginary part of X is in radians.

Examples. EXP (0.0) is 1.0 with default real kind; EXP ((1.0, 2.0d0)) is $-1.13 + 2.47i$ with double precision kind.

EXPONENT (X) Elemental Function

Exponent of a real value.

X Of type real.

Result Characteristics. Of type default integer.

Result Value. The result is the exponent e of the model representation of X as determined by the model for real numbers (13.2.3) of the kind of X. If X is zero, the result is zero. If X is an IEEE infinity or NaN, the result is HUGE (0).

Examples. If b=2 in the real model (13.2.3), the values 1.0 and 0.125 are represented as $2^1 \times f$ and $2^{-2} \times f$ where f is the fraction one half. EXPONENT (1.0) and EXPONENT (0.125) have the values 1 and –2, respectively, of default integer kind.

EXTENDS_TYPE_OF (A, MOLD) — Inquiry Function

True if the dynamic type of the first argument is an extension type (4.4.12) of the dynamic type of the second argument.

A, MOLD Of any extensible type. If either is a pointer, it must have a defined association status.

Result Characteristics. Scalar of type default logical.

Result Value. Except when A or MOLD is unlimited polymorphic, and either A or MOLD is a disassociated pointer or unallocated allocatable, the result is true if and only if the dynamic type of A is an extension type of the dynamic type of MOLD. In the exceptional cases:

- if MOLD is unlimited polymorphic and is either a disassociated pointer or unallocated allocatable, the result is true. Otherwise,
- if A is unlimited polymorphic and is either a disassociated pointer or unallocated allocatable, the result is false.

Note that the dynamic type of a disassociated pointer or unallocated allocatable is its declared type.

Examples. Consider the example of 7.5.4 where the types painted_line_type and labeled_line_type are extensions of the type line_type and the following declarations:

```
type( line_type ) :: line, divider
type( painted_line_type )  :: a
type cartesian; real :: x, y; end type cartesian
type( cartesian) :: point
```

EXTENDS_TYPE_OF (a, line), EXTENDS_TYPE_OF (a, a), and EXTENDS_TYPE_OF (divider, line) are all true, but EXTENDS_TYPE_OF (line, a) and EXTENDS_TYPE_OF (point, line) are false; see 4.4.12 for the definition of type extension.

FLOOR (A {, KIND}) — Elemental Function

Greatest integer less than or equal to a value.

A Of type real.

KIND A scalar integer initialization expression whose value is an integer kind value.

Result Characteristics. Of type integer with kind KIND if KIND is present, or otherwise, with the kind of default integer.

Result Value. The result is the greatest integer less than or equal to A.

Examples. FLOOR (3.7) is 3; FLOOR (3.0) is 3; and FLOOR (–4.1, P) is –5 of integer kind P.

FRACTION (X) Elemental Function

Fractional part of a real value.

X Of type real.

Result Characteristics. Same as X.

Result Value. The result is the fraction $X \times b^{-e}$ as determined by the model for real numbers (13.2.3). If X is zero, the result is zero. If X is an IEEE infinity, the result is X; if X is a IEEE NaN, the result is a NaN.

Examples. If b=2 in the real model, the values 1.0 and –0.125 are represented by $2^1 \times f$ and $-2^{-2} \times f$ where f is the fraction one half. FRACTION (1.0) and FRACTION (–0.125) are 0.5 and –0.5, respectively, of kind default real.

GET_COMMAND ({COMMAND, LENGTH, STATUS}) Subroutine

Obtain the entire command initiating the program.

COMMAND Scalar of type default character with INTENT (OUT). If COMMAND is present, COMMAND is assigned the entire command that invoked the program or blanks if the command cannot be determined; the entire command includes all arguments.

LENGTH Scalar of type default integer with INTENT (OUT). If LENGTH is present, LENGTH is assigned the significant length of the command, defined as the number of characters specifying the command, including any significant trailing blanks if the processor supports significant trailing blanks. This length is determined by the command and not the length of the COMMAND argument of this procedure. If the command length cannot be determined, LENGTH is assigned zero.

STATUS Scalar of type default integer with INTENT (OUT). If STATUS is present, STATUS is assigned the value –1 if the COMMAND argument is present and has a length less than the significant length. It is assigned a processor-dependent positive value if the command retrieval fails and zero otherwise.

Example. Consider the code segment:

```
integer :: stat, leng
character(len=100) :: cmd
call  get_command (cmd, leng, stat)
print *, "command:", trim (cmd)
print *, "length:", leng
print *, "status", stat
```

Executing this code segment (called `sample`) with the command:

```
sample xx.f90 < yy
```

might print the following output:

```
command:sample xx.f90
length: 13
status: 0
```

It is processor-dependent whether redirection (that is, < yy) is part of the command The relationship between what is written as a command and what is interpreted as the command by this procedure is processor dependent; redirection is one example, and use of wildcards is another.

GET_COMMAND_ARGUMENT (NUMBER {, VALUE, LENGTH, STATUS}) — Subroutine

Obtain a specified command argument.

NUMBER — Scalar of type default integer with INTENT (IN). It specifies the ordinal number of the argument whose properties are to be determined. To be useful it should have a value between 0 and the number of arguments of the command; if there is an argument numbered 0, it is the command name that invoked the program if there is one. Except for argument number 0, the ordering of the arguments is processor dependent.

VALUE — Scalar of type default character with INTENT (OUT). If VALUE is present, it is assigned the command argument specified by NUMBER. If the value of the command argument cannot be determined, VALUE is assigned blanks.

LENGTH — Scalar of type default integer with INTENT (OUT). If LENGTH is present, it is assigned the significant length of the command argument specified by NUMBER, defined as the number of characters specifying the command argument, including any significant trailing blanks if the processor supports significant trailing blanks. This length is determined by the command argument and not the length of the VALUE argument of this procedure. If the command argument length cannot be determined, zero is assigned to LENGTH.

STATUS Scalar of type default integer with INTENT (OUT). STATUS is assigned the value –1 if the VALUE argument is present and has a length less than the significant length of the command argument. It is assigned a processor-dependent positive value if the argument retrieval fails and zero otherwise.

Example. Consider the code segment:

```
integer :: stat, leng
character(len=100) :: val
call get_command_argument (1, val, leng, stat)
print *, "argument 1:", trim (val)
print *, "length:", leng
print *, "status", stat
```

Executing this code segment (called `sample`) with the command:

```
sample xx.f90 < yy
```

might print the following output:

```
argument 1:xx.f90
length: 6
status: 0
```

GET_ENVIRONMENT_VARIABLE (NAME {, VALUE, LENGTH, STATUS, TRIM_NAME}) Subroutine

Obtain the value of a system environment variable.

NAME Scalar of type default character with INTENT (IN). The interpretation of the case of its value is processor dependent. It specifies the name of the environment variable whose value is to be determined.

VALUE Scalar of type default character with INTENT (OUT). If VALUE is present, it is assigned the character value of the environment variable specified by NAME or blanks if the processor can not determine the value or the value of the environment variable NAME does not exist.

LENGTH Scalar of type default integer with INTENT (OUT). If LENGTH is present, it is assigned the length of the value of the environment variable if the environment variable NAME exists and 0 otherwise.

STATUS Scalar of type default integer with INTENT (OUT). It is assigned values as shown in Table A-1.

TRIM_NAME Scalar of type logical with INTENT (IN). If TRIM_NAME is true or is not present, the trailing blanks are not considered part of the name of the environment variable; if TRIM_NAME is present with the value false, it specifies that trailing blanks in NAME are considered significant if the processor supports trailing blanks in the name of an environment variable.

Table A-1 Values assigned to STATUS

Value	Condition
–1	The VALUE argument is present and has a length less than the significant length of the environment variable value.
0	The environmental variable NAME exists and either has no value or its value was successfully retrieved.
1	The specified environment variable NAME does not exist.
2	The processor does not support environment variables.
>2	Some other error condition occurs.

Example. Consider the code segment:

```
integer :: stat, leng
character(len=100) :: val
call  get_environment_variable ("SSH_ASKPASS, val, leng, stat)
print *, "value of SSH_ASKPASS:", trim (val)
print *, "length:", leng
print *, "status", stat
```

Executing this code segment (called `sample`) with the command:

```
sample
```

might print the following output:

```
value of ASK_PASS:/usr/libexec/openssh/gnome-ssh-askpass
length: 38
status: 0
```

HUGE (X) — Inquiry Function

Largest number in the real or integer model.

X — Scalar or array of type integer or real.

Result Characteristics. Scalar of the type and kind of X.

Result Value. The result is the number r^q-1 for integers (13.2.2) or $(1-b^{-p})b^{e_{max}}$ for real numbers (13.2.3) of the type and kind of X, namely the largest model number of the type and kind of X.

Examples. If r=2 and q=31 in the integer model (13.2.2), HUGE (0) has the value $2^{31}-1$ of default integer type; if b=2, p=24 and e_{max}=128 in the real model (13.2.3), HUGE (0.0) has the value $(1-2^{-24})2^{128}$ or 0.3403×10^{39}, of default real type.

IACHAR (C {, KIND}) — Elemental Function

Position of a specified character in the ASCII character set.

C — Of type character and of length one.

KIND Scalar integer initialization expression whose value is an integer kind value.

Result Characteristics. Of type integer with kind KIND if KIND is present, or otherwise, with the kind of default integer.

Result Value. If C is a character in the ASCII collating sequence, the result is the position of the character C in the ASCII collating sequence and is an integer value in the range [0, 127]; otherwise, the result is a processor-dependent integer value. The results are consistent with the results returned by the character intrinsic functions LGE, LGT, LLE, and LLT; that is, for example, if LLT(C1,C2) is true, IACHAR (C1) < IACHAR (C2).

Examples. IACHAR ("a") is 97; IACHAR ("a", P) is 97 with integer kind P.

IAND (I, J) Elemental Function

Logical AND of two integers.

I, J Of type integer with the same kind.

Result Characteristics. Same as I.

Result Value. The result is the value obtained by "and-ing" corresponding bit positions of I and J; that is, the *k*th bit position of the result is 1 if the *k*th bit positions in both I and J are 1, and 0 otherwise.

Examples. Using the particular bit model for default integers described in 13.2.1, IAND (16, 8) is 0. IAND (INT (B"1010"), INT (B"1100")) is 8, which is the same value as INT (B"1000").

IBCLR (I, POS) Elemental Function

Value with a specified bit set to zero.

I Of type integer.

POS Of type integer. Its value must be in the range [0, BIT_SIZE (I)−1].

Result Characteristics. Same as I.

Result Value. The result is the integer obtained by setting the bit position POS of I to zero.

Example. Using the particular bit model for default integers described in 13.2.1, IBCLR ([24,16], 4) is [8 0] of kind default integer; IBCLR (INT (B'1111'), 0) is 14 which has the same value as the BOZ literal constant B'1110'.

IBITS (I, POS, LEN) Elemental Function

Specific bits extracted from an integer value.

I Of type integer.

POS Of type integer. Its value must be in the range [0, BIT_SIZE (I)−LEN−1].

LEN Of type integer with a nonnegative value.

Result Characteristics. Same as I.

Result Value. The result is the integer value obtained by extracting the sequence of LEN bits from I starting in bit position POS, right adjusting them in the result, and setting all other bits in the result to zero.

Example. Using the particular bit model for default integers described in 13.2.1, IBITS (16, 4, 2) is 1; IBITS (INT (B'10110'), 1, 2) is 3 which has the same value as the BOZ literal constant B'11'.

IBSET (I, POS) Elemental Function

Value with a specified bit set to one.

I Of type integer.

POS Of type integer. Its value must be in the range [0, BIT_SIZE (I) – 1].

Result Characteristics. Same as I.

Result Value. The result is the integer value obtained by setting the bit position POS of I to one.

Examples. Using the particular bit model for default integers described in 13.2.1, IBSET (24, [4, 2]) is [24 28] of kind default integer; IBSET (INT (B"0000"), 2) is 4 which has the same value as the BOZ literal constant B"0100".

ICHAR (C {, KIND}) Elemental Function

Position of a specified character in a character set.

C Of type character and of length one. It must be a character capable of being represented in the character type and kind of C.

KIND Scalar integer initialization expression whose value is an integer kind value.

Result Characteristics. Of type integer with kind KIND if KIND is present, or otherwise, with the kind of default integer.

Result Value. The result is the position of the character C in the collating sequence for characters of the kind of C. Additionally, for any two characters C1 and C2 with the same kind capable of being represented by the processor, C1≤C2 if and only if ICHAR (C1) ≤ ICHAR (C2), and C1=C2 if and only if ICHAR (C1) = ICHAR (C2).

Examples. ICHAR ("a") is 97, if the processor uses the ASCII representation in its default character kind. On the same processor, ICHAR ("a", P) is the integer 97 of integer kind P.

IEOR (I, J) Elemental Function

Logical exclusive-OR of two integers.

I, J Of type integer with the same kind.

Result Characteristics. Same as I.

Result Value. The result is the value obtained by "exclusive or-ing" corresponding bit positions of I and J; that is, the *k*th bit position of the result is 1 if exactly one of the *k*th bit positions in I and J is 1, and 0 otherwise.

Examples. Using the particular bit model for default integers described in 13.2.1, IEOR (24, 16) is 8; IEOR (INT (B"1100"), INT (B"0110")) is 10, which has the same value as the BOZ literal constant B"1010".

INDEX (STRING, SUBSTRING {, BACK, KIND}) Elemental Function

Location of a given substring in a character string.

STRING, SUBSTRING Of type character with the same kind.

BACK Of type logical.

KIND Scalar integer initialization expression whose value is an integer kind value.

Result Characteristics. Of type integer with kind KIND, if KIND is present, or otherwise, with the kind of default integer.

Result Value. The result is the beginning index position in STRING of the substring SUBSTRING, or 0 if there is no such position. If BACK is present with the value false or is absent, the result is:

- the beginning index position of the first occurrence of substring SUBSTRING in STRING.
- 0 if the length of STRING is less than the length of SUBSTRING.
- 1 if the length of SUBSTRING is 0.

If BACK is present with the value true, the result is:

- the beginning index position of the last occurrence of substring SUBSTRING in STRING.
- 0 if the length of STRING is less than the length of SUBSTRING.
- LEN (STRING) + 1 if the length of SUBSTRING is 0.

Examples. INDEX ("input_file.f90", "f") is 7; INDEX ("input_file.f90", "f", .true.) is 12; and INDEX ("input_file.f90", "i", BACK=.true., KIND=P) is 8 of integer kind P; INDEX ("input_file.f90", "pt") is 0 of default integer kind.

When the length of the substring is greater than the length of the string, the substring is not present in the string; consequently, zero is returned. For example, INDEX ("f90", "input_file.f90") is zero. The empty string is at both ends of any string; thus, INDEX ("file", "") is 1 and INDEX ("file", "", .true.) is 5.

INT (A {, KIND}) Elemental Function

Truncated integer value.

A — Of type integer, real, or complex, or a BOZ literal constant.

KIND — Scalar integer initialization expression whose value is an integer kind value.

Result Characteristics. Of type integer with kind KIND if KIND is present, or otherwise, of the kind of default integer.

Result Value. The result is the truncated value of A or of the real part of A if A is complex, represented as an integer. If |A|<1, the truncated value is 0; otherwise, the truncated value is the largest integral value in magnitude, smaller than or equal to A in magnitude and with the same sign as A. If A is a BOZ literal constant, it is treated as if it were a constant of the integer type with the largest decimal exponent range supported by the processor and whose bit pattern is that given by the BOZ literal constant; the interpretation of the bit pattern is processor dependent.

Examples. INT ((3.6,1)) is 3; INT (–3.6) is –3. INT (0.5, P) is 0 of integer kind P. INT(–5) and INT (–5.0) are –5; INT (5.0) is 5. INT (B'101') is 5.

IOR (I, J) — Elemental Function

Logical inclusive-OR of two integers.

I, J — Of type integer with the same kind.

Result Characteristics. Same as I.

Result Value. The result is the value obtained by "inclusive or-ing" corresponding bit positions of I and J; that is, the *k*th bit position of the result is 1 if either one or both of the *k*th bit positions in I and J are 1, and 0 otherwise.

Examples. Using the particular bit model for default integers described in 13.2.1, IOR (24, 16) is 24; IOR (INT (B"1100"), INT (B"0110")) is 14 which has the same value as the BOZ literal constant B"1110".

ISHFT (I, SHIFT) — Elemental Function

Logical end-off shift of an integer.

I — Of type integer.

SHIFT — Of type integer. Its absolute value must be less than BIT_SIZE (I).

Result Characteristics. Same as I.

Result Value. The result is the value of I shifted left by the SHIFT bits if SHIFT is positive, or shifted right by |SHIFT| bits if SHIFT is negative. The bits shifted out are lost and the bits shifted in are zero. No shift is performed if SHIFT is zero.

Examples. Using the particular bit model for default integers described in 13.2.1, ISHFT (24, 2) is 96; ISHFT (24, –1) is 12; ISHFT (INT (B'0011'), 2) is 12 which has the same value as the BOZ literal constant B'1100'.

ISHFTC (I, SHIFT {, SIZE}) Elemental Function

Logical circular shift in a field of an integer.

I	Of type integer.
SHIFT	Of type integer with absolute value less than or equal to the value of SIZE.
SIZE	Of type integer with a positive value less than or equal to BIT_SIZE (I). If it is not present, it is treated as if it were present with the value BIT_SIZE (I).

Result Characteristics. Same as I.

Result Value. The result is the rightmost SIZE bits of I circularly shifted left by SHIFT bits if SHIFT is positive, or circularly shifted right by |SHIFT| bits if SHIFT is negative. The bits shifted out of the field are brought in to the opposite end of the field of size SIZE; the unshifted bits are not altered.

Examples. Using the particular bit model for default integers described in 13.2.1 with $z = 32$, ISHFTC (24, 2) is 96; ISHFTC (24, –1) is 12; ISHFTC (1, –1, 10) is 512; ISHFTC (1, –2) is 2^{30}; and ISHFTC (INT (B'10110111'), 2, 4) has the value of the BOZ literal constant B'10111101'.

IS_IOSTAT_END (I) Elemental Function

True if a value indicates an end-of-file IOSTAT condition.

I	Of type integer.

Result Characteristics. Of type default logical.

Result Value. The result is true if the value of I is the value of the IOSTAT specifier that indicates an end-of-file condition and false otherwise.

Examples. IS_IOSTAT_END (–1) is true on some Fortran processors but IS_IOSTAT_END (1) is always false.

IS_IOSTAT _EOR (I) Elemental Function

True if a value indicates an end-of-record IOSTAT condition.

I	Of type integer.

Result Characteristics. Of type default logical.

Result Value. The result is true if the value of I is the value of the IOSTAT specifier that indicates an end-of-record condition and false otherwise.

Examples. IS_IOSTAT_EOR (–2) is true on some Fortran processors, but IS_IOSTAT_EOR (2) is always false.

KIND (X) Inquiry Function

Kind parameter.

X — Scalar or array of any intrinsic type.

Result Characteristics. Scalar of type default integer.

Result Value. The result is the kind type parameter value of X.

Examples. If the value 1 is a supported kind value for a real entity and X is declared by the statement:

```
real(kind=1)  X
```

KIND (X) has the value 1. If 1 is also the kind value of the default real type, KIND (0.0) is 1.

LBOUND (ARRAY {, DIM, KIND}) — Inquiry Function

Lower bound(s) of an array or a dimension of an array.

ARRAY	An array of any type. If it is allocatable, it must be allocated; if it is a pointer, it must be associated.
DIM	Scalar of type integer with a value in the range [1, *n*] where *n* is the rank of ARRAY. The corresponding actual argument must not be an optional dummy argument.
KIND	Scalar integer initialization expression whose value is an integer kind value.

Result Characteristics. Scalar or rank-one array of type integer with kind KIND if KIND is present, or otherwise, with the kind of default integer.

Result Value. If DIM is present, the result is a scalar integer representing the lower bound of ARRAY in the DIM dimension as defined in 5.4, 6.6, and 7.2.4. If DIM is absent, the result is a rank-one array of the lower bounds of each dimension of ARRAY.

Examples. For the following statements:

```
real, target :: A(2:3, 7:10)
real, pointer, dimension(:,:) :: B, C, D
B => A
C => A(:,:)
allocate ( D(-3:3,-7:7) )
```

LBOUND (A) is [2 7], LBOUND (A, DIM=2) is 7, LBOUND (B) is [2 7], LBOUND (C) is [1 1], and LBOUND (D) is [–3 –7].

LEN (STRING {, KIND}) — Inquiry Function

Length of a character string.

STRING	Scalar or array of type character. If it is an unallocated allocatable or a pointer that is not associated, it must not have a deferred length type parameter.
KIND	Scalar integer initialization expression whose value is an integer kind value.

Result Characteristics. Scalar of type integer with kind KIND if KIND is present, or otherwise, with the kind of default integer.

Result Value. The result value is the length of STRING as defined in 7.2.2 if STRING is scalar, and of an element of STRING if STRING is an array.

Examples. If STR and STR_ARRAY are declared by the statement:

```
character(kind=1,len=25) :: STR, STR_ARRAY(10,10)
```

LEN (STR) and LEN (STR_ARRAY) both return the value 25. LEN ("ABbb") and LEN ("AB" // "CD") both return the value 4.

LEN_TRIM (STRING {, KIND}) Elemental Function

Length of a string after trailing blanks have been removed.

STRING Of type character.

KIND Scalar integer initialization expression whose value is an integer kind value.

Result Characteristics. Of type integer with kind KIND if KIND is present, or otherwise, with the kind of default integer.

Result Value. The result is the length of STRING not counting trailing blanks.

Examples. LEN_TRIM ("*bb*input*bbbb*") is 7; LEN_TRIM ("*bb*input", P) is 7 with kind P; LEN_TRIM ("*bbbb*") is 0; LEN_TRIM (GREEK_"αβ*bb*") is 2, where GREEK is a named integer constant whose value is the character kind value of a Greek character set if the processor supports this character set.

LGE (STRING_A, STRING_B) Elemental Function

Greater than or equal to comparison based on the ASCII collating sequence.

STRING_A, STRING_B

Of type default character.

Result Characteristics. Of type default logical.

Result Value. The result is true if STRING_A follows or is equal to STRING_B in the ASCII collating sequence, or otherwise, false. The shorter string is padded on the right with blanks to the length of the longer string before the comparison is performed. The result is processor dependent if either string contains a non-ASCII character.

Examples. LGE ("a", "k") is false; LGE ("x*b*", "x") is true; and LGE ("ak", "") is true. Note that the result is true if both STRING_A and STRING_B are of zero length.

LGT (STRING_A, STRING_B) Elemental Function

Greater than comparison based on the ASCII collating sequence.

STRING_A, STRING_B

Of type default character.

Result Characteristics. Of type default logical.

Result Value. The result is true if STRING_A follows STRING_B in the ASCII collating sequence, or otherwise, false. The shorter string is padded on the right with blanks to the length of the longer string before the comparison is performed. The result is processor dependent if either string contains a non-ASCII character.

Examples. LGT ("a", "k") is false; LGT ("x*b*", "x") is false; and LGT ("ak", "") is true. Note that the result is false if both STRING_A and STRING_B are of zero length.

LLE (STRING_A, STRING_B) — Elemental Function

Less than or equal to comparison based on the ASCII collating sequence.

STRING_A, STRING_B

Of type default character.

Result Characteristics. Of type default logical.

Result Value. The result is true if STRING_A precedes or is equal to STRING_B in the ASCII collating sequence, or otherwise, false. The shorter string is padded on the right with blanks to the length of the longer string before the comparison is performed. The result is processor dependent if either string contains a non-ASCII character.

Examples. LLE ("a", "k") is true; LLE ("x*b*", "x") is true; and LLE ("ak", "") is false. Note that the result is true if both STRING_A and STRING_B are of zero length.

LLT (STRING_A, STRING_B) — Elemental Function

Less than or equal to comparison based on the ASCII collating sequence.

STRING_A, STRING_B

Of type default character.

Result Characteristics. Of type default logical.

Result Value. The result is true if STRING_A precedes STRING_B in the ASCII collating sequence, or otherwise, false. The shorter string is padded on the right with blanks to the length of the longer string before the comparison is performed. The result is processor dependent if either string contains a non-ASCII character.

Examples. LLT ("a", "k") is true; LLT ("x*b*", "x") is false; and LLT ("ak", "") is false. Note that the result is false if both STRING_A and STRING_B are of zero length.

LOG (X) — Elemental Function

Natural logarithm.

X — Of type real or complex. If it is of type real, its value must be positive. If it is of type complex, it must not be zero.

Result Characteristics. Same as X.

Result Value. The result is the natural logarithm of X. If X is of type complex, the result is the principal value, where the imaginary part of the result is in the range $[-\pi, \pi]$; if, in addition, the real part of X is negative and the imaginary part of X is zero, the imaginary part of the result is π in magnitude, either π if the imaginary part of X is positive zero or the processor cannot determine the sign of zero, or $-\pi$ if the imaginary part of X is negative zero.

Examples. LOG (1.0) is 0.0; LOG ((–1.0d0, 0.0)) is πi.

LOG10 (X) — Elemental Function

Logarithm to the base 10.

X — Of type real; it must be a positive.

Result Characteristics. Same as X.

Result Value. The result is the base 10 logarithm of X.

Examples. LOG10 (1.0) is 0.0; LOG10 (10.0d0) is 1.0 with kind double precision; LOG10 ([1.0_Q, 10.0_Q, 100.0_Q]) is [0.0 1.0 2.0] with kind Q.

LOGICAL (L {, KIND}) — Elemental Function

Logical value.

L — Of logical type.

KIND — Scalar integer initialization expression whose value is a logical kind value.

Result Characteristics. Of type logical with kind KIND if KIND is present, or otherwise, with the kind of default logical.

Result Value. The result is L.

Example. LOGICAL (.true., T) is the value true of kind T, where T is a named integer constant with a logical kind value.

MATMUL (MATRIX_A, MATRIX_B) — Transformational Function

Matrix multiplication.

MATRIX_A, MATRIX_B — Arrays of type integer, real, complex or logical of rank one or two, with at least one of them of rank two. The last (or only) dimension of MATRIX_A must have the same extent as the first (or only) dimension of MATRIX_B. They must be both of numeric type or both of logical type.

Result Characteristics. Rank-one or rank-two array of the type and kind of the expression in the column headed **Value** and of shape specified by the column headed **Shape of Result** in the table below.

Result Value. The result is the matrix (linear algebra) product of the arguments, treated as vectors (when rank one) or matrices (when rank two). The shape and value of the result MATMUL (A, B) is:

Shape of A	Shape of B	Shape of result	Result subscript	Value of element result	
				Arithmetic	Logical
$[m\ n]$	$[n\ p]$	$[m\ p]$	(i, j)	SUM (A(i, :)*B(:, j))	ANY (A(i, :).AND.B(:, j))
$[m]$	$[m\ p]$	$[p]$	(j)	SUM (A(:)*B(:, j))	ANY (A(:).AND.B(:, j))
$[m\ n]$	$[n]$	$[m]$	(i)	SUM (A(i, :)*B(:))	ANY (A(i, :).AND.B(:))

Examples. Let A and B be the matrices $\begin{bmatrix} 1 & 2 & 3 \\ 2 & 3 & 4 \end{bmatrix}$ and $\begin{bmatrix} 1 & 2 \\ 2 & 3 \\ 3 & 4 \end{bmatrix}$; let X and Y have the values [1 2] and [1 2 3].

The result of MATMUL (A, B) is the matrix-matrix product AB with the value $\begin{bmatrix} 14 & 20 \\ 20 & 29 \end{bmatrix}$.

The result of MATMUL (X, A) is the vector-matrix product XA with the value [5 8 11].

The result of MATMUL (A, Y) is the matrix-vector product AY with the value [14 20].

MAX (A1, A2 {A3, ... }) Elemental Function

Maximum of specified values.

A1, A2, ... Of type integer, real, or character with all of the same type and kind type parameter. If the arguments are of type character, they may be of differing lengths.

Result Characteristics. Same as A1. For A1 of type character, the length of the result is the length of the longest argument.

Result Value. The result is the maximum of all objects A1, A2, ... For type character arguments, the comparisons are made using the intrinsic character relational operators. If the maximum argument is shorter in character length than the longest argument, the result is the maximum argument padded with blanks on the right to equal the length of the longest argument.

Examples. MAX (1.0, 2.0, 30.0) is 30.0; MAX (1, 2, –3) is 2; and MAX ([–1.0d0, –10.0d0], [–2.0d0, 20.1d0]) is [–1 20.1] with double precision kind. MAX ("a", "b*b*", "c") is "c*b*".

MAXEXPONENT (X) Inquiry Function

Maximum value of the model exponent.

X Scalar or array of type real.

Result Characteristics. Scalar of default integer type.

Result Value. The result is the integer e_{max} for real numbers (13.2.3) of the type and kind of X, namely the maximum exponent for model numbers of the type and kind of X.

Examples. If e_{max}=128 in the real model (13.2.3), MAXEXPONENT (0.0) is the value 128 of default integer kind.

MAXLOC (ARRAY, DIM {, MASK, KIND}) or
MAXLOC (ARRAY {, MASK, KIND}) Transformational Function

Location of the first maximum element of an array.

ARRAY	Array of type integer, real, or character.
DIM	Scalar of type integer in the range [1, n] where n is the rank of ARRAY. The corresponding actual argument must not be an optional dummy argument.
MASK	Of type logical, conformable with ARRAY.
KIND	Scalar integer initialization expression whose value is an integer kind value.

Result Characteristics. Of type integer with kind KIND if KIND is present, or otherwise, with the kind of default integer. If DIM is absent, it is a rank-one array of size equal to the rank n of ARRAY; if DIM is present and ARRAY is of rank one, it is a scalar; otherwise, it is a rank-one array of size n–1.

Result Value. The result is:

- if MASK and DIM are not present, a rank-one array whose element values are the subscripts of the first maximum in array element order of all elements of ARRAY provided ARRAY is of nonzero size. The ith subscript is in the range [1, e_i] where e_i is the extent of ith dimension of ARRAY. If ARRAY is zero sized, all elements of the result are zero.
- if MASK is present and DIM is not present, a rank-one array whose element values are the subscripts of the first maximum in array element order of all elements of ARRAY corresponding to a true element of MASK. The ith subscript is in the range [1, e_i] where e_i is the extent of ith dimension of ARRAY. If ARRAY is zero sized or all elements of MASK are false, all elements of the result are zero.
- if DIM is present and ARRAY has rank one, MAXLOC (ARRAY[, MASK = MASK]). If DIM is present and the rank of ARRAY is at least two, the value of the element (s_1, s_2,..., $s_{\mathrm{DIM}-1}$, $s_{\mathrm{DIM}+1}$,..., s_n) of the result is equal to MAXLOC (ARRAY (s_1, s_2,..., $s_{\mathrm{DIM}-1}$, :, $s_{\mathrm{DIM}+1}$,..., s_n), DIM=1 [, MASK = MASK (s_1, s_2,..., $s_{\mathrm{DIM}-1}$, :, $s_{\mathrm{DIM}+1}$,..., s_n)]).

- If ARRAY is of type character, the comparisons to determine the maximum value are the same as those used by the intrinsic relational operators for operands of the same type as ARRAY.

Examples. MAXLOC ([1, 2, 4, 3]) is the rank-one array [3] of size 1; MAXLOC ([4, 2, 4, 3]) is [1]; MAXLOC ([1, 2, 4, 3], 1) is the scalar 3; MAXLOC ([1, 2, 4, 3], [.true., .true., .false., .true.]) is the rank-one array [4] of size 1; and MAXLOC ([1, 2, 4, 3], 1, [.true., .true., .false., .true.], P) is the scalar 4 of integer kind P.

Given the following statements:

```
logical, parameter :: T = .true., F = .false.
integer, dimension(11:12,31:33) :: a
logical, dimension(21:22,21:23) :: m
integer, dimension(2) :: r
r = maxloc(a,dim=2,mask=m)
```

with the values shown in Table A-2.

Table A-2 Values for MAXLOC

a	m	a's subscripts	maxloc's subscripts
1 **4** 7	T T F	(11, 31) **(11, 32)** (11, 33)	(1, 1) **(1, 2)** (1, 3)
2 **5** 8	F T F	(12, 31) **(12, 32)** (12, 33)	(2, 1) **(2, 2)** (2, 3)
3 6 **9**	T F T	(13, 31) (13, 32) **(13, 33)**	(3, 1) (3, 2) **(3, 3)**

When executed, MAXLOC returns [2 2 3] which is the list of indices indicating the maximum values of those being selected along the rows of a, thus indicating the 4 in row 1, the 5 in row 2, and the 9 in row 3 of array a (indicated in bold font).

MAXVAL (ARRAY, DIM {, MASK}) or MAXVAL (ARRAY {, MASK}) — Transformational Function

Maximum value of array elements.

ARRAY — Array of type integer, real, or character.

DIM — Scalar of type integer with a value in the range [1, n] where n is the rank of ARRAY. The corresponding actual argument must not be an optional dummy argument.

MASK — Of type logical, conformable with ARRAY.

Result Characteristics. Same type and type parameters as ARRAY. If DIM is absent or ARRAY has rank one, it is a scalar; otherwise, it is a rank n–1 array whose shape is that of ARRAY with the DIM dimension removed.

Result Value. The result is:

- if MASK and DIM are not present, the maximum of all elements of ARRAY provided ARRAY is of nonzero size. If ARRAY is zero sized and is of a numeric type, the result is the negative value of largest magnitude supported by the processor of the same type and kind as ARRAY; if ARRAY is zero sized and of type character, the result is the value of the string of the length of ARRAY, with each character equal to CHAR (0, KIND = KIND (ARRAY)).

- if MASK is present and DIM is not present, MAXVAL (PACK (ARRAY, MASK)).
- if DIM is present and ARRAY has rank one, MAXVAL (ARRAY, MASK = MASK). If DIM is present and the rank of ARRAY is at least two, the value of the element (s_1, s_2, ..., s_{DIM-1}, s_{DIM+1}, ..., s_n) is the value MAXVAL (ARRAY (s_1, s_2, ..., s_{DIM-1}, :, s_{DIM+1}, ..., s_n) [, MASK = MASK (s_1, s_2, ..., s_{DIM-1}, :, s_{DIM+1}, ..., s_n)]).
- if ARRAY is of type character, the comparisons to determine the maximum value are the same as those used by the intrinsic relational operators for operands of the same type as ARRAY.

Examples. MAXVAL ([–1, 5, 2]) is 5; MAXVAL (A, 1), where A = $\begin{bmatrix} 1 & 2 \\ -2 & -3 \end{bmatrix}$ is [1 2];

MAXVAL (A, 2, M) where M = $\begin{bmatrix} \textit{false} & \textit{true} \\ \textit{true} & \textit{true} \end{bmatrix}$ is [2 –2]; and MAXVAL (A, M) is 2.

MERGE (TSOURCE, FSOURCE, MASK) Elemental Function

Selection of values under control of a mask.

TSOURCE, FSOURCE
Of any type but with the same type and type parameters.

MASK Of type logical.

Result Characteristics. Same as TSOURCE.

Result Value. The result is TSOURCE if MASK is true and FSOURCE if MASK is false.

Examples. MERGE (1, 2, .false.) is 2; MERGE ([1, 2, 3], [4, 5, 6], .true.) is [1 2 3]; MERGE ([1, 2, 3], [4, 5, 6], [.false., .true., .false.]) is [4 2 6]; if TSOURCE is $\begin{bmatrix} 1 & 6 & 5 \\ 7 & 4 & 6 \end{bmatrix}$,

FSOURCE is $\begin{bmatrix} 0 & 3 & 2 \\ 7 & 4 & 8 \end{bmatrix}$, and MASK is $\begin{bmatrix} \textit{true} & \textit{false} & \textit{true} \\ \textit{false} & \textit{false} & \textit{true} \end{bmatrix}$, MERGE (TSOURCE, FSOURCE,

MASK) is $\begin{bmatrix} 1 & 3 & 5 \\ 7 & 4 & 6 \end{bmatrix}$.

MIN (A1, A2 {A3, ... }) Elemental Function

Minimum of specified values.

A1, A2, ... Of type integer, real, or character with the same type and kind type parameter. If the arguments are of type character, they may be of differing lengths.

Result Characteristics. Same as A1. For A1 of type character, the length of the result is of the length of the longest argument.

Result Value. The result is the minimum of all objects A1, A2, ... For type character arguments, the comparisons are made using the intrinsic character relational operators. If the minimum argument is shorter in character length than the length of the longest argument, the result is the minimum argument padded with blanks on the right to equal the length of the longest argument.

Examples. MIN (1.0, 2.0, 30.0) is 1.0; MIN (1, 2, –3) is –3; and MIN ([–1.0d0, –10.0d0], [–2.0d0, 20.1d0]) is [–2 –10] with double precision kind; MIN ("a", "b*b*", "c") is "a*b*".

MINEXPONENT (X) — Inquiry Function

Minimum value of the model exponent.

X Scalar or array of type real.

Result Characteristics. A scalar of default integer type.

Result Value. The result is the integer e_{min} for real numbers (13.2.3) of the type and kind of X, namely the minimum exponent for model numbers of the type and kind of X.

Examples. If e_{min}=–1021 in the integer model (13.2.3), MINEXPONENT (0.0) is the value –1021 of default integer kind.

MINLOC (ARRAY, DIM {, MASK, KIND}) or MINLOC (ARRAY {, MASK, KIND}) — Transformational Function

Location of the first minimum element of an array.

ARRAY Array of type integer, real, or character.

DIM Scalar of type integer with a value in the range [1, *n*] where *n* is the rank of ARRAY. The corresponding actual argument must not be an optional dummy argument.

MASK Of type logical, conformable with ARRAY.

KIND Scalar integer initialization expression whose value is an integer kind value.

Result Characteristics. Of type integer with kind KIND if KIND is present, or otherwise, with kind of default integer. If DIM is absent, it is a rank-one array of size equal to the rank *n* of ARRAY; if DIM is present and ARRAY is of rank one, it is a scalar; otherwise, it is a rank-one array of size *n*–1.

Result Value. The result is:

- if MASK and DIM are not present, a rank-one array whose element values are the subscripts of the first minimum in array element order of all elements of ARRAY provided ARRAY is of nonzero size. The ith subscript is in the range $[1, e_i]$ where e_i is the extent of ith dimension of ARRAY. If ARRAY is zero sized, all elements of the result are zero.
- if MASK is present and DIM is not present, a rank-one array whose element values are the subscripts of the first minimum in array element order of all elements of ARRAY corresponding to a true element of MASK. The ith subscript is in the range $[1, e_i]$ where e_i is the extent of ith dimension of ARRAY. If ARRAY is zero sized or all elements of MASK are false, all elements of the result are zero.
- if DIM is present and ARRAY has rank one, MINLOC (ARRAY[, MASK = MASK]). If DIM is present and the rank of ARRAY is at least two, the value of the element $(s_1, s_2, \ldots, s_{DIM-1}, s_{DIM+1}, \ldots, s_n)$ of the result is equal to MINLOC (ARRAY $(s_1, s_2, \ldots, s_{DIM-1}, :, s_{DIM+1}, \ldots, s_n)$, DIM=1 [, MASK = MASK $(s_1, s_2, \ldots, s_{DIM-1}, :, s_{DIM+1}, \ldots, s_n)$]).
- If ARRAY is of type character, the comparisons to determine the minimum value are the same as those used by the intrinsic relational operators for operands of the same type as ARRAY.

Examples. MINLOC ([1, 2, 4, 3]) is the rank-one array [1] of size 1; MINLOC ([4, 2, 4, 2]) is [2]; MINLOC ([1, 2, 4, 3], 1) is the scalar 1; MINLOC ([1, 2, 4, 3], [.false., .true., .false., .true.]) is the rank-one array [2] of size 1; and MINLOC ([1, 2, 4, 3], 1, [.false., .true., .false., .true.], P) is the scalar 2 of integer kind P.

Given the following statements:

```
logical, parameter :: T = .true., F = .false.
integer, dimension(11:12,31:33) :: a
logical, dimension(21:22,21:23) :: m
integer, dimension(2) :: r
r = minloc(a,dim=2,mask=m)
```

with the following values:

Table A-3 Values for MINLOC

a	m	a's subscripts	minloc's subscripts
1 4 7	T T F	**(11, 31)** (11, 32) (11, 33)	**(1, 1)** (1, 2) (1, 3)
2 **5** 8	F T F	(12, 31) **(12, 32)** (12, 33)	(2, 1) **(2, 2)** (2, 3)
3 6 9	T F T	**(13, 31)** (13, 32) (13, 33)	**(3, 1)** (3, 2) (3, 3)

When executed, MINLOC returns [1 2 1] which is the list of indices indicating the minimum values of those being selected along the rows if a, thus indicating the 1 in row 1, the 5 in row 2, and the 3 in row 3 of array a (indicated in bold font).

MINVAL (ARRAY, DIM {, MASK}) or
MINVAL (ARRAY {, MASK}) Transformational Function

Minimum value of array elements.

ARRAY Array of type integer, real, or character.

DIM Scalar of type integer with a value in the range [1, n] where n is the rank of ARRAY. The corresponding actual argument must not be an optional dummy argument.

MASK Of type logical and conformable with ARRAY.

Result Characteristics. Same type and type parameters as ARRAY. If DIM is absent or ARRAY has rank one, it is a scalar; otherwise, it is a rank n–1 array whose shape is that of ARRAY with the DIM dimension removed.

Result Value. The result is:

- if MASK and DIM are not present, the minimum of all elements of ARRAY provided ARRAY is of nonzero size. If ARRAY is zero sized and is of a numeric type, the result is the positive value of largest magnitude supported by the processor of the same type and kind as ARRAY; if ARRAY is zero sized and of type character, the result is the value of the string of the length of ARRAY, with each character equal to CHAR (n–1, KIND = KIND (ARRAY)) where n is the number of characters in the collating sequence for characters of the kind of ARRAY.
- if MASK is present and DIM is not present, MINVAL (PACK (ARRAY, MASK)).
- if DIM is present and ARRAY has rank one, MINVAL (ARRAY, MASK = MASK). If DIM is present and the rank of ARRAY is at least two, the value of the element (s_1, s_2, ..., s_{DIM-1}, s_{DIM+1}, ..., s_n) is the value MINVAL (ARRAY (s_1, s_2, ..., s_{DIM-1}, :, s_{DIM+1}, ..., s_n) [, MASK = MASK (s_1, s_2, ..., s_{DIM-1}, :, s_{DIM+1}, ..., s_n)]).
- if ARRAY is of type character, the comparisons to determine the minimum value are the same as those used by the intrinsic relational operators for operands of the same type as ARRAY.

Examples. MINVAL ([–1, 5, 2]) is –1; MINVAL (A, 1), where A = $\begin{bmatrix} 1 & 2 \\ -2 & -3 \end{bmatrix}$ is [–2 –3];

MINVAL (A, 2, M) where M = $\begin{bmatrix} \mathit{false} & \mathit{true} \\ \mathit{true} & \mathit{true} \end{bmatrix}$ is [2 –3]; and MINVAL (A, M) is –3.

MOD (A, P) Elemental Function

Remainder function, having the sign of the first argument.

A, P Of type integer or real; A and P must have the same type and kind type parameter. P must not be zero.

Result Characteristics. Same as A.

Result Value. The result is the remainder of A when divided by P; that is, the result is A–INT (A/P)*P.

Examples. MOD (5, 2) is 1; MOD (–3.1, 2.0) is –1.1; MOD (3.1, –2.0) is 1.1; and MOD (–6.2d0, –2.1d0) is –2.0 with double precision kind.

MODULO (A, P) Elemental Function

Remainder function, having the sign of the second argument.

A, P Of type integer or real; A and P must have the same type and kind type parameter. P must not be zero.

Result Characteristics. Same as A.

Result Value. For A of type integer, the result is the modulo R of A with respect to P; that is, where Q is an integer, the integer result R satisfies the requirement that $A = Q \times P + R$ where $0 \le R < P$ if P is positive, and $P < R \le 0$ if P is negative. For A of type real, the result is the same as the Fortran expression A–FLOOR(A/P)*P.

Examples. MODULO (5, 2) is 1; MODULO (–3.1, 2.0) is 0.9; MODULO (3.1, –2.0) is –0.9; and MODULO (–6.2d0, –2.1d0) is –2.0 with double precision kind. Note that despite the different forms of the definitions for real and integer values, the definitions are consistent; for example, MODULO (–3, 2) and MODULO (–3.0, 2.0) both have the value 1.

MOVE_ALLOC (FROM, TO) Pure Intrinsic Subroutine

Transfer an allocation from one object to another of the same type.

FROM Allocatable scalar or array of any type with INTENT (INOUT).

TO Allocatable scalar or array, type compatible with FROM, with the same rank as FROM and with INTENT (OUT). If FROM is polymorphic, TO must be polymorphic. Any nondeferred parameter of the declared type of TO must have the same value as the corresponding parameter of the declared type of FROM.

Result Value. The allocation is moved from the allocatable object FROM to the allocatable object TO and FROM is deallocated. If FROM is unallocated on invocation of MOVE_ALLOC, TO becomes unallocated; otherwise, TO is allocated with the same dynamic type, type parameters, and array bounds, and is given the same value as FROM had before MOVE_ALLOC was invoked; the allocation status of FROM becomes unallocated. If TO has the TARGET attribute, any pointer associated with FROM becomes associated with TO after MOVE_ALLOC is invoked; otherwise, any pointer associated with FROM when MOVE_ALLOC is invoked becomes undefined.

It is expected that allocatable objects involve the use of descriptors to locate allocatable storage. Using descriptors, MOVE_ALLOC can be implemented by transferring the descriptor of FROM to that of TO and clearing the descriptor of FROM, and thus no target data will move.

Example. Suppose more data is collected than will fit into an initially allocated array. In the following code sequence, an allocatable array TMP is allocated twice the size of the array SYMBOLS; the value of SYMBOLS is copied to TMP, and then the array TMP becomes SYMBOLS by using the subroutine MOVE_ALLOC; in this last step, no data will be moved.

```
character( len=len(SYMBOLS)), allocatable, dimension(:) :: TMP
   . . .
allocate (TMP(2*size (SYMBOLS))
TMP(1:size (SYMBOLS)) = SYMBOLS
TMP(size(SYMBOLS)+1:) = ""
call MOVE_ALLOC ( TMP, SYMBOLS)
```

After these statements are executed, the array SYMBOLS is now twice its original size with the first half of it having the same values as it had originally.

MVBITS (FROM, FROMPOS, LEN, TO, TOPOS) — Elemental Subroutine

Copy a sequence of bits from one integer to another.

FROM	Of type integer with INTENT (IN).
FROMPOS	Of type integer with INTENT (IN) in the range [0, BIT_SIZE (FROM)–LEN].
LEN	Of type integer with INTENT (IN) and with a nonnegative value.
TO	Of type integer with INTENT (INOUT) and with the same kind as FROM. It may be associated with the FROM argument. TO, starting at position TOPOS for LEN bits, is set to LEN bits from FROM, starting at position FROMPOS (13.2.1).
TOPOS	Of type integer with INTENT (IN). It must have a nonnegative value such that TOPOS+LEN ≤ BIT_SIZE (TO).

Example. Consider the following code segment:

```
integer :: T
T = 16
call  MVBITS (31, 0, 2, T, 1)
```

Using the particular bit model for default integers described in 13.2.1, the value of T is 22 after the call to MVBITS. In place of the last two statements above, the statements

```
T = INT (B"10000")
call  MVBITS (INT(B"11111"), 0, 2, T, 1)
```

sets T to the same value 22 which is the same value as the BOZ literal constant B'10110'.

NEAREST (X, S) — Elemental Function

Nearest machine-representable number in a given direction.

X	Of type real.
S	Of type real. It may be of a different kind than X and must not be zero.

Result Characteristics. Same as X.

Result Value. The result is the machine-representable value nearest X toward the direction of the infinity of the sign of S. Note that the result is described in terms of machine-representable values rather than model numbers.

Examples. If a processor uses for its default real type values the values described by the real model (13.2.3) with b=2, p=24, and e_{min}=–125, the values of NEAREST (1.0, –1.0) and NEAREST (0.0, 0.125) are $1 - 2^{-24}$ and 2^{-126}, respectively.

NEW_LINE (A) Inquiry Function

New line character for the character kind of the argument.

A Scalar or array of type character.

Result Characteristics. Scalar of type character, length 1, and the kind of A.

Result Value. The result is the new line character for the character set specified by the kind of A, as follows:

Kind of A	Condition	Result
Default	ACHAR (10) is a representable character	ACHAR (10)
ASCII, ISO 10646	None	CHAR (10, KIND (A))
Other kinds	A new line character *ch* in files connected for formatted stream output exists	*ch*
	No new line character in file connected for formatted stream output exists	Blank character

Example. The statements:

```
print "(/)"
print "(a)",  NEW_LINE ("a")
```

will produce the same result on most systems.

NINT (A {, KIND}) Elemental Function

Real value rounded to the nearest integer.

A Of type real.

KIND Scalar integer initialization expression whose value is an integer kind value.

Result Characteristics. Of type integer with kind KIND if KIND is present, or otherwise, with the kind of default integer.

Result Value. The result is the nearest integer value to A; if there are two such nearest integers, the result is the one of greater magnitude.

Examples. NINT (3.1) is 3; NINT (–3.5, P) is –4 of kind P.

NOT (I) Elemental Function

Logical complement of an integer.

I Of type integer.

Result Characteristics. Same as I.

Result Value. The result is the value obtained by complementing the bit positions of I; that is, I is complemented bit-by-bit.

I	NOT (I)
1	0
0	1

Example. Using the particular bit model for default integers described in 13.2.1, NOT (4) has integer value represented as a bit string 11111111111111111111111111111011, which has the first bit set to 1. For all arguments, the result value as an integer is processor dependent because the standard does not specify the representation of negative integers.

NULL ({MOLD}) Transformational Function

A disassociated pointer or unallocated allocatable component of a structure constructor.

MOLD It must be a pointer or allocatable and may be of any type or a procedure pointer. If a pointer, its association status may be associated, disassociated, or undefined. If it is allocatable, its allocation status may be allocated or unallocated. Its value may be undefined.

MOLD must be present in the following cases:

- any type parameter of the contextual entity in Table A-4 is assumed.
- NULL appears as an actual argument of a generic procedure and type, type parameters, or rank of this actual argument is needed to determine which specific procedure is to be referenced.
- NULL appears as an actual argument corresponding to a dummy argument with assumed character length.
- in all contexts other than those listed in Table A-4.

Table A-4 Result characteristics of NULL

Appearance of NULL ()	Type, type parameters, and rank of the result
Right side of a pointer assignment	The pointer on the left side
Initialization of an object in a declaration	The object
Default initialization for a component	The component
A value in a structure constructor	The corresponding component
An actual argument of a procedure	The corresponding dummy argument
A data value in a DATA statement	The corresponding pointer data object

Result Characteristics. The results characteristics are the same as MOLD if it is present; otherwise, the characteristics are determined by the context as specified in Table A-4. In addition, if the contextual entity has deferred-type parameters, those type parameters of the result are deferred.

Result Value. The result is a disassociated pointer or an unallocated allocatable entity.

Examples. If A is a pointer, NULL (A) is a disassociated pointer with the characteristics of A. Consider the following code segment:

```
type node_type
  integer, allocatable, dimension(:) :: vals
  type(node_type), pointer :: nxt
end node type
type(node_type) :: node
   . . .
node = node_type (null (), null ())
```

The assignment statement with the structure constructor node_type sets the component vals to an unallocated allocatable entity and the component nxt to a disassociated pointer.

PACK (ARRAY, MASK {, VECTOR}) Transformational Function

Masked array packed into a vector.

Arguments.

ARRAY	Array of any type.
MASK	Of type logical, conformable with ARRAY.
VECTOR	Rank-one array of the same type and type parameters as ARRAY. If MASK is an array, VECTOR must have a size at least as large as the number of true elements of MASK. If MASK is a scalar with the value true, VECTOR must have a size at least as large as the size of ARRAY.

Result Characteristics. Rank-one array of the type and type parameters of ARRAY. If VECTOR is present, the result size is the size of VECTOR; otherwise, if MASK is an array, the result size is the number of true elements in MASK. If MASK is a scalar with the value true, the result size is the size of ARRAY; if MASK is a scalar with the value false, the result size is zero.

Result Value. The result consists of the elements of ARRAY corresponding to the true elements of MASK in array element order. If VECTOR is present and is larger in size than the number of true elements in MASK, the remaining elements of the result are the corresponding remaining elements of VECTOR.

Examples. If ARRAY is $\begin{bmatrix} 4 & 2 \\ 3 & 1 \end{bmatrix}$, PACK (ARRAY, .true.) is [4 3 2 1]; PACK (ARRAY, .false., [1, 2, 3, 4, 5, 6]) is [1 2 3 4 5 6]; and if MASK is $\begin{bmatrix} true & false \\ false & true \end{bmatrix}$, PACK (ARRAY, MASK, [1, 2, 3, 4, 5, 6]) is [4 1 3 4 5 6].

PRECISION (X) Inquiry Function

Decimal precision of a model number.

X Scalar or array of type real or complex.

Result Characteristics. Scalar of default integer type.

Result Value. The result is the integer part of $((p-1)\times\log_{10}(b))+k$ for model parameters p and b (13.2.3) for the real type of X and k is 1 if b is an integral power of 10 and 0 otherwise, namely the decimal precision of real numbers of the kind of X.

Example. If a processor supports the double precision kind using the real model (13.2.3) with p=53 and b=2, the value PRECISION (0.0d0) is 15.

PRESENT (A) Inquiry Function

True if an actual argument of a procedure is present.

A Scalar or array optional dummy argument name of any type. The actual argument may be a dummy procedure or a pointer. A has no INTENT attribute. The actual argument corresponding to A must be an accessible optional dummy argument in the subprogram that invokes the PRESENT function (12.6.2).

Result Characteristics. Scalar of type default logical.

Result Value. The result is true if A is present (12.6.2) and false otherwise.

Example. In the following code segment:

```
print "(es7.1)", define_small ()
print "(es7.1)", define_small (1.0e10)
contains
```

```
real function define_small (x)
   real, optional :: x
   if( present(x) )  then
       find_small = abs (x)*epsilon (x)
   else
       find_small = epsilon (x)
   endif
end function define_small
```

the printed output might be: 1.2e–07 and 1.2e+03.

PRODUCT (ARRAY, DIM {, MASK}) or PRODUCT (ARRAY {, MASK})

Transformational Function

Product of array elements.

ARRAY — Array of type integer, real, or complex.

DIM — Scalar of type integer with a value in the range [1, n] where n is the rank of ARRAY. The corresponding actual argument must not be an optional dummy argument.

MASK — Of type logical, conformable with ARRAY.

Result Characteristics. Same as ARRAY. If DIM is not present or ARRAY has rank one, it is a scalar; otherwise, it is a rank n–1 array whose shape is that of ARRAY with the DIM dimension removed.

Result Value. The result is:

- if MASK and DIM are not present, the product of all elements of ARRAY provided ARRAY is of nonzero size. If ARRAY is zero sized, the result is one.
- if MASK is present and DIM does not appear, the product of the elements of ARRAY corresponding to the true elements of MASK and one if there are no true elements.
- if DIM is present and ARRAY has rank one, PRODUCT (ARRAY, MASK = MASK). If DIM is present and the rank of ARRAY is at least 2, the value of the element $(s_1, s_2, ..., s_{DIM-1}, s_{DIM+1}, ..., s_n)$ is equal to PRODUCT (ARRAY $(s_1, s_2, ..., s_{DIM-1}, :, s_{DIM+1}, ..., s_n)$ [, MASK = MASK $(s_1, s_2, ..., s_{DIM-1}, :, s_{DIM+1}, ..., s_n)$]).
- if ARRAY is of zero size or no elements of MASK are true, the result is 1.

Examples. PRODUCT ([–1, 5 ,2]) is –10. If A = $\begin{bmatrix} 4 & 2 \\ -2 & -3 \end{bmatrix}$, PRODUCT (A, 1) is [–8 –6];

and if M = $\begin{bmatrix} false & true \\ true & true \end{bmatrix}$, PRODUCT (A, 2, M) is [2 6] and PRODUCT (A, A>0) is 8.

RADIX (X)

Inquiry Function

Base of a model number.

X Scalar or array of type integer or real.

Result Characteristics. Scalar of default integer type.

Result Value. The result is the value *r* of the integer model (13.2.2) or the value *b* of the real model (13.2.3) of the type and kind of X, namely the radix of numbers of the type and kind of X.

Examples. If *r*=2 in the integer model (13.2.2), RADIX (0) has the value 2 of default integer kind; if *b*=2 in the real model (13.2.3), RADIX (0.0) has the value 2 of default integer kind.

RANDOM_NUMBER (HARVEST) Subroutine

Generate pseudorandom scalar or array of real type.

HARVEST Scalar or array of type real. It is an INTENT (OUT) argument that is assigned a scalar or array of uniformly distributed pseudorandom real values in the interval [0, 1).

Examples. The code segment:

```
real :: S
real( kind=KIND (0.0d0) ) :: D
real, dimension(3) :: A

call  RANDOM_NUMBER (S)
call  RANDOM_NUMBER (D)
call  RANDOM_NUMBER (A)
```

produces a set of pseudorandom numbers, probably different for each execution of the code segment. Upon completion of this code segment, the value for S is a scalar of type default real, such as 0.5587673; the value of D is a real scalar with double precision kind such as 0.2024475895811094, and the value of A is a default real rank-one array of three numbers, such as [0.5366381 0.2763737 .012461195].

RANDOM_SEED ({SIZE, PUT, GET}) Subroutine

Retrieve or set the seed of the pseudorandom number generator.

SIZE Scalar of default integer type. If it is present, it is an INTENT (OUT) argument that is assigned the size of the array used by the processor to hold the seed for the pseudorandom number generator.

PUT Rank-one array of default integer type. If it is present, it is an INTENT (IN) argument that is the value used to set the seed of the pseudorandom number generator.

GET Rank-one array of default integer type. If it is present, it is an INTENT (OUT) argument that is assigned the current seed for the pseudorandom number generator.

There must be zero or one argument. If no argument is present, the processor assigns a processor-dependent value to the seed. If the PUT argument is used to set the seed with a particular value, the same sequence of pseudorandom numbers must be generated when that particular seed is specified a second time.

Examples. Consider the following program:

```
real :: S
real( kind=KIND (0.0d0) ) :: D
real, dimension(3) :: A
integer :: sz
integer, dimension(:), allocatable :: seed

call  RANDOM_SEED (size = sz)  ! Finds the size sz of the seed
allocate (seed(sz))
sz = [(i,i=1,sz)]               ! Establishes a user seed
call  random_seed (put = seed)  ! Sets the array of seeds

call  RANDOM_NUMBER (S)    ! A scalar default real random number
call  RANDOM_NUMBER (D)    ! A scalar double precision random number
call  RANDOM_NUMBER (A)    ! An array of random numbers
```

This program segment produces a set of pseudorandom numbers, the same for each execution of the code segment; see the similar example in the description of RANDOM_NUMBER. and 13.3.4.1 for further discussion of the properties of the procedures RANDOM_SEED and RANDOM_NUMBER.

RANGE (X) — Inquiry Function

Decimal exponent range of a model number.

X	Scalar or array of type integer, real, or complex.

Result Characteristics. Scalar of default integer type.

Result Value. The result value is:

for integer X	INT (LOG10 (HUGE (X)))
for real X	INT (MIN (LOG10 (HUGE (X)), –LOG10 (TINY(X))))
for complex X	RANGE (REAL (X))

Examples. If r=2 and q=31 in the integer model (13.2.2), RANGE (0) has the value 9 of default integer; if b=2, p=24, e_{min}=–125, and e_{max}=128 in the real model (13.2.3), RANGE (0.0) has the value 37 of default integer kind.

REAL (A {, KIND}) — Elemental Function

Real value.

A	Scalar or array of type integer, real, or complex type, or a BOZ literal constant.
KIND	Scalar integer initialization expression whose value is a real kind value.

Result Characteristics. Of type real with kind KIND if KIND is present, or otherwise, with the kind of default real.

Result Value. The result is A if A is of integer or real type, or of the real part of A if A is of complex type. If A is a BOZ literal constant, the result is that value of type real with the kind of the result whose bit pattern is that given by the BOZ literal constant; the interpretation of the bit pattern is processor dependent.

Examples. REAL (3) is 3.0; REAL ((4.0, 1.0)) is 4.0. REAL (4.1, KIND (0.0d0)) is 4.1, converted from real kind to double precision kind; in general, it is not equal to 4.1d0.

REPEAT (STRING, NCOPIES) Transformational Function

Concatenation of several copies of a character string.

STRING Scalar of type character.

NCOPIES Scalar of type integer. Its value must not be negative.

Result Characteristics. Scalar of the type and kind of STRING and with length NCOPIES*LEN (STRING).

Result Value. The result is the string consisting of NCOPIES copies of STRING concatenated together.

Example. REPEAT ("BAD*b*", 3) is the string BAD*b*BAD*b*BAD*b*; REPEAT ("BAD*b*", 0) is the empty string.

RESHAPE (SOURCE, SHAPE {, PAD, ORDER}) Transformational Function

Rank-one array reshaped to an array of a specified shape.

SOURCE Array of any type. If PAD is absent or of zero size, the size of SOURCE must be greater than or equal to PRODUCT (SHAPE).

SHAPE Rank-one array of type integer and of a positive size n less than 8. No element may have a negative value. The size n must be determinable at compile time; what is determinable at compile time is open to interpretation.

PAD Array of the same type and type parameters as SOURCE

ORDER Rank-one array of type integer with any integer kind and of size n. It must have a value that is a permutation of the integers from 1 to n. If it is not present, it is as if it is present with the value [1 2 ... n].

Result Characteristics. Array of shape SHAPE with the type and type parameters of SOURCE.

Result Value. The elements of the result, taken in permuted subscript order ORDER(1), ORDER(2), ..., ORDER(n), are those of the array SOURCE in array element order, followed, if needed to complete the elements of the result, by the elements of PAD in array element order, followed if needed, by further copies of PAD in array element order.

Examples. RESHAPE ([1, 2, 3, 4, 5, 6], [2, 3]) has the value $\begin{bmatrix} 1 & 3 & 5 \\ 2 & 4 & 6 \end{bmatrix}$.

RESHAPE ([1, 2, 3, 4, 5, 6], [2, 4], [0, 0], [2, 1]) has the value $\begin{bmatrix} 1 & 2 & 3 & 4 \\ 5 & 6 & 0 & 0 \end{bmatrix}$.

Consider the following program segment:

```
real, dimension(2,2) :: a
integer, dimension(2) :: shp = [ size(a,1), size(a,2) ]
a = reshape( [1,2,3,4], shp, order = [2,1] )
```

a is the array $\begin{bmatrix} 1.0 & 2.0 \\ 3.0 & 4.0 \end{bmatrix}$.

RRSPACING (X)

Elemental Function

Reciprocal of model relative spacing near a specified value.

X Of type real.

Result Characteristics. Same as X.

Result Value. The result is the value $|Xb^{-e}|b^p$ as determined by the model for real numbers (13.2.3). If X is an IEEE infinity or NaN, the result is zero or that NaN, respectively.

Examples. If a processor uses for its default real type values the values described by the real model (13.2.3) with b=2 and p=24, the values of RRSPACING (1.0) and RRSPACING (0.0) are 2^{23} and 0.0, respectively.

SAME_TYPE_AS (A, B)

Inquiry Function

True if two objects are of the same dynamic type.

A, B Of any extensible type. Either or both may be pointers that have a defined association status.

Result Characteristics. Scalar of type default logical.

Result Value. The result is true if the dynamic type of A is the same as the dynamic type of B. Note that the dynamic type of a disassociated pointer or unallocated allocatable is its declared type.

Example. Using the declarations given for the example of the intrinsic function EXTENDS_TYPE_OF, SAME_TYPE_AS (a, line), SAME_TYPE_AS (a, divider), and SAME_TYPE_AS (point, divider) are each false whereas SAME_TYPE_AS (line, divider) is true.

SCALE (X, I) Elemental Function

Value scaled by a power of the radix.

X	Of type real.
I	Of type integer.

Result Characteristics. Same as X.

Result Value. The result is the value Xb^I as determined by the model for real numbers (13.2.3), provided the value is in range, and otherwise, is processor dependent.

Examples. If a processor uses for its default real type values the values described by the real model (13.2.3) with b=2, the values of SCALE (1.0, –2) and SCALE (0.0, 10) are 0.25 and 0.0, respectively.

SCAN (STRING, SET {, BACK, KIND}) Elemental Function

Position in a string of any one of a given set of characters.

STRING, SET	Of type character, but both with the same kind.
BACK	Of type logical.
KIND	Scalar integer initialization expression whose value is an integer kind value.

Result Characteristics. Of type integer with kind KIND if KIND is present, or otherwise, with the kind of default integer.

Result Value. The result is the index position in STRING of a character in the string SET, or zero if no character in SET appears in STRING. If BACK is present with the value false or is not present, the index position is that of the first occurrence in STRING of any character in SET; otherwise, if BACK is present with the value true, the index position is that of the last occurrence in STRING of a character in SET. If the length of STRING or SET is zero, zero is returned.

Examples. SCAN ("input*b*string", "if") is 1; SCAN ("input*b*string", "if", .true.) is 10; and SCAN ("input*b*string", "f", KIND=P) is 0 of integer kind P.

SELECTED_CHAR_KIND (NAME) Transformational Function

Kind parameter of a specified character set.

NAME	Scalar of type default character.

Result Characteristics. Scalar of type default integer.

Result Value. The result is the kind parameter value of the character type whose name is the value of NAME. If the named character type is not supported by the processor, the value is –1. The value of NAME is interpreted without regard to case and trailing blanks. The names used for character kinds that support the default, ASCII, and ISO/IEC 10646-1:2000 UCS-4 character sets are the strings DEFAULT, ASCII and ISO 10646, respectively.

Examples. If a processor supports the default character and the ISO/IEC 10646-1:2000 UCS-4 character types with kind values 1 and 2, respectively, the kind values for X, Y, and Z declared as follows:

```
character(kind=kind("ABC"))  X
character(kind=selected_char_kind("ISO_10646"))  Y
character(kind=selected_char_kind("default"))  Z
```

are 1, 2, and 1, respectively; in this case, the first and third statements declare the same character kind.

SELECTED_INT_KIND (R) — Transformational Function

Kind parameter of an integer data type, specified by a minimum decimal range.

R — Scalar of type integer.

Result Characteristics. Scalar of type default integer.

Result Value. The result is the kind parameter value of an integer type that supports integers n in the range $-10^R < n < 10^R$. If such an integer kind does not exist, the value is –1. If there is more than one kind available, the kind corresponding to the smallest decimal exponent range is returned. If there are more than one of those, the smallest kind value of those is returned.

Examples. If q=63 and r=2 in the integer model (13.2.2) for integers with kind 2, SELECTED_INT_KIND (15) has the value 2 of default integer kind.

SELECTED_REAL_KIND ({P, R}) — Transformational Function

Kind parameter of a real data type, specified by a minimum decimal precision and/or exponent range.

P, R — Scalars of type integer. At least one of them must be present. If either is absent, it is as if it were present with the value zero.

Result Characteristics. Scalar of type default integer.

Result Value. The result is the kind parameter value of a real type that supports real numbers x of decimal precision at least P digits and a decimal range of at least R as defined by the intrinsic functions PRECISION and RANGE, respectively. If more than one kind type parameter meets the criterion, the kind value returned is the one that has the least decimal precision; if there is more than one of these, the one with the smallest kind value is returned. If the processor does not support such a real kind, the values shown in Table A-5 are returned.

Examples. If a processor supports the default real type modeled with e_{min}=–125, e_{max}=128, p=24, and r=2 and a double precision real type modeled with e_{min}=–1021, e_{max}=1024, p=53 and r=2 using the real model (13.2.3) with kind values 4 and 8, respectively, the kind values for SELECTED_REAL_KIND (6) and SELECTED_REAL_KIND (R=100) are 4 and 8, respectively.

Table A-5 Negative values of SELECTED_REAL_KIND

Value	Condition
–1	No kind with precision P, but a kind with range R
–2	No kind with range R, but a kind with precision P
–3	No kind with precision P or range R
–4	A kind with precision P and another with range R, but not one kind that supports both

See Appendix B for the requirements of the similar IEEE module procedure IEEE_SELECTED_REAL_KIND.

To illustrate the negative returned values, consider one processor that has two real kinds, whose precisions and ranges returned by the functions PRECISION and RANGE are 6 and 36, and 15 and 307 respectively, and a second processor whose precisions and ranges for its two kinds are 10 and 30, and 4 and 40, respectively. Table A-6 illustrates the returned values for these two processors with various requested precisions P and ranges R:

Table A-6 Returned values of SELECTED_REAL_KIND (P, R)

Processor's two P, R, pairs	Requested P, R	Returned Value
(6, 37), (15, 307)	20, 37	–1
	10, 400	–2
	20, 400	–3
(10, 30), (4, 40)	6, 37	–4

Note that the processor corresponding to the last row of Table A-6 supports two real kinds; neither of these real kinds satisfy the requirement for at least 6 decimal digits of precision and decimal exponent range of 37, but each satisfies one of the requirements; thus, the value of SELECTED_REAL_KIND (6, 37) must be –4 for this processor.

SET_EXPONENT (X, I) — Elemental Function

Value with its exponent set to a specified value.

X — Of type real.

I — Of type integer.

Result Characteristics. Same as X.

Result Value. The result is the value Xb^{I-e} as determined by the model for real numbers (13.2.3), provided the value is in range, and otherwise, is processor dependent. If X is zero, the result is zero.

Examples. If a processor uses for its default real type values the values described by the real model (13.2.3) with b=2, the values of SET_EXPONENT (1.0, 3) and SET_EXPONENT (0.0, 120) are 4.0 and 0.0, respectively.

SHAPE (SOURCE [, KIND]) — Inquiry Function

Number of elements in each dimension of an array.

SOURCE — Scalar or array of any type. If it is allocatable, it must be allocated; if it is a pointer, it must be associated. It must not be an assumed-size array.

KIND — Scalar integer initialization expression whose value is an integer kind value.

Result Characteristics. Rank-one array of integer type with kind KIND if KIND is present, or otherwise, with the kind of default integer. Its size is equal to the rank of SOURCE.

Result Value. The result is the shape of SOURCE. Note that if SOURCE is a scalar, the result is a zero-sized rank-one array.

Examples. For the following code segment:

```
real, allocatable, dimension(:) :: vector
real, dimension(0:100, 1:1000) :: table
allocate (vector(10:20))
```

SHAPE (vector), SHAPE (vector(15:)), and SHAPE (table) are [11], [6,] and [101 1000], respectively. SHAPE(3) is a rank-one array of size zero.

SIGN (A, B) — Elemental Function

Value with a specified sign.

A, B — Of type integer or real with the same type and kind type parameter.

Result Characteristics. Same as A.

Result Value. The result is |A| if B is positive, –|A| if B is negative. If B is positive zero or the processor cannot distinguish the sign of zero or if B is of type integer and is zero, the result is |A|; otherwise, the result is –|A|.

Examples. SIGN (1, –10) is –1; SIGN (10.1, 2.0) is 10.1; and SIGN (1.0d0, –0.0d0) is –1 with double precision kind on a processor that can distinguish positive and negative zero, and 1 with double precision kind otherwise.

SIN (X) — Elemental Function

Sine.

X — Of type real or complex.

Result Characteristics. Same as X.

Result Value. The result is the sine of X. If X is real, X is in radians. If X is complex, the real part of X is in radians.

Examples. SIN (0.0) is 0.0; SIN (X) where X is $\pi/2$ is 1.0; and SIN ((1.0_P, 1.0_P)) is 1.29846 + 0.63496i of kind P.

SINH (X)

Elemental Function

Hyperbolic sine.

X Of type real.

Result Characteristics. Same as X.

Result Value. The result is the hyperbolic sine of X.

Examples. SINH (0.0) is 0.0; and SINH (–0.5_P) is –0.521095 of real kind P.

SIZE (ARRAY {, DIM, KIND})

Inquiry Function

Number of elements of an array or a dimension of an array.

ARRAY Array of any type. If it is allocatable, it must be allocated. If it is a pointer, it must be associated. If it is an assumed-size array, DIM must be present with a value that is not equal to the rank of ARRAY.

DIM Scalar of type integer with a value in the range [1, n] where n is the rank of ARRAY.

KIND Scalar integer initialization expression whose value is an integer kind value.

Result Characteristics. Scalar of integer type with kind KIND if KIND is present, or otherwise, with the kind of default integer.

Result Value. If DIM is present, the result is the extent of the DIM dimension of ARRAY; if absent, the result is the number of elements in ARRAY

Examples. Consider the following code segment, in a subprogram with dummy argument D:

```
real, allocatable, dimension(:) :: vector
real, dimension(10,20,*) :: D
allocate (vector(10:20))
```

Table A-7 shows the values of various references to SIZE.

Table A-7 Values of SIZE

Reference	Value
size(vector)	11
size(vector(15:))	6
size(D,1)	20
size(D(:,:10))	200

Note that D(:,:10) is not an assumed-size array even though D is an assumed-size array.

SPACING (X)

Elemental Function

Model absolute spacing near a specified value.

X Of type real.

Result Characteristics. Same as X.

Result Value. The result is the value $b^{max((e-p),(e_{min}-1))}$ as determined by the model for real numbers (13.2.3), provided X is neither zero, IEEE infinity, nor NaN. If X is zero, the result is TINY (X); if X is an IEEE infinity, the result is positive IEEE infinity; if X is NaN, the result is that NaN.

Examples. If a processor uses for its default real type values the values described by the real model (13.2.3) with b=2 and e_{min}=–125, the values of SPACING (1.0) and SPACING (0.0) are 2^{-26} and $2^{e_{min}-1}$, respectively, or 1.19×10^{-7} and 1.18×10^{-38}, respectively.

SPREAD (SOURCE, DIM, NCOPIES)

Transformational Function

Array replicated by adding a dimension.

SOURCE Scalar or array of any type. If it is an array, its rank must not exceed 6.

DIM Scalar of type integer with a value in the range [1, n+1] where n is the rank of SOURCE.

NCOPIES Scalar of type integer.

Result Characteristics. Array of the type and type parameters of SOURCE and of rank n+1. If SOURCE is a scalar, the shape of the result is [MAX (NCOPIES,0)]; if SOURCE is an array, the shape of the result is [d_1, d_2, ..., $d_{\mathrm{DIM}-1}$, MAX (NCOPIES, 0), d_{DIM}, ..., d_n] where [d_1, d_2, ..., d_n] is the shape of SOURCE.

Result Value. If SOURCE is a scalar, every element of the result is that scalar; if SOURCE is an array, the element (r_1, r_2, ..., r_{n+1}) has the value SOURCE (r_1, r_2, ..., $r_{\mathrm{DIM}-1}$, $r_{\mathrm{DIM}+1}$, ..., r_{n+1}).

Examples. SPREAD ("A", 1, 3) is the character array [A A A]. If B is the array [1, 3, 7], SPREAD (B, DIM=1, NCOPIES=NC) is the array $\begin{bmatrix} 1 & 3 & 7 \\ 1 & 3 & 7 \\ 1 & 3 & 7 \end{bmatrix}$ if NC has the value 3 and is a zero-sized array if NC has the value 0.

SQRT (X)

Elemental Function

Square root.

X Of type real or complex. If it is of type real, its value must be nonnegative.

Result Characteristics. Same as X.

Result Value. The result is the square root of X. If X is of type real, the result is nonnegative; if X is of type complex, the result is the principal value; namely, the real part is nonnegative, and if the real part of the result is zero, the sign of the imaginary part of the result is the sign of the imaginary part of X.

Examples. SQRT (4.0) is 2.0; SQRT ((–25.0d0, 0.0d0)) is (–3.0d0, 4.0d0) is 1+2*i* of double precision kind. If A = $\begin{bmatrix} 1.0 & 9.0 & 25.0 \\ 4.0 & 16.0 & 36.0 \end{bmatrix}$, SQRT (A) is $\begin{bmatrix} 1.0 & 3.0 & 5.0 \\ 2.0 & 4.0 & 6.0 \end{bmatrix}$.

SUM (ARRAY, DIM {, MASK}) or SUM (ARRAY {, MASK}) — Transformational Function

Sum of array elements.

ARRAY	Array of type integer, real, or complex.
DIM	Scalar of type integer with a value in the range [1, *n*] where *n* is the rank of ARRAY. The corresponding actual argument must not be an optional dummy argument.
MASK	Of type logical, conformable with ARRAY.

Result Characteristics. Same as ARRAY. If DIM is absent or ARRAY has rank one, it is a scalar; otherwise, it is a rank *n*–1 array whose shape is that of ARRAY with the DIM dimension removed.

Result Value. The result is:

- if MASK and DIM are absent, the sum of all elements of ARRAY provided ARRAY is of nonzero size. If ARRAY is zero sized, the result is zero.
- if MASK is present and DIM is absent, the sum of the elements of ARRAY corresponding to the true elements of MASK and zero if there are no true elements.
- if DIM is present and ARRAY has rank one, SUM (ARRAY, MASK = MASK). If DIM is present and the rank of ARRAY is at least 2, the value of the element (s_1, s_2, ..., s_{DIM-1}, s_{DIM+1}, ..., s_n) is equal to SUM (ARRAY (s_1, s_2, ..., s_{DIM-1}, :, s_{DIM+1}, ..., s_n) [, MASK = MASK (s_1, s_2, ..., s_{DIM-1}, :, s_{DIM+1}, ..., s_n)]).
- if ARRAY is of zero size or no elements of MASK are true, the result is 0.

Examples. SUM ([–1, 5, 2]) is 6. If A = $\begin{bmatrix} 1 & 2 \\ -2 & -3 \end{bmatrix}$, SUM (A, 1) is [–1 –1]; and if M = $\begin{bmatrix} false & true \\ true & true \end{bmatrix}$, SUM (A, 2, M) is [2 –5] and SUM (A, A>0) is 3.

SYSTEM_CLOCK ({COUNT, COUNT_RATE, COUNT_MAX}) — Subroutine

Obtain data from the system clock.

COUNT Scalar of type integer with INTENT (OUT). If COUNT is present, COUNT is assigned a processor-dependent value based on the processor clock, representing time in terms of counts of the clock. The value is in the range [0, COUNT_MAX]. If the processor clock is not available, it is assigned –HUGE (COUNT).

COUNT_RATE Scalar of type integer or real with INTENT (OUT). If COUNT_RATE is present, COUNT_RATE is assigned the number of clock counts per second or zero if there is no processor clock.

COUNT_MAX Scalar of type integer with INTENT (OUT). If COUNT_MAX is present, COUNT_MAX is assigned the maximum value of COUNT, if a processor clock is available, or otherwise, zero.

Examples. The code segment:

```
integer :: cnt, cnt_rate, cnt_max
call system_clock (cnt, cnt_rate, cnt_max)
print *, "clock count:", cnt
print *, "clock counts per second:", cnt_rate
print *, "maximum clock count:", cnt_max
```

might produce output such as:

```
clock count: 50110264
clock counts per second: 1000
maximum clock count: 86399999
```

This output might be produced on a system with a 24-hour clock (8640000 is the number of seconds in 24 hours and 5011064 is the number of seconds from when the clock was started).

TAN (X) Elemental Function

Tangent.

X Of type real.

Result Characteristics. Same as X.

Result Value. The result is the tangent of X where X in radians.

Examples. TAN (0.0) is 0.0; TAN (X) where X is $\pi/4$ is 1.0; and TAN (0.5d0) is 0.54630 with double precision kind.

TANH (X) Elemental Function

Hyperbolic tangent.

X Of type real.

Result Characteristics. Of the type of X.

Result Value. The result is the hyperbolic tangent of X.

Examples. TANH (0.0d0) is 0.0 with double precision kind; TANH (–0.5_Q) is –0.46212 of real kind Q.

TINY (X) Inquiry Function

Smallest positive number in the real model.

X Scalar or array of type real.

Result Characteristics. Scalar of the type and kind of X.

Result Value. The result is the value $b^{e_{min}-1}$ (13.2.3), for real numbers of the type and kind of X, namely the smallest positive model number of the type and kind of X.

Example. If b=2 and e_{min}=−125 in the real model (13.2.3), the value TINY (0.0) has the value 2^{-126} or 1.175×10^{-38} with default real kind. Note that the returned values for TINY are defined in terms of the model and may not be the smallest number in magnitude nor the smallest positive number; for example, denormalized numbers, if supported by the processor, are smaller than the value returned by TINY.

TRANSFER (SOURCE, MOLD [, SIZE]) Transformational Function

Value transferred from an object to the result without conversion.

SOURCE Scalar or array of any type.

MOLD Scalar or array of any type. If it is a variable, it need not be defined.

SIZE Scalar of integer type. The corresponding actual argument must not be an optional dummy argument.

Result Characteristics. Same as MOLD. It is a scalar if SIZE is absent and MOLD is a scalar or, if MOLD is an array and SIZE absent, it is a rank-one array of a size that is the smallest size that is not shorter than the physical representation of SOURCE. If SIZE is present, the result is a rank-one array of size SIZE.

Result Value. The result is a value with the physical representation of SOURCE, interpreted as an entity of the type and type parameters of MOLD. If the sizes of SOURCE and the result are the same, the result is SOURCE; if the size of SOURCE is smaller than that of the result, the leading part of the result is SOURCE and the remaining part is processor dependent; if the size of SOURCE is larger than the size of the result, the result is the leading part of SOURCE.

Examples. Suppose FFT is a default complex array of three elements and X is a default real array of size six. TRANSFER (FFT, X) is a rank-one array of six elements, representing the real and imaginary components of the three elements of FFT in order, with the elements of the odd indices of the result being the real parts and the elements with the even indices being the imaginary parts.

Suppose STRUCTURE is a sequence derived type with five default real components. TRANSFER (STRUCTURE, 0.0, 3) is a rank-one array of size three and type default real consisting of the first three components of STRUCTURE, in order.

TRANSFER ([1.1, 2.2, 3.3], [(0.0,0.0)]) is a complex rank-one array of length two whose first element has the value 1.1 + 2.2i and whose second element has a real part with value 3.3. The imaginary part of the second element is processor dependent.

TRANSPOSE (MATRIX) Transformational Function

Matrix transpose.

MATRIX A two-dimensional array of any type.

Result Characteristics. Array of the type and type parameters of MATRIX and of shape [*m n*] where MATRIX has shape [*n m*].

Result Value. The (i, j) element of the result is MATRIX (j, i) for all $1 \le j \le m$ and $1 \le i \le n$.

Example. If MATRIX = $\begin{bmatrix} 1 & 3 \\ 2 & 4 \end{bmatrix}$, TRANSPOSE (MATRIX) is $\begin{bmatrix} 1 & 2 \\ 3 & 4 \end{bmatrix}$.

TRIM (STRING) Transformational Function

String without trailing blanks.

STRING Scalar of type character.

Result Characteristics. Of the type and kind of STRING and length equal to the length of STRING less the number of trailing blanks.

Result Value. The result is the value of STRING with all trailing blanks removed.

Example. TRIM ("*bbb*string*bbb*") is *bbb*string.TRIM ("*bbbb*") is the null string.

UBOUND (ARRAY {, DIM, KIND}) Inquiry Function

Upper bound(s) of an array or a dimension of an array.

ARRAY Array of any type. If it is allocatable, it must be allocated. If it is a pointer, it must be associated. If it is an assumed-size array, DIM must be present with a value not equal to the rank of ARRAY.

DIM Scalar of type integer with a value in the range [1, *n*] where *n* is the rank of ARRAY. The corresponding actual argument must not be an optional dummy argument.

KIND Scalar integer initialization expression whose value is an integer kind value.

Result Characteristics. Scalar or rank-one array of type integer with kind KIND if KIND is present, or otherwise, with the kind of default integer.

Result Value. If DIM is present, the result is a scalar integer representing the upper bound of ARRAY in the DIM dimension as defined in 5.4, 6.6, and 7.2.4. If DIM is absent, the result is a rank-one array of the upper bounds of each dimension of ARRAY.

Examples. Consider the following code in a subprogram where D is a dummy argument:

```
real, allocatable, dimension(:) :: vector
real, dimension(0:100, -10:30) :: array
real, dimension(20,*) :: D
allocate (vector(10:20))
```

UBOUND (vector), UBOUND (array,1), and UBOUND (array,2) are [20], 100, and 30, respectively; UBOUND (D,1) and UBOUND (D(:,21:35),2) are 20 and 15, respectively.

UNPACK (VECTOR, MASK, FIELD) — Transformational Function

Array unpacked from a vector under mask control.

VECTOR — Rank-one array of any type. Its size must be at least as large as the number of true elements in MASK.

MASK — Array of type logical.

FIELD — Of the type and type parameters of VECTOR and conformable with MASK.

Result Characteristics. Array of the type and type parameters of VECTOR with the shape of MASK.

Result Value. The elements of the result, where MASK is true, taken in array element order, are those of VECTOR in array element order, and where MASK is false, have the value FIELD if FIELD is a scalar or have the value of the corresponding element of FIELD if FIELD is an array.

Examples. Particular values may be "scattered" to particular positions in an array by using UNPACK. If M is the array $\begin{bmatrix} 1 & 0 & 0 \\ 0 & 1 & 0 \\ 0 & 0 & 1 \end{bmatrix}$, V is the array [1 2 3], and U is the logical mask $\begin{bmatrix} false & true & false \\ true & false & false \\ false & false & true \end{bmatrix}$, the result of UNPACK (V, MASK = U, FIELD = M) has the value $\begin{bmatrix} 1 & 2 & 0 \\ 1 & 1 & 0 \\ 0 & 0 & 3 \end{bmatrix}$ and the result of UNPACK (V, MASK = U, FIELD = 0) has the value $\begin{bmatrix} 0 & 2 & 0 \\ 1 & 0 & 0 \\ 0 & 0 & 3 \end{bmatrix}$.

VERIFY (STRING, SET {, BACK, KIND}) — Elemental Function

Position in a string of a character that is not one of a given set.

STRING — Of type character.

SET — Of type character of the same kind as STRING.

BACK Of type logical.

KIND A scalar integer initialization expression whose value is an integer kind value.

Result Characteristics. Of type integer with kind KIND if KIND is present, or otherwise, with the kind of default integer.

Result Value. The result is the index position in STRING of a character that is not in the string SET. If BACK is present with the value false or is not present, the index position is that of the first occurrence of a character not in SET; otherwise, if BACK is present with the value true, the index position is that of the last occurrence of a character not in SET. If all characters in SET appear in the string or the length of STRING is zero, zero is returned.

Examples. VERIFY ("input*b*string", "if") is 2; VERIFY ("input*b*string", "if", .true.) is 12; VERIFY ("input*b*string", "xy", KIND=P) is 0 of integer kind P; and VERIFY ("","abc") is 0.

B IEEE Module Procedures

This appendix contains detailed specifications of the module procedures in the IEEE intrinsic modules IEEE_ARITHMETIC and IEEE_EXCEPTIONS. The title of each description gives the name of the procedure and the names of its dummy arguments, with an indication of which arguments are optional. On the right side of each title line, the classification of each procedure is given; it indicates whether the procedure is an inquiry, transformational or elemental function, or an elemental or nonelemental subroutine. Also, on the right side of each title line is the name of the IEEE module that defines the module procedure.

Whether the actual argument is optional is indicated by braces { } enclosing the arguments in the section titles; the braces indicate that each argument within the braces is optional and may appear as an argument with or without the other optional arguments.

Some of the procedure headings show an alternative interface which make it appear as though the last argument is optional; as explained in 14.3, the argument is not optional in the sense of an OPTIONAL attribute for the argument; rather, there are two distinct versions of the procedure.

If the specification for an argument specifies a type but not a kind, any acceptable kind for that type is permitted. Sometimes the kind is required to be one for which IEEE_SUPPORT_DATATYPE function returns true; this is a program restriction and is indicated by the phrase "restricted kind" in the description of the arguments.

For inquiry functions, the arguments need not have defined values, pointer arguments need not have a defined association status, and allocatable arguments need not be allocated. For all functions, their arguments have intent IN but this intent is not stated explicitly. If a function returns an array, the array has lower bounds of 1 and strides of 1 in each dimension. For all subroutines, the intent of each dummy argument is specified in the description; if those arguments have intent OUT or INOUT, the value returned by the subroutine is specified in the description of the argument.

In all examples, it is assumed that, unless otherwise stated, USE statements referring to the appropriate IEEE modules appear in the scoping unit of the example codes, and IEEE_SUPPORT_STANDARD (0.0) is true; note that if this support standard function is true for default real, it implies that it IEEE_SUPPORT_HALTING and IEEE_SUPPORT_ROUNDING are true for any valid exception flag and rounding mode.

IEEE_CLASS (X) — Elemental Function

IEEE_ARITHMETIC

The IEEE class of the argument.

X — Of type real with restricted kind.

Result Characteristics. Of type IEEE_CLASS_TYPE.

Result Value. The result value is:

- IEEE_SIGNALING_NAN or IEEE_QUIET_NAN if IEEE_SUPPORT_NAN (X) is true and X is a signaling or quiet NaN, respectively;
- IEEE_NEGATIVE_INF or IEEE_POSITIVE_INF if IEEE_SUPPORT_INF (X) is true and X is a negative or positive infinity, respectively;
- IEEE_NEGATIVE_DENORMAL or IEEE_POSITIVE_DENORMAL if IEEE_SUPPORT_DENORMAL (X) is true and X is a negative or positive denormal value, respectively;
- IEEE_NEGATIVE_NORMAL, IEEE_POSITIVE_NORMAL, IEEE_NEGATIVE_ZERO, or IEEE_POSITIVE_ZERO if X is a negative normal, positive normal, negative zero, or positive zero value, respectively; or otherwise,
- IEEE_OTHER_VALUE; this value is provided for processors that support the IEEE arithmetic in some partial way and need to specify a value that can be created but is not supported. For example, a processor may write in an unformatted file a denormalized number when gradual underflow mode is enabled but when the file is read, the underflow mode is abrupt in which case denormalized numbers are not supported.

The invocation of this function never raises an exception, even if X is a signaling NaN.

Example. IEEE_CLASS (–TINY(1.0)/4.0) is the value IEEE_NEGATIVE_DENORMAL.

IEEE_COPY_SIGN (X, Y)

Elemental Function

IEEE_ARITHMETIC

Value with specified sign.

X, Y Of type real with restricted kind.

Result Characteristics. Same as X.

Result Value. The result is the value of X with the sign of Y. This is true even if X is a exceptional value, provided the processor supports the particular exceptional value.

Examples. IEEE_COPY_SIGN (X, 1.0) is |X|; IEEE_COPY_SIGN (X, –1.0) is –|X| for X of type default real; these references are standard conforming giving the specified results even if X is an exceptional value, such as an infinity or NaN.

IEEE_GET_FLAG (FLAG, FLAG_VALUE)

Elemental Subroutine

IEEE_EXCEPTIONS

Obtain a specified exception flag.

FLAG Of type IEEE_FLAG_TYPE with INTENT (IN). It specifies an exception, which may be IEEE_INVALID, IEEE_OVERFLOW, IEEE_DIVIDE_BY_ZERO, IEEE_UNDERFLOW, or IEEE_INEXACT.

FLAG_VALUE

Of type default logical with INTENT (OUT). The value is true if the corresponding IEEE exception is signaling, or false otherwise.

Example. The following program segment determines the current condition of the overflow exception flag, clears the flag, performs a computation that possibly raises the overflow exception flag, and then restores it to its initial value:

```
logical :: initial_fv, computation_fv
call  IEEE_GET_FLAG (IEEE_OVERFLOW, initial_fv)
call  IEEE_SET_FLAG (IEEE_OVERFLOW, .false.)
! Perform some computation.
  . . .
call  IEEE_GET_FLAG (IEEE_OVERFLOW, computation_fv)
if( computation_fv )  then
  ! Fix up the computation to address the overflow.
  . . .
endif
! Restore the initial overflow flag value.
call  IEEE_SET_FLAG (IEEE_OVERFLOW, initial_fv)
```

IEEE_GET_HALTING_MODE (FLAG, HALTING) Elemental Subroutine

IEEE_EXCEPTIONS

Obtain the halting mode for a specified exception.

FLAG Of type IEEE_FLAG_TYPE with INTENT (IN). It specifies the exception, which may be IEEE_INVALID, IEEE_OVERFLOW, IEEE_DIVIDE_BY_ZERO, IEEE_UNDERFLOW, or IEEE_INEXACT.

HALTING Of type default logical with INTENT (OUT). The value is true if an IEEE exception corresponding to FLAG will cause halting, or false otherwise.

Example. The following program segment determines the current condition of the halting mode for a divide-by-zero exception, sets the halting mode for divide-by-zero to the "continue computation" mode, performs a computation that possibly raises this exception, and then restores it to its initial value:

```
logical :: initial_dv_halting, computation_dv
call  IEEE_GET_HALTING_MODE (IEEE_DIVIDE_BY_ZERO, initial_dv_halting)
call  IEEE_SET_HALTING_MODE (IEEE_DIVIDE_BY_ZERO, .false.)
! Perform some computation.
  . . .
call  IEEE_GET_FLAG (IEEE_DIVIDE_BY_ZERO, computation_dv)
if( computation_dv )  then
  ! Fix up the computation to address the divide-by-zero.
  . . .
endif
! Restore the initial overflow flag value.
call  IEEE_SET_HALTING_MODE (IEEE_DIVIDE_BY_ZERO, initial_dv_halting)
```

IEEE_GET_ROUNDING_MODE (ROUND_VALUE)

Subroutine

IEEE_ARITHMETIC

Obtain the IEEE rounding mode.

Argument.

ROUND_VALUE

Scalar of type IEEE_ROUND_TYPE with INTENT (OUT). The value is the current IEEE rounding mode, which is either IEEE_NEAREST, IEEE_TO_ZERO, IEEE_UP, IEEE_DOWN, or IEEE_OTHER.

Example. The following program segment determines the current rounding mode, sets the rounding mode to round up, performs some computation where all floating-point results are rounded up, and then restores the rounding mode to its initial value:

```
type (IEEE_ROUND_TYPE) :: initial_rnd_value
call  IEEE_GET_ROUNDING_MODE (initial_rnd_value)
call  IEEE_SET_ROUNDING_MODE (IEEE_UP)
! Perform some computation, rounding all floating-point
! computations up.
  . . .
! Restore the initial rounding mode.
call  IEEE_SET_ROUNDING_MODE (initial_rnd_value)
```

IEEE_GET_STATUS (STATUS_VALUE)

Subroutine

IEEE_EXCEPTIONS

Obtain the processor's IEEE floating-point status.

STATUS_VALUE

Scalar of type IEEE_STATUS_TYPE with INTENT (OUT). The value is the IEEE floating-point status.

Example. Assume a module my_status_module has established a floating-point status that has the desired modes and exception flags set; this has been saved in a variable clean_status of type IEEE_STATUS_TYPE. The following program segment determines the current floating-point status, sets the status to the "clean" status, performs some computation where the three modes may be changed or floating-point exceptions may be signaled, and then restores the floating-point status to its initial value:

```
use my_status_module, only: clean_status
type (IEEE_STATUS_TYPE) :: initial_status
call  IEEE_GET_STATUS (initial_status)
call  IEEE_SET_STATUS (clean_status)
! Perform some computation, for which any of the three modes
! may be set or any of the exceptions may be raised.
  . . .
! Restore the initial status.
call  IEEE_SET_STATUS (initial_status)
```

IEEE_GET_UNDERFLOW_MODE (GRADUAL)

Subroutine

IEEE_ARITHMETIC

Obtain the IEEE underflow mode.

GRADUAL Scalar of type default logical with INTENT (OUT). The value is true if the underflow mode is gradual underflow or false if the underflow mode is abrupt underflow.

Example. The following program segment determines the current underflow mode, sets the underflow mode to gradual underflow, performs some computation where gradual underflow is used to maintain some accuracy involving small values, and then restores the underflow mode to its initial value:

```
logical :: initial_underflow_mode
call  IEEE_GET_UNDERFLOW_MODE (initial_underflow_mode)
call  IEEE_SET_UNDERFLOW_MODE (.true.)
! Perform some computation, using gradual underflow.
   . . .
! Restore the initial underflow mode.
call  IEEE_SET_UNDERFLOW_MODE (initial_underflow_mode)
```

IEEE_IS_FINITE (X)

Elemental Function

IEEE_ARITHMETIC

True if a value is finite.

X Of type real with restricted kind.

Result Characteristics. Of type default logical.

Result Value. The result is true if X is finite; that is, X is a positive or negative zero, normal value, or denormal value.

Examples. IEEE_IS_FINITE (1.0) is true; IEEE_IS_FINITE (1.0/0.0) is false.

IEEE_IS_NAN (X)

Elemental Function

IEEE_ARITHMETIC

True if a value is a NaN.

X Of type real with IEEE_SUPPORT_NAN (X) being true.

Result Characteristics. Of type default logical.

Result Value. The result is true if X is a quiet or signaling NaN.

Examples. IEEE_IS_NAN (SQRT (–1.0)) is true whereas IEEE_IS_NAN (SQRT (1.0)) is false.

IEEE_IS_NEGATIVE (X)

Elemental Function

IEEE_ARITHMETIC

True if a value is negative.

Argument.

X Of type real with restricted kind.

Result Characteristics. Of type default logical.

Result Value. The result is true if X is a negative infinity, negative normal (including negative zero), or negative denormal value.

Examples. IEEE_IS_NEGATIVE (–1.0) and IEEE_IS_NEGATIVE (–0.0) are true whereas IEEE_IS_NEGATIVE (1.0) is false.

IEEE_IS_NORMAL (X)

Elemental Function

IEEE_ARITHMETIC

True if a value is a normal number.

X Of type real with restricted kind.

Result Characteristics. Of type default logical.

Result Value. The result is true if X is a negative or positive normal value (including either zero).

Examples. IEEE_IS_NORMAL (3.14159) and IEEE_IS_NORMAL (–0.0) are true whereas IEEE_IS_NORMAL (1.0/0.0) is false.

IEEE_LOGB (X)

Elemental Function

IEEE_ARITHMETIC

Unbiased exponent (14.2.1) in the IEEE format.

X Of type real with restricted kind.

Result Characteristics. Same as X.

Result Value. If X has a nonzero finite normal value, the result is the unbiased IEEE exponent *e* of X, namely EXPONENT (X)–1 if the real model in 13.2.3 is used; if X is zero, the result is –inf if IEEE_SUPPORT_INF (X) is true and –HUGE (X) otherwise. If X is zero, the divide-by-zero and inexact exceptions are raised. If X is positive or negative infinity, the result is positive infinity with no exception raised. If X is a NaN, the result is a NaN with no exception raised.

Example. IEEE_LOGB (2.1) is 1.

IEEE_NEXT_AFTER (X, Y)

Elemental Function

IEEE_ARITHMETIC

The next representable neighbor toward a specified value.

X, Y Of type real with restricted kind.

Result Characteristics. Same as X.

Result Value. If X equals Y, the result is X and no exception is signaled. Otherwise, the result is the nearest representable neighbor of X toward Y. If X is zero and Y is not zero, the result is nonzero. The result may be a denormalized number in which case both the underflow and inexact exceptions are raised. The result may be an infinity if X is finite in which case both the overflow and inexact exceptions are raised.

Examples. IEEE_NEXT_AFTER (1.0, 2.0) is 1 + EPSILON (1.0) with no exceptions raised and IEEE_NEXT_AFTER (HUGE (0.0), IEEE_VALUE (1.0, IEEE_POSITIVE_INF)) is positive infinity with the overflow and inexact exceptions raised. If the underflow mode is gradual, IEEE_NEXT_AFTER (0.0, 2.0) is the smallest denormalized positive number with the underflow and inexact exceptions raised; if the underflow mode is abrupt, IEEE_NEXT_AFTER (0.0, 2.0) is TINY (0.0) with only the inexact exception raised.

IEEE_REM (X, Y)

Elemental Function

IEEE_ARITHMETIC

IEEE remainder.

X, Y Of type real with restricted kind.

Result Characteristics. Real with the kind of the argument with the largest precision.

Result Value. For all rounding modes, the result is exactly X – Y×N where N is the nearest whole number to the exact quotient X/Y; if there are two nearest whole numbers (that is, |X/Y – N| = 1/2), N is chosen to be even. If the result is zero, it has the sign of X.

Examples. IEEE_REM (–2.1, 1.05) is –0; IEEE_REM (2.0, 1.05) is +0; and IEEE_REM (4.1, –2.0) is 0.1.

IEEE_RINT (X)

Elemental Function

IEEE_ARITHMETIC

A value rounded to a whole number using the current IEEE rounding mode.

X Of type real with restricted kind.

Result Characteristics. Same as X.

Result Value. The result is the value of X rounded to the whole number determined by the current rounding mode. If the result is zero, it has the sign of X.

Examples. Table B-1 gives the value of IEEE_RINT (X) where X is given in column 1 for the four rounding modes IEEE_NEAREST, IEEE_TO_ZERO, IEEE_UP, and IEEE_DOWN.

Table B-1 IEEE_RINT (X)

X	NEAREST	TO_ZERO	UP	DOWN
0.9	1	0	1	0
0.1	0	0	1	0
–0.1	–0	–0	–0	–1
–0.9	–1	–0	–0	–1

IEEE_SCALB (X, I)

Elemental Function

IEEE_ARITHMETIC

A value scaled by a specified power of 2.

X Of type real with restricted kind.

I Of type integer.

Result Characteristics. Same as X.

Result Value. The result is $X \times 2^I$ if the result is representable and a normal number. If X is finite and $X \times 2^I$ is too large, the overflow and inexact exceptions are raised; the result is infinity of the sign of X if IEEE_SUPPORT_INF (X) is true, or is SIGN (HUGE(X), X) otherwise. If $X \times 2^I$ is too small to be a normal number and there is a loss of accuracy to represent it, the underflow and inexact exceptions are raised; the result is the nearest representable value to $|X \times 2^I|$ of the same sign as X. If X is infinite, the result is the infinity of the same sign as X with no exceptions raised.

Examples. IEEE_SCALB (1.0, 2) is 4.0; on a processor that uses IEEE single precision and supports the IEEE standard, IEEE_SCALB (1.0, 200) is positive infinity with both the inexact and overflow exceptions signaled; on a processor that supports the IEEE standard and gradual underflow, IEEE_SCALB (1.0, –130) is the denormalized number 2^{-130} with the underflow but not the inexact exception signaled.

IEEE_SELECTED_REAL_KIND ({P, R})

Transformational Function

IEEE_ARITHMETIC

Kind parameter of an IEEE real data type, specified by a minimum decimal precision and minimum exponent range.

P, R Scalars of type integer. At least one of P or R must be present. If P or R is absent, it is as if they were present with the value 0.

Result Characteristics. Scalar of type default integer.

Result Value. The result is a kind parameter value that specifies an IEEE real type whose precision, as measured by the intrinsic function PRECISION, is at least as large as P, and whose decimal range, as measured by the intrinsic function RANGE, is at least as large as R. This kind value K will have the property that IEEE_SUPPORT_DATATYPE (1.0_K) is true. If there is more than one kind that satisfies this requirement, the one of least precision is returned; if there is more than one with the same precision, the one with the smallest kind value is selected. If no such kind is available, the result is –1 if the requested precision is not available, –2 if the requested range is not available, or –3 if both the requested precision and range are not available.

Example. IEEE_SELECTED_REAL_KIND (15, 200) gives the same kind as KIND (0.0d0) on a processor that supports the IEEE 64-bit format as the double precision type.

IEEE_SET_FLAG (FLAG, FLAG_VALUE)

Pure Subroutine

IEEE_EXCEPTIONS

Assign a value to a specified exception flag.

FLAG — Scalar or array of type IEEE_FLAG_TYPE with INTENT (IN). A value is set to one of the exceptions IEEE_INVALID, IEEE_OVERFLOW, IEEE_DIVIDE_BY_ZERO, IEEE_UNDERFLOW, or IEEE_INEXACT. No two elements of FLAG may have the same value.

FLAG_VALUE — Scalar or array of type default logical, conformable with FLAG, and with INTENT (IN). The exception in FLAG corresponding to any true element of FLAG_VALUE is set to signaling, and otherwise is set to quiet.

Example. CALL IEEE_SET_FLAG (IEEE_DIVIDE_BY_ZERO, .true.) sets the IEEE divide-by-zero flag to signaling.

IEEE_SET_HALTING_MODE (FLAG, HALTING)

Pure Subroutine

IEEE_EXCEPTIONS

Assign a value to a specified halting mode.

FLAG — Scalar or array of type IEEE_FLAG_TYPE with INTENT (IN). A value is set to one of the exceptions IEEE_INVALID, IEEE_OVERFLOW, IEEE_DIVIDE_BY_ZERO, IEEE_UNDERFLOW, or IEEE_INEXACT. No two elements of FLAG may have the same value. This procedure must not be invoked if IEEE_SUPPORT_HALTING (FLAG) is false.

HALTING — Scalar or array of type default logical, conformable with FLAG, and with INTENT (IN). The halting mode of the exception in FLAG corresponding to any true element of HALTING is set to cause halting, and otherwise is set to continue execution, when the exception is signaled.

Although the exact point where the execution halts is processor dependent, it is expected to be close to where the exception occurred.

IEEE_SET_HALTING_MODE (FLAG, HALTING) must not be invoked unless IEEE_SUPPORT_HALTING (FLAG) is true.

Example. CALL IEEE_SET_HALTING_MODE (IEEE_ALL, .true.) sets the halting mode for all exceptions so that the processor halts subsequent to the signaling of any of the IEEE exceptions. At program startup, the halting mode is processor dependent. The halting of the processor may occur at any point convenient for the processor after the particular exception has been signaled but the point in the execution sequence where the processor halts is expected to be near to the place where the exception occurred.

IEEE_SET_ROUNDING_MODE (ROUND_VALUE)

Subroutine

IEEE_ARITHMETIC

Assign a value to the rounding mode.

ROUND_VALUE

Scalar of type IEEE_ROUND_TYPE with INTENT (IN). The value is an IEEE rounding mode, which is either IEEE_NEAREST, IEEE_TO_ZERO, IEEE_UP, IEEE_DOWN, or IEEE_OTHER. The rounding mode is set to the value of ROUND_VALUE.

This procedure must not be invoked unless IEEE_SUPPORT_ROUNDING_MODE (ROUND_VALUE, X) is true for some X for which IEEE_SUPPORT_DATATYPE (X) is true.

Examples. CALL IEEE_SET_ROUNDING_MODE (IEEE_DOWN) sets the processor's rounding mode to round down all floating-point operations, except for formatted input/output conversion. The initial rounding mode is processor-dependent. See the description of IEEE_GET_ROUNDING_MODE for another example of the use of this subroutine.

IEEE_SET_STATUS (STATUS_VALUE)

Subroutine

IEEE_EXCEPTIONS

Assign a value to the floating-point status.

STATUS_VALUE

Scalar of type IEEE_STATUS_TYPE with INTENT (IN). Its value must be an IEEE floating-point status obtained from a previous invocation of the procedure IEEE_GET_STATUS. The current IEEE floating-point status is set to the value of STATUS_VALUE.

Examples. See IEEE_GET_STATUS for an example of the use of this subroutine.

IEEE_SET_UNDERFLOW_MODE (GRADUAL)

Subroutine

IEEE_ARITHMETIC

Assign a value to the underflow mode.

GRADUAL Scalar of type default logical with INTENT (IN). If the value is true, the current underflow mode is set to gradual underflow, or if it is false, the current underflow mode is set to abrupt underflow. This procedure must not be invoked unless IEEE_SUPPORT_UNDERFLOW_CONTROL (X) is true for some X.

Examples. CALL IEEE_SET_UNDERFLOW_MODE (.false.) sets the processor's underflow mode so that underflow is abrupt. See IEEE_GET_UNDERFLOW_MODE for another example of the use of this subroutine.

IEEE_SUPPORT_DATATYPE ( ) or IEEE_SUPPORT_DATATYPE (X)

Inquiry Function

IEEE_ARITHMETIC

True if the processor supports IEEE arithmetic.

X Scalar or array of type real.

Result Characteristics. Scalar of type default logical.

Result Value. The result is true if the processor supports IEEE arithmetic for all real types if X does not appear, or for the real data type of the kind of X if X appears. Supporting IEEE arithmetic is defined in 14.1.

Example. Consider a processor that supports IEEE arithmetic but for the underflow mode, supports only gradual underflow in single precision, supports only abrupt underflow in the double precision type, and there are no other real kinds. For such a processor, IEEE_SUPPORT_DATATYPE ( ) is true.

IEEE_SUPPORT_DENORMAL ( ) or IEEE_SUPPORT_DENORMAL (X)

Inquiry Function

IEEE_ARITHMETIC

True if the processor supports denormalized numbers.

X Scalar or array of type real.

Result Characteristics. Scalar of type default logical.

Result Value. The result is true if the processor supports arithmetic operations with denormalized values and assignment of denormalized values for all real types if X does not appear, or for the real data type of the kind of X if X appears. Denormalized numbers are defined in 14.1.

Examples. For the processor described in the example for IEEE_SUPPORT_DATATYPE, IEEE_SUPPORT_DENORMAL ( ) and IEEE_SUPPORT_DENORMAL (1.0d0) are false whereas IEEE_SUPPORT_DENORMAL (1.0) is true. Because IEEE_SUPPORT_DENORMAL (1.0d0) is false, there must be some operation where the processor should return a denormalized value but does not.

IEEE_SUPPORT_DIVIDE ( ) or IEEE_SUPPORT_DIVIDE (X) — Inquiry Function

IEEE_ARITHMETIC

True if the processor supports the IEEE divide.

X — Scalar or array of type real.

Result Characteristics. Scalar of type default logical.

Result Value. The result is true if the processor supports the accuracy specified by the IEEE standard [13] for the divide operation for all real types if X does not appear, or for the real data type of the kind of X if X appears. See 14.1 for a specification of supporting the IEEE divide operation.

Example. IEEE_SUPPORT_DIVIDE ( ) is true for a processor that provides the correctly rounded result for a divide operation for all rounding modes and signals IEEE_DIVIDE_BY_ZERO returning the correctly signed infinite value for a division by zero if the dividend is finite and nonzero.

IEEE_SUPPORT_FLAG (FLAG) or IEEE_SUPPORT_FLAG (FLAG, X) — Inquiry Function

IEEE_EXCEPTIONS

True if the processor supports the specified exception.

FLAG — Scalar of type IEEE_FLAG_TYPE. Its value is one of the exceptions IEEE_INVALID, IEEE_OVERFLOW, IEEE_DIVIDE_BY_ZERO, IEEE_UNDERFLOW, or IEEE_INEXACT.

X — Scalar or array of type real.

Result Characteristics. Scalar of type default logical.

Result Value. The result is true if the processor supports the detection of the exception specified by FLAG for all real types if X does not appear, or for the real data type of the kind of X if X appears.

Example. IEEE_SUPPORT_FLAG (IEEE_INVALID) is true if the processor supports the invalid exception for all real kinds.

IEEE_SUPPORT_HALTING (FLAG) — Inquiry Function

IEEE_EXCEPTIONS

True if the processor supports the halting mode for a specified exception.

FLAG Scalar of type IEEE_FLAG_TYPE. Its value is one of the exceptions IEEE_INVALID, IEEE_OVERFLOW, IEEE_DIVIDE_BY_ZERO, IEEE_UNDERFLOW, or IEEE_INEXACT.

Result Characteristics. Scalar of type default logical.

Result Value. The result is true if the processor supports the ability to specify the halting mode for the exception specified by FLAG, by invoking the procedure IEEE_SET_HALTING_MODE.

Example. IEEE_SUPPORT_HALTING (IEEE_OVERFLOW) is false for a processor that does not permit the continue mode after an overflow exception is raised.

IEEE_SUPPORT_INF ( ) or IEEE_SUPPORT_INF (X) Inquiry Function

IEEE_ARITHMETIC

True if the processor supports IEEE infinities.

X Scalar or array of type real.

Result Characteristics. Scalar of type default logical.

Result Value. The result is true if the processor supports positive and negative infinities for all real types if X does not appear, or for the real data type of the kind of X if X appears.

Example. IEEE_SUPPORT_INF ( ) is true if the processor supports infinities for all real kinds.

IEEE_SUPPORT_IO ( ) or IEEE_SUPPORT_IO (X) Inquiry Function

IEEE_ARITHMETIC

True if the processor supports the IEEE base conversion rounding during formatted input/output.

X Scalar or array of type real.

Result Characteristics. Scalar of type default logical.

Result Value. The result is true if the processor supports IEEE input and output as specified in 14.1 for all real types if X does not appear, or for the real data type of the kind of X if X appears.

Example. IEEE_SUPPORT_IO (1.0) is true if the processor supports IEEE base conversion for the default real data type.

IEEE_SUPPORT_NAN ( ) or IEEE_SUPPORT_NAN (X) Inquiry Function

IEEE_ARITHMETIC

True if the processor supports IEEE NaNs.

X Scalar or array of type real.

Result Characteristics. Scalar of type default logical.

Result Value. The result is true if the processor supports IEEE NaNs for all real types if X does not appear, or for the real data type of the kind of X if X appears.

Example. IEEE_SUPPORT_NAN ( ) is false if the processor does not support NaNs for the real type of kind double precision.

IEEE_SUPPORT_ROUNDING (ROUND_VALUE) or IEEE_SUPPORT_ROUNDING (ROUND_VALUE, X)

Inquiry Function

IEEE_ARITHMETIC

True if the processors supports the specified IEEE rounding.

ROUND_VALUE
: Scalar or array of type IEEE_ROUND_TYPE. The value is an IEEE rounding mode, which is either IEEE_NEAREST, IEEE_TO_ZERO, IEEE_UP, IEEE_DOWN, or IEEE_OTHER.

X
: Scalar or array of type real.

Result Characteristics. Scalar of type default logical.

Result Value. The result is true if the processor supports the rounding mode specified by the ROUND_VALUE for all real types if X does not appear, or for the real data type of the kind of X if X appears. If the result is true, support includes the ability to specify and change the rounding mode using the procedure IEEE_SET_ROUNDING_MODE.

Example. IEEE_SUPPORT_ROUNDING (IEEE_NEAREST, 1.0) is true if the processor supports the nearest rounding mode for default real type.

IEEE_SUPPORT_SQRT ( ) or IEEE_SUPPORT_SQRT (X)

Inquiry Function

IEEE_ARITHMETIC

True if the processor's intrinsic SQRT satisfies the IEEE square root specification.

X
: Scalar or array of type real.

Result Characteristics. Scalar of type default logical.

Result Value. The result is true if the processor supports the IEEE specification of the square root operation for all real types if X does not appear, or for the real data type of the kind of X if X appears. See 14.1 for the specification of supporting the IEEE square root operation.

Example. IEEE_SUPPORT_SQRT ( ) is false if the square root intrinsic function SQRT aborts when it is given the negative argument –1.0.

IEEE_SUPPORT_STANDARD ( ) or IEEE_SUPPORT_STANDARD (X)

Inquiry Function

IEEE_ARITHMETIC

True if the processor supports the IEEE facilities specified in the Fortran 2003 standard.

X Scalar or array of type real.

Result Characteristics. Scalar of type default logical.

Result Value. The result is true if the functions:

- IEEE_SUPPORT_DATATYPE ( ),
- IEEE_SUPPORT_DENORMAL ( ),
- IEEE_SUPPORT_DIVIDE ( ),
- IEEE_SUPPORT_FLAG (FLAG) for a valid FLAG,
- IEEE_SUPPORT_HALTING (FLAG) for a valid FLAG,
- IEEE_SUPPORT_INF ( ),
- IEEE_SUPPORT_NAN ( ),
- IEEE_SUPPORT_ROUNDING (ROUND_VALUE) for a valid ROUND_VALUE, and
- IEEE_SUPPORT_SQRT ( )

are all true for all real types if X does not appear, or for the real data type of the kind of X if X appears.

Example. IEEE_SUPPORT_STANDARD ( ) is false if the processor supports both IEEE and non-IEEE real types.

IEEE_SUPPORT_UNDERFLOW_CONTROL ( ) or IEEE_SUPPORT_UNDERFLOW_CONTROL (X)

Inquiry Function

IEEE_ARITHMETIC

True if the processor supports the IEEE underflow control during execution.

X Scalar or array of type real.

Result Characteristics. Scalar of type default logical.

Result Value. The result is true if the processor supports the control of the underflow modes gradual and abrupt for all real types if X does not appear, or for the real data type of the kind of X if X appears. Such control is exercised by invoking the procedure IEEE_SET_UNDERFLOW_MODE.

Example. IEEE_SUPPORT_UNDERFLOW_CONTROL (0.0) is true if the processor supports the control of the underflow mode for the default real type.

IEEE_UNORDERED (X, Y)

Elemental Function

IEEE_ARITHMETIC

True if the arguments are unordered.

X, Y Of type real with restricted kind.

Result Characteristics. Of type default logical.

Result Value. The result is true if either X or Y or both are NaNs.

Example. IEEE_UNORDERED (SQRT (–1.0), 1.0) is true.

‘IEEE_VALUE (X, CLASS)

Elemental Function

A value of a specified type from a specified class.

IEEE_ARITHMETIC

X — Of type real with restricted kind.

CLASS — Of type IEEE_CLASS_TYPE. Its value is one of:

- IEEE_SIGNALING_NAN if IEEE_SUPPORT_NAN (X) is true;
- IEEE_QUIET_NAN if IEEE_SUPPORT_NAN (X) is true;
- IEEE_NEGATIVE_INF if IEEE_SUPPORT_INF (X) is true;
- IEEE_POSITIVE_INF if IEEE_SUPPORT_INF (X) is true;
- IEEE_NEGATIVE_DENORMAL if IEEE_SUPPORT_DENORMAL (X) is true;
- IEEE_POSITIVE_DENORMAL if IEEE_SUPPORT_DENORMAL (X) is true;
- IEEE_NEGATIVE_NORMAL;
- IEEE_NEGATIVE_ZERO;
- IEEE_POSITIVE_ZERO; or
- IEEE_POSITIVE_NORMAL.

Note that CLASS cannot have the value IEEE_OTHER_VALUE.

Result Characteristics. Same as X.

Result Value. The result is an IEEE value as specified by class. The value in general is processor dependent except that the processor must produce the same value in all contexts for the same kind type parameter specified by X.

Examples. IEEE_VALUE (0.0, IEEE_POSITIVE_NORMAL) and IEEE_VALUE (2.0, IEEE_POSITIVE_NORMAL) must be the same value in all contexts; the value may be 1.0, for example.

C Language Evolution

New features are defined in each standard revision cycle. These features sometimes replace those that have become archaic in the language and rarely used. In order to alert the user to redundant and seldom used features, they are declared obsolescent. Features from the obsolescent list of a current standard are candidates for deletion in a future revision. There are three classes of features that determine the language evolution of Fortran.

1. New
2. Obsolescent
3. Deleted

Compilers must implement all new and obsolescent features. They are not required to implement deleted features, but most do in order to be able to process existing programs.

C.1 New Features

The following is a summary of the major new features of Fortran 2003. They are described throughout this book.

1. Derived-type enhancements, such as parameterized derived types and mixed accessibility
2. Features for object-oriented programming, such as extensible types (inheritance), polymorphic variables, type-bound procedures, and pointers to functions
3. Interoperability with the C programming language
4. Support for IEEE arithmetic and exceptions
5. User control of derived-type input/output
6. Stream and asynchronous input/output
7. Environment inquiries and command line arguments
8. Support for international usage, for example, character sets and decimal symbol usage

C.2 Obsolescent Features

The obsolescent features are those features of Fortran 95 that are redundant and for which better methods are available in Fortran 2003. The first three items on the following list were obsolescent in Fortran 90 and remain so. Six items were added to the list in Fortran 95. No new obsolescent features were added in Fortran 2003. These features are candidates for deletion in the next revision of the standard and their use is discouraged. The obsolescent features in Fortran 2003 are:

1. Arithmetic IF statement(8.8.4)
2. Some forms of DO termination (8.7.5)
3. Alternate return (12.6.9)
4. Computed GO TO statement (8.8.5)
5. Statement function (12.4.4)
6. DATA statement among executable statements (2.6)
7. Assumed-length character function (4.3.5.1(4d))
8. Fixed form source (3.3.2)
9. CHARACTER* form for a character declaration (4.3.5.1)

C.2.1 Arithmetic IF Statement

The arithmetic IF statement is redundant. The functionality of the arithmetic IF statement is achieved with the IF statement or IF construct.

C.2.2 DO Termination

Shared DO termination or DO termination with other than an END DO or CONTINUE statement is potentially confusing and error prone. It is safer and better programming practice to use a block DO construct with an unlabeled END DO statement.

C.2.3 Alternate Return

The effect of an alternate return can be achieved with a return code that is used in a CASE construct after returning. For example,

```
CALL SUBR_NAME (X, Y, Z, *100, *200, *300)
```

may be replaced by

```
CALL SUBR_NAME (X, Y, Z, RETURN_CODE)
SELECT CASE (RETURN_CODE)
   CASE (1)
      ...
   CASE (2)
      ...
```

```
        CASE (3)
           ...
        CASE DEFAULT
           ...
     END SELECT
```

C.2.4 Computed GO TO Statement

The CASE construct is a more general facility than the computed GO TO statement; a case selector may be of type character or logical, as well as integer. Use of the CASE construct produces more structured code and is more readable.

C.2.5 Statement Functions

An internal function is a more general replacement for a statement function. A statement function statement is easily confused with an assignment statement.

C.2.6 DATA Statements among Executable Constructs

When a DATA statement appears in the execution part, it gives the impression that a DATA statement causes assignment during program execution rather than initialization before execution.

C.2.7 Assumed-Length Character Functions

The assumed-length character function is an irregularity in the language; it requires the declaration of the function length in the calling program. Instead, a function with an explicit interface or an automatic character length can be used.

C.2.8 Fixed-Form Source

Fixed-form source was dependent on the card column restrictions of punched card input. It has been replaced by a free-form source.

C.2.9 CHARACTER* Form of Syntax in Declarations

Declarations that use CHARACTER* are irregular. There are several other ways to declare character length, such as:

```
CHARACTER (LEN=17) :: C1, C2*6
```

C.3 Deleted Features

The deleted features are those features of earlier versions of Fortran that are redundant and considered largely unused. The list of deleted features for Fortran 90 was empty; there were none. The following obsolescent features of Fortran 90 were deleted from Fortran 95 and are not features of either Fortran 95 or Fortran 2003:

1. Real and double precision DO variables

2. Branching to an END IF from outside the block

3. PAUSE statement

4. ASSIGN statement, assigned GO TO statement, and related features
5. *n*H edit descriptor

There are no deleted features in Fortran 2003.

C.4 Other Compatibility Issues

All standard Fortran 95 programs are standard Fortran 2003 programs. However, there are some cases where the program might be interpreted slightly differently.

1. Fortran 95 had the concept of printing (9.4.6) and the character printed in column one might (or might not) be interpreted to control the printer carriage. There is no such concept in Fortran 2003. However, some devices may still interpret the character in column one to control the printer carriage.
2. The list-directed and namelist output format for real zero values is different from Fortran 95 in Fortran 2003 (10.10.2, 10.11.2.1).
3. If a processor can distinguish between positive and negative zero, the result of ATAN2 (Y, X) when X < 0 and Y is negative real zero is different from that of Fortran 95. The result of LOG (X) and SQRT (X) when X is complex, REAL (X) < 0, and the imaginary part of X is negative zero is different from that of Fortran 95.

Index of Examples

D

E

F

G

H

I

S

T

U

V

W

Z

Index

Symbols

A

B

C

D

E

O

P

Q

R

S

T

U

V

W

X

Z

Printed in the United States of America